ZOO and AQUARIUM HISTORY

Ancient Animal Collections to Zoological Gardens

ZOO and AQUARIUM HISTORY

Ancient Animal Collections to Zoological Gardens

Edited by
Vernon N. Kisling, Jr.

CRC PRESS

Boca Raton   London   New York   Washington, D.C.

COVER PHOTOGRAPHS

Background Photograph: African Panorama exhibit at Hagenbeck's Tierpark, Stellingen (Hamburg), Germany. Panorama exhibits revolutionized the exhibition of animals during the twentieth century (see Chapter 3). (Photograph by Harro Strehlow.)

Front Cover: Camel exhibit during the mid-1920s at Ueno Zoo, Tokyo, Japan (see Chapter 9). (Photograph courtesy of Ueno Zoo.)

Back Cover:

Upper Left: Heron painted as an illustration for the Indian Natural History Project, from a specimen at the Barrackpore menagerie, Calcutta, India, early to mid-1800s (see Chapter 8). (© The British Library.)

Lower Right: Tiger painted from a specimen in the Raja Serfagee menagerie (early to mid-1880s), the precursor to the Shivaganga Gardens Zoo, Tamil Nadu, India. From the Raja Serfagee Collection (see Chapter 8). (© The British Library.)

Library of Congress Cataloging-in-Publication Data

Zoo and aquarium history : ancient animal collections to zoological gardens / edited by Vernon N. Kisling, Jr.
      p. cm.
    Includes bibliographical references (p.).
    ISBN 0-8493-2100-x
    1. Zoos—History. 2. Menageries—History. 3. Aquariums, Public—History. I. Kisling, Vernon N.

QL76 .Z727 2000
590′.7′309—dc21
                                                                                                     00-030362
                                                                                                                    CIP

This book contains information obtained from authentic and highly regarded sources. Reprinted material is quoted with permission, and sources are indicated. A wide variety of references are listed. Reasonable efforts have been made to publish reliable data and information, but the author and the publisher cannot assume responsibility for the validity of all materials or for the consequences of their use.

Neither this book nor any part may be reproduced or transmitted in any form or by any means, electronic or mechanical, including photocopying, microfilming, and recording, or by any information storage or retrieval system, without prior permission in writing from the publisher.

The consent of CRC Press LLC does not extend to copying for general distribution, for promotion, for creating new works, or for resale. Specific permission must be obtained in writing from CRC Press LLC for such copying.

Direct all inquiries to CRC Press LLC, 2000 N.W. Corporate Blvd., Boca Raton, Florida 33431.

Trademark Notice: Product or corporate names may be trademarks or registered trademarks, and are used only for identification and explanation, without intent to infringe.

Visit the CRC Press Web site at www.crcpress.com

© 2001 by CRC Press LLC

No claim to original U.S. Government works
International Standard Book Number 0-8493-2100-x
Library of Congress Card Number 00-030362
Printed in the United States of America   2 3 4 5 6 7 8 9 0
Printed on acid-free paper

Preface

History is a human invention, so it tends to have an anthropocentric perspective that excludes other species. As historian Max Oelschlaeger points out, "The wild plants and animals, the web of life with which our humanity is bound, and without which the human drama could not be enacted, become bit players."[1] This history of zoos and aquariums is about some of those bit players and the forgotten roles they have played in our human drama. Much of our past has been an integral part of nature: our response to environmental changes, our need for natural resources, and our need for suitable land have influenced past social, economic, and political activities usually considered strictly human endeavors. Animals and plants have been significant parts of this history, but they are often overlooked, or, if mentioned, their importance has not been fully appreciated. Animals have been important for many reasons, particularly in the past when our survival depended upon them. This history, however, is about animals as nonutilitarian resources, about animals maintained in collections for many reasons throughout the past 5,000 years — as symbols of power and prestige, as luxury and diplomatic gifts, as objects of personal pleasure, for recreational use, for educational purposes, to increase zoological knowledge, and for conservation purposes.

Animal collections of the past especially (but even modern zoos and aquariums) have not been well studied. Although the published information may be sufficient for an overview, it does not provide an in-depth understanding of these collections and the many facets of their complex evolution. Interestingly, more research has been undertaken on the history of botanical gardens, natural history museums, and circuses. There are journals specifically devoted to the history of these institutions, as well as numerous books and academic studies. In comparison, very little widely distributed, easily accessible information has been published on the history of zoos and aquariums — no journals, few books, few academic studies.

To conduct original research, one must still visit zoos and aquariums to find the necessary historical material. And even if the researcher does this, the material is often minimal or is no longer at the institution. It may be in municipal archives or in regional historical society archives. Very often, it no longer exists because it was thrown out or destroyed to make room for new paperwork. Just trying to determine the year of establishment of an institution can be difficult. Zoo and aquarium personnel often do not know where to look for historical information, letters of inquiry go unanswered, and there is no consensus on what criteria should be used to establish this date (often there is an "official" opening, but other criteria are also used, such as when the first animals were acquired).

Most zoo and aquarium histories are institutional ones with limited distribution. Relatively comprehensive histories have occasionally been published, but they have had limited distribution and use. Most are now out of print, so copies are difficult to find.[2-7] A more comprehensive history of the world's zoos and aquariums from ancient times to the present has been a needed, but daunting, task. Historical research on zoos and aquariums is still at a rudimentary stage, with much of it concerned with basic information: what collections existed, when were they established, when were the structures built, how were they managed, who managed them, and what animals were collected. Nevertheless, more in-depth, analytical studies are now examining the cultural, institutional, and environmental context of zoos and aquariums. So the time is ripe for a comprehensive review of what is known about zoos and aquariums, one that provides an introduction to the subject and also highlights the published and archival resources for those who want to know more.

It is doubtful that a single individual could compile a comprehensive worldwide history of zoos and aquariums. To overcome the diversity of languages involved with such a task, to gain access to the regional information, and to understand the varied cultural influences on zoo and aquarium development, contributions for this history were sought from individuals who live in, or are from, the countries and regions about which they write. Initially, some 8 years ago, trying to find these contributors appeared to be an impossible task; however, cooperative contacts, persistence, and good fortune produced knowledgeable individuals who were able to contribute to this history.

Every effort has been made to publish a book that is as comprehensive and authoritative as possible. Such an endeavor, however, is destined to fall short. A single volume cannot completely cover 5,000 years of wild animal collections, nor can it cover every country or detail every collection. It is also difficult to provide a balance among regions, and among countries within a region. Each contributor is more familiar with particular countries and collections and less familiar with others. Even within a single country, much less within a large region, there may be an imbalance in coverage because of difficulties obtaining information.

Although the historical study of zoos and aquariums needs to be approached from many analytical and disciplinary perspectives, it still requires a practical review of basic information. *Zoo and Aquarium History* is intended as an overview of the current state of our knowledge, with references to the more detailed information. It is hoped that this history will stimulate the growing interest in this subject and that it will encourage further research. New research, which is very much needed, will undoubtedly produce additional facts and new insights that will improve the material in this volume. This research will also help fill the gaps that are evident in this volume. Much still needs to be learned about the collections, the individuals who managed the collections, past institutions no longer extant, and existing institutions throughout the world.

A growing interest in the history of zoos and aquariums has already become apparent. The 1980s and 1990s have been significant decades for this fledgling field of study. The Bartlett Society was founded in England on October 27, 1984 as an international society for the study of zoos and wild animal husbandry history. In 1989, the National Zoological Park held an international zoo history symposium in Washington, D.C. and later published the proceedings as *New Worlds, New Animals: From Menagerie to Zoological Park in the Nineteenth Century*.[7] Additional institutional histories have also appeared, three professional zoo associations (American Zoo and Aquarium Association, American Association of Zoo Veterinarians, and American Association of Zoo Keepers) in the United States have appointed association historians, and the number of zoo history theses and dissertations at academic institutions has increased. If this awakening continues its momentum, the future looks bright — but only if zoos and aquariums archive their records, recognize the interest in their histories, and understand the value of their past.

Origins and Definitions

What constitutes a zoo and which zoos were first are points of contention that continue to be debated. The approach taken in this history is a broad one, beginning with the first efforts to keep wild native animals. This period (ca. 10,000–3000 B.C.) was dominated by the gathering of wild animals for what turned out to be utilitarian purposes, regardless of whether this was the original intention. Since domestication is a biological process requiring many generations, these early efforts involved keeping animals that remained wild for quite some time. Some species continued to remain wild and were never domesticated. These wild species may have formed the precursors of later collections, but they could not be considered collections themselves (at best, they may have been proto-collections). Rather, these early experiences with wild animals were simply the important first steps leading to animal collections.

These first efforts evolved into keeping wild native and exotic animals for nonutilitarian purposes. During this period (ca. 3000 B.C.–A.D. 1456) Mesopotamia, Egypt, China, and possibly India were the first societies known to have animal collections. The epicenter of these collecting activities then shifted to the Greco-Roman regions, to the Persian and Arabic regions, and later to Medieval Europe. Meanwhile, collections continued to exist in China, India, and other Asian countries. Large collections also existed in Central America (the Aztec collections) and South America (the Inca collections).

Animal collections evolved into menageries during and after the European Renaissance period (1456–1828), and then into zoological gardens beginning in the nineteenth century (1828 to present). In hindsight, however, earlier collections may also be considered menageries or zoos. Modern usage of these words applies to any collection of wild animals, including those collections existing in the past. The idea that collections evolved first into menageries and then into zoological gardens has generated a great deal of discussion. As collections changed from private to public entities, as they shifted from the domain of the wealthy to that of the general public, as individual ownership switched to government or society ownership, as individual collections became cultural institutions, and as animal husbandry and exhibition standards improved, collections have certainly become different kinds of places. These evolutionary changes have prompted the use of *menagerie* and *zoo* to acknowledge the differences, but there is no consensus on the criteria to be used for defining the differences. In contrast, aquarium evolution is less complicated since it sprang forth as a relatively modern concept during the 1850s.

There is no precise definition for a menagerie, but in determining that animal collections evolved into menageries, certain characteristics may be recognized. In a menagerie, as many species as possible are exhibited, animals are exhibited in taxonomically arranged rows of barred cages, staff is somewhat knowledgeable about animals, and there are limited education and science programs; the main emphasis is on recreation or entertainment. *Menagerie*, as a word, did not enter the European vocabulary until it was first used in France in the early eighteenth century, and it was some time before it was used to refer to a collection of wild animals. However, *menagerie* is not a well-defined word and the above characteristics are not precise criteria. Menagerie tends to be a concept that individuals, including zoo historians, view differently.

Neither is there any precise definition for a zoological garden, but in determining that menageries became zoological gardens, certain characteristics may again be recognized. In a sense, zoological gardens are simply sophisticated menageries. Nevertheless, they have more naturalistic animal exhibits arranged ecologically or zoogeographically, staff that is increasingly knowledgeable about animals, and improved education, research, and conservation programs. Conservation parks (or bioparks) are similar to zoological gardens, but with an increased emphasis on immersion exhibits that re-create natural habitats and on conservation (*in situ* field programs, as well as *ex situ* captive management programs).

Moving along the continuum from menageries to zoological gardens (and on to conservation parks), it is difficult to pinpoint any clearly defined transition points. However, it can be said with some degree of certainty that particular institutions led the transition from menageries to zoological gardens, such as Schönbrunn (Vienna), the Jardin des Plantes (Paris), the London Zoological Garden, and the Philadelphia Zoological Garden. Other institutions of the world have performed similar roles for their regions.

Whatever definition one chooses, modern zoos may include a variety of facilities: zoological parks, conservation parks, aviaries, herpetariums, safari parks, insectariums, butterfly parks, and endangered species rehabilitation centers. Aquariums and oceanariums are unique forms of zoological gardens and are here distinguished from the other terrestrially oriented facilities (as the aquarium profession generally prefers). All of these variations are considered in this history under the umbrella term *zoological gardens*. In addition, other modern institutions are merging with the zoological garden concept. National parks and wildlife reserves are becoming so intensively managed that they are becoming zoogeographic megazoos.[8] Examples of this trend are presented in some of the chapters. Although they are not bona fide zoos, they do resemble the ancient royal animal parks, and as natural habitats decrease and management of the remaining park areas increases, these park areas may one day be included in the zoological garden concept.

Historical Trends

A great deal of pride is taken in historical "firsts," such as which zoo was the first established in a particular country, which zoo had the first exhibition or first birth of a particular species, which zoo developed certain kinds of exhibits first, and so on. Pride aside, what is more important are the trends, of which the firsts are just the beginning. Institutions, such as zoological gardens, do not begin fully developed and, in fact, are never fully developed. Zoological gardens are still evolving and today's state-of-the-art facilities will appear crude to future generations. Some individuals do not recognize this and prefer to disparage earlier collections rather than understand them, criticizing them based on today's standards rather than on standards contemporary with the period. Institutions must be understood within their historical context and are, at any particular time, merely snapshots of broader trends contributing to that historical context.

"We have a responsibility to our captive animals, brought from their native wilds to minister to our pleasure and instruction.... Much as has been done in this direction, we must all admit that there is still more required. The buildings of today will ... some day seem to our successors what the former ones seem to us."[9] The "today" of this statement could easily be 1987, but it is not — the today of this statement is 1887. And the truth of this century-old statement has become evident with the often-heard, disparaging remarks about menageries. Our successors in 2087 will, no doubt, also agree with this statement as they look back on the zoological gardens of the twentieth century. One hopes they will be more understanding and appreciative of our efforts than we are of our predecessors' efforts.

This comparative view of zoos over time can be undertaken because of many institutional trends that have occurred. Private, privileged collections have evolved into public, cultural institutions. An emphasis on private, personal pleasure first gave way to public entertainment and recreation, and then to educational, scientific, and conservation concerns. Improved knowledge and technology were used to improve the husbandry and exhibition of animals in the collections. Workforce diversity, both in job responsibilities (based on improved knowledge and skills) and in types of jobs, transformed what was essentially agricultural work into a zoo profession. The regional uniqueness of collections has turned into global conformity.

Zoos and aquariums have also been a part of other, broader trends in wildlife biology, conservation, veterinary medicine, technology, education, park and recreation development, human sensibilities regarding nature, and many other facets of cultural change. Reflecting these many influences, zoos have evolved from mere collections (with individuals interested in wildlife as managers), to menageries (with naturalists as managers in an era of natural history), to zoological gardens (with zoologists or veterinarians as managers in an era of specialized science), to conservation parks (with conservationists as managers in an era of endangered species), to what is now emerging at the beginning of the twenty-first century — the environmental center (with business administrators as managers in an era of marketing, public relations, and fund-raising). Zoos are returning to the integrated, environmental animal parks of ancient Mesopotamia, but in a more sophisticated manner. And at the same time, wild nature itself is becoming a megazoo — just a different kind of captivity.

All of the trends affecting zoos and aquariums need to be studied as we move beyond the accumulation of basic information about these institutions. However, a great deal of basic information remains to be discovered in this neglected arena of history. Many good institutions, animals, and people have made our existing zoos and aquariums what they are today, and many more have labored at institutions that did not survive. We still know very little about many of them.

Overview

Zoo and Aquarium History begins with a chronologically arranged global perspective in Chapter 1 and then, about the time of the European Renaissance, changes to a geographic perspective. Chapter 1 provides an introduction and covers the first keeping of wild animals, the ancient collections, and the menageries. It covers the world from roughly 10,000 B.C. (the end of the last Ice Age and the beginning of domestication) to about A.D. 1456 (the beginning of the

European Renaissance). However, there is some chronological overlap between the end of this chapter and the beginnings of the other chapters, since the first chapter ends with the development of menageries and their transformations into zoological gardens. This was an evolutionary process that took time, that cannot be well defined, and that occurred at different times in different countries.

Chapters 2 through 11 are geographically arranged histories of zoological gardens, usually beginning with the menageries upon which these zoological gardens were founded. Some of the chapters cover individual countries, and others cover regions with several countries. The chapters provide information on the major zoological gardens of Great Britain, Europe (based on its historic division into Western Europe and Eastern Europe), the United States, Australia, Asia, India, Japan, Africa, and South America. Each chapter contains a list of references, which also serves as bibliography.

An appendix presents a chronological listing of the world's zoos and aquariums. It is not intended to be a directory of zoos and aquariums, as only those zoos and aquariums with known years of establishment are listed. The list is arranged geographically, as are the chapters, but not all the zoos and aquariums listed are discussed in the text because of space limitations.

References

1. Oelschlaeger, Max, *The Idea of Wilderness: From Prehistory to the Age of Ecology*, Yale University Press, New Haven, CT, 1991, 7–8.
2. Baratay, Eric and Hardouin-Fugier, *Zoos — Histoire des jardins zoologiques en occident (XVIe –XXe siècles)*, La Découverte, Paris, 1998 [*Zoos — History of Western Zoological Gardens (16th–20th Centuries)*].
3. Fisher, James, *Zoos of the World: The Story of Animals in Captivity*, Natural History Press, Garden City, NY, 1967.
4. Loisel, Gustave, *Histoire des ménageries de l'antiquité à nos jours*, Octave Doin et Fils and Henri Laurens, Paris, 1912.
5. Lukaszewicz, K., *Ogrody Zoologiczne–Wczoraj–Dzis–Jutro*, Wiedza Powszechna, Warszawa, 1975 [*Zoological Gardens–Yesterday–Today–Tomorrow*].
6. Hediger, H., *Zoologische Gärten. Gestern–Heute–Morgen*, Hallwag Verlag, Bern, Switzerland, 1977 [*Zoological Gardens: Yesterday, Today, Tomorrow*].
7. Hoage, R. J. and Deiss, William A., Eds., *New Worlds, New Animals: From Menagerie to Zoological Park in the Nineteenth Century*, Johns Hopkins University Press, Baltimore, MD, 1996.
8. Sullivan, Arthur L. and Shaffer, Mark L., "Biogeography of the megazoo: biogeographic studies suggest organizing principles for a future system of wild lands," *Science*, 189, 13, 1975.
9. Flower, William, "Jubilee address," in *Annual Report*, Zoological Society of London, London, 1887.

Contributors

Catherine de Courcy is a resident of Melbourne, Australia, where she is a librarian at the State Library of Victoria. Her master's thesis was on the history of the Melbourne Zoological Gardens. Her publications include two on the history of Australian zoos.

James F. Ellis, Jr. and Georgeann A. Ellis. Dr. James Ellis is the Associate Director of International Programs at Washington State University, Pullman. He has worked at the Oklahoma City Zoo, Peoria Zoo, and Santa Fe Community College Teaching Zoo. Together with his wife, Georgeann Ellis, a licensed microbiologist, he has spent many years in South America and the Caribbean consulting with zoos in Brazil, Guatemala, and Jamaica, as well as the Museum Paraense Emilio Goeldi. The two have written numerous articles on zoos, wildlife conservation, and conservation education in developing countries.

Ken Kawata is General Curator of the Staten Island Zoo, New York. Born in Japan, he interned at the Ueno Zoo, Tokyo, before moving to the United States and working at several U.S. zoos including those in Topeka (Kansas), Indianapolis (Indiana), and Tulsa (Oklahoma) as a keeper and a curator. He has published numerous articles in various American, British, European, and Japanese zoo journals and translated articles from Japanese to English for the *International Zoo Yearbook* and *International Zoo News*.

Clinton H. Keeling, a resident of Shalford, England, is a self-taught zoologist and a former zoological garden curator who is primarily concerned with education and wild animal husbandry. He has published more than 40 animal and zoo books, both for adults and children. He is the founder of the Bartlett Society (1984), an international society for the study of zoos and wild animal husbandry.

Vernon N. Kisling, Jr. is a collection management coordinator at the Marston Science Library, University of Florida at Gainesville, and a past curator at the Atlanta Zoo and at the Crandon Park Zoo (Miami). He has written numerous articles on wildlife conservation and zoo history. He is the chair of the American Zoo and Aquarium Association History Committee and is the North American representative of the Bartlett Society.

Wilhelmus Labuschagne, a resident of Pretoria, South Africa, has been Director of the National Zoological Garden of South Africa since 1985. For 13 years prior to this he was a zoologist and then curator at the Johannesburg Zoo. He is a member of the International Union of Directors of Zoological Gardens and is the Chairman of the Pan African Association of Directors of Zoological Gardens, Aquariums and Botanical Gardens.

Leszek Solski began his professional career as a veterinarian in Szczecinek (northern Poland). He was General Manager of the Polish Fauna Garden in Bydgoszcz (central Poland) from 1981 to 1982 and then worked as a scientific assistant at the Institute of Biomaterials Research, within the Medical Academy, Wroclaw. He has written more than a dozen scientific publications, and 60 articles for more general audiences, on zoos and wildlife conservation.

Harro Strehlow holds a Diplom-Biologist from the Freie Universität Berlin and a Dr. rer. nat. from the Technische Universität Berlin. He has been a biologist and a teacher, and has worked as an animal keeper. He has written more than 90 articles on animal behavior, wildlife conservation, the history of biology, and the history of zoos.

Sally Walker established Friends of Mysore Zoo, Zoo Outreach Organisation, the Society for Promotion of History of Zoos and Natural History in India, and the Asian Regional Network of Zoo Educators. She is the regional representative for the Captive Breeding Specialists Group and other wildlife conservation organizations. She founded two monthly zoo magazines now in their fifteenth year of regular publication and she has published more than 500 articles in India and abroad, primarily about zoos and conservation science in India and Southeast Asia.

Acknowledgments

A few acknowledgments are necessary. First, and foremost, this book has been possible only because of the contributors who have written about the zoos and aquariums in their regions and countries. No one individual could know, or even gather, this historical information from throughout the world. Too much time, money, language skills, and effort would be required to accomplish what these contributors have achieved. Although some guidance was given to the contributors to establish some degree of conformity, contributors were encouraged to write in their unique ways from their cultural viewpoints. An editorial effort was made to balance consistency with uniqueness. Several other individuals were unable to contribute chapters, but suggested others who could, and these gestures were greatly appreciated. Without such cooperative assistance, the necessary contacts could not have been made.

This project has not enjoyed the usual array of grants, secretaries to do the typing, research assistants, or the like. Some eight years of voluntary efforts by the contributors and the editor got the work done. In addition to the contributors, I would like to thank Ross Arnett, publisher, editor, writer, scientist, and friend. Ross, who passed away just as the manuscript was finished, had faith in the book from the beginning, and helped get it into the publisher's hands, a daunting task for an author to accomplish unaided. I would also like to thank John Sulzycki, our editor at CRC Press, for his keen interest, constructive criticism, advice, and help. The book, for whatever it is worth, is far better than it would have been without John's work. While the contributors made the book possible, Ross and John made the book publishable.

Contributors for most chapters have relied upon numerous individuals who provided them with information. It would be difficult to list all of them here, but they have been listed in the chapter references as personal communications. Reviewers, some known to me and some not known, provided constructive criticism on individual chapters, for which I am grateful. In particular, I would like to thank Herman Reichenbach for reviewing several of the chapters and for sharing his expertise on zoos, natural history, European history, and the Chinese language. Without this array of contacts, both within and outside the zoo profession, the chapters would have been far less informative and less accurate.

The final result, this particular history, has been a complex effort driven to completion with the help of individuals with a keen interest, and a passion, for the history of zoos and aquariums. I have enjoyed the opportunity to work with them, and to learn from them, on this project.

Vernon Kisling
Gainesville, Florida

Contents

1   Ancient Collections and Menageries
Vernon N. Kisling, Jr.

- 1.1 Introduction 1
- 1.2 Keeping Wild Animals 2
  - 1.2.1 Environmental Knowledge 2
  - 1.2.2 Domestication of Wild Animals 5
  - 1.2.3 Beyond Domestication — Collecting Wild Animals 7
- 1.3 Ancient Collections 8
  - 1.3.1 Mesopotamian Collections 8
  - 1.3.2 Ancient Egyptian Collections 12
  - 1.3.3 Ancient Asian Collections — India and China 16
  - 1.3.4 Greek and Roman Collections 17
  - 1.3.5 Persian and Arab Collections 21
  - 1.3.6 Medieval Collections 21
  - 1.3.7 Aztec and Inca Collections 25
- 1.4 Menageries 28
  - 1.4.1 European Menageries 28
  - 1.4.2 Colonial Menageries 33
  - 1.4.3 Evolution of the Zoo and Aquarium 37
- References 42
  - Additional Sources 47

2   Zoological Gardens of Great Britain
Clinton H. Keeling

- 2.1 Introduction 49
- 2.2 The Tower Menagerie 50
- 2.3 Estate Collections and the Windsor Great Park 56
- 2.4 The Exeter 'Change 58
- 2.5 Three Nineteenth-Century Private Collections 60
- 2.6 Zoological Gardens at Surrey and Liverpool 61

2.7	Manchester Zoological Gardens and the Belle Vue Zoo	65
2.8	Zoological Gardens at Edinburgh, Bristol, Hull, and Preston	66
2.9	Zoological Gardens at Paignton, Chester, and Dudley	67
2.10	London Zoological Garden	68
2.11	Zoological Gardens at Whipsnade and Jersey	72
	References	73

3 Zoological Gardens of Western Europe
Harro Strehlow

3.1	Introduction	75
3.2	Post-Medieval Collections to Modern Zoos	75
	3.2.1 Game Parks, Falconries, and Pheasantries	76
	3.2.2 Deer Moats, Bear Pits, and Lion Cages	78
	3.2.3 Menageries	80
	3.2.4 Transition from Menagerie to Modern Zoo — Schönbrunn and Berlin	83
	3.2.5 Transition from Menagerie to Modern Zoo — Jardin des Plantes	89
3.3	Early Modern Zoos	90
	3.3.1 The Exotic Style and the Systematic Zoo	97
	3.3.2 The Hagenbeck Revolution	102
3.4	Modern Zoos	104
	3.4.1 Zoos of the Early Twentieth Century	104
	3.4.2 Zoos of the Late Twentieth Century	108
	References	112
	Additional Sources	116

4 Zoological Gardens of Central-Eastern Europe and Russia
Leszek Solski

4.1	Introduction	117
4.2	Poland	117
	4.2.1 Wild Animal Keeping in Poland through the Nineteenth Century	117
	4.2.2 Origin of Modern Polish Zoos to 1939	119
	4.2.3 Polish Zoos during and after World War II	122
	4.2.4 General Characteristics and Comments on Polish Zoos	130

	4.3	Russia	132
		4.3.1 Wild Animal Keeping in Russia through the Nineteenth Century	132
		4.3.2 Origin of Modern Zoos in Russia to 1917	133
	4.4	Soviet Union	135
		4.4.1 Soviet Zoos — Political and Economic Realities	135
		4.4.2 Russian Zoos in the Soviet Union	137
		4.4.3 Other Zoos in the Soviet Union	139
		4.4.4 General Characteristics and Comments on Soviet Zoos	140
	4.5	Czech and Slovak Republics	140
	4.6	Hungary	142
	4.7	Bulgaria	143
	4.8	General Comments on Central-Eastern European Zoos	145
	References		145
		Additional Source	146

5 Zoological Gardens of the United States
Vernon N. Kisling, Jr.

	5.1	Introduction	147
	5.2	Eighteenth- and Nineteenth-Century Menageries	147
	5.3	Nineteenth-Century Zoos and Aquariums	150
	5.4	Twentieth-Century Zoos and Aquariums	164
	5.5	Summary	177
	References		177
		Additional Sources	180

6 Zoological Gardens of Australia
Catherine de Courcy

	6.1	Introduction	181
	6.2	Foundation and Development, 1857 to 1920	182
		6.2.1 Origins	182
		6.2.2 Enclosure Design	186
		6.2.3 The Animal Collections	188
		6.2.4 Early Development of the Four Objectives	192
		6.2.5 Visitors	197

	6.3	Survival, 1920–1960 198
		6.3.1 Enclosure Design and the Animal Collection................198
		6.3.2 Hallstrom and Taronga Zoo...200
		6.3.3 Survival of the Four Objectives.......................................201
	6.4	Modernization, 1960 to the Present 202
		6.4.1 Modernization of the Four Objectives............................205
		6.4.2 Visitors 211
	References 211

7 Zoological Gardens of Asia
Sally Walker

7.1	Introduction 215
7.2	Southwest Asia 215
	7.2.1 Afghanistan 217
	7.2.2 Cyprus 218
	7.2.3 Iran 218
	7.2.4 Iraq 219
	7.2.5 Israel 219
	7.2.6 Jordan 221
	7.2.7 Kuwait 221
	7.2.8 Oman 221
	7.2.9 Saudi Arabia 222
	7.2.10 Turkey...222
	7.2.11 United Arab Emirates...223
	7.2.12 Yemen..223
7.3	South Asia...223
	7.3.1 Bangladesh 223
	7.3.2 Bhutan 224
	7.3.3 Nepal 224
	7.3.4 Pakistan 225
	7.3.5 Sri Lanka 226
7.4	Southeast Asia 228
	7.4.1 Brunei 229
	7.4.2 Cambodia 229
	7.4.3 Laos..229
	7.4.4 Malaysia 229
	7.4.5 Myanmar 231
	7.4.6 Philippines 231

	7.4.7	Singapore	232
	7.4.8	Thailand	233
	7.4.9	Vietnam	235
7.5	Indonesia		236
	7.5.1	Java	237
	7.5.2	Sumatra	238
7.6	East Asia		239
	7.6.1	Hong Kong	239
	7.6.2	Korea	240
	7.6.3	Macau	240
	7.6.4	Taiwan	240
	7.6.5	China	241
References			244
	Additional Sources		250

8 Zoological Gardens of India

Sally Walker

8.1	Introduction	251
8.2	Ancients and Invaders — Vedics, Guptas, Moguls, Europeans	252
	8.2.1 The Vedic Period — Spiritual and Mystical Values	252
	8.2.2 The Gupta and Mogul Periods	253
	8.2.3 The European Period — Utilitarian or Mechanistic Values	256
8.3	Nineteenth-Century Indian Zoos	257
	8.3.1 Calcutta's Wild Animal Collections — Four Early Ones	257
	8.3.2 Old Madras State Zoos	265
	8.3.3 Kerala Trivandrum (Old Travancore State) and Trichur Zoos	269
	8.3.4 Sakkarbaug Zoo and the Gir Lions	270
	8.3.5 Maharastra State and the Bombay Zoo	270
	8.3.6 Princely Zoos	271
	8.3.7 Old Mysore's Zoos	274
	8.3.8 Early Collections Not Normally Mentioned	276
8.4	Twentieth-Century Indian Zoos	277
	8.4.1 Modern Zoos and the National Zoological Park	277
	8.4.2 Indian Crocodile Project and Specialist Zoos	283

    8.5   Indian Zoos and Wildlife 284
            8.5.1   Indian Board for Wildlife and Wildlife Protection Act... 285
            8.5.2   Management of Zoos in India..286
            8.5.3   The Future of Indian Zoos 289
    References 290

9   Zoological Gardens of Japan
Ken Kawata

    9.1   Introduction 295
    9.2   Historical Overview 295
            9.2.1   Pre-Restoration Era 295
            9.2.2   The Emergence of Modern Zoos 296
            9.2.3   World War II and Beyond 298
    9.3   Institutional Overview...302
            9.3.1   The Setting...302
            9.3.2   Zoos 302
            9.3.3   Aquariums 303
            9.3.4   Traveling Menageries and Safari Parks 304
            9.3.5   Japanese Association of Zoological Gardens
                      and Aquariums..305
    9.4   Administrative Overview 306
            9.4.1   Governing Authorities 306
            9.4.2   Marketing and Events 311
            9.4.3   Philanthropy 312
            9.4.4   Animals as Commodities...312
    9.5   Animal Collections 313
            9.5.1   Marveling at Giraffe 313
            9.5.2   Animal News Makers 313
            9.5.3   Exhibits..315
            9.5.4   Breeding Programs 317
            9.5.5   Research Activities 322
    9.6   Internationalization and Cultural Uniqueness 323
            9.6.1   The Sakoku Factor...323
            9.6.2   Environmental Awareness 324
            9.6.3   Penchant for Group Acceptance...325
            9.6.4   Uncharted Waters 325
    References 327
            Additional Sources 329

10 Zoological Gardens of Africa
Wilhelmus Labuschagne and Sally Walker

10.1	Introduction	331
10.2	Arab Republic of Egypt	331
	10.2.1 Giza Zoological Gardens	331
	10.2.2 Other Egyptian and Sudanese Territory Zoos	339
10.3	Sub-Saharan Africa	342
	10.3.1 Kenya	342
	10.3.2 Republic of South Africa	342
	10.3.3 Other African Nations	346
10.4	African Region	347
	10.4.1 Malagasy Republic (Madagascar)	347
	10.4.2 Mauritius	347
References		348
	Additional Sources	349

11 Zoological Gardens of South America
James F. Ellis, Jr. and Georgeann A. Ellis

11.1	Introduction	351
11.2	Brazilian Zoos	351
	11.2.1 The North (Norte)	353
	11.2.2 The Northeast (Nordeste)	355
	11.2.3 The Center-West (Centro-Oeste)	355
	11.2.4 The Southeast (Sudeste)	356
	11.2.5 The South (Sul)	358
	11.2.6 Summary	358
11.3	Parque Zoobotânico Museu Paraense Emilio Goeldi	359
11.4	South American Zoos in Other Countries	362
	11.4.1 Argentina	362
	11.4.2 Bolivia	363
	11.4.3 Colombia	364
	11.4.4 Venezuela	364
	11.4.5 Summary	365
References		365

Appendix   Zoos and Aquariums of the World

Zoological Gardens of Great Britain 369
Zoological Gardens of Western Europe 371
    Austria 371
    Belgium 371
    Denmark 371
    Finland 371
    France 371
    Germany 372
    Italy 373
    The Netherlands 373
    Portugal 373
    Spain 373
    Sweden 373
    Switzerland 373
Zoological Gardens of Eastern Europe 374
    Bulgaria 374
    Czech and Slovak Republics 374
    Hungary 374
    Poland 374
    Russia and Former Soviet Union 375
Zoological Gardens of the United States 375
Zoological Gardens of Australia 380
Zoological Gardens of Asia 380
    Southwest Asia (Middle East) 380
    South Asia 381
    Southeast Asia 381
    Indonesia 382
    East Asia 382
Zoological Gardens of India 382
Zoological Gardens of Japan 384
Zoological Gardens of Africa 387
Zoological Gardens of South and Central America 387
    Argentina 387
    Bolivia 388
    Brazil 388

Chile	389
Colombia	389
Guyana	389
Peru	389
Venezuela	389
Central America	390
Index	391

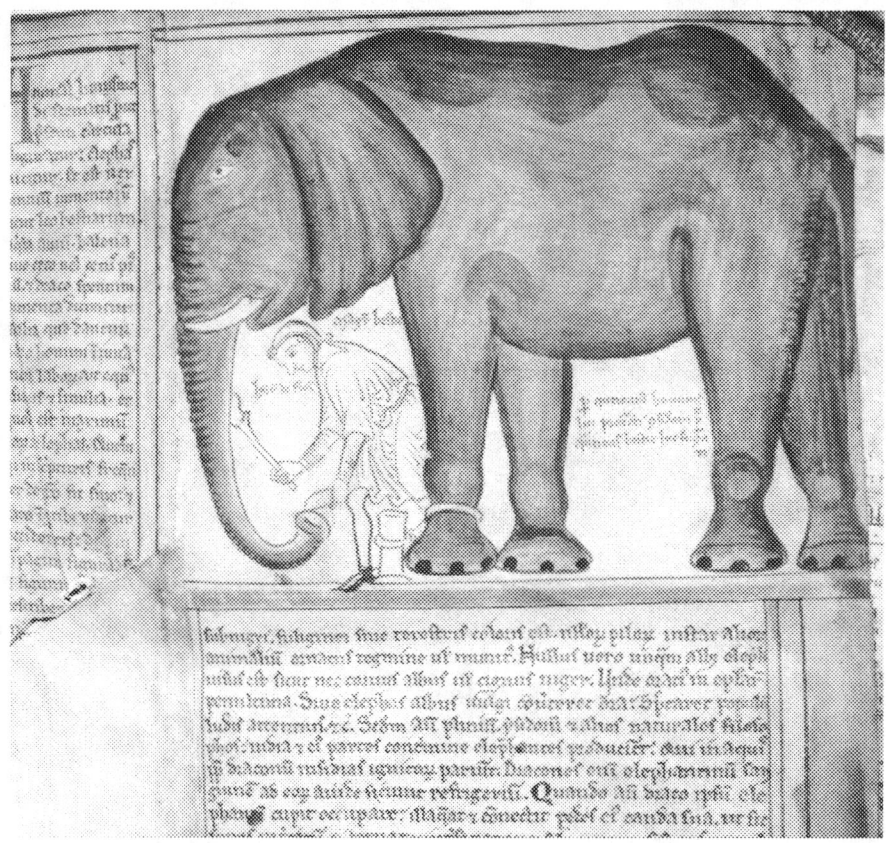

Elephant King Louis IX of France brought with him when he returned from the Crusades during the mid-1200s. From the *Chronicles of Matthew Paris: Monastic Life in the Thirteenth Century.* © The Master and Fellows of Corpus Christi College, Cambridge.

1 Ancient Collections and Menageries

Vernon N. Kisling, Jr.

1.1 Introduction

Exotic animals have long been the ultimate collectibles. Exotic animals, alive and active, have been more fascinating and exciting than natural history (museum) specimens, plants, or cultural artifacts — in part, because animals are less common, more difficult to acquire, and more expensive to maintain. And then, there is the fascination, both emotional and scientific. Since ancient times the passion for possessing wild animals from distant lands has overcome the great difficulties and expense of capturing, transporting, and maintaining them. To paraphrase a proverbial saying, if there were no zoo, someone would invent one. And many have done so over the past 5,000 years, in various ways.

Cultural institutions, like the cultures that foster them, evolve over time. This evolution is certainly true for zoos and aquariums, which have evolved in parallel with the diversity of cultures that have nurtured them. Keeping wild animals is as old as the first attempts at domestication, which began about 10,000 B.C.; however, "collections" of wild animals were not assembled until the earliest urbanized civilizations began about 3000 B.C. These early collections, within the context of their times, were in effect the earliest zoos, even though they were not then referred to as zoos. Zoo-related terminology, as it now exists, did not come into use until the modern zoo concept developed during the eighteenth and nineteenth centuries, a period when animal collections significantly changed and became the cultural institutions that are now familiar to us. Because an etymology for zoos and aquariums analogous to etymologies for natural history museums and botanical gardens[1,2] has not been developed, a preliminary etymology for zoos and aquariums is provided at the end of this chapter, following the discussion of the historical context within which these terms emerged.

Past cultures, ancient through early modern, viewed nature as an integrated whole even while attempting to categorize its many parts. Ancient collections began as more than just gatherings of these parts; animals were kept within a natural setting. Early trade in exotic products also included both animals and plants, and rarely seen species were coveted acquisitions for those who could afford such extravagances. Collecting plants was useful and important because plants had food and medicinal values, while some were popular for their ornamental uses. Plant collections were widespread because plants were easy to transport (as seeds, bulbs, or cuttings) and were economical to maintain and display. Live animals, on the other hand, were more difficult and expensive to transport, maintain, and exhibit, and these difficulties made animals more coveted and exclusive. While gardens were pervasive throughout the social strata of societies, living animal collections were, for most of their history, restricted to royalty and the wealthy classes.

Animal collections, and the cultural activities associated with these collections, have not been as well studied as plant collections. No doubt the paucity of information on zoos is partly because of the scarcity of historical documentation and partly because, in the broad sweep of past cultural activities, the gathering of wild animals for collections has played a minor role. This situation is not surprising, but what is surprising is how the study of botanical gardens and natural history museums has overshadowed the study of zoos. Greater historical interest in plants and animal naturalia than in living animals is not easily explained, although the lack of historical interest in the natural world in general is less difficult to explain. The past is, of course, distorted when viewed through the lens of history. Some distortion is unavoidable since the historical record is fragmentary and pertinent information is often difficult to find, as the subject of zoos so well illustrates. Especially important to the study of zoos and aquariums, historical viewpoints are anthropocentric, with interpretations of past events focused on interactions among humans. Historian Max Oelschlaeger has concisely stated this way of thinking:

> The world becomes merely a stage upon which the human drama is enacted. The wild plants and animals, the web of life with which our humanity is bound, and without which the human drama could not be enacted, become bit players. The modern viewpoint thus impels us to relentlessly subjugate the wilderness, since things wild and free are alien to sensibilities nurtured so carefully in the garden of civilization.[3]

To test the veracity of this statement one need only look at the numerous collecting activities that have existed in virtually every civilization throughout history. Gathering animals for collections in ancient times was not as rare an occurrence as one might suspect and became steadily more common during each succeeding century. Ancient animal collections developed initially from the convergence of two ancient trends: the keeping of native wild animals for utilitarian purposes and the evolution of societies into civilizations with attributes favorable to collecting. Essential to both trends were the relationships of individuals to wild animals in particular, and to nature in general. These relationships, in turn, had their foundations in earlier preurbanized and preliterate societies, when pertinent environmental knowledge had developed out of necessity.

1.2 Keeping Wild Animals

1.2.1 Environmental Knowledge

In the Mesolithic period (10,000–8000 B.C.), which followed the end of the last Ice Age, humans were preliterate food gatherers, hunters, and fishers living in small social groups. During this era these groups developed distinct cultures, improved their tool technology, and exploited natural resources. They advanced socially, intellectually, and technologically as they adapted to the challenges of a changing environment, caused by the ending Ice Age with its climatic changes and accompanying changes in local flora and fauna distributions.[4,5] It was also a time when humans began refining their aesthetic appreciation for music, art (figurines, drawings, and decorative pottery) and luxury items (jewelry).[6–9]

While still rudimentary, aesthetic and intellectual interests in nature were parts of a broader environmental knowledge that, among other things, laid the foundation for keeping animals. Humans accumulated knowledge about their natural surroundings over many generations and each social group became increasingly familiar with the local animals, plants, habitats, and

	Historical Periods and the Environmental Activities Characterizing Them
10,000–8000 B.C. (Mesolithic period)	Small social groups Environmental knowledge (an integration of science, religion, and magic) Environmental changes due to the end of last Ice Age Minimal impact on natural resources Food gathering, hunting, and fishing Folk systematics, ecology, and medicine
8000–3000 B.C. (Neolithic period)	Villages (kin groups, tribes with chiefs) *Domestication* (animal and plant) *Agricultural gardens and animal yards* Exploitation of natural resources More complex myths and religious rituals
3000 B.C.–A.D. 1500	Urbanization and city-states *Gardens, parks, game reserves, fishponds, animal collections* Comprehensive exploitation of natural resources Knowledge of environmental change, but not its effects Scientific knowledge beyond observation Writing and history
A.D. 1500–2000	Nation-states *Menageries, zoological gardens, and aquariums* Cabinets and natural history museums, botanical gardens Conservation awareness and environmental ethics Human-caused extinctions and extensive natural resources depletion Scientific specialization (natural sciences from natural history)

weather essential to its survival. This kind of knowledge has since been categorized as "folk systematics" and "folk ecology."

Folk systematics concerns the identification and classification of animals (and plants) important to the social groups, based on gross morphological similarities or differences among the animals as well as on the utility of the animals to the group. Each distinct kind of animal had a name and was part of hierarchically arranged categories, much as they are today. For the most part, observations of biological discontinuities used in preliterate societies have been shown to correspond closely to those now recognized in modern classification schemes, and a large portion of their individually named animals and plants correspond with currently recognized taxa.[10–13]

Preliterate societies could recognize an average of 520 plants and 390 animals, based on studies of more modern, yet still primitive, societies. The memory and verbal abilities of individuals, rather than the biological diversity of the region, limited the number of animals and plants that could be identified. With the advent of writing, literate societies expanded upon this basic system of identification and classification, which continued to suffice until 1758, when Carolus Linnaeus (Carl von Linné) codified the folk systematics of Europe into a scientific system using Latin binomial (genus and species) names for each animal and plant.[11,13]

In addition to information on specific kinds of animals and plants, preliterate societies were intimately knowledgeable about the habitats of these animals and plants. This folk ecology included information about the weather and seasons, the geography of the region, the location of water sources, animal behavior, and the germination and growth of plants — environmental information that was essential to the emergence of domestication and agriculture.[14–16]

FOLK SYSTEMATICS TO SCIENTIFIC SYSTEMATICS

An inclination to categorize has always been a part of human nature. With regard to animals and plants, this process is known as systematics. Studies of several "primitive" societies that exist relatively unchanged today indicate their forebearers developed accurate biological classification systems and common organizing principles. Organisms are placed in naturally occurring groupings, with each group having easily recognized characteristics distinguishable from characteristics of other groups (what is now referred to as a natural system of classification). These groupings are further merged into several increasingly larger units arranged in a hierarchical order. The extent to which individual genera or species are named depends on the importance of the animals to that society.

Within each society studied, the number of animal taxa recognized (with an average of 390) was fewer than the number of plant taxa (with an average of 520); the number of plant taxa recognized was higher in agricultural societies. Throughout history, prior to the Renaissance, the number of taxa recognized remained relatively constant. Western systematics did not evolve out of these primitive classification schemes, however, because the schemes of the primitive groups studied were unknown to the ancient Greeks, who developed the foundations of modern Western systematics independently. Medieval European systematics, in turn, was derived primarily from the work of Aristotle for animals, Dioscorides for medicinal plants, and Theophrastus for all other plants.

This classification work was considered inadequate during the early Renaissance and afterward when an increasing number of new species were discovered. In 1700, the French botanist, Joseph P. de Tournefort, published a botanical list of 698 genera, not much more than the average number of genera recognized by primitive societies and the Greeks. Within a half century, however, this number almost doubled with Linnaeus' description of 1,239 genera in 1764.

Carolus Linnaeus (Carl von Linné, 1707–1778) built upon the folk systematics of Europe as it existed in the first half of the eighteenth century, but with substantial improvements, including the use of binomial nomenclature (using a genus and species to identify each distinct organism). This replaced the polynomial nomenclature (where the name of each organism

This systematic and ecological knowledge was transmitted through folklore and folk rituals, which both aided and impeded the progress of these societies. Although folklore and rituals contained practical information about animals, plants, and the environment, they also perpetuated myths. Fact and fiction frequently merged into an inseparable body of lore, with religion, magic, medicine, and science indistinguishable. Similarly, humans, nature, and the gods were one interdependent whole. Nevertheless, this rudimentary base of knowledge was as rational and scientific as it could have been at the time and provided preliterate populations with a way of looking at, analyzing, and understanding a difficult-to-comprehend world.[9,17–19]

Social complexity increased, as did intellectual and technological solutions to the increasingly complex problems faced by these societies. As rudimentary knowledge expanded beyond simple natural curiosity and need-to-know, humans began to progress as a unique species and began

> Number of Animal and Plant[a] Generic Taxa Discerned by Various Societies
>
> | Animals | | Plants | |
> |---|---|---|---|
> | *Primitive Societies* | | | |
> | Ndumba (New Guinea) | 186 | Primitive societies range | 238–956 |
> | Sahaptin (United States) | 236 | Primitive societies average | 520 |
> | Piaroa (Venezuela) | 305 | | |
> | Tzeltal (Mexico) | 335 | | |
> | Kalam (New Guinea) | 345 | *Greek Naturalists*[b] | |
> | Anindilyakwa (Australia) | 417 | Theophrastus | 550 |
> | Tobelo (Indonesia) | 420 | Dioscorides | 537 |
> | Hanunoo (Philippines) | 461 | | |
> | Wayampi (French Guiana) | 589 | *European Naturalists* | |
> | Aguaruna (Peru) | 606 | Tournefort (1700) | 698 |
> | Primitive societies average | 390 | Linnaeus (1764) | 1,239 |
>
> [a] Plant studies are more numerous.
> [b] Aristotle, while often mentioned, did not provide a clear set of genera.
>
> consisted of a string of descriptive words). Another important contribution was the use of Latin for these scientific names, a "dead" language known to naturalists and that did not favor one nationality over the others. The Linnaeus system of classification was an artificial one, relying on an arbitrary set of characteristics considered more important than others and, often, a set of characteristics that could not be easily discerned by the field naturalist in nature.
>
> *Systema Naturae*, published by Linnaeus in 1758 (the tenth and most important edition), is considered the starting point of modern (Western) systematics. It is the foundation upon which modern international codes of nomenclature have been built. Classification schemes continue to be artificial and increasingly sophisticated, relying on chemical, genetic, and molecular characteristics. After remaining stable for most of history, the number of genera exploded to several thousand in the eighteenth century, to a few million in the nineteenth century, to several million in the twentieth century.
>
> Sources: Based on Berlin, Brent, *Ethnobiological Classification: Principles of Categorization of Plants and Animals in Traditional Societies*, Princeton University Press, Princeton, NJ, 1992; Raven, Peter H., Berlin, Brent, and Breedlove, Dennis E., "The origins of taxonomy," *Science*, 174, 1210, 1971; and Morton, A. G., *History of Botanical Science: An Account of the Development of Botany from Ancient Times to the Present Day*, Academic Press, New York, 1981.

to distinguish between themselves and other species. Preliterate humans, however, continued to be part of the natural order, dependent upon an uncontrolled, and still pervasive, wilderness. Humans were not able to escape from this situation until they became relatively self-sufficient producers with the advent of domestication and agriculture.

1.2.2 Domestication of Wild Animals

As early humans developed into a socially and intellectually unique species, they intensified their dominion over other species, while distancing themselves from other species through development of socially elaborate cultures. These developments accelerated during the Neolithic period (8000–3000 B.C.) as environmental conditions emerged that were favorable for the settlement of new kinds of organized, and relatively permanent, communities. This in turn provided the impetus for domesticating animals.[4,5]

Some early societies remained groups of gatherers, hunters, and fishers, entirely dependent upon nature. Other societies became self-sufficient producers by extending their control over the animals and plants near their settlements. This control involved farming plants (agriculture) through controlled seeding, harvesting, storage, and maintenance, as well as farming animals (domestication) through controlled breeding, feeding, and maintenance.

Initially, animal husbandry evolved from the practice of herding or penning wild animals, perhaps those animals that raided the crops, those that were easily corralled, or young animals kept as pets. However it might have begun, gaining this initial control of wild animals was a difficult, but important, first step. Some wild species were more compatible with the social environment of humans than were others. Compatible wild animal species had attributes that favored the process of domestication: hardiness and the ability to adapt to new surroundings, social gregariousness that permitted herding and overcoming natural flight behavior, dominance hierarchies that recognized humans as the alpha species, reproductive behavior that adapted to captive breeding conditions, temperaments that facilitated tending with minimum effort, and utilitarian value.[20–22]

Domesticating animals was the first effort at keeping wild animals. Although it is now taken for granted, domestication was a long-term biological process. As a biological process, domestication required the keeping of wild animals through many generations and was achieved only after significant changes were made in the behavioral, physical, and genetic attributes of the captive species. These necessary changes could not have been anticipated initially, so domestication was not known as such until after it had occurred.

Evolution of Human Social Organizations

Social Groups
— Continued as social groups
Dependent Societies (gatherers, hunters, and/or fishers)
— Continued as societies dependent on wild nature
Simple Producing Societies (domestication along with gathering, hunting, and fishing)
— Continued as simple producing societies
Complex Producing Societies
  New World
  Maize Group: Mesoamerican and Andean
Old World
  Grain Group: Alluvial River Valley Societies
    Mesopotamian
      Arabic and Orthodox (Russian)
        Western (Greco-Roman/European)
    Egyptian
    Indic (Indian)
    Pre-Sinic/Rice Group (Chinese, Japanese, Far Eastern Asia)

Source: Based on Quigley, Carroll, *The Evolution of Civilizations*, 2nd ed., Liberty Press, Indianapolis, IN, 1979.

Domesticated Species[20–22]

Species	First Domestication, B.C.	Range
Dog	10,000–8000	Global; various species of wild dogs, wolves, and jackals
Goat	8000–7000	Mesopotamia and western Asia initially; various species
Sheep	8000–7000	Mesopotamia, western Asia, and southeastern Europe initially; various species
Reindeer	8000	Scandinavia, Greenland, and polar North America; wild reindeer
Pig	7000	Europe and western Asia initially; wild boar
Cattle	6400	Mesopotamia, western Asia, and southeast Europe initially; various species
	2500–1500	Northwest India; humped cattle
Llama	5500–4200	South America; guanaco and vicuna
Horse	4000	Europe and Asia (horses); Mesopotamia and western Asia (onagers); western Asia and Africa (donkey, asses)
Camel	2600	North Africa and western Asia (one-humped camel); central Asia (two-humped)
Elephant	2500	Indus valley (India); Asian elephants
Ferret	1800–400	Europe; European polecats
Cat	1600–500	Egypt, Europe, Mesopotamia, India; various species small wild cats
Rabbit	?	Europe; wild European rabbit
Guinea Pig	?	Peru; South American cavy
Bird	?	India (chicken and peacock), Asia (pheasant), Africa (guinea fowl), Central America (turkey), Mesopotamia (pigeon, goose, and duck)

Palaeozoological evidence from archaeological sites indicates that only a few species were kept for domestication purposes, although other species were present. Untamable species were eventually killed for food or released, while those that did not present difficulties were maintained over time and eventually domesticated. Regardless of how many species were kept, only a few were successfully domesticated and an unknown number remained wild. It is possible that the still-wild species, rather than being killed for food or released, were retained and kept as protocollections. Since changes occur in incremental steps, this kind of experience may possibly have been a precursor to later collecting activities.

Domestication enabled human population growth and affected the social structure of societies. Societies, previously dependent upon nature, became producing societies, some maintaining simple social structures, others developing more complex structures. Both simple and complex producing societies kept domesticated livestock; however, other factors existed that propelled the societies leaning toward complexity into an era of urbanized and literate civilizations. Regardless of whether protocollections of native wild species existed previously, it was at the newer urbanized level of socioeconomic development that humankind entered into the collecting of wild species for aesthetic, nonutilitarian reasons.[23]

1.2.3 Beyond Domestication — Collecting Wild Animals

Complex producing societies evolved into urbanized, literate civilizations between 3000 and 1500 B.C. The transition from society to civilization was both a physical and a philosophical phenomenon. Physical aspects of urbanization accommodated increasingly large populations, providing social order, civil administration, common defense, cooperative and specialized labor, and foreign trade to obtain needed materials not available locally. Although the bulk of the population had little money or time for leisure activities, royalty and a wealthy class of individuals — which included government officials, priests, merchants, and landowners — prospered and took advantage of cultural luxuries that became available.

For these privileged classes, the urbanized lifestyle provided a relatively stable social environment conducive to long-term endeavors such as collecting. These wealthy individuals were able to accumulate large tracts of land, not all of which had to be developed for economic reasons and could, therefore, be set aside as animal parks for their pleasure. Expansion of foreign trade increased exposure of these wealthy individuals to exotic lands and the wildlife from these lands. Wealth to buy what they wanted, leisure time to do what they wanted, the availability of luxury items, and heightened aesthetic sensibilities provided the newly emerging upper class with the opportunity to plant gardens, build reserves and parks, and collect animals.

Philosophical aspects of urbanization involved attitudes that focused on the distinction between the city and the country, as well as intellectual issues associated with the dichotomy. For the first time, urbanized human environments differed substantially from the rural and wilderness environments. As urban areas increased in number and size, and urban citizens were more removed from the country, these individuals began to feel the loss of their natural environments. The need for a connection with nature became stronger as urbanized populations continued to grow and their lifestyles became increasingly complex. This need contributed to the upper-class desire to re-create natural settings and to collect animals, a desire that existed in tandem with their newfound ability to pursue these activities.

Ancient civilizations first emerged in Mesopotamia along the Tigris and Euphrates Rivers of western Asia in what is now Iraq; in Egypt along the Nile River in northeastern Africa; among the Indus society of India along the Indus River in what is now Pakistan; and in China, along the Yellow River (the Huang Ho). There were other, less complex societies throughout Europe, Asia, and Africa; however, they probably did not develop the social structures that were favorable to collecting. Over time, a complex evolution of animal collections occurred as the epicenter of power shifted throughout these regions and finally settled in Europe. Animal collections in Mesopotamia and Egypt became less prominent but nevertheless continued, while those in India disappeared with the Indus society and reemerged with India's Indo-Aryan societies. Collections in China and the Americas developed in isolation, and newer collections appeared in Greece, Rome, Persia, the Arab regions, and eventually Medieval Europe.

1.3 Ancient Collections

1.3.1 Mesopotamian Collections

Mesopotamian societies developed as riverine city-states along the Tigris and Euphrates Rivers. Sumer was the first, flourishing from ca. 3000 to 2800 B.C. in the lower and middle portions of the Mesopotamian region. Akkad society superseded the Sumer society between 2800 and 2200 B.C. The Akkad, located in the middle and upper portions of the Mesopotamian region, eventually ca. 2200–330 B.C. split into Babylonia (the middle region) and Assyria (the upper region). Eventually, these societies were conquered by the Persians in 539 B.C., and later by Alexander the Great of Greece in 330 B.C.[15, 24–27]

Mesopotamian city-states had to protect themselves from the ravages of a harsh environment with unpredictable rivers (floods and periodic changes in the courses of these rivers). This region's environment was perceived as a place of uncontrollable forces, fierce animals, and wicked demons. Mesopotamian gods and religion were centered on the forces of nature, the heavens and its planets, the visible earth, and the invisible interior earth. Every natural animate and inanimate object, as well as everything manufactured, was assigned to the sphere of a particular god. Divination, magic, medicine, and the sciences were often indistinguishable.

Knowledge consisted of the ability to produce quoted phrases of established information appropriate to a given situation, requiring memorization rather than original thought. This information was written on clay tablets and stored in libraries, which most Mesopotamian kings maintained.

The natural sciences, less important than mathematics and astronomy, were of practical importance and consisted of observing and classifying animals, plants, and minerals. The practice of medicine was first recorded by the Sumerians, but was more fully developed during the Babylonian and Assyrian period. Physicians practiced human medicine, whereas veterinarians known as "ox and ass doctors" practiced animal medicine. Knowledge about animals and plants consisted of being aware of, but not necessarily understanding, their occurrence, properties, and habits. Based on this knowledge, animals and plants were listed in groupings that were rudimentary classification schemes. Boundaries between animals, plants, and minerals, as well as the groupings within each, were recognized. These natural resources were all given names and some thought was given to their interrelationships. By ca. 2500 B.C. the natural world around Mesopotamia was categorized into domestic animals, wild animals, wild birds, fishes, insects, plants, trees, vegetables, and minerals.

Mesopotamians believed that nothing existed without a name, and that once named, the namer held power over the named. They also thought that putting something in writing gave it, and the knowledge about it, permanence. This had the unfortunate effect that texts, once established, were used uncritically, stifling original thinking on a given subject. Thus, efforts to develop knowledge from practical observations of nature and wildlife had begun, but did not progress very far. Nevertheless, the natural world was beginning to be controlled and understood, albeit slowly.

At the same time, the natural world and its wildlife were also increasingly utilized. Much of the area around the rivers was suitable for agriculture but lacked resources necessary to sustain other advancing technologies. This situation encouraged the establishment of trade, which became well developed over time. Mesopotamian merchants traded with other civil societies, as well as with the "barbarians" and their nomadic caravans. Of primary concern was the acquisition of essential raw materials lacking in the Tigris–Euphrates region; however, as commercial ventures progressed and the wealthy prospered, trade in luxury items increased. Included in this trade in luxury items were animals, which were also obtained through confiscations in conquered lands and as tribute from other societies. Early trade in wild exotic animals was undoubtedly negligible since it only fed the passion of a small class of wealthy individuals; however, over time the trade grew as royalty and an increasing number of wealthy individuals attempted to surpass the collections of one another and their predecessors.

Although all classes of society had kitchen gardens and fishponds, royalty and the wealthy landowning class had shade gardens, ornamental gardens, and parks. Sumerian shade and ornamental gardens were often small and combined with vegetable plots and orchards. Purely ornamental gardens developed later during the Babylonian and Assyrian period when the wealthy class became more prosperous. For the most part, these gardens and the smaller parks were strictly botanical; the larger parks contained the animal collections.

One of the earliest references to gardens is in the Gilgamesh epic, a story about the adventures of Gilgamesh, King of Uruk (Sumeria) ca. 2750 B.C.; one city he encountered in his adventures was proud that one third of its territory consisted of garden-orchards. Although references to gardens or garden-orchards between 3000 and 2000 B.C. are scarce, later references between 2000 and 1000 B.C. to house, temple, and royal gardens are more numerous. By the later Babylonian and Assyrian period, between 1000 and 330 B.C., references to gardens and the larger royal parks become even more common. Likewise, land records during this later period indicate the extent to which gardens had become common features of the wealthy citizen's property holdings.

Property holdings included both domesticated and wild animals. Animals kept included household pets, fish in ponds, birds in flight cages, falcons for sport, lions in cages, and wild game in parks. Individuals used blunt arrows to stun wild animals, and traps (usually concealed pits) were used to catch these wild animals alive for pets, collections, and trade. Some animals, particularly those species rarely seen, were valuable luxury items. Royalty frequently kept tame lions as pets, and other lions were used for hunting or fighting. Lions and other animals were kept for exhibit purposes to impress and entertain local guests and foreign dignitaries. Royal lions were kept in cages and pits during the Ur III period (beginning ca. 2100 B.C.). It is conceivable, therefore, that cages were constructed to hold other dangerous or rare species as well.

Bas-reliefs from Assyrian royal palaces show monkeys, antelopes, camels, elephants, and other species brought to the Assyrian kings as tribute. Although they usually viewed nature from a practical perspective, the kings of Sumeria, Babylonia, and Assyria were proud of their animal collections, which were symbols of power, wealth, and authority. The kings were especially proud of rare specimens their subjects and foreign dignitaries sent after diligent searches and difficult transport. Animals came from Asia via trade with the Indus society and from Africa via trade with the Egyptians.

Royalty and wealthy individuals constructed fishponds that also served the economic purpose of keeping fish fresh for the table. They kept wild birds such as ibis, cranes, herons, peacocks, and pelicans as pets and in flight cages. Initially, these bird collections also served the economic purposes of being handy food sources and commodities for sale. Falconry and wild beast (primarily lion) hunts were royal sports conducted in the wild and in the royal parks. Royal park collections of elephants, wild bulls, lions, apes, ostrich, deer, gazelle, ibex, and other species were combination protomenageries, hunting reserves, and garden-parks. The royal families also used the parks for entertaining guests and for personal pleasure.

Eventually, the ability to maintain animals and plants in large park areas was taken to a new level of sophistication with the re-creation of entire habitats. Sennacherib (Assyria, 704–681 B.C.) simulated a marsh environment of southern Babylonia to exhibit rarely seen marsh species

Lion released from its transport crate into the animal park of Ashurbanipal, King of Assyria (668–627 B.C.) at Nineveh, Mesopotamia. This scene is part of a relief that was on one of the palace walls. © The British Museum.

Royal Parks and the Hanging Gardens of Babylon–Nineveh

Babylonian and Assyrian royal parks and hanging gardens were the result of Mesopotamian garden evolution. Some of these parks and gardens may have been public parks for the benefit of the cities in which they were established. However, for the most part, they were for the use and enjoyment of the royal family. Royal parks and gardens were often the site of royal hunts, a place to entertain guests, and a place to keep animals.

Tiglath-Pileser I (Assyria, 1114–1076 B.C.) kept herds of deer, gazelle, and ibex from conquered territories in his park. He was proud that some of these animals were rare and had never before been seen in Assyria. Ashurnasirpal II (Assyria, 883–859 B.C.) had herds of wild bulls, lions, ostriches, and apes, along with many species of imported trees and fruiting plants. Sargon II (Assyria, 721–705 B.C.) was particularly fond of lions and falcons, and laid out several parks around his capital city. Merodach-Baladan II (Babylonia, 721–710 B.C.) had extensive gardens that may have been the predecessors to the fabled hanging gardens. Sennacherib (Assyria, 704–681 B.C.) laid out several parks around his capital (Nineveh) and imported trees and other plants. He also re-created a southern Babylonian marsh environment when he had a swamp created and populated with animals and plants imported from the actual marsh habitat he admired. Sennacherib, Esarhaddon (Assyria, 680–669 B.C.) and Ashurbanipal (Assyria, 668–627 B.C.) re-created habitats of the Amanus mountains in Syria. The conquering Achaemenid (Persian) kings (539–331 B.C.) and Greek rulers that followed continued this tradition of extensive gardens, parks, and animal collections. Some of these collections still existed when the Roman armies invaded the region in A.D. 363.

The hanging gardens have become the most famous of these collections. They were urban gardens planted on the terraces of ziggurats, and from a distance they appeared to be vegetation-covered mountains. Surprisingly, these spectacular gardens were not mentioned in Mesopotamian cuneiform texts, which described virtually all aspects of life, business, administration, and royal activities. Their description instead comes to us from the writings of Greek travelers, such as Berossos, Strabo, Quintus Curtius Rufus, and Diodorus Siculus. The particular hanging garden made famous in these Greek accounts was thought to be one Nebuchadnezzar II (Babylonia, 604–562 B.C.) planted at Babylon to remind Nebuchadnezzar's wife of her mountain meadow homeland, which she missed living in urbanized Babylon.

New research, however, indicates the hanging gardens of Babylon were actually at the palace garden of Sennacherib located at Nineveh, also known at the time (ca. 700 B.C.) as Old Babylon. It was one of Sennacherib's re-creations of habitats that he enjoyed, in this case mountain scenes. The gardens were not unique, as these were only one of many efforts Sennacherib, and his predecessors and successors made to create magnificent gardens, parks, and animal collections. But they were likely the hanging gardens the Greek travelers saw and wondered at.

Source: Based on Dalley, Stephanie, "Ancient Mesopotamian gardens and the identification of the hanging gardens of Babylon resolved," *Garden History*, 21, 1, 1993 and Finkel, Irving L., "The hanging gardens of Babylon," in *The Seven Wonders of the Ancient World*, Clayton, Peter A. and Price, Martin J., Eds., Routledge, London, 1988, chap. 2.

from that region of Mesopotamia. He also re-created mountain habitats, one of which is now thought to have been the site of the fabled hanging gardens of Babylon (actually in Nineveh).[28,29] His successors, Esarhaddon (Assyria, 680–669 B.C.) and Ashurbanipal (Assyria, 668–627 B.C.), also provided mountain habitats that resembled the nearby Amanus mountains. Within the cities, the terraces of the monumental, pyramid-shaped, terraced buildings known as ziggurats sometimes were planted with trees, shrubs, and vines to give a mountain-like appearance, similar to the hanging gardens of Babylon.

One aspect of civilization was specialized labor, and this included work related to the keeping of wildlife. Fishermen, bird keepers–fowlers, and shepherds cared for the domestic stocks. Servants (animal keepers, perhaps) captured animals in the royal park for release at the appropriate time and place for royal hunts. These servant/animal keepers probably cared for the animals as well. Veterinarians, the ox and ass doctors, dealt primarily with domestic and military livestock, and their fees were preset in King Hammurabi's (Babylonia, 1728–1686 B.C.) Code of Laws.

Animal collections in this region of the world were not unique to Mesopotamian societies. Neighboring Hittite societies also kept pets, domestic livestock, and wild animals; however, their wild animal collections were less extensive than those in Mesopotamia. Their collections also served more than one purpose: secular uses, including food, hunting, and pleasure; religious uses, including rituals that required animals; and symbolic uses, for which certain species, such as lions and eagles, were reserved for royalty as emblems of their power and authority. As with Mesopotamian collections, Hittite collections contained native and exotic species, including lions, tigers, leopards, wolves, foxes, bears, deer, wild goats, boars, bison, elephants, hares, mice, eagles, snakes, frogs, bees, ants, spiders, and several other animals that have not yet been identified.[30]

Persian kings, who conquered Mesopotamia in 539 B.C., continued the Mesopotamian collections using their own traditions of garden design and animal collecting. Greek rulers, conquering the Persians in 330 B.C., did the same, as did the Muslim Arab rulers who followed. Collections in this region still existed in A.D. 363 when Roman armies conquered Babylonia.

1.3.2 Ancient Egyptian Collections

Egyptian civilization began with the unification of lower (the northern delta region of the Nile River) and upper (the southern riverine region) Egypt into the Old Kingdom (ca. 2700–2200 B.C.). Several periods of fragmentation and unification followed: the First Intermediate period and Middle Kingdom (ca. 2200–1786 B.C.), the Second Intermediate period and New Kingdom (ca. 1786–1087 B.C.), the Late Dynastic period (ca. 1087–332 B.C.), the Ptolemaic period (under Greek rule, 332–30 B.C.), and the Roman province period (beginning in 30 B.C.). As time went by, the southern region of upper Egypt was extended farther southward along the Nile River until it covered approximately the same area as does modern Egypt.[15,25,31–35]

This riverine area the Egyptians inhabited was a long, narrow valley dependent upon the seasonally predictable Nile River. There were numerous cities and villages, but these settlements were essentially rural communities with no sharp distinctions between the urban and countryside environments. Kings and priests, with the help of an extensive bureaucracy, controlled government and daily life. Royalty, priests, government officials, and certain other individuals formed the wealthy upper class, artisans and shop owners formed the middle class, and serfs (workers) and slaves formed the lower class. Over time, the society in general, and the upper class in particular, became more prosperous and their luxuries became more extravagant. In the later dynasties, whoever could afford it indulged in a beautiful villa, a fine carriage, a boat, numerous slaves, rich clothing, costly food, good wine, a large herd of cattle, elaborate gardens, and exotic animals.

Egyptian belief in magic and superstition hindered significant advances in learning. Nevertheless, outstanding achievements were accomplished in astronomy, geography, biology, and technology. Egyptians particularly understood most aspects of basic mathematics, the only science their magic and superstition did not contaminate. Physicians, including veterinarians, developed useful medical and botanical knowledge; however, this knowledge was heavily laced with magic and superstition. These two areas of expertise and belief were closely linked, for many of the drugs and potions the physicians used were of plant origin. Ingredients of animal origin in these drugs and potions were rare. Beyond this, zoology and botany apparently did not become well developed although native and exotic animals and plants were known to the Egyptians, domestication of wild animals and plants was one of their most important activities, and the keeping of animals and plants was one of their great passions.

Representations of animals and plants were pervasive throughout Egyptian culture. Many Egyptian gods assumed animal forms, while other kinds of animals were worshiped without being transformed into gods. Symbols for the lower and upper regions of the Nile River Valley were flowers (the papyrus and the flowering rush, respectively). Animals and plants were graphically rendered as much as any other subject. Domestication of a large variety of birds, ruminants, and carnivores was attempted.

Physically, the extent of the Egyptian's natural world was not great. The Nile Valley was relatively narrow, with deserts bordering it on both sides. It had an almost harborless Mediterranean coast to the north and the unknown tropics of Africa to the south. While the valley was pleasant, it was also monotonous, with a limited variety of wildlife, few trees, and few wildflowers. Wood was used extensively in the building of structures and boats during the Old Kingdom period, resulting in a gradual decrease in trees and vegetation. Orchards and trees became valuable in the Middle and New Kingdom dynasties, perhaps because so many were cut down during the Old Kingdom. Trees represented shade and protection in an otherwise hot and harsh environment. Deities were thought to live in trees and several groves in each district were considered sacred. The Egyptians believed that trees were essential in cemeteries and near water wells, tombs, and temples. They were valued in every kind of garden.

Animal husbandry was primarily concerned with cattle, but the Egyptians experimented with many species. In addition to several breeds of cattle, there were breeds of sheep, goats, donkeys, and pigs. Horses and mules came into use between the Middle and New Kingdoms, and the camel during the Ptolemaic period. Egyptian domestication attempts included many different kinds of native wild ruminants and carnivores. These animals were fattened on bread dough, as were many kinds of birds. A variety of birds were kept in domestic flocks, particularly geese and ducks, but also swans, doves, ibis, cranes, herons, and other marsh and waterbirds. Bees were kept both in the desert and in private gardens, for their honey.

Animal-related jobs in Egyptian society included herdsmen, fowlers–bird catchers, beekeepers–honey gatherers, and veterinarians. No doubt there were also animal keepers, but as yet no specific information is known about the individuals who maintained the animal collections. Veterinarians were "physicians" — no separate term existed for animal doctors — who cared for cattle and whose practice was based on human medicine. There are no records to indicate clearly how valued the veterinarian, or any of these animal-related positions, was in Egyptian society. Agriculturists worked harder than anyone at what was considered one of the most important jobs, but these individuals apparently received little recognition or appreciation.

Fowlers came from the lower class, as bird catching was considered one of the dirtiest jobs. Bird hunting, on the other hand, was a sport of the upper class. As one Egyptologist has observed:

> Much of the country formerly covered by marshes and tropical forests was already arable land. At the same time old river beds remained ... [and] the greatest delight perhaps that the Egyptian knew was to row in a light boat between the beautiful waving tufts of the papyrus reeds, to pick the lotus flowers, to start the wild birds and then knock them over with the throw-stick, to spear the great fish of the Nile and even the hippopotamus, with the harpoon. Pictures of all periods exist representing these expeditions, and we have but to glance at them in order to realize how much the Egyptians loved these wild districts.[36]

Both bird hunting and spearfishing, as opposed to commercial bird catching and fishing, were sports of the wealthy. Other sports included harpooning hippopotamus and crocodile and desert hunts for gazelle, antelope, hyena, jackal, fox, leopard, lion, and numerous small mammals.

In addition to being a sport of the wealthy, desert hunting was a professional occupation of archers who patrolled the deserts and who protected the honey and resin gatherers working far into these deserts from marauding tribes. Royal hunts ventured not only to the desert regions, but also to more remote regions outside of Egyptian territory. Kings of the New Kingdom dynasties traveled up the Nile River into African territory and along the Euphrates River into western Asia to hunt elephant, rhinoceros, lion, and wild bull. Although hunting in the wild deserts and other remote regions predominated, "there is hardly a decorated tomb without its scene of the owner launching showers of arrows with unfailing accuracy at gazelles and antelopes in a fenced-in park rather like a zoo."[37]

> Wealthy Egyptians ... at all times kept menageries, in which they brought up the animals taken by the lasso or by the dogs in the desert, as well as those brought into Egypt by way of commerce or as tribute. From the neighboring deserts they obtained the lion and the leopard (which were brought to their masters in great cages), the hyena, gazelle, ibex, hare, and porcupine, were also found there; from the incense countries and from the upper Nile came the pard, the baboon, and the giraffe; and from Syria the bear and the elephant.[38]

Egyptians particularly liked to tame (and eventually domesticate, as has been mentioned) as many of these species as possible. Tuthmosis IV (1425–1408 B.C.) was accompanied by two tame lions used to hunt antelope, and Ramesses II (1298–1235 B.C.) had a tame lion that not only accompanied him into battle, but also guarded the royal tent at night.

Monkeys and birds were often kept as pets. Birdcages in the houses of royalty and wealthy citizens became popular in the New Kingdom, in addition to the more common birdhouses that had always been used for keeping fowl for food, tribute, and sacrifice. Egyptians took special delight in displaying wild birds from the marshes, birds of prey, and imported exotic species. Fishponds and beehives were also maintained, and were often used to complement gardens. Pools became microhabitats containing fish, birds, papyrus, lotus, and other aquatic plants. Grander houses and palaces had rooms opening onto gardens in such a way that painted gardens on the walls and floor blended with the real garden just outside the rooms. Paintings of plants, birds, and marsh life, together with live birds in cages, blurred the boundary between house and garden as the garden was entered from these rooms.

Trading expeditions were made to obtain needed resources, exotic materials, and animals. These expeditions were made to the Holy Land in the east extending up through Palestine, Syria, and Mesopotamia for wood, gold, silver, precious stones, horses, bears, elephants, fish, and cattle. They extended to Nubia in the south via the Nile River for gold, ebony, ivory, acacia wood, precious stones, dogs, cattle, ostrich feathers and eggs, panthers, panther skins, giraffes, and monkeys. They extended to the land of Punt farther south in Africa via the Red Sea for incense, myrrh, other plants, and exotic African animals. Little is known about these trading expeditions other than those to the land of Punt.[39,40]

When the Greek Ptolemaic dynasties (323–30 B.C.) replaced the Egyptian New Kingdom dynasties, the city of Alexandria boasted many great amenities. One of these included the largest and most varied animal collection in the ancient world. When the Romans took over Egypt in 30 B.C., this animal collection in Alexandria continued and was probably the source of supply for African animals used in Roman spectacles. Since the collection during these later periods was more properly Greco-Roman than Egyptian, it will be discussed later.

EGYPT'S PUNT EXPEDITIONS

Ancient Egyptians did not often undertake long cross-country journeys; however, trading expeditions to the land of Punt, an African country in what is now the Ethiopia area, had been made since at least the fourth dynasty of the Old Kingdom. These long and difficult expeditions were motivated by the desire of the Egyptians for Punt incense and myrrh, which were used in Egyptian religious services. Even so, there was scant mention of these expeditions until the New Kingdom rule of Hatchepsut (during the eighteenth dynasty, ca. 1520–1480 B.C.). Hatchepsut, the only female pharaoh (something that was generally forbidden — all other female rulers were queens), accomplished much during her well-organized administration. Under her rule, the most ambitious and famous expeditions were made to Punt between ca. 1510 and 1490 B.C. These expeditions were proudly recorded pictographically in her Deir el-Bahari temple.

Expeditions to Punt under Hatchepsut's sponsorship were significant because of their size and the large amount and variety of items brought back to Egypt, including live myrrh trees for the first time. These trees were planted on the terraced grounds of Hatchepsut's temple to resemble the myrrh terraces found in Punt, thus providing Egypt with its own supply of myrrh. Other items of importance included incense (frankincense), gold, silver, precious gems, ivory, ebony, slaves, animal skins, and live animals.

Animals brought to Egypt on the Punt expeditions included birds, greyhounds, cattle, monkeys, apes, leopards, rhinoceroses, and giraffes. Some of these animals may have been seen in Egypt for the first time, and they formed the largest known animal collection in Egypt to that time. Unfortunately, although Hatchepsut's temple paintings proclaim the greatness of these acquisitions, they do not depict where the animals were housed or how they were maintained.

Hatchepsut's successors, including the pharaohs of the eighteenth through twentieth dynasties (ca. 1580–1085 B.C.) continued her Punt expeditions. Of particular significance were those of Tuthmosis III, Amenhotep II, Tuthmosis IV, Amenhotep III, Ramesses II, and Ramesses III. These pharaohs, along with Hatchepsut, had extensive animal collections, primarily because they were supplied with exotic African species from Punt.

Source: Based on Montet, Pierre, *Everyday Life in Egypt*, Edward Arnold, London, 1962 [translated by A. R. Maxwell-Hyslop and Margaret S. Drower]; Naville, Edouard H., *The Temple of Deir el-Bahari*, Trubner & Co., London, 1894–1908. [issued as the 12–14th, 16th, 19th, 27th, 29th Memoir of the Egypt Exploration Fund]; and Tyldesley, Joyce, *Hatchepsut: The Female Pharaoh*, Viking, New York, 1996.

1.3.3 Ancient Asian Collections — India and China

Observations of the earliest European explorers and merchants to reach Asia indicate royal animal collections existed throughout the region. These observations, made during the sixteenth and seventeenth centuries, are brief morsels scattered among many observations in voluminous travel accounts. Collections mentioned in these accounts were probably continuations of older royal collections, but until more is known about them, the age of these collections remains undetermined. Considering what is known about collections in India and China, however, these other Asian collections may date to ancient times.

An Indus civilization that existed from ca. 2500 to 1500 B.C. was centered around two principal cities, Harappa and Mohenjo-daro, on the Indus River in what is now Pakistan. To date, little is known about the zoological activities of this Indus civilization; however, animals were domesticated, most importantly the elephant. Contact existed with Mesopotamia, as evidenced by the authenticating seals used on traded commodities. These seals depicted wild animals, such as elephants, tigers, rhinoceroses, buffalo, antelope, and gharials. Animals such as peacocks and apes, as well as ivory, were exported.[15,25,41–48]

A wealthy class of people in Indus society enjoyed gold, silver, jewels, and other marks of wealth. Because of their wealth, their artistic depiction of wild animals, and their contacts with other ancient societies (which had animal collections), it is probable that animal collections of some kind were part of Indus society. However, no direct evidence indicates that any structures were used for animal collections.[49] Since the Indus written language has not been deciphered, their attitudes toward nature or animals are not known.

There was little continuity between the ancient Indus civilization and the later Indo-Aryan societies that prospered throughout India between ca. 1500 B.C. and A.D. 1500. Indo-Aryan animal collections are much better known because there is more documentation and because these collections still existed when the Europeans arrived. Indo-Aryan societies, along with their animal collections, may be representative of the kinds of collections that existed throughout Asia during this same period. These collections had uniquely Asian characteristics that are best described by Walker in Chapter 8 on the zoological gardens of India.

Another region in Asia with a highly developed civilization was China. Like Egypt, China's history has been marked by alternating periods of unification and fragmentation. The first period of unification began with the Shang dynasty (ca. 1500–1000 B.C.) along the Yellow River.[50–56] Periods of unification represented stable times under the most influential rulers. It was during these periods that significant animal collections often developed. Some dynasties are better known for their animal collections and gardens, the importation of exotic species, and the study of the natural sciences; nevertheless, it seems they all had animal collections.

Ancient Chinese attitudes toward wild animals and nature vacillated from ethical concern to uncaring use, but for the most part they were unfriendly and fearful. The Chinese viewed animals and landscapes they did not control as menacing and alien, as much, if not more so, than did the Mesopotamians or Egyptians. These attitudes moderated in later dynasties but never completely subsided. Intellectual interest in the natural sciences was minimal except for plants, animals, and minerals used for medicinal purposes. Nevertheless, wild animals, especially exotic species, held the interest of rulers and the wealthy class, as they did in other ancient societies.

From the beginning, royal animal collections served both practical and aesthetic purposes. Starting with the founder of the Shang dynasty, China's rulers built animal reserves; however, it was Wen Wang, founder of the Zhou dynasty, who built the first well-known animal reserve. This reserve and similar royal and baronial parks (parks owned by the wealthy class) of the Zhou period (ca. 1000–200 B.C.) were large, walled-in natural areas that required their own staffs of administrators, keepers, and veterinarians.

Han dynasty (ca. 200 B.C.–A.D. 220) rulers continued this tradition of large royal parks. Emperor Wu Di enlarged an established reserve around 140 B.C. The reserve was a large wilderness area, but it also had some 70 palace buildings and water gardens. It contained exhibits of gems and minerals, native and exotic plants, and animals, including swans, geese, ducks, waterbirds, shorebirds, great bustards, herons, cormorants, turtles, alligators, sturgeons, various aquatic species, camels, asses, mules, horses, yaks, deer, elephants, rhinoceroses, and possibly giant pandas. Animals were obtained locally and from Mesopotamia, India, and other regions of Asia. In addition to keeping animals in their parks, Han emperors kept birds, bears, tigers, and other mammals in palace annexes, in what appear to have been private menageries kept for their own personal pleasure. These animals, as well as dangerous animals in the parks, were kept in cages.

Royal and baronial parks served multiple purposes. Collections in these parks were used to provide food and as sources of specimens for religious ceremonies, hunting, entertainment, and leisurely enjoyment. The parks were also the stages for combat spectacles between animals, as well as between men and animals. Chinese combat between men and animals usually involved unarmed men who fought lions, leopards, bears, yaks, elephants, or rhinoceroses. The death of the animal in the course of combat was apparently not the purpose, for it seems the combat itself was the primary interest.

Following the Han dynasty, the Qin, Tang, and Song dynasties (ca. 265–1279) continued the fashion of extravagant parks. Generic terms were used to describe different kinds of parks and gardens, for example, *yuan* (a royal park, hunting park, or forest reserve, originally used to describe a large open park) and *yu* (a fenced animal reserve or wildlife enclosure, originally meant to connote a fenced *yuan*). Many of these parks and gardens were given individual names by their owners. One park most often known by its personal name was the famous animal reserve of Wen Wang, which he called "Lingyou," commonly referred to as the "Garden of Intelligence"; however, a more accurate meaning would be "Garden for the Promotion (Encouragement) of Knowledge."

Unlike the Mesopotamian and Egyptian collections, which continued under the influence of other societies, the Chinese collections evolved without outside interference. The story of the Chinese collections of the Yuan and Ming dynasties (1200s–1600s), made famous through the travels of Marco Polo and the trading expeditions of Zheng He, respectively, is told below within the discussion of the medieval period (Section 1.3.6).

1.3.4 Greek and Roman Collections

Greco-Roman societies (ca. 1100 B.C.–A.D. 476) existed alongside those of Mesopotamia and Egypt and eventually overshadowed and conquered them both. Greek intellectual inquiries went beyond what could be learned simply through observation, the factor limiting the growth of knowledge in other previous societies. However, the Greeks' curiosity was still primarily practical and descriptive. Although they had an appreciation for nature, the Greeks were intent on controlling nature, creating what Cicero later called "another nature" out of the wilderness that surrounded them. A primary concern of the Greek city-states was the creation of this second nature, an agricultural and man-made environment. Associations between certain gods and trees were particularly strong, with places of worship located in sacred groves.

Roman natural science was derived from Greek science and made few independent, original contributions. Nevertheless, Roman efforts to copy texts (that were later lost) and compile what was known into encyclopedic tomes carried ancient knowledge about animals forward to the Medieval period. Similar to the Greek view, Roman attitudes toward nature were practical and utilitarian. While Romans enjoyed the beauty of nature and preferred rural life over urban life, they felt the land and its natural resources should be used productively. Extensive agriculture, deforestation, mining, quarrying, and hunting modified the environment significantly.[15,57–60]

By the end of the Greco-Roman period, a number of animal populations within their region of influence had been eliminated and the environment was deforested and degraded. This damage was so extensive that evidence was still obvious some 1,400 years later, prompting George Marsh to write *Man and Nature: Or, Physical Geography as Modified by Human Action*, one of the first modern books (published in 1864) concerning the need for environmental conservation.[59] To understand how the use of wildlife during this period impacted wild populations so severely, one needs to first examine the early years of the period.

While expressing a desire to enjoy and protect their environment, the early Greeks (up through 323 B.C.) depleted many of their natural resources. Agriculture and deforestation in particular modified the environment. As a result, the Greeks took an interest in forestry, controlling the cutting, exporting, and importing of lumber and experimenting with replanting some areas. Mining and quarrying also presented environmental problems, as did hunting, which brought some animal populations to extinction and decimated others. Grazing domestic stocks also took their toll on the countryside and the wild animals that lived there.

Greek curiosity, travel, and trade provided favorable conditions for developing animal collections; however, the ruling city-states did not have enough wealth or influence to develop large collections. Pets included various kinds of birds, monkeys, weasels (probably the polecat, domesticated as the ferret to catch vermin), hedgehogs, and harmless snakes. Temple collections maintained animals for processionals, during which wild animals appeared in cages, while tame ones walked on leashes or pulled vehicles. Showmen and professional animal trainers exhibited wild animals for entertainment. Some animals, such as bears and lions, were common; others, such as tigers, were rare. Bears and lions were common because they could still be found in Greece and its neighboring areas at the time. Rare animals were ones from other lands that did not exist locally and were not often seen within Greece. Whether an individual animal was considered native or exotic depended on where it was captured. An animal from another land was considered exotic, even if it was the same species as one found locally.

During the Hellenistic Greek period (323–27 B.C.), from the death of Alexander the Great to the reign of the Roman Empire, the number of exotic animals brought into Greece increased, but this does not seem to have influenced the development of collections. Alexander's campaign into Persia and elsewhere in Asia was the most significant Greek expansion of power and brought the Greeks into contact with little-known lands and animals. Alexander had many of these animals sent back for Aristotle to study, but there is as yet no known information on whether the animals remained alive for an appreciable period of time or if they were maintained in collections. Hellenistic monarchies established after Alexander's death in Macedonia, Asia, and Egypt maintained animal collections, but there is little information to indicate these were very extensive, except for the collection at Alexandria in Egypt.

Ptolemy II (283–246 B.C.) established the largest and most varied of the Hellenistic collections at Alexandria. After taking over the Egyptian dynasties, Ptolemaic rulers, the wealthiest, most powerful and most influential of the Hellenistic monarchies established many excellent cultural institutions at Alexandria, making the city one of the centers of ancient scholarship. Unfortunately, while ancient scholars and travelers often mentioned the animal collection of Alexandria, no details were provided. This collection still existed when Egypt became a Roman province in 30 B.C.

The Roman Republic (ca. 509–27 B.C.) began as a city-state at Rome and eventually occupied the entire Italian peninsula. These early Romans were content with their native soil and wildlife for many years, having no desire to conquer other territories or to import exotic wildlife. During this time, hunting and baiting native wildlife were popular. In baiting, animals (usually dogs) were used to attack another animal (such as a bear) that was tethered. This form of "entertainment" remained popular through the Medieval period. Deer, wild goats, boars, bears, and bulls were often hunted and used in baiting. The first known exotics in the Republic were four Indian elephants captured in a battle at Heraclea against King Pyrrhus, and later marched in the procession celebrating the victory (280 B.C.). Thirty years later some 100 African elephants, along with their mahouts, were taken in a battle with the Carthaginians and marched to Rome. These elephants were exhibited to the citizens of the Roman Republic at towns along the route.[61,62]

Although there is no direct evidence of elephant collections during this time, the large numbers of elephants used in military battles indicate that these collections must have existed. Elephants used for military purposes had mahouts who controlled, trained, and cared for them, all of which needed to be done at "collection" sites. Throughout the late Republic period and the Empire period, Roman legions faced elephants in battle, their Eastern enemies using Indian elephants and their Carthaginian enemies using African elephants. Despite the use of elephants in battle by the opponents of Rome and the large number of elephants that Rome captured and kept, the Roman army itself rarely used elephants. The Romans believed their enemies' elephants could be successfully opposed by well-disciplined Roman troops and that elephants were just as likely to trample their own troops, if routed, as they were the enemies' troops.

Early Republic hunts and processionals evolved into increasingly elaborate spectacles. These included religious ceremonies, triumphant marches, special events, and the *venationes* (public games with animal versus animal and human versus animal combats). Extravagant spectacles began in the late Republic period, but did not reach full notoriety until the Empire period (27 B.C.–A.D. 476). Historians and others have written a great deal about these spectacles because of the large number of animals displayed and killed; however, little has been written about the collections where these animals were kept.[63,64]

In addition to these rather well-known uses of animals, more humane, aesthetic uses of animals existed among the wealthy class, civil administrators, and emperors. Both native and exotic wild animals could be found in villa gardens, ponds, bird enclosures, cages, large parks, and hunting reserves. The largest of these collections belonged to the emperors, and these collections grew as Roman imperialism spread throughout Europe, Africa, and Asia. Imperial expansions brought the Romans into contact with new species of exotic wildlife and with existing foreign collections. As previously noted, Ptolemy II's collection still existed when the Romans overtook Alexandria in 30 B.C. Roman armies also encountered a large collection in Babylonia when they conquered that region in A.D. 363.[65]

Public entertainment included itinerant acts with performing animals, but the greatest interest was in the more exciting processionals and public games that emperors, provincial administrations, and sometimes wealthy citizens provided. After the initial appearance of elephants in the third century B.C., exotics began replacing native species in the spectacles. Some of these early exotic species included lions, leopards, and ostriches, and by the first century B.C., hippopotamuses and crocodiles were included as well. An exhibit at Pompey in 55 B.C. had lions, leopards, monkeys, elephants, and an Indian rhinoceros. A few years later, the first incidence of a giraffe in Roman territory was recorded in 46 B.C. These uses of exotic species for exhibition purposes in spectacles increased significantly during the Empire period.

Emperors of the Roman Empire surpassed the Republic's grandeur in every way, including their private collections and public spectacles. These spectacles were presented throughout the Empire, but the most extravagant were those the emperors provided at Rome. Emperors and administrators had the best access to wild animals, although the upper class had some access as well. The ability to obtain wild animals, especially exotic species from distant lands, depended on an individual's wealth and influence, and exotic animals provided the most status.

Hunters, collectors, and dealers existed throughout the Roman Empire to obtain animals for private collections. Emperors, however, simply had to make a request to obliging officials and military officers to have animals delivered. Soldiers often captured wild animals for the emperors; the commanders considered it good training for battle. Large military units and municipalities had vivaria for keeping wild animals after they were caught, and there was a state-owned vivarium at Laurentum outside Rome where the imperial elephant herd, managed by a *procurator ad elephantos*, was kept during the first and second century A.D. *Vivarium* is one of several Latin terms that have given rise to modern zoo terminology. Although popular through the nineteenth century, *vivarium* is seldom used today, while other terms, such as *aviary*, are still used.

A variety of capture techniques were used, as one might expect, but the most common were the pit and the netted corral, into which wild animals were driven. Once caught, the animals were transported in solid wood crates, or in barred crates if the animals had calmed down sufficiently. They were transported by wagon, as well as by boat; many animals did not survive the difficult journey. Animals needed for a procession or spectacle were sometimes requested as far as two years in advance.

Animals used in public spectacles were kept in stockyards (a crude form of vivarium) located on the outskirts of town. These holding areas were purposely located away from the main activity centers of town as a safety precaution in case an animal escaped. Public games featuring animal versus animal combats, or human versus animal combats, were usually held in the morning, with the gladiatorial combats following in the afternoon. The latter were considered more important and were preferred by the cultured classes. After the games, any animals still alive were caught, crated, and moved back to the stockyards.

Various professions were associated with keeping wild animals, including dealers, performers, animal slayers (those who fought animals in public games), trainers, keepers, and veterinarians. In addition to the state-owned stockyards, collections were maintained in a variety of small enclosures, large parks, hunting reserves, bird flight cages, and fishponds. Ordinary citizens commonly maintained freshwater fishponds, whereas saltwater fishponds, which were expensive to create and maintain, were a hobby of the nobility.

Roman bird and mammal collections contained a wide variety of species. In the bird collections were waterfowl, poultry, nightingales, goldfinches, thrushes, parrots, peafowl, pheasants, cranes, storks, and flamingos. The most common mammals kept as pets or in collections were dogs, cats, monkeys, weasels (tame ferrets used to control vermin), deer, gazelles, wolves, foxes, lynxes, caracals, hyenas, and camels. Lions, leopards, and bears were sometimes kept in cages or as pets in the houses of nobles and in the imperial palaces. Rare species, such as tiger, cheetah, zebra, giraffe, rhinoceros, hippopotamus, and crocodile, were usually only found in the emperor's collections. Others, such as elephants, were only exhibited during the spectacles, and no more than one or two elephants are known to have been privately owned.

As the Western Roman Empire declined, so did the public spectacles; however, they continued in the Eastern (Byzantine) Roman Empire. But the spectacles of Constantinople were not the extravagant affairs they had been in Rome, and, in time, they declined as well. The number of animals used in public spectacles increased from hundreds to thousands toward the end of the Roman Empire, if the Roman chroniclers can be believed. Even taking some degree of exaggeration into account, public spectacles involved an enormous number of animals during the existence of the Empire. While almost nothing is known about where the animal collections were housed or how they were managed, the types of animals in the collections are known because of their appearances in these spectacles.

Disintegration of the Roman infrastructure included the loss of Roman collections; however, monarchs, monasteries, and municipalities continued to keep wild animals in post–Roman Empire Europe. These European collections continued through the Medieval period to emerge as the menageries of the Renaissance period. An intervening period of Arabic influence also included animal collections.

1.3.5 Persian and Arab Collections

Alexander the Great did not conquer Persia, now Iran, until 331 B.C. But the Greeks already knew of their *paradeisos*, the great paradise gardens of the Persians that had existed since about 546 B.C. These gardens were re-creations of the Garden of Eden, a state of supreme bliss, an ideal place. Essential elements of these gardens were geometric designs, water, trees, and flowers; animals were not essential to these gardens, but they were sometimes included, especially in the larger parklike gardens. The influence of these gardens was widespread, impressing those societies that conquered Persian territories and spreading to those societies the Persians conquered. These gardens also influenced Islamic garden design as it spread throughout Asia, northern Africa, and Spain (ca. 622–1492).[66]

After Roman Egypt became a province of the Islamic caliphate (in 641), the Egyptian caliph demanded tribute from its southern regions — camels, elephants, giraffes, and other animals. Animals were also captured on the caliph's hunting trips and military excursions in these areas. These animals were kept in the caliph's animal collection at Cairo, occasionally to be used in processions or presented as gifts to foreign dignitaries of Persia, India, China, and Europe.

During this same period, Arabic science was probably the most advanced in the world, especially in astronomy, mathematics, alchemy, and medicine. The natural sciences, however, were not considered important. Despite this lack of interest in the natural world, Arabic animal collections were part of the Muslim court. Abderrahman III (912–961) established an animal park as part of his new city, Zahra (north of Cordova, Spain), in 936. Here he kept animals in cages and fenced enclosures.[67] Large collections also existed in Baghdad, Constantinople, and Cairo. These collections continued through the Medieval and into the Renaissance period. During both periods, animals were exchanged between European and Muslim rulers. Ottoman Turkish rulers continued the collections under their control and took part in the exchange of animals with European rulers, exchanges that continued through the early nineteenth century when, for example, giraffes were sent from the Cairo collection to England, France, and Germany.[68,69]

1.3.6 Medieval Collections

The Medieval period (476–1453) is a construct of European history; the era extended from the decline of the Western Roman Empire in Europe to the emergence of the European Renaissance with its accompanying age of European exploration and influence. For other regions of the world, this time period was merely a continuation of their ancient civilizations and remained so until the Europeans arrived on their doorsteps.

With the disintegration of the Roman infrastructure, Europe developed in relative autonomy. Roman provincial animal collections were dismantled or abandoned when the Romans withdrew, and European monarchs, monasteries, and municipalities and their collections filled the vacuum resulting from the collapse of the Roman infrastructure. Monarchs throughout Europe, along with wealthy barons and nobles, provided structure to civilian life and government; monasteries were the spiritual and political centers of the Church and the community; and villages and towns became important urban municipalities.[67,70–76]

The most important of these monarchies, monasteries, and municipalities maintained animal collections. Emperor Charlemagne (Charles the Great, 742–814), founder of the Holy Roman Empire, had royal collections at several of his estates. These collections included elephants, lions, bears, camels, monkeys, and birds, especially falcons, reflecting the popularity of falconry as a sport of royalty during the Medieval period. Some of these animals were sent to Charlemagne from collections held by other monarchs in Europe and elsewhere, as such gifts were a form of recognition and influence; for example, Haroun-el-Raschid, Caliph of Baghdad (765–809), sent two elephants to Charlemagne.

Monks and nuns at the monasteries usually had a keen interest in gardening and small animal husbandry. These activities had practical purposes, but sometimes served ornamental or aesthetic purposes as well. Monks at St. Gallen, Switzerland (early to mid-800s) maintained one of the better known animal collections, which had animal houses, outdoor paddocks, and keepers.

When William the Conqueror swept over the isle of Great Britain (1066), he seized the existing game reserves, forming the first British hunting forests and parks. At his manor, Woodstock, he began a collection of exotic animals about 1100. His son, Henry I, enclosed Woodstock and enlarged the collection, which included lions, leopards, lynx, camels, and an owl considered to be "rare." In 1235, during the reign of Henry III, animals from this collection were moved to the Tower of London.

Frederick II (1194–1250), Emperor of the Holy Roman Empire, had an extensive collection at his Palermo residence in Italy. He had a particularly fine collection of falcons and was an authority on falconry and bird biology. Based on his experience with this collection he wrote an authoritative text on the subject, *Über die Kunst, mit Vögeln zu Jagen* (*On the Art of Hunting with Birds*, or *The Art of Falconry*), which has been reprinted many times to the present day.[77] Frederick II also had an animal collection that included elephants, giraffes, and white bears. Some of his animals were obtained in exchanges with other European monarchs and with Muslim rulers in India (from whom he acquired an elephant), Egypt (with whom he exchanged a white bear for a giraffe), Spain, and Constantinople.

Sixteenth-century European noble in falconer's regalia. From an illustration in a later edition of *The Art of Falconry* by Frederick II of Hohenstaufen, translated and edited by Casey A. Wood and F. Marjorie Fyfe. © Stanford University Press.

Louis IX of France (1214–1270) brought back an elephant when he returned from the Crusades. Philip VI (1293–1350) created an area at his Chateau de Louvre for lions and leopards. Charles V (1337–1380) moved this collection to St. Pol and enlarged it. He particularly liked birds and kept many of them in gold and silver cages throughout every room of his chateau. He also kept caged nightingales in his garden trees and had a large flight cage for other birds.

Many of the French nobility had animal collections as well. A collection of René, Count of Anjou (1440s), was one of the largest of these collections. The count's estate included a lion house, a small mammal house, a flight cage for birds, a pond for waterbirds, as well as ostriches, camels, and elephants.

In Florence, a new lion house was built in 1293 to replace a deteriorating structure the Romans had constructed. By the early 1300s there were three keepers to oversee the 24 lions in the house. On special occasions these lions were set loose in arenas to do battle with other animals, much like scaled-down versions of the earlier Roman spectacles. Medici rulers of Florence, beginning in 1434, maintained animal collections for many generations. Some of the Popes had collections housed at the Vatican. An early collection of Pope Benedict XII (1285–1342) contained ostriches. Pope Leo X (1475–1521) kept tropical birds, lions, leopards, other cats, and other mammals. His prize possession was an elephant named Hanno, which the King of Portugal, Dom Manuel I (Emanuel I, 1469–1521), had given him. Collections of the Italian princes were the first to flourish as the Renaissance began to sweep across Europe. By the fifteenth century, animal parks (as well as the earliest botanical gardens) were found throughout Italy.

Giraffe the Sultan of Egypt presented to Lorenzo de Medici of Florence, along with a lion and other animals, in 1487. From Sigismondo Tizio's *Historiae Senenses*. © Città del Vaticano, Biblioteca Apostolica Vaticana, Chigi G II 36, fol 148v. Photograph © Biblioteca Apostolica Vaticana.

Sweyn II, King of Denmark (1047–1075), was known to have an animal collection. William IV, Count of Holland (1350s), kept lions, bears, and falcons on his estate at The Hague. About the same time, the Dukes of Guelre had a collection at their castle, Rozendaal, near Arnhem. Prosperous Dutch citizens also maintained collections that, whenever possible, included royal animals (those considered emblematic symbols) such as lions and eagles.

Throughout Europe the population at large, although not allowed to view collections of the privileged, saw animals when they appeared in itinerant animal acts during Medieval fairs. Municipalities also provided their citizens and visitors with views of animals, which were kept in moats no longer needed for defensive purposes, as well as in pits and cages. Municipalities often kept animals that appeared on the coats-of-arms of the municipalities, such as eagles, lions, and hoofed stock.

Medieval moat at Berne, Switzerland used to display deer. From *Wild Animals in Captivity* by Heini Hediger. Courtesy of Dover Publications, Inc.

Animal collections at Cairo and Constantinople continued to thrive during the Medieval period. Justinian I, Roman Emperor at Constantinople (483–565), maintained one of the largest animal collections of the time. This collection still existed when the Turks overthrew the Eastern Roman Empire in 1453 and it was continued by the Turkish rulers, who also maintained the collections at Cairo when they gained control of Egypt. Travelers fortunate enough to visit the Holy Land and Egypt to see the antiquities and historical sights often stopped to view the menageries at Cairo and Constantinople.

In China, the Yuan and Ming dynasties (ca. 1271–1644) continued their tradition of animal reserves, parks, and gardens during this period. Breeding goldfish, as the gold-colored Crucian carp became known, was popular during this period and resulted in a variety of forms, shapes, and colors. Kept in ornamental ponds and earthen jars, goldfish enhanced many Chinese gardens.[78]

Marco Polo's travels, which supposedly took place from 1271 to 1295, made the Yuan (Mongol) dynasty well known in Europe (whether Polo actually traveled in China is still debated).[52-54] While in China, Marco Polo was in the service of Kublai Khan, which allowed him ample opportunity to observe the emperor's estates and to travel throughout China. Every once in a while, Polo provides us with a glimpse of the animal collections. As was popular with earlier Chinese emperors, the largest collections existed in the royal parks. The Great Khan's royal park was rich with "beautiful meadows, watered by many rivulets, where a variety of animals of the deer and goat kind are pastured, to serve as food for the hawks and other birds employed in the chase, whose pens are also in the grounds. The number of these birds is upwards of two hundred, without counting the hawks [and] when [the Great Khan] rides about this enclosed forest, he has one or more small leopards carried on horseback, behind their keepers."[79]

The Great Khan also maintained an elaborate palace where spaces between the buildings were filled with meadows, trees, hoofed stock, and other animals. He also kept a herd of some 5000 elephants for use in processions, as well as a herd of camels. His collections included many leopards and lynxes (now considered to have been cheetahs) used for hunting deer. There were lions said to be larger than Babylonian lions with white, black, and red stripes (if by red, Polo meant orange, these "lions" were probably tigers). These cats, transported to the hunt in cages on wagons, were used to seize boars, wild oxen, asses, stags, deer, and bear for sport. The Great Khan also had a large collection of falcons for falconry, requiring 10,000 falconers to care for the birds. Polo also saw the great parks of regional rulers, which were stocked with animals, but unfortunately he says very little about them.[53]

In turn, the Grand Eunuch Zheng He made China's presence better known in Asia and Africa with his Indian Ocean trading expeditions (1405–1433). These trading expeditions were short-lived, however, because Chinese royal court officials did not hold the merchants, who encouraged these commercial ventures, in high esteem and the expedition costs were felt to be an unnecessary expense. Zheng He was the admiral of these expedition flotillas, which involved 62 ships, over 100 auxiliary vessels, and almost 30,000 people. These expeditions called at ports in Indochina, Indonesia, India, Ceylon, and the eastern coast of Africa. Included in the fleet were "horse-ships," ships normally used to transport horses and domestic stock, but also used to carry wildlife obtained during the expeditions. Spectacular exotic species obtained during these expeditions included giraffe, lion, rhinoceros, oryx, leopard, ostrich, and zebra.[55]

Giraffes were particularly important to the Chinese because they thought giraffes were the legendary unicorn, a sacred animal in Confucian tradition representing the virtue of utmost benevolence. A unicorn would appear only if a ruler possessed this virtue. It was therefore with great joy that the newly throned Ming ruler Yongle (1359–1424) received two giraffes a year apart in 1414 and 1415. These events were signals to the Chinese people that Yongle was the ruler of a perfect government, something that had not happened since the great emperors of the past had ruled China. The Chinese royal court prized the other exotic animals brought to China as well, but what could compete with a unicorn?[55]

1.3.7 Aztec and Inca Collections

Civilizations in the Americas may be as old as those of the Old World, but they did not reach their heights until the Medieval period. Aztec and Inca civilizations, the largest and most powerful of the New World civilizations when the Europeans arrived, were continuations of earlier civilizations. They had a long history of established urban settlements, monumental architecture (including huge stone pyramids), extensive road systems, and an extensive system of irrigated agriculture. There were well-organized and structured trading systems that extended throughout the Americas, and there were impressive animal collections.

Aztec civilization, which flourished between 1345 and 1521, was at its height prior to the arrival of Hernando Cortes in 1519. When Cortes arrived, Montezuma had just taken power from his rivals, Nezahualpilli and Nezahualcoyotl, and had moved the capital from Tezcuco to Tenochtitlan, a Venice-like city at the center of Lake Texcoco (where Mexico City is now located).[80–86]

The Aztecs were pictographically literate, artistic, agricultural (although few animals were domesticated), and skilled, but they did not use iron, the wheel, or the plow. A favorite science of the Aztec dynasties was astronomy, although civil engineering was clearly well developed in view of the quality of their communities, houses, and pyramid temples. They also possessed a good understanding of animals, plants, and minerals. Metals, gemstones, silver, gold, and other materials, in addition to their practical uses, were made into figurines resembling animals familiar to the Aztecs. Their live collections were supplemented with well-made, detailed figurines. These replicas amazed Cortes: "There was not a living thing on land or sea of which Muteczuma [sic] could have knowledge which was not so cunningly represented in gold, silver, precious stones or featherwork as almost to seem the thing itself."[87]

There were extensive gardens throughout Tezcuco and the surrounding countryside. These gardens contained water basins stocked with various kinds of fish, flight cages containing a variety of birds, and other animals. Terraced hills were filled with hanging gardens, streams, lakes, and waterfalls. Royal forests were extensive and the Aztecs had laws to ensure their preservation. On the march to Tenochtitlan, Cortes came upon these gardens and well-tended groves. Surrounding the capital were groves of oak, sycamore, and cedar intermingled with farms, orchards, and flower gardens. Streets into the capital were lined with buildings that had gardens on the ground and on their upper levels.

Montezuma had several residences, each of which had gardens, fishponds, and birds. His most elaborate residence was at the new capital, Tenochtitlan. This aggregation of several thousand buildings had, among other things, a birdhouse, a wild animal house (with mammals, birds of prey, and reptiles), and a collection of deformed humans. The animal houses and birdhouses were contained within extensive gardens of flowering plants, fragrant shrubs, and medicinal plants. These gardens also contained freshwater and saltwater ponds filled with fish and waterfowl.

Describing the birdhouse Cortes observed, "There were also ten pools of water in which were kept every kind of waterfowl known in these parts, fresh water being provided for the river birds, salt for those of the sea, and the water itself being frequently changed to keep it pure: every species of bird, moreover, was provided with its own natural food, whether fish, worms, maize or the smaller cereals." Impressed with this collection, Cortes noted, "It was the whole task of three hundred men to look after these birds. Others likewise were employed in ministering to those who were ill."[88]

A separate collection contained mammals, birds of prey, and reptiles, perhaps because these animals required a building with sturdy cages.

> He had also another very beautiful house in which there was a large courtyard, paved very prettily with flagstones in the manner of a chessboard. In this palace there were cages some nine feet high and six yards round: each of these was half covered with tiles and the other half by a wooden trellis skillfully made. They contained birds of prey, and there was an example of every one that is known in Spain, from kestrel to eagle, and many others which were new to us. Other large rooms on the ground floor were full of cages made of stout wood very firmly put together and containing large numbers of lions, tigers, wolves, foxes and wild cats of various kinds.

The size of this collection required a workforce equal to that needed at the birdhouse, as "there were likewise another three hundred men to look after these animals and birds."[89]

One of Cortes's men, Bernal Diaz del Castillo, recorded his observations on this mixed collection of birds and mammals: "Many kinds of carnivorous beasts of prey, tigers and two kinds of lions, and animals something like wolves and foxes, and other smaller carnivorous animals, and all these carnivores they feed with flesh, and the greater number of them breed in the house." In addition, he mentions the reptiles housed in this collection: "They also have in that cursed house many vipers and poisonous snakes which carry on their tails things that sound like bells. These are the worst vipers of all, and they keep them in jars and great pottery vessels with many feathers and there they lay their eggs and rear their young."[90]

While Cortes considered these collections, along with many other things he saw, to be as good as anything in his native Spain, and although his awe of the Aztec civilization is evident in his writings, he still considered the Aztecs barbarians. By 1521, Cortes had conquered the Aztecs, decimated their population, and destroyed their cities, gardens, and animal collections.

Inca civilization, which reached its pinnacle between 1440 and 1533, was weakening when Francisco Pizarro arrived in 1531. The Incas were less cultured than their predecessors and neighbors, but incorporated the best aspects of these cultures into their own as they assumed power. The Incas eventually established a well-organized government and a complex social system that stretched from present-day Ecuador to central Chile. It had about 10,000 miles of roads with runners communicating information orally or with the *quipu* (a set of colored cords with coded knots). They had no written language, did not use the wheel or iron, and domesticated very few animals. Royal Inca were the ruling caste with absolute control over other castes and societies; however, they integrated these others into their own system and were tolerant of conquered societies. They were less brutal than the Aztecs and attempted to mold conquered societies into one nation.[85,91]

The Incas were not scientifically oriented, but they did have a refined knowledge of astronomy and various aspects of nature, particularly medicinal plants. Like the Aztecs, the Incas had animal collections and gardens, both real and artificial. The royal palace at Cuzco contained one such collection fabricated in gold and silver. The chronicler Garcilaso de la Vega noted, "It contained many herbs and flowers of various kinds, small plants, large trees, animals great and small, tame and wild, and creeping things such as snakes, lizards, and snails, butterflies and birds, each placed in an imitation of its natural surroundings." In this extensive royal collection, "There were birds of all kinds, some perched on the trees as if they were singing, while others were flying and sucking honey from the flowers. There were, too, deer and stags, lions, tigers, and all the other animals and birds that bred in the country, each being set in its natural surroundings to give greater similitude."[92] This palace collection was imitated in the estates of the provincial lords and in the temples.

The Temple of the Sun was covered with "many gold and silver figures copied from life — of men and women, birds of the air and waterfowl, and wild animals such as tigers, bears, lions, foxes, hounds, mountain cats, deer, guanacos and vicunas, and domestic sheep — were placed round the walls in spaces and niches." In addition, "they imitated herbs and such plants as grow on buildings and placed them on the walls so that they seemed to have grown on the spot. They also scattered over the walls lizards, butterflies, mice, and snakes, large and small, which seemed to be running up and down."[93]

There were also real gardens filled with orchards, groves of trees, flower beds, and sweet-smelling herbs that were all native to Peru. And there were real animal collections as well. Royal Inca and provincial lords conducted hunts and captured a variety of wild animals, along with guanacos and vicunas that were caught for the domesticated herds. Some wild animals that regional lords sent the Royal Inca as tribute were killed. Captured wild animals were kept in collections maintained at a variety of locations. Although some of these animals were kept at the royal court, others were kept in various districts of the capital, districts that were named for the kinds of animals kept there, such as the *amarus* district (where large snakes were kept), the *pumacurcu* and *pumapchupan* districts (where large cats and bears were kept), and the *surihualla* (the ostrich field). Wild animals were kept for aesthetic purposes, for demonstrating the ruler was pleased with the lords who presented the animals, and for ceremonial uses. More specific information on these collections is lacking since, unlike the conquerors of the Aztecs, Pizarro and his officials did not leave detailed eyewitness descriptions of the Inca animal collections. By 1533, Pizarro had succeeded in subduing the Inca, destroying the animal collections in the process.

One other major Central American civilization, the Mayan civilization (ca. 250–925), still existed when the Europeans arrived, but was greatly diminished in stature. Much of what we know about the Mayans therefore comes from archaeological evidence rather than colonial documentation. Mayans had gardens, fishponds, and pets — such as parrots, other birds, monkeys, coatis, and kinkajous. They hunted wild animals, were familiar with wild animals and plants, and domesticated a few species. Folk systematics of their descendants, the Tzeltal society, indicates a knowledge of animals and plants that no doubt existed in the past as well.[94,95] Long-distance trade was used to obtain exotic luxury items, including feathers, jaguar hides and teeth, shark teeth, stingray spines, and shells. Royalty and the wealthy imported these items for status purposes. Based on their socioeconomic characteristics and zoological activities, it is conceivable that they kept wild animals in addition to animals for pets and for domestication. Unfortunately, as with the ancient Indus society of India, no conclusive evidence proves that animal collections existed. Knowledge of this aspect of their lives has thus far eluded the historian.[94–98]

While Aztec and Inca animal collections were quite impressive, they were destroyed and had no influence on the emerging European Renaissance collections. Exploration of the Americas and the discovery of many New World species, however, did have profound consequences for these European collections.

1.4 Menageries

1.4.1 European Menageries

Europe evolved into a continent of nation-states with increasing power, wealth, and influence during its Renaissance. Relatively small and scattered royal, monastic, and municipal collections from the Medieval period began to increase in size and numbers. These collections were to become known as *menageries* (although this term did not come into use until about 1712, it is often used to describe earlier collections, especially from this period). Proliferation of fifteenth- and sixteenth-century European menageries coincided with expanding European exploration. Most animals came to these collections from the three

Advertising poster dated 1751 for the Blaauw Jan menagerie, Amsterdam. From a steel engraving attributed to C. F. Fritzsch. © Artis Library, University of Amsterdam.

THE WORLD FOR A RHINO

Soon after Columbus discovered the Americas for Spain in 1492, Portugal established key bases on the coast of India and competed with Spain for other colonial bases. In 1493, Pope Alexander VI established a demarcation line that separated Portuguese and Spanish territories. Soon thereafter, in 1494, the Treaty of Tordesillas extended this line farther west, allowing Portugal to claim Brazil and leaving Spain to claim the other American territories. These newly acquired territories were important to the king of Portugal, Dom Manuel I (Emanuel I, 1469–1521), especially the ports established in India and the New World.

Portuguese authorities, however, were not satisfied and wanted the Pope to move the demarcation line even farther west, encroaching on Spain's empire. Wanting to be on good terms with Pope Leo X (1475–1521) and hoping to surpass efforts made by other European countries to win the Pope's favors, Dom Manuel I sent the Pope unusual African and Asian animals, including an elephant in 1514. These animals were from the king's royal palace menageries and were chosen because they were animals from Portugal's colonies that the Pope did not already have in his Vatican menagerie. Portuguese kings had maintained menageries since at least the time of Dom Diniz in the late 1200s. By the time of King Dom Manuel I, the royal palace menagerie at Ribeira, Lisbon included birds, gazelles, antelope, lions, and a trained cheetah, and the royal palace menagerie at Estãos maintained a herd of elephants and other large animals.

The Pope's elephant, to be named Hanno, was considered of particular interest since there had not been one seen in Rome since the time of the Roman Empire. The Pope enjoyed this prize possession, but the Portuguese king soon had an opportunity to present an even more prestigious gift. In 1515, Sultan Muzafar presented an Indian rhinoceros from his menagerie at Champanel to Afonso Albuquerque, governor of Portuguese India. Albuquerque, in turn, sent this rare animal to Dom Manuel I. It was the first rhinoceros seen in Europe since Roman times, and it is the one that appears in the famous Albrecht Dürer drawing of an armor-plated rhinoceros, as well as the drawings of several other lesser known contemporary artists.[*] Another live rhinoceros would not arrive in Europe until one reached Madrid in 1579. Such a rare beast could have brought Dom Manuel I great fame among his European peers; however, he did not keep it. Instead, he sent it to Pope Leo X, no doubt expecting such a rare gift to influence the Pope greatly on behalf of the Portuguese empire. Such was the value of rarely seen exotic animals in Renaissance Europe.

Unfortunately for Dom Manuel I and Portugal, the ship carrying the rhinoceros to Italy was caught in a storm and sank, taking its cargo with it. Except for a few minor adjustments in the 1700s, the demarcation line remained where it was and Portugal had to rely on itself to gain new territories.

Sources: based on Loisel, Gustave, *Histoire des ménageries de l'antiquité à nos jours*, Octave Doin et Fils and Henri Laurens, Paris, 1912; Clarke, T. H., *The Rhinoceros from Dürer to Stubbs, 1515–1799*, Sotheby, London, 1986; Rookmaaker, L. C., *Bibliography of the Rhinoceros: An Analysis of the Literature on the Recent Rhinoceroses in Culture, History and Biology*, A. A. Balkema, Rotterdam, 1983; Bedini, Silvio A., *The Pope's Elephant*, J. S. Sanders, Nashville, 1998. *Albrecht Dürer's now famous illustration of this 1515 Indian rhinoceros is in the British Museum and is illustrated in many books.

parts of the world then familiar to collectors: Europe, Africa, and Asia.[99] Collectors knew these animals through books, art, and other collections. Along with familiar animals, collectors obtained animals that had not been seen in Europe since ancient times, such as the rhinoceros brought to Portugal in 1515 for King Dom Manuel I. A new region also joined the three parts of the world then familiar to Europe. As the German cartographer who named this new region, Martin Waldseemüller, pointed out in 1507, the "fourth part of the earth, which ... we may call ... America," was different and unexpected.[100] New World fauna provided many familiar species (the North American), as well as many unknown species (the South American).

This unexpected New World fauna, in addition to new species being discovered in Africa and Asia, created difficulties for European naturalists. In considering impacts of the New World on the Old, historian John Elliott observes, "By the later sixteenth century, as a result of the vast amount of fresh observation during the preceding decades, the problem of classification was becoming acute in every field of knowledge."[101] As Elliott further notes, "The superiority of direct personal observation over traditional authority was proved time and time again in the new environment of America. And on each fresh occasion, another fragment was chipped away from the massive rock of authority."[102] Collections were essential to this process; proceeding beyond description and taxonomy required live animals.

Jacob Burckhardt, in his classic history of Renaissance Italy, describes the importance of collections during this period. "By the end of the fifteenth century ... true menageries, ... now reckoned part of the suitable appointments of a court, were kept by many of the [Italian Renaissance] princes. It belongs to the position of the great ... to keep horses, dogs, mules, falcons and other birds, court-jesters, singers, and foreign animals."[103] It was the wealth of these Italian princes, together with their interest in natural history and their contacts with Arab animal traders, that caused their collections to be the earliest in Renaissance Europe. Living collections were few in comparison with natural history cabinets (the precursors to museums, which contained stuffed specimens and naturalia), and they could be found in Florence, Palermo, and Naples. These living collections included lions, tigers, leopards, bears, elephants, camels, giraffes, zebras, ostriches, and crocodiles. It is generally claimed that some species in these collections, like the giraffe and zebra, had not been seen in Europe since Roman times, although there are indications (such as when Frederick II sent a white bear to Cairo in exchange for a giraffe) that these animals had in fact been in Medieval collections.[68,104]

In the sixteenth century, Francis Bacon wrote in his drama, *Gesta Grayorum*, that a knowledgeable ruler should have four institutions established under his authority: a library, a combined botanical and zoological garden, a cabinet, and a museum of instruments and machinery. The garden was "to be built about with Rooms, to stable in all rare Beasts, and to cage in all rare Birds; with two Lakes adjoining, the one of fresh Water, and the other of salt, for like variety of Fishes: And so you may have, in a small Compass, a Model of Universal Nature made private."[105]

Information on 13 seventeenth-century cabinets and two seventeenth-century menageries (at the Tower of London and Versailles) provides some insight into the collections of this period. An analysis of these collections indicates, as one might expect, that the overall number of live specimens was considerably fewer than the number of cabinet specimens. The number of exotic specimens was also fewer than the number of native European specimens. The exotic specimens from Africa and South America held the most interest, and those from North America the least.[106]

By the seventeenth century, European collections were so prevalent that one explanation given for how animals arrived in the Americas after the Biblical flood was that the animals had

Ancient Collections and Menageries

Versailles menagerie, Paris. From an engraving by A. Perelle as it appeared in *Histoire des ménageries de l'antiquité à nos jours* by Gustave Loisel.

escaped from ships transporting them to European menageries. Athanasius Kircher (1602–1680), a German Jesuit priest interested in natural philosophy (as natural history was then known) proposed this explanation. Like many natural philosophers, he was also a Biblical scholar interested in reconciling the Bible with the discovery of the New World — a place, and a new group of animals, not mentioned in the Bible. For Kircher and many Europeans, the story of Noah's Ark explained animal biodiversity, and the descent from Mount Ararat after

Analysis of Seventeenth-Century Zoological Collections

	Europe and North Asia	India and Siam	East Indies	Africa	South America	North America
Cabinets	58	7	6	10	15	4
London menagerie	49	5	5	23	10	8
Paris menagerie	61	4	3	18	13	1

Cabinets:
All 13 cabinets, with a total number of specimens = 1,657
England: J. Tradescant, R. Hubert, J. Petiver, Royal Society
France: P. Borel, Sainte Genevieve
Italy: F. Calceolari, F. Cospi, A. Kircher
Germany: B. Besler, M. Besler
Denmark: O. Worm
Netherlands: J. van de Mere, J. Swammerdam

Menageries:
London (Tower of London), with a total number of specimens = 39
Paris (Versailles), with a total number of specimens = 188

Note: Thirteen cabinets and two menageries in seventeenth-century Europe showing the geographic sources of their specimens as percentages of the total number of specimens in each collection.
Source: Based on George, Wilma, in *The Origins of Museums: the Cabinet of Curiosities in Sixteenth- and Seventeenth-Century Europe*, Imprey, O. and MacGregor, A., Eds., Oxford University Press, Oxford, 1985.

the flood was the explanation for zoogeography. As new species were discovered, Kircher and other scholars devised increasingly elaborate blueprints for the Ark, the stalls labeled with the names of the numerous new occupants. But it became more difficult to believe in the literal truth of these Biblical stories as the number of new species increased, especially after the discovery of an entirely new fauna from the New World.[107]

In similar fashion, many of the wealthier Renaissance animal collectors attempted to gather all known species into their menageries. After all, if Noah fit them into the Ark, the collector should be able to fit them into his menagerie. But, like the Biblical scholars, animal collectors eventually succumbed to the realization that the number of species exceeded their grasps. A modified version of this obsession with having complete collections continued, however, as later menageries attempted to acquire "postage stamp" collections containing specimens representing each of the taxa — a philosophy that lingered at many zoos until only recently.

Bringing back specimens for collections was one thing; bringing them back alive was another matter. As early as the 1660s Robert Boyle, John Woodward, and others published instructions on the observation, collecting, preservation, and transport of natural history specimens.[108–110] These instructions, however, were not concerned with collecting and transporting living specimens for menageries. Living specimens were not often shipped because knowledge about these animals and their captive care was rudimentary. For practical reasons, it was far easier to ship preserved animals, skins, and other animal naturalia than it was to ship living animals.

Wilma George, in her study of the representation of animals in European culture, describes the reality of early transportation as it applied to animal shipments. "Some specimens were brought back to Europe alive or, at least, started their journey alive. Live animals were normal ship's cargo. Sheep, pigs, and poultry taken on board for food on the outward journey could be replaced by exotic animals on the return." George further points out that animals often did not survive and ended up in the cooking pot with their skins and skeletons saved for the museums.[106]

Even domestic animals, which were shipped from Europe with regularity, suffered on long voyages. The horse, for example, was a domestic animal whose needs were well known, and yet the death rate on long sea voyages was high, sometimes exceeding 50%.[111–114] Shipping animals with unknown behaviors and diets, and that were not comfortable around people or on the high seas, was far more difficult. Nevertheless, animal shipments were occasionally attempted in the sixteenth and seventeenth centuries, more frequently ventured in the eighteenth century, and became commonplace in the nineteenth century. However, despite improvements in transportation, even nineteenth-century professional animal traders had difficulties with the loss of animals during transport.[115]

Early Renaissance collections obtained Asian and African species through northern and central Italy. In the fifteenth century, Arabs were active purveyors of menagerie animals, using southern and eastern Mediterranean ports to transport animals to Europe via Italy.[116] At that time, Arabs were still the primary seafarers throughout the Indian Ocean, trading among the Asian and West African regions.[117] Later sources for European collections usually depended on the country's influence in a particular region of the world; hence, the European colonial sphere of influence was an important factor in determining a collection's inventory. However, other factors affected collection inventories, such as the collector's individual preferences, available access to field collectors, and exotic qualities of the species. Animal collectors of the time were almost always influential men. As time passed, animal providers became more varied and included explorers, sailors, travelers, colonial officials, other collectors, and eventually professional traders.

Animal shipments increased between the sixteenth and nineteenth centuries as new lands were settled, as scientific knowledge evolved from a generalized natural history into specialized disciplines, as transportation improved, as the popularity of animals increased among the public, and as the number of collections increased. The nineteenth century became the heyday of professional collections and popular natural history. It was an era of *omnium gatherum*, when everything in nature was deemed worthy for someone to collect and study.[118]

1.4.2 Colonial Menageries

Europe's growth and power had an impact beyond its borders as it explored and settled other regions of the world. European settlement in newly discovered regions evolved into possession, and along with "ownership" came a desire to know about these new worlds. Early on, colonial officials understood that the "first step for a nation to take is to recognize the lands where the people live, what they contain, what they produce, and what they are capable of."[119] Animals from these newly settled regions were of commercial value, but they held Europe's interest for other popular and scientific reasons.

Cartographers used exotic faunas to fill barely explored, empty voids on their maps. On maps and in works of art, the elephant, rhinoceros, lion, zebra, giraffe, ostrich, gazelle, and monkey represented things African; the elephant, rhinoceros, tiger, anteater, deer, crocodile, parrot (short tailed), and peacock represented things Asian; the jaguar, llama, tapir, peccary, and parrot (long tailed) represented things South American; the bear, puma, fox, wolf, opossum, and beaver represented things North American; the kangaroo, bird of paradise, and pouched animals represented things Australian; and the white bear represented things polar.[99]

It is difficult to imagine what these new species meant to the Europeans, especially now that these animals are so familiar. And it was not just the Europeans, for unfamiliar, exotic animals were exciting for everyone at that time. This widespread interest in exotic species resulted from a basic emotional kind of response, and in some cases a scientific curiosity, or a passionate fascination. The idea that European menageries exhibited colonial animals as a form of domination over their colonial possessions does not explain colonial menageries exhibiting wildlife from Europe and the other colonies. And as is evident from the discovery of the Americas, interest in newly discovered exotic animals was immediate. When Columbus's first New World expedition returned to Europe, it carried live animals home.

By the seventeenth century, animals were regular cargo at major European and colonial ports. The Dutch East India Company, for example, had special warehouses and stables on the Amsterdam docks to receive animal shipments, as it did at Cape Town and other colonial ports.[73,106,109] As the Portuguese, Spanish, Dutch, British, French, and German empires expanded, they established botanical collecting stations, which were also used to house animals. Some of these holding facilities developed into colonial menageries and eventually into national zoological parks following the independence of the colonies.

Europeans encountered local menageries as they explored India, Indonesia, China, Egypt, Central America, and South America. When the first ships of the English East India Company landed at Aceh (a city-state on the northwest tip of Sumatra) in June 1602, the crew was treated to royal entertainments that included buffalo, tiger, and elephant combats.[120] Obviously, the Sultan of Aceh, Ala-uddin Shah, had a collection of animals at his disposal but, as usual, the spectacle, not the collection, is described. Other examples of this kind of collection will likely come to light as the logbooks and diaries of other early explorers and colonial administrators are adequately researched. These observations, however, tend to be brief comments widely

> ## Earliest Live New World Animals Sent to European Menageries
>
> Sources and dates for the earliest known importation of various New World animals (and the bases of the information) are as follows:
>
> **Parrots.** Columbus returned from his first expedition in 1493 with a new kind of parrot, the long-tailed macaw. His second expedition brought back more than 60 parrots for Queen Isabella and King Ferdinand of Spain in 1494 and 1496. (Morison, Samuel E., *Admiral of the Ocean Sea: A Life of Christopher Columbus*, Book of the Month Club, New York, 1992 [Reprint of 1942 edition].)
>
> **Opossum.** Vicente Pinzon shipped an opossum from Brazil to Queen Isabella and King Ferdinand of Spain in 1500. (Archer, Michael, Ed., *Carnivorous Marsupials*, Royal Zoological Society of New South Wales, Mosman, 1982 and Struik, D. J., "Early colonial science in North America and Mexico," *Quipu*, 1, 25, 1984.)
>
> **Iguana.** Gonzalo Fernandez de Oviedo y Valdes, in the early 1500s, took "every precaution for the safe dispatch of a live iguana from Hispaniola to his friend Ramusio in Venice, but omitted to inform himself adequately about its dietary habits. He gave it a barrel of earth for its sustenance, and the unfortunate creature died on the voyage." (Elliott, John H., *The Old World and the New, 1492–1650*, Cambridge University Press, London, 1970, 37.)

scattered among many pages of documentation — the proverbial menagerie needles in the historical haystack.

Colonizing Europeans reacted to local menageries in remote regions with either indifference or destruction. As with other products of those remote societies, Europeans viewed these native menageries as inferior to their own. Colonial administrations and naturalists often took over local menageries or replaced them with their own collections. Many private menageries did not last beyond the lifetimes of the owners, but some did evolve into public menageries; for example, the private colonial menagerie of Cecil Rhodes, which was kept at his estate, Groote Schuur (Cape Town, Republic of South Africa). Other colonial officials too numerous to mention, and whose names have long since lost their importance in world events, established collections. However, if these officials are still mentioned in colonial histories, their listed activities generally do not include "founder of a menagerie." Colonial acclimatization and natural history societies also established menageries, as did some municipal governments. Origins for collections extant in the 1950s in countries that were soon to be former colonies can be found in the first directory of zoos published in the *International Zoo Yearbook*. Later editions of the *Yearbook* and other zoo directories do not provide this kind of information.[121]

Some of the earliest colonial menageries started as acclimatization farms at colonial botanical-collecting stations. These collecting stations were important to the agricultural and economic development of the colonies and were usually introduced shortly after a colony was established. Portuguese collecting stations were founded between 1415 and 1487 at Madeira, São Tomé, and Fernando Po (West African coast), and were followed by Dutch stations at Malabar (India) sometime during the 1600s, Batavia (Java) sometime during the 1600s, and Cape Town (South Africa) in 1652. Several others were founded during the 1700s at Mauritius,

> **Bison.** Andre Thevet, in the mid-1500s, mentioned two bison that were brought to Spain, but they were probably not exhibited and did not survive long. (Krumbiegel, Ingo and Sehm, Gunter G., "The geographic variability of the plains bison: a reconstruction using the earliest European illustrations of both subspecies," *Archives of Natural History*, 16, 169, 1989.)
> **Guinea pig.** The guinea pig arrived in Europe directly from South America, or indirectly from West Africa, between about 1550 and 1580. (Weir, Barbara J., "Notes on the origin of the domestic guinea-pig," in *The Biology of Hystricomorph Rodents*, Rowlands, I. W. and Weir, Barbara J., Eds., Academic Press, London, 1974 [Symposia of the Zoological Society of London 34].)
> **Turkey.** Turkeys were brought to Spain, possibly as early as 1500, but definitely between then and the 1520s, as well as to Italy in the 1520s, France in 1538, and England in 1541. (Schorger, A. W., *The Wild Turkey: Its History and Domestication*, University of Oklahoma Press, Norman, 1966.)
> **Muscovy duck.** The Muscovy duck arrived in Spain, Portugal, and France between about 1520 and 1530. (Donkin, R. A., *The Muscovy duck,* Cairina moschata domestica*: Origins, Dispersal, and Associated Aspects of the Geography of Domestication*, A. A. Balkema, Rotterdam, 1989.)
> **Miscellaneous birds and mammals.** Various individuals continued importing turkeys, macaws, guinea pigs, and llamas throughout the 1500s. In the 1600s, the Dukes of Montmorency imported American wood ducks, Muscovy ducks, South American parrotlets, and South American guans and curassows to their aviaries at Chateau de Chantilly, France. (Fisher, James, *Zoos of the World: The Story of Animals in Captivity*, Natural History Press, Garden City, NY, 1967.)

India, St. Vincent, Reunion Island, Jamaica, St. Helena, the Canary Islands, Mexico, and Brazil.[109,122,123] The tremendous increase in the number of these collecting stations continued throughout the 1800s.

Collecting stations functioned as nurseries for native colonial plants, as agricultural acclimatization gardens, and as medicinal gardens. Several had animal-holding areas for European domestic stock being exported to the colonies and for colonial wildlife being shipped to Europe or to other colonies. As collecting stations evolved into botanical gardens, their animal-holding areas evolved into menageries. Collecting stations also formed botanical and agricultural networks. Colonial natural history museums developed networks in the eighteenth and nineteenth centuries as well. Networking, which facilitated the efforts of colonial gardens and museums to hire directors, train staff, procure supplies, and exchange specimens, was self-contained within each European empire. Control of these activities resided in each empire's major European gardens and museums.[109,119,122–126] Colonial menageries had a weaker network than did the gardens or museums, and the European menageries exerted less influence over them.[127]

A variety of individuals were responsible for animal transactions, including diplomats, colonial officials, explorers, ship captains and sailors, collectors, and, later, ordinary citizens. The exchange of exotic specimens among colonial menageries, therefore, may have been the least important activity of the menageries since there were so many other sources for animal acquisitions. The extent to which animal exchanges had increased by the nineteenth century is evident from an inventory survey of the Calcutta Zoological Gardens covering 1875 to 1891. An analysis of the geographic sources of its mammal and bird collections shows these animals came from all regions of the world to reside in this colonial zoo. Only

Analysis of Geographic Sources for the Mammal and Bird Collections Exhibited at the Calcutta Zoological Gardens, 1875–1891

Region	Mammal Species		Bird Species	
	No.	%	No.	%
India/Ceylon/Himalayan Region	94	39	212	53
Asia/Indonesia	49	20	46	11.5
China/Japan	4	2	20	5
Europe	10	4	40	10
Africa/Madagascar	32	13	17	4
North America	8	3	5	1
Central/South America	23	10	22	5
Australia/New Zealand/New Guinea	12	5	39	10
Polar	1	1	0	0
Domestic	8	3	1	0.5
Total	241	100	402	100

Source: Based on Samyal, R. B., *A Handbook of the Management of Animals in Captivity in Lower Bengal*, Natraj Publishers, Dehra Dun, India, 1995.

39% of the mammal species and 53% of the bird species came from India, Ceylon, or the Himalayan region.[128]

The acclimatization of economically valuable animals was a primary function of many colonial menageries. This activity was important to the well-being of the colonies, although not to the same extent as economic botany and the acclimatization of economically valuable plants. European menageries, such as those of the Zoological Society of London and Société Zoologique d'Acclimatation, established acclimatization programs of their own. The Zoological Society of London did not actively pursue acclimatization for very long (ca. 1826–1834) because the program could not compete with the exhibition of exotic wildlife and because the acclimatization farm was too expensive to maintain.[129,130]

British colonies, however, made more extensive attempts at acclimatization since these efforts were important to their struggling agricultural economies. Acclimatization societies and collections in the colonies also eventually ceased, or evolved into zoological societies and zoological gardens. Sometimes agricultural societies continued the acclimatization efforts, but none was particularly successful. As Sir William Flower, President of the Zoological Society of London noted in his 1887 review of the society's programs, "No addition of any practical importance has been made to our stock of truly domestic animals since the commencement of the historic period of man's life upon the earth."[131]

France and its colonies took a more active role in acclimatization. The Société Zoologique d'Acclimatation (established 1854) and Jardin Zoologique d'Acclimatation (established 1860) were begun to populate French fields, forests, and streams with new animals, to increase the number of domestic animals, to increase their nutritional resources, and to endow their agriculture. Société efforts applied to France and its colonies, and its source of animals was global. Jardin exhibits promoted fauna from the French colonies, as well as from non-European countries with which France had commercial or diplomatic relationships. Initially (1854–1860), about 22% of the members in the society were foreign, mostly from Algeria.[132,133]

Toward the latter part of the nineteenth century, and into the early twentieth century, the emphasis of colonial agriculture shifted from small farms (with their experimentation and need for a diverse source of animals and plants) to plantation economies (with their need for a few

Ancient Collections and Menageries

Jardin Zoologique d'Acclimatation, Paris. This nineteenth-century postcard shows the importance of family recreation in menageries, even this menagerie, which was established to further work in economic (applied) natural history. © National Zoological Park, Smithsonian Institution.

established animals and crops). This diminished need for acclimatization experiments, along with the difficulty of developing new domestic breeds and the desire of the public to see exotic wildlife, transformed acclimatization collections into exotic fauna collections.

1.4.3 Evolution of the Zoo and Aquarium

Menageries were well established throughout the world during the eighteenth- and nineteenth-centuries. A shift from royal and private menageries to public menageries occurred between the late 1700s and early 1800s. This shift did not occur in a vacuum, but was part of a larger cultural shift. Prior to this time only the privileged few had the interest, leisure time, and financial ability to support animal collections. During the 1800s, the privileged few grew in number, supporting menageries through societies funded with membership fees or through governments funded with taxes. Support for menageries then shifted to the general public, although in some places, such as the United States, support began with the public.[134]

The transition from private collection to public institution is the point of departure for the chapters to follow covering regional histories of zoological gardens (with some variation in the chronological starting points and some overlap with the history of menageries presented in this chapter). But before proceeding to these chapters, the transition to zoological gardens and the appearance of aquariums must be considered.

Within a century of referring to collections as menageries (which began about 1712 in Europe), a proposal was made to establish a zoological garden. A Zoological Society of London prospectus dated March 1, 1825 stated:

> It has long been a matter of deep regret to the cultivators of Natural History, that we possess no great scientific establishment either for teaching or elucidating Zoology; and no public menageries or collections of living animals where their nature, properties and habits may be studied. In almost every other part of Europe, except in the metropolis of the British Empire, something of this kind exists.

PRELIMINARY ETYMOLOGY FOR ZOOS AND AQUARIUMS

Little is known about the terminology used to describe past animal collections and those who managed the collections. Many terms used in the past have been lost or are unknown because they have been translated literally or figuratively into modern terms. An etymology for zoos does not yet exist as it does for natural history museums and botanical gardens. [1,2] Aquariums do not present a problem because the aquarium concept and terminology developed more recently in England during the 1850s.

Mesopotamian and **Egyptian** terminology existed, but the words are usually translated literally. For example, a veterinarian in Mesopotamia was an "ox and ass doctor." Also, many of their animal collections were an integral part of their gardens and parks and therefore zoo-related terms are intertwined with garden- and park-related terminology.

China had several terms to describe its animal collections. There are also descriptions of annex-courts, which were private apartments within the late Han dynasty palaces where rulers kept caged wild mammals and birds. In addition, owners gave individual names to their parks and gardens, such as the *Guangcheng Yuan* (Broadly Perfected Park) or *Lingyou* (Garden for the Promotion/Encouragement of Knowledge). [54,71]

yu: A fenced animal reserve or wildlife enclosure; originally a fenced *yuan*.
yuan: A royal park, hunting park, or forest reserve; originally a large open park.

Persia had one particular term to describe its passion for walled gardens and its use of them as sanctuaries from the harsh environment and everyday life. The Greeks, Romans, and other Arab cultures adopted its use, and it was then passed on to form the modern European word, *paradise*.[66]

paradeisos: A walled garden, with or without animals, that resembles a heavenly paradise on earth.

This prospectus also asserted that the society would "offer a collection of living animals, such as never yet existed in ancient or modern times" and that it would benefit Britain to offer "a very different series of exhibitions to the population of her metropolis," one that could be used for the "objects of scientific research, not of vulgar admiration."[135]

A leading journal then devoted to science and literature, the *Literary Gazette*, considered these objectives of the proposed society and asked, "Is it not altogether visionary?"[135] This remark was meant critically, as well as being complimentary. This new direction, or new emphasis on existing directions, had considerable support. As menageries proliferated and a new age of sensibility developed, higher expectations were requiring higher standards of animal husbandry. Increased knowledge about animals and improved technology made achieving these standards possible.

Collections established during the 1800s began calling themselves zoological gardens, zoological parks, or simply zoos. In some cases this was simply fashionable because zoological gardens were considered professionally managed facilities, whether they were or not. In other cases there was a distinct difference in facilities, staff, programs, and budgets, as well as an emphasis on education and science rather than on entertainment. Differences that occurred

Greco-Roman terminology has provided a number of terms for animal collections and the individuals who manage these collections. Some of these terms have become root words for more modern European terms.[63,64]

amphitheatrales magistri: Wild animal trainers.
aviarium/aviaria: A place/places where birds were kept.
bestiarii: Performers who work with wild animals in the spectacles, such as runners, dodgers, polejumpers, and others.
magistri: Common keepers and trainers (often slaves or foreigners).
negotiator ursorum: Bear dealers; middle men who bought and sold bears. No doubt the term negotiator applied to other kinds of animal dealers as well.
piscinae: Freshwater (*dulces*) and saltwater (*maritimae*) fish ponds.
piscinarii: A fishpond owner or keeper.
procurator ad elephantos: Individual in charge of the state-owned elephant herd.
therotrophium: Later known as a chase (in Europe, ca. 1460), it was a wooded area surrounded by a park wall to keep wild animals for hunting and other purposes. It differed from a *leporarium* or *roboraria*, later known as a warren (in Europe, ca. 1377), which was used to keep hares and sometimes other "wild, but tame," animals such as deer and wild goats.
veterinarius: Animal doctor, also known as *mulomedicus*, whose primary function was treating horses.
vivarium/vivaria: An enclosed private park/parks for keeping wild animals. It also referred to the cruder stockyards and arenas where wild animals were kept for public spectacles during the late Roman Republic and Roman Empire periods.

European terminology evolved into our modern Western terminology.[151] This list is in chronological order of the introduction of the term.

aviary: 1577 (L.), an enclosure for keeping birds that can fly freely.
vivarium: 1600 (L.), an enclosure for live animals, especially tanks for small animals in a natural setting.
menagerie (or less commonly, *menagery*): 1712 (F.), management of a group of animals, originally the household or domestic stock, but later this included exotic animals. Eventually, the term was used to describe a collection of wild animals managed primarily for public exhibition.

continued

in animal husbandry, exhibition, veterinary medicine, education, research, and conservation may be attributed to improvements in knowledge and technology; however, by purposely taking advantage of this knowledge and technology, zoo and aquarium professionals transformed their collections into new institutions (zoological gardens) that were recognizably different from the older collections (menageries).

The zoological garden of the nineteenth century eventually evolved into the biopark, or conservation park, of the late twentieth century.[136,137] Some features of the conservation park concept had existed previously, such as the exhibition of multispecies natural habitats, which were found in tropical colonial zoos during the nineteenth century. Other design and exhibition features of the conservation park concept are only now possible because of improved animal husbandry knowledge, animal health care, and modern technology. Still other features of this concept are now feasible because of the increasingly popular emphasis on conservation. Thus, as with earlier transitions, the trend toward conservation parks is a reflection of related social, environmental, and technological changes. The conservation park concept is quickly being superseded with an even newer one, the environmental center of the twenty-first century.

> ## Preliminary Etymology
>
> *zoological garden* (or *zoological park*): 1829 (B.), a garden or park setting in which wild animals are kept for public exhibition with an emphasis on education, science, and conservation. The Zoological Society of London first used the term to describe its collection at Regent's Park (although this collection was simultaneously referred to as a menagerie). Zoos often use this term to distinguish themselves from menageries, although this is less true in Europe. European languages use variations, such as *Tiergarten* or *zoologischer Garten* (German), *giardino zoologico* (Italian), *dierenpark* (Dutch), *zoologiske have* (Danish), *jardin zoologique* or *parc zoologique* (French), *zoologicka zahrada* (Czech), *miejski ogród zoologiczny* (Polish), and *jardim zoológico* or *parque zoológico* (Portuguese). The American Zoo and Aquarium Association (AZA) has developed a definition for zoological gardens and aquariums as part of its accreditation standards: "A permanent cultural institution which owns and maintains captive wild animals that represent more than a token collection and, under the direction of a professional staff, provides its collection with appropriate care and exhibits them in an aesthetic manner to the public on a regularly scheduled basis. They shall further be defined as having as their primary business the exhibition, conservation, and preservation of the earth's fauna in an educational and scientific manner." [152] Other accreditation standards and definitions may also exist within other national and regional associations.
>
> *zoo*: 1847 (B.), this abbreviation of the term zoological garden was first used in Britain as a popular nickname. It first appeared in print to describe the Clifton Zoo, but was popularized in the famous contemporary music hall song, "Walking in the Zoo on Sunday."[153]
>
> *aquarium*: 1854 (L.), a tank for keeping aquatic plants and animals that is self-contained and self-sustained as a natural system. Originally, aquariums were also referred to as *aquatic vivariums*. Warington and Gosse coined the term *aquarium* in their original British publications on the subject.[138–144]
>
> *oceanarium*: 1938, a public aquarium with at least one very large aquarium tank for keeping sea mammals, such as dolphins or killer whales. It was first used to describe Marineland (St. Augustine, Florida), which opened in 1938.
>
> *conservation park* (or *biopark*): 1990s, relatively new terms for a zoo that emphasizes ecological immersion exhibits and conservation programs. No clear definition exists and some aspects of this concept appeared previously; however, the concept only became popular in the 1990s.

Aquarium history, on the other hand, is essentially a modern one with fewer changes. The aquarium concept, using self-contained and self-sustaining systems, originated in England during the 1850s. Prior to this development there were ornamental fishponds, which were open systems supplied with water from nearby seas, lakes, or streams. There were also fish bowls, which were closed systems, but not self-sustaining. Robert Warington discussed the modern concept of freshwater aquariums in articles published betweeen 1850 and 1854, and saltwater aquariums in articles published between 1853 and 1854. Some of the earliest aquariums in Britain were known as Warrington [sic] Cases. It was Philip H. Gosse who popularized the aquarium. Unlike Warington, who wrote articles for scientific journals, Gosse wrote books for the public. With several books on natural history and the microscope to his credit, he was already popular when he wrote his books on aquariums, *The Aquarium* (1854) and *Handbook to the Marine Aquarium* (1855). Both of these men coined the term *aquarium*, which was consciously chosen over other terms.[138–144]

The popularity of glass during the early 1800s was one factor in the popularity of the new aquariums in the mid-1800s. Technology for making glass had improved at the time and, as a result, glass was used for a variety of new ornamental purposes. In 1829, Nathaniel B. Ward accidentally discovered that plants could be self-maintained in closed glass containers. His published findings came at a time when the restrictive glass tax in Britain was about to be repealed (in 1845), making glass more affordable. Glass tanks called Wardian Cases became popular in the 1800s for displaying miniature self-sustaining habitats, then known as vivariums, but now known as terrariums.[145]

Among the most popular plants kept in Wardian Cases were ferns. During the Victorian period, many animal and plant fads came and went. Collecting ferns was one of these fads, and it was waning just as goldfish keeping and aquariums were becoming popular in the 1850s and early 1860s. In effect, aquariums were simply upside-down Wardian Cases, and the aquatic environments made the cases even more interesting than the previously popular terrestrial environments. Keeping aquariums gave individuals vacationing at the beaches a mission — looking for miniature animals and plants to bring home for the aquarium. Soon after the aquarium fad hit England, it spread to continental Europe and the United States. Americans initially relied on British publications for their knowledge of aquariums. The first American book on aquariums may have been *Life Beneath the Waters; or, the Aquarium in America*, by Arthur M. Edwards (1858). In this book Edwards indicates that aquariums, and aquarium specimens, were already sold at stores in New York and other large cities.[146]

As aquariums became popular for the parlor, Gosse assisted the Zoological Society of London with the building of London Zoo's Aquarium, the first zoo aquarium open to the public (1853). This event was followed by the opening of a number of public aquariums from 1853 to 1899 in Europe (for example, Paris 1859, Hamburg 1864, Hanover 1866, Brussels 1868, Cologne 1868, Berlin 1869, Brighton 1872, Manchester 1874, Southport 1874, Yarmouth 1876, Westminster 1876, Edinburgh 1878, Amsterdam 1880, Sevastopol 1897) and the United States (Boston 1859, Washington 1873, San Francisco 1894, New York 1896). Numerous other public aquariums and marine research stations opened during the twentieth century.

In 1938, Marineland (St. Augustine, Florida) was the first oceanarium (aquariums centered around a large central tank used to exhibit marine mammals). It was initially used for filming underwater movies and only later became a public attraction. Most public aquariums established during the early to mid-1900s were exhibits or buildings within zoological gardens. Marineland and other early oceanariums were the exceptions and were the forerunners of the large, independent aquariums. Between the 1970s and the 1990s, technological advances in design and materials and the public's increasing interest in aquatic environments, prompted a shift to large, independent aquariums with a focus on native habitat exhibits, information on aquatic systems, and environmental education. Improved technology provided the ability to maintain crystal-clear water and the ability to display huge aquatic vistas without obstructions. These improvements in exhibits made the large, independent aquariums immensely popular.

Zoos and aquariums continue to evolve. Their roles as cultural institutions are changing in many ways. At the same time they remain true to their traditional commitments concerning recreation, education, research, and conservation. How these roles and commitments were met in the past was determined by the state of knowledge and technology, as well as societies' expectations. How these roles and commitments will be met in the future will be determined by the same factors.

The world has become a megazoo, with parks representing scattered ecosystem "exhibits." Zoological gardens are becoming small parks and parks are becoming large zoological gardens. Some parks are already centers for the rehabilitation of endangered species and their management programs deal with populations that can no longer survive on their own.[147–150] The zoo and aquarium story continues to reveal itself. To understand the changes now occurring and those yet to come requires an understanding of the changes past generations have experienced and the reasons they occurred.

References

1. Findlen, Paula, "The museum: its classical etymology and Renaissance genealogy," *Journal of the History of Collections*, 1, 59, 1989.
2. van Erp-Houtepen, Anne, "The etymological origin of the garden," *Journal of Garden History*, 6, 227, 1986.
3. Oelschlaeger, Max, *The Idea of Wilderness: From Prehistory to the Age of Ecology*, Yale University Press, New Haven, CT, 1991, 7–8.
4. Sulman, Felix G., *Short and Long-Term Changes in Climate*, CRC Press, Boca Raton, FL, 1982.
5. Thomas, William L., Jr., Ed., *Man's Role in Changing the Face of the Earth*, University of Chicago Press, Chicago, 1956.
6. Clark, Grahame and Piggott, Stuart, *Prehistoric Societies*, Knopf, New York, 1965.
7. Dance, S. Peter, *A History of Shell Collecting*, E. J. Brill, Leiden, 1986.
8. Faul, Henry and Faul, Carol, *It Began with a Stone: A History of Geology from the Stone Age to the Age of Plate Tectonics*, Wiley, New York, 1983.
9. Levi-Strauss, Claude, *The Savage Mind*, Chicago University Press, Chicago, 1966.
10. Berlin, Brent, "Folk systematics in relation to biological classification and nomenclature," in *Annual Review of Ecology and Systematics*, Johnston, Richard F., Frank, Peter W., and Michener, Charles D., Eds., Annual Review, Palo Alto, CA, 4, 259, 1973.
11. Berlin, Brent, Breedlove, Dennis E., and Raven, Peter H., "General principles of classification and nomenclature in folk biology," *American Anthropologist*, 75, 214, 1973.
12. Berlin, Brent, *Ethnobiological Classification: Principles of Categorization of Plants and Animals in Traditional Societies*, Princeton University Press, Princeton, NJ, 1992.
13. Raven, Peter H., Berlin, Brent, and Breedlove, Dennis E., "The origins of taxonomy," *Science*, 174, 1210, 1971.
14. Atran, Scott, *Cognitive Foundations of Natural History*, Cambridge University Press, New York, 1990.
15. Hughes, J. Donald, *Ecology in Ancient Civilizations*, University of New Mexico Press, Albuquerque, 1975.
16. Morton, A. G., *History of Botanical Science: An Account of the Development of Botany from Ancient Times to the Present Day*, Academic Press, New York, 1981.
17. Bernal, J. D., *Science in History*, MIT Press, Cambridge, MA, 1971.
18. Ronan, Colin A., *Science: Its History and Development among the World's Cultures*, Facts on File, New York, 1982.
19. Singer, Charles, Holmyard, E. J., and Hall, A. R., Eds., *A History of Technology*, Oxford University Press, New York, 1954.
20. Clutton-Brock, Juliet, *Domesticated Animals from Early Times*, University of Texas Press, Austin, 1981.
21. Ucko, Peter J. and Dimbleby, G. W., Eds., *The Domestication and Exploitation of Plants and Animals*, Aldine, Chicago, 1969.
22. Zeuner, Frederick E., *A History of Domesticated Animals*, Hutchinson, London, 1963.
23. Leeds, Anthony and Vayda, Andrew P., *Man, Culture, and Animals: The Role of Animals in Human Ecological Adjustments*, American Association for the Advancement of Science, Washington, D.C., 1965.
24. Contenau, Georges, *Everyday Life in Babylon and Assyria*, Edward Arnold, London, 1964 [Reprint of 1954 edition, translated by K. R. and A. R. Maxwell-Hyslop].
25. Hawkes, Jacquetta, *The First Great Civilizations: Life in Mesopotamia, the Indus Valley, and Egypt*, Knopf, New York, 1973.
26. Luckenbill, Daniel D., *Ancient Records of Assyria and Babylonia*, University of Chicago Press, Chicago, 1926–1927.
27. Oppenheim, A. Leo, *Ancient Mesopotamia: Portrait of a Dead Civilization*, University of Chicago Press, Chicago, 1964.

28. Dalley, Stephanie, "Ancient Mesopotamian gardens and the identification of the hanging gardens of Babylon resolved," *Garden History*, 21, 1, 1993.
29. Finkel, Irving L. The hanging gardens of Babylon, in *The Seven Wonders of the Ancient World*, Clayton, Peter A. and Price, Martin J., Eds., Routledge, London, 1988, chap. 2.
30. Collins, Billie Jean, *The Representation of Wild Animals in Hittite Texts*, Ph.D. dissertation, Yale University, New Haven, CT, 1989.
31. Breasted, James H., *Ancient Records of Egypt*, University of Chicago Press, Chicago, 1906–1907.
32. Erman, Adolf, *Life in Ancient Egypt*, Dover, New York, 1971 [Reprint of 1894 edition, translated by H. M. Tirard].
33. Kees, Hermann, *Ancient Egypt: A Cultural Topography*, University of Chicago Press, Chicago, 1961 [edited by T. G. H. James, translated by Ian F. D. Morrow].
34. Montet, Pierre, *Everyday Life in Egypt*, Edward Arnold, London, 1962 [translated by A. R. Maxwell-Hyslop and Margaret S. Drower].
35. Wilkinson, Alix, "Gardens in ancient Egypt: their locations and symbolism," *Journal of Garden History*, 10, 199, 1990.
36. Erman, 1971, 235.
37. Montet, 1962, 130.
38. Erman, 1971, 243.
39. Naville, Edouard H., *The Temple of Deir el-Bahari*, Trubner & Co., London, 1894–1908. [issued as the 12–14th, 16th, 19th, 27th, 29th Memoir of the Egypt Exploration Fund].
40. Tyldesley, Joyce, *Hatchepsut: The Female Pharaoh*, Viking, New York, 1996.
41. Allchin, Bridget and Allchin, Raymond, *The Rise of Civilization in India and Pakistan*, Cambridge University Press, Cambridge, U.K., 1982.
42. Auboyer, Jeannine, *Daily Life in Ancient India*, Macmillan, New York, 1965 [translated by Simon W. Taylor].
43. *Cambridge History of India*, Cambridge University Press, Cambridge, 1922 ff.
44. Fairservis, Walter A., Jr., *The Roots of Ancient India*, Macmillan, New York, 1971.
45. Kosambi, D. D., *Ancient India: A History of Its Culture and Civilization*, Meridian Books, New York, 1969.
46. Possehl, Gregory L., Ed., *Harappan Civilization: A Contemporary Perspective*, Oxford and IBH Publishing, New Delhi, 1982.
47. Rao, S. R., *Lothal and the Indus Civilization*, Asia Publishing House, New York, 1973.
48. Wheeler, R. E. Mortimer, *The Indus Civilization: Supplementary Volume to the Cambridge History of India*, 3rd ed., Cambridge University Press, Cambridge, U.K., 1968.
49. Dales, George F., University of California archaeologist involved with the excavation of Indus Valley sites, personal communication, 1989.
50. Graham, Dorothy, *Chinese Gardens*, Dodd Mead, New York, 1938.
51. Menzies, Nicholas K., *Forest and Land Management in Imperial China*, St. Martin's Press, New York, 1994.
52. Needham, Joseph, *Science and Civilisation in China*, Cambridge University Press, London, 1954 ff.
53. Polo, Marco, *The Travels of Marco Polo*, Heritage Press, New York, 1934 [edited and translated by Manuel Komroff].
54. Schafer, Edward H., "Hunting parks and animal enclosures in ancient China," *Journal of the Economic and Social History of the Orient*, 11, 318, 1968.
55. Snow, Philip, *The Star Raft: China's Encounter with Africa*, Weidenfeld and Nicolson, New York, 1988.
56. Wong, K. Chimin and Lien-Teh, Wu, *History of Chinese Medicine*, AMS Press, New York, 1973 [Reprint of 1936 edition].
57. Clagett, Marshall, *Greek Science in Antiquity*, Abelard-Schuman, New York, 1955.
58. Glacken, Clarence J., *Traces on the Rhodian Shore: Nature and Culture in Western Thought From Ancient Times to the End of the Eighteenth Century*, University of California Press, Berkeley, 1967.

59. Marsh, George P., *Man and Nature: Or, Physical Geography as Modified by Human Action*, Belknap Press–Harvard University Press, Cambridge, 1965 [Reprint of 1864 edition, edited by David Lowenthal].
60. Perlin, John, *A Forest Journey: The Role of Wood in the Development of Civilization*, Norton, New York, 1989.
61. Gowers, William, "The African elephant in warfare," *African Affairs*, 46, 42, 1947.
62. Scullard, Howard H., *The Elephant in the Greek and Roman World*, Cornell University Press, Ithaca, NY, 1974.
63. Jennison, George, *Animals for Show and Pleasure in Ancient Rome*, Manchester University Press, Manchester, 1937.
64. Toynbee, Jocelyn M. C., *Animals in Roman Life and Art*, Cornell University Press, Ithaca, NY, 1973.
65. Jennison, 1937, 134–135.
66. Moynihan, Elizabeth B., *Paradise as a Garden: In Persia and Mughal India*, Braziller, New York, 1979.
67. Huff, Toby E., *The Rise of Early Modern Science: Islam, China, and the West*, Cambridge University Press, Cambridge, U.K., 1993.
68. Laufer, Berthold, *The Giraffe in History and Art*, Field Museum of Natural History, Chicago, 1928 [FMNH Anthropology Leaflet 27].
69. Allin, Michael, *Zarafa: a Giraffe's True Story, from Deep in Africa to the Heart of Paris*, Walker, New York, 1998.
70. Bennett, Edward T., *The Tower Menagerie*, R. Jennings, London, 1829.
71. Fisher, James, *Zoos of the World: The Story of Animals in Captivity*, Natural History Press, Garden City, NY, 1967.
72. Lloyd, Joan B., *African Animals in Renaissance Literature and Art*, Clarendon Press, Oxford, U.K., 1971.
73. Loisel, Gustave, *Histoire des ménageries de l'antiquité à nos jours*, Octave Doin et Fils and Henri Laurens, Paris, 1912.
74. Maguire, Henry, "A description of the Aretai Palace and its garden," *Journal of Garden History*, 10, 209, 1990.
75. Rybot, Doris, *It Began before Noah*, Michael Joseph, London, 1972.
76. Wilson, Derek, *The Tower 1078–1978*, Hamish Hamilton, London, 1978.
77. Frederick II, *The Art of Falconry*, Stanford University Press, Stanford, CA, 1943 [translated and edited by Casey A. Wood and F. Marjorie Fyfe].
78. Matsui, Yoshiichi, *Goldfish*, Hoikusha, Osaka, 1971 [translated by Don Kenny].
79. Polo, 1934, 142.
80. Cortes, Hernan, *Five Letters, 1519–1526*, Norton, New York, n. d. (196?) [translated by J. Bayard Morris].
81. Diaz del Castillo, Bernal, *The Discovery and Conquest of Mexico, 1517–1521*, Farrar, Straus and Cudahy, New York, 1956 [translated by A. P. Maudslay].
82. Duran, Fray D., *The Aztecs*, Orion Press, New York, 1964 [translated by Doris Heyden and Fernando Horcasitas].
83. López de Gómara, Francisco, *The Conquest of the Weast India (1578)*, Scholars Facsimiles and Reprints, New York, 1940.
84. Nuttall, Zelia, "The gardens of ancient Mexico," in *Annual Report of the Board of Regents of the Smithsonian Institution … for the Year Ending June 30 1923*, U.S. Government Printing Office, Washington, D.C., 1925.
85. Prescott, William H., *History of the Conquest of Mexico and History of the Conquest of Peru*, Modern Library, New York, 1936 [Reprints of *History of the Conquest of Mexico*, Harper, New York, 1843 and *History of the Conquest of Peru*, Harper, New York, 1847].
86. Sahagun, Bernardino de, *General History of the Things of New Spain: Florentine Codex*, School of American Research, University of Utah, Salt Lake City, 1950–1982.
87. Cortes, n. d. (196?), 84.

88. Cortes, n. d. (196?), 95.
89. Cortes, n. d. (196?), 95–96.
90. Diaz del Castillo, 1956, 213.
91. Vega, Garcilaso de la, *Royal Commentaries of the Incas*, University of Texas Press, Austin, 1966 [translated by Harold V. Livermore from 1609 Lisbon edition].
92. Vega, 1966, 188, 315.
93. Vega, 1966, 313–314.
94. Berlin, Brent, Breedlove, Dennis E., and Raven, Peter H., *Principles of Tzeltal Plant Classification: An Introduction to the Botanical Ethnography of a Mayan-Speaking People of Highland Chiapas*, Academic Press, New York, 1974.
95. Hunn, Eugene S., *Tzeltal Folk Zoology*, Academic Press, New York, 1977.
96. DiPeso, Charles C., "Macaws ... crotals ... and trumpet shells," *Early Man*, 2, 4, 1980.
97. Morley, Sylvanus G. and Brainerd, George W., *The Ancient Maya*, 4th ed., Stanford University Press, Stanford, CA, 1983 [revised by Robert J. Sharer].
98. Wood, W. Raymond, "Plains trade in prehistoric and protohistoric intertribal relations," in *Anthropology on the Great Plains*, Wood, W. Raymond and Liberty, Margot, Eds., University of Nebraska Press, Lincoln, 1980.
99. George, Wilma, *Animals and Maps*, University of California Press, Berkeley, 1969.
100. Waldseemuller, Martin, *The Cosmographiae Introductio*, U.S. Catholic Historical Society, New York, 1907, 63, 68–70 [Facsimile of 1507 edition, U.S. Catholic Historical Society Monograph 4, edited by Charles G. Herbermann].
101. Elliott, John H., *The Old World and the New, 1492–1650*, Cambridge University Press, London, 1970, 37.
102. Elliott, 1970, 40.
103. Burckhardt, Jacob, *The Civilisation of the Renaissance in Italy*, Macmillan, New York, 1928, 294–295 [Reprint of 1878 edition].
104. Findlen, Paula, *Possessing Nature: Museums, Collecting, and Scientific Culture in Early Modern Italy*, University of California Press, Berkeley, 1994.
105. Bland, Desmond, Ed., *Gesta Grayorum*, Liverpool University Press, Liverpool, 1968, 47–48 [Reprint of 1594 edition ascribed to Francis Bacon, English Reprints Series 22].
106. George, Wilma, "Alive or dead: zoological collections in the seventeenth century," in *The Origins of Museums: the Cabinet of Curiosities in Sixteenth- and Seventeenth-Century Europe*, Impey, Oliver and MacGregor, Arthur, Eds., Oxford University Press, Oxford, 1985, 184.
107. Browne, Janet, *The Secular Ark: Studies in the History of Biogeography*, Yale University Press, New Haven, CT, 1983.
108. Boyle, Robert, "General heads for a natural history of a countrey, great or small," *Philosophical Transactions [of the Royal Society]*, 1, 186, 1666.
109. Grove, Richard H., *Green Imperialism: Colonial Expansion, Tropical Island Edens and the Origins of Environmentalism, 1600–1860*, Cambridge University Press, Cambridge, U.K., 1995.
110. Woodward, John, *Brief Instructions for Making Observations in All Parts of the World*, Society for the Bibliography of Natural History, London, 1973 [Facsimile of 1696 edition, introduced by V. A. Eyles, Sherborn Fund Facsimile 4].
111. Crosby, Alfred W., *The Columbian Exchange: Biological and Cultural Consequences of 1492*, Greenwood Press, Westport, CT, 1972.
112. Crosby, Alfred W., *Ecological Imperialism: The Biological Expansion of Europe, 900–1900*, Cambridge University Press, New York, 1986.
113. Denhardt, Robert M., *The Horse of the Americas*, University of Oklahoma Press, Norman, 1975.
114. Graham, R. B. C., *The Horses of the Conquest*, William Heinemann, London, 1930.
115. Hagenbeck, Carl, *Beasts and Men*, Longmans, Green and Co., London, 1909 [abridged translation of German edition, translated by H. S. R. Elliot and A. G. Thacker].
116. Laufer, 1928, 79.

117. Hourani, George F., *Arab Seafaring in the Indian Ocean in Ancient and Early Medieval Times*, rev. ed., Princeton University Press, Princeton, NJ, 1995.
118. Barber, Lynn, *The Heyday of Natural History, 1820–1870*, Doubleday, New York, 1980, 152.
119. Segawa, H., "Brazilian colonial gardens and the Rio de Janeiro Passeio Publico," *Journal of Garden History*, 13, 213, 1993, 214.
120. Keay, John, *The Honourable Company: A History of the English East India Company*, Macmillan, New York, 1991.
121. Morris, Desmond and Jarvis, Caroline, Eds., *International Zoo Yearbook*, Volume 1, Zoological Society of London, London, 1959.
122. Brockway, Lucile H., *Science and Colonial Expansion: The Role of the British Royal Botanic Gardens*, Academic Press, New York, 1979.
123. McCracken, D. P., *Gardens of Empire: Botanical Institutions of the Victorian British Empire*, Leicester University Press, London, 1997.
124. Pinar, S., "Little-known travellers and natural systems: Francisco Norona's exploratory voyage through the islands of the Indian Ocean (1784–1788)," *Archives of Natural History*, 24, 127, 1997.
125. Orosz, Joel J., *Curators and Culture: The Museum Movement in America, 1740–1870*, University of Alabama Press, Tuscaloosa, 1990.
126. Sheets-Pyenson, Susan, *Cathedrals of Science: The Development of Colonial Natural History Museums during the Late Nineteenth Century*, McGill-Queen's University Press, Montreal, 1988.
127. Kisling, Vernon N., Jr., "Colonial menageries and the exchange of exotic faunas," *Archives of Natural History*, 25, 303, 1998.
128. Sanyal, R. B., *A Handbook of the Management of Animals in Captivity in Lower Bengal*, Natraj Publishers, Dehra Dun, India, 1995 [Reprint of 1892 edition].
129. Mitchell, P. Chalmers, *Centenary History of the Zoological Society of London*, Zoological Society of London, London, 1929.
130. Lord Zuckerman, Ed., *The Zoological Society of London 1826–1976 and Beyond*, Zoological Society of London–Academic Press, London, 1976 [Symposia of the Zoological Society of London 40].
131. Anonymous, "The Zoological Society of London," *Nature*, 36, 186, 1887, 188.
132. Osborne, Michael A., *The Societe Zoologique d'Acclimatation and the New French Empire: the Science and Political Economy of Economic Zoology During the Second Empire*, Ph.D. dissertation, University of Wisconsin, Madison, 1987.
133. Osborne, M. A., *Nature, the Exotic, and the Science of French Colonialism*, Indiana University Press, Bloomington, 1994.
134. Hoage, R. J. and Deiss, William A., Eds., *New Worlds, New Animals: From Menagerie to Zoological Park in the Nineteenth Century*, Johns Hopkins University Press, Baltimore, MD, 1996.
135. Mitchell, 1929, 10–13.
136. Conway, William, "The conservation park: a new zoo synthesis for a changed world," in *The Ark Evolving: Zoos and Aquariums in Transition*, Wemmer, Christen M., Ed., Smithsonian Institution–National Zoological Park, Front Royal, VA, 1995, chap. 13.
137. Robinson, Michael H., "Beyond the zoo: the biopark," *Defenders*, 62, 10, 1987.
138. Warington, Robert, "Notice of observations on the adjustments of the relations between the animal and vegetable kingdoms, by which the vital functions of both are permanently maintained," *Chemical Society Journal*, 3, 52, 1851.
139. Warington, Robert, "Observations on the natural history of the water-snail and fish kept in a confined and limited portion of water," *Annals of Natural History*, 10, 273, 1852.
140. Warington, Robert, "On preserving the balance between the animal and vegetable organisms in sea-water," *Annals of Natural History*, 12, 319, 1853.
141. Warington, Robert, "Memoranda of observations made in small aquaria, in which the balance between the animal and vegetable organisms was permanently maintained," *Annals of Natural History*, 14, 366, 1854.
142. Warington, Robert, "On artificial sea-water," *Annals of Natural History*, 14, 419, 1854.

143. Gosse, Philip H., *The Aquarium*, London, 1854.
144. Gosse, Philip H., *Handbook to the Marine Aquarium*, London, 1855.
145. Ward, Nathaniel B., *On the Growth of Plants in Closely Glazed Cases*, John Van Vorst, London, 1842.
146. Edwards, Arthur M., *Life beneath the Waters: or the Aquarium in America*, Bailliere, New York, 1858.
147. Budiansky, Stephen, *Nature's Keepers: the New Science of Nature Management*, Free Press, New York, 1995.
148. Schaller, George B., *The Last Panda*, University of Chicago Press, Chicago, 1993.
149. Sullivan, Arthur L. and Shaffer, Mark L., "Biogeography of the megazoo: biogeographic studies suggest organizing principles for a future system of wild lands," *Science*, 189, 13, 1975.
150. Western, David, *In the Dust of Kilimanjaro*, Island Press, Washington, D.C., 1997.
151. *Oxford English Dictionary*, 2nd ed., Oxford University Press, Oxford, 1989.
152. American Zoo and Aquarium Association, *Accreditation of Zoos and Aquariums*, American Zoo and Aquarium Association, Wheeling, WV, 1991, 3.
153. Blunt, Wilfrid, *The Ark in the Park: The Zoo in the Nineteenth Century*, Hamish Hamilton, London, 1976.

Additional Sources

1. Archer, Michael, Ed., *Carnivorous Marsupials*, Royal Zoological Society of New South Wales, Mosman, 1982.
2. Bedini, Silvio A., *The Pope's Elephant*, J. S. Sanders, Nashville, TN, 1998.
3. Clarke, T. H., *The Rhinoceros from Dürer to Stubbs, 1515–1799*, Sotheby, London, 1986.
4. Donkin, R. A., *The Muscovy duck,* Cairina moschata domestica*: Origins, Dispersal, and Associated Aspects of the Geography of Domestication*, A. A. Balkema, Rotterdam, 1989.
5. Krumbiegel, Ingo and Sehm, Gunter G., "The geographic variability of the plains bison: a reconstruction using the earliest European illustrations of both subspecies," *Archives of Natural History*, 16, 169, 1989.
6. Morison, Samuel E., *Admiral of the Ocean Sea: A Life of Christopher Columbus*, Book of the Month Club, New York, 1992 [Reprint of 1942 edition].
7. Quigley, Carroll, *The Evolution of Civilizations*, 2nd ed., Liberty Press, Indianapolis, IN, 1979.
8. Rookmaaker, L. C., *Bibliography of the Rhinoceros: An Analysis of the Literature on the Recent Rhinoceroses in Culture, History and Biology*, A. A. Balkema, Rotterdam, 1983.
9. Schorger, A. W., *The Wild Turkey: Its History and Domestication*, University of Oklahoma Press, Norman, 1966.
10. Struik, D. J., "Early colonial science in North America and Mexico," *Quipu*, 1, 25, 1984.
11. Weir, Barbara J., "Notes on the origin of the domestic guinea-pig," in *The Biology of Hystricomorph Rodents*, Rowlands, I. W. and Weir, Barbara J., Eds., Academic Press, London, 1974 [Symposia of the Zoological Society of London 34].

Heron painted as an illustration for the Indian Natural History Project, from a specimen at the Barrackpore menagerie, Calcutta, India (early to mid-1800s). From the Buchanan-Hamilton Collection. © The British Library.

2 Zoological Gardens of Great Britain

Clinton H. Keeling

2.1 Introduction

Looking at wild animals on public exhibition has always been a popular pursuit in England and throughout Great Britain. Britons are country folks and naturalists at heart, with a national characteristic of individuality that could account for the many private zoological collections that have existed in England.[1,2] As far as it is possible to ascertain, the first exotic specimen to be seen in England was a walrus exhibited during the reign of King Alfred the Great, who ruled from A.D. 871 to 899. During the sixteenth century, Shakespeare complained — possibly with tongue in cheek — that the English would rather pay to admire a painted fish than give a dolt to relieve a lame beggar. This statement was made at a time when exotic animals were beginning to be exhibited as attractions at traveling fairs, brief interludes of recreation in an otherwise hard and dreary life, which often became rowdy and out of hand from merriment that involved too much drinking. This tradition of traveling fairs showing animals existed in England into the 1940s and a more modern version of these fairs remains popular today. During the mid-eighteenth century John Hunter, the father of modern surgery, bravely frequented some of the roughest and toughest fairs to see species he would otherwise never have encountered. He also made arrangements to collect any casualties suffered by the proprietors. Some of these animals can still be seen in an excellent state of preservation at the Royal College of Surgeons' Museum in Lincoln's Inn Fields, London.

There were many opportunities for obtaining exotic animals in England. One of the prime sources for the ready demand of naturalists and showmen was the mercantile shipping trade, which by the eighteenth century was worldwide. Ship captains knew full well how lucrative it was to obtain unusual animals from their agents in exotic ports, and to endeavor to ensure the animals arrived in London, Liverpool, or Portsmouth alive, if not in good order.

The first permanent English zoological collection, for which there is irrefutable proof, is the collection of King Henry I (1068–1135) at Woodstock, Oxfordshire, some 60 miles west of London. In this collection, as recorded by William of Malmesbury, there were lions, lynx, leopards, camels, porcupines and a "remarkable owl," which William of Montpellier presented to the king. Most unfortunately, from the historian's point of view, not a word was written about the way the animals were kept. Certainly it would be most interesting to know the methods of housing in particular, but one should not lose sight of the ultramaterialism of the times in question. Recording of such details by the collectors would have been trivia of the highest, or lowest, order, and clearly not the sort of thing these gentlemen cared to spend time doing. As is so frequently the case in zoological garden history, the student can only speculate based on what little information is available.

2.2   The Tower Menagerie

At the entrance to the famed Tower of London, in front of the Middle Tower, are the foundations of a semicircular building. A plaque informs the visitor it was the site of the Lion Tower. Looking upon it, the visitor can easily visualize unfamiliar animals from the world's tundras, forests, plains, and mountains that once lived, bred, fought, and died there as part of the Tower menagerie (ca. 1235–1832).

This menagerie, in effect the royal collection for many centuries, was by all accounts established by, for, or in the reign of King Henry III (1216–1272) when the Holy Roman Emperor Frederick II presented him with three leopards (those still depicted on the Royal Arms of England). King Henry ordered the sheriffs of London and the counties of Bedfordshire and Buckinghamshire to provide for the maintenance of the animals and their keepers. Thus, in 1252, funds were provided for the upkeep of a white bear, and in the following year a muzzle and chain were provided to hold it while it fished in the Thames. Many historians have related this story, but frankly, doubts need to be expressed as (1) it would have to have been an abnormally tractable bear to be restrained in this way while disporting itself in the water; (2) the muzzle would have been extremely incapacitating had the bear been expected to eat its catch; and (3) while the Thames was then largely unpolluted and probably swarming with fish, a polar bear (assuming the white bear was of this species) would not have been interested in the fish as this species rarely eats this type of food. There is little doubt some (probably young) bear was accustomed to being led about the grounds, but its alleged riverine activities were probably the stuff of legend.

Four years later, England's first elephant of historic times arrived when King Louis IX of France presented a specimen to King Henry III. Henry sent a precept to the sheriffs stating, "We command that you, of the farm of our city, to cause, without delay, to be built at our Tower of London, one house of forty feet long, and twenty feet deep, for our elephant." [3] The king must have been a realist, as another report states he specified that the house must be practicable for some other purpose on the death of its inhabitant. This animal must have developed into an impressive tusker, as it gradually wore holes in the stone wall with these exaggerated incisors. It would insert them into the wall to support its head when sleeping in the characteristic standing position of the adult animal. Whoever had charge of the elephant fulfilled his duties more than satisfactorily, as clearly it lived long enough to gouge holes in solid rockwork. The elephant's longevity is impressive, considering no one knew anything about the proper care of African elephants until the middle of the nineteenth century. The elephant was no doubt African, as is evident from a contemporary drawing by Matthew Paris, a monk from St. Albans, in his thirteenth-century book, *Chronicles of Matthew Paris*.[4]

During the reigns of the first three Edwards (1272–1377), the collection consisted almost entirely of lions, one of which was traditionally named after the current monarch. It would appear the lions did quite well and one, named Pompey, is alleged to have reached the incredible age of 75. This assertion, however, is utter nonsense as the world longevity record (which the Cologne Zoological Garden holds) for this comparatively short-lived species is 27 years, while the average is nearer 15. Accuracy is always a major problem when researching old notes and records, where fact and fiction are interwoven with impartiality.[5]

According to the historian John Strype in 1708 there were lions, two leopards or tigers, three eagles, two owls, two "cats of the mountain" (these could have been the European wild cat, which was then more common in England than it is now, although at that period the puma of the New World was often known as the catamount), and a jackal at the Tower. From about 1750 the collection began to improve, and in a description of the Tower published in 1800, John Bayely included a species list that makes fascinating reading. It is reproduced here in his own words, with the present author's comments in parentheses.[3]

1. "Fanny. Lioness from Africa."
2. "Young Nero. He lion bred at the Tower. Now eight years."
3. "Miss Fanny. Bred at the Tower at same time as Nero."
4. "Miss Howe and Miss Fanny Howe. Whelped on 1st June 1794 and named after the gallant conqueror of the French on that day." (This would have been the Glorious First of June, a naval encounter in which Admiral Howe defeated the French fleet.)
5. "Lion and lioness newly arrived from Bussora on the Gulf of Persia." (Virtually all authorities now agree the lion is extinct in this region.)
6. "Male elephant from Calcutta."
7. "Royal tiger — China."
8. "Duchess — young leopardess from Malabar coast. 2 or 3 leopards besides this."
9. "Jack — a black leopard presented by Governor Hastings."
10. "Hyena — Saltset near Bombay." (This would have been the striped hyena, *Hyena hyena*, the only Asiatic member of the family.)
11. "Jumbo — large baboon." (This name is most interesting, as the famous Jumbo, an African elephant that lived at the London Zoological Garden, 1865–1882, was so well known his name has gone down in history to describe anything of large dimensions — yet here, over half a century before he was born, is another outsized animal with the same name. Most odd, unless it is an incredible coincidence.)
12. "Monkeys — various species."
13. "Tiger cat from river Gambia." (This animal could have been the blotched genet, a catlike relative of the mongoose and civet, which is common in that part of the world.)
14. "Coatimundi — Honduras." (An early misuse of the species name *coati*; a coatimundi is an individual animal that has left the group to live on its own.)
15. "2 raccoons bred at the Tower." (A reasonable achievement.)
16. "Jackal."
17. "Eagle of the sun." (A complete mystery, unless it was the bald eagle that some writers have said has a habit of sitting looking at the sun, which may or may not be true.)
18. "Wolves and other animals."

The collection was open to the public, with an admission fee of one shilling, which was decidedly expensive for the time. By 1822, the Tower menagerie had dwindled to an elephant, a grizzly bear, and two or three birds. According to the long-defunct *Library of Entertaining Knowledge:*[6]

It had gradually declined in value for half a century in some degree perhaps from the force of popular prejudice which was accustomed to consider it only an occupation and amusement for children to make a visit to the "Lions at the Tower." The beasts of prey are in nearly every case sent to the Tower, but His present Majesty has, during the last ten years, formed a very fine collection of such quadrupeds as are capable of domestication in Windsor Great Park at a lodge called Sand Pit Gate.

Alfred Cops arrived at the Tower in 1822 after working at the Exeter 'Change, where he had received injuries from Chunee, the elephant. Just who Cops really was is something of a mystery, but his chosen calling was that of wild animal keeper. By all accounts he was a very good one indeed, as his charges seemed to thrive under his care. It is said of him in the preface to *The Tower Menagerie* by E. T. Bennett,[7] "So excellent is the management of Mr. Cops, especially regarding cleanliness, that essential security of animal health, that not a single death has occurred from disease, and only one from an accidental cause, the secretary bird having incautiously introduced its long neck into the den of the hyena, was deprived of it and its head at one bite."[8] As Bennett wrote his book seven years after Cops' arrival, it is extremely difficult to believe this unfortunate bird was the only death during that time. However, this event in no way decries the skill and knowledge of Cops, doubtless of humble origin, who pioneered modern wild animal husbandry in England. Within a year or two he had built a collection hitherto unknown for its sheer variety of species, some of which were the first of their kind to be seen in England. Again, it is said of him, "By his spirited and judicious exertion the empty dens have been filled and new ones have been constructed; and the whole of them now being permanently tenanted, the menagerie affords a really interesting and attractive spectacle to the numerous visitors who are drawn thither either from motives of curiosity or a love of science."[9]

The Tower Menagerie, published in November 1829, contains much valuable information about the contemporary collection, but one cannot help wishing it had also described the way the collection was housed. Also, it is not informative whether there was some definite policy in carrying out extensive improvements after Cops' appointment, or whether he was given *carte blanche* to select new arrivals as he chose. Bennett was Assistant Secretary to the infant Zoological Society of London and, to judge by what he wrote, he was a good all-around zoologist, apart from where herpetology was concerned, as will be seen in due course.

The book was illustrated with the usual misshapen, crude, and inaccurate woodcuts of the time and was a chatty volume, but it was packed with information about many of the individual animals described. At the beginning is a curiously prophetic observation, decades in advance of its time. It must be borne in mind that during this period animals in menageries and shows were merely kept, with little or no attempt made to pursue scientific policies, although the barely launched Zoological Society of London was making earnest and largely successful attempts to remedy matters. Bennett observed:

> But as civilization advanced and the progress of society favored the development of the mind, when those who were no longer compelled by necessity to labor for their daily bread found leisure to look abroad with expanded views upon the wonders of creation, the animal kingdom created new attractions and awakened ideas which had before laid dormant. What was at first a mere sentiment of curiosity became a love of science; known objects were explained with more minute attention, and whatever was rare or novel was no longer regarded with a stupid stare of astonishment, but became an object of careful investigation and philosophic mediation.[10]

Rather strangely, there is an apology for the fact that some of the animals described therein were no longer on exhibition. These animals included a cape lion, cheetahs, a Himalayan bear, and a deep blue macaw (the Hyacinthine species?) which "have passed into foreign hands and are now on the continent of Europe." Others, such as the wolves, a Javan civet, and a white antelope were sent to form the nucleus of the collection at Regent's Park. With these exceptions all the animals listed were stated to be in the menagerie.

The first species to be described was the Asiatic lion, and the reader was informed that George, no doubt named after King George IV, had been captured in Bengal during 1823. It appears that General Watson was out riding when a lion and lioness attacked him, but he was able to kill them both with his double-barreled rifle. On casting about he found two cubs some three days old — the cause, no doubt, of the animals' aggressiveness — which he took back to camp where they were reared with the help of a goat. In September of that year, they were brought to England and presented to the Tower, where for the first year they were allowed to wander about almost at will, as they were so tame. It is interesting to note that these animals were obtained in Bengal (in eastern India), whereas today, little more than a century and a half later, the Asiatic lion (*Panthera leo persica*) is restricted to a small area almost on the west coast of India. The female gave birth to three cubs on October 20, 1828.

There were also two cape lions, brown and yellow, respectively, which were of particular interest as this subspecies was by then becoming decidedly rare, finally becoming extinct when the last known specimen was shot in Natal in 1865. The curious fact about this very distinctive race was the cranial difference from other lions; the head is more massive and oblong in profile, as can clearly be seen on the mounted specimen that is still in the Natural History Museum, London. It is believed there are only five other preserved examples.

There were two tiger specimens, a Malayan male and an Indian female. It is reasonable to assume they were approximately the same size as the Malayan subspecies is generally smaller than the *forma typica*. The male, said to be very tractable, had been caught as a cub and reared on boiled food, which is surprising as normally it is difficult in the extreme to induce big cats to eat cooked meat. At the Tower the tiger was persuaded to take raw food, but "he has by no means lost his appetite for soup."

Lord Exmouth obtained the jaguar which, most unexpectedly, was present at the bombardment of Algiers in 1825 during the Barbary wars. On arrival in England it was given to the Marchioness of Londonderry, who gave it to the King, who in turn ordered it to the Tower. This bewildering lifestyle could well have been too much for it, as "it has recently died," which decidedly contradicts the earlier assertion that all species listed were currently thriving. There was also a puma, and it was stated in passing that Mr. Kean, the tragedian, possessed one so tame it would follow him about and was frequently introduced into a filled drawing room at complete liberty.

The menagerie held cheetahs, too, but as their diet consisted of boiled meat and meal, and in winter, hot mashes and gruel, one is left to ponder whether they managed to survive more than one winter on this entirely unsuitable fare. Along these lines, it would be tempting to disparage some of the methods of feeding and general management then practiced, but it must be borne in mind that these were the pioneer days of wild animal husbandry in England. Valuable lessons were being learned through trial and error from which others to come could benefit gratis. Menageries like the Tower came and went, and if their standards are no longer approved, one should at least keep in mind that standards for professions at that time were much lower than are those of the present day.

Little was said about the caracal, other than that it came from Bengal. This beautiful reddish lynx is found over a wide area of southern Asia and Africa and takes its name from the Turkish *kara kulac* (black ears), a reference to the long tufts on this part of the body, while its Arab name *um rashat* means "mother of tassels."

There was a cape hunting dog housed with two spotted hyenas and a striped hyena, which must have made an interesting, if curious, association. It mentioned that William Burchell, the naturalist, imported the first cape hunting dog seen in England. He kept it chained in a yard for over a year. There were also African bloodhounds that Major Denham, an early African explorer, imported. Apparently they were kept for hunting gazelles, and although no description was given, they were probably the variety now known as the Rhodesian ridgeback, or, although less likely, the basenji. Wolves were represented by an adult pair with five cubs.

A gray mongoose from the collection was once turned loose in a room 16 feet square and infested with rats, and within a minute and a half had accounted for a dozen of them. The mongoose proved so valuable in its ability to control rodent populations that, during the time between the two world wars, many London firms imported them to replace cats in controlling vermin in warehouses.

The black bear (probably the American rather than the Asiatic species) was housed with a spotted hyena with which it was on friendly terms, although the latter was dominant at feeding time. There was also a grizzly bear that the Hudson Bay Company had presented 17 years earlier, obviously the bear listed among the few animals inherited by Cops. Originally known simply as Martin, with the passage of time it had been afforded the prefix of "old" and was indeed the menagerie's oldest inhabitant. The Bornean bear (clearly a sun bear) was said to be excessively voracious and "appeared disposed to eat almost without cessation." Its principal food was bread, an extremely unbalanced diet for an omnivorous mammal. It was so tame strangers could pet and rub it. The bear's greed finally cost it its life, having gorged itself at breakfast during hot weather and dying within 10 minutes. So here is yet another casualty not mentioned in the preface of Bennett's book.

Monkeys were represented by the bonnet species from southern India, an anubis baboon from East Africa and a chacma or, as Bennett called it, a pig-faced baboon. There was also a white-handed mongoose, which must have been the mongoose lemur, a primitive primate from Madagascar, which is not the easiest species to keep even today.

An animal whose description sounded very much like the great gray kangaroo had been obtained from the royal collection at Windsor. These kangaroos obviously thrived on their diet of greenstuff and hay as they had bred on at least one occasion. What was described as a "zebra of the plains" was a mare, which had been in the collection two years and was so tame that a boy was able to lead her about the yard. On occasions she could be left to roam about the Tower with a man at her side "who she does not attempt to quit except to run to the canteen, where she is occasionally indulged with a draught of ale, of which she is particularly fond." This kind of relationship between man and captive wild animal is the very stuff of good animal keeping.

There was a very casual mention of a three-year-old elephant, presumably Asiatic, and two llamas which were said to prefer carrots to any other food. There was also a deer, which the king sent from Windsor and which was described as the Malayan rusa deer. In the next sentence the reader is informed it was also called "the samboo deer, as the present species is known to its keepers," all of which could be thoroughly confusing were it not for the fact that here was one of the few species whose scientific name was given, *Cervus equinus*. This was one of the older scientific names, long since gone in one of the periodic taxonomic shake-ups, for the sambar of Southeast Asia. So it is fairly safe to assume it was a sambar, particularly as someone might well have misheard "samboo" for the correct name. *Rusa* is simply Malayan for any species of deer.

An Indian antelope, which Lord Liverpool had presented six years before, was also graced with its scientific name, *cervicapra*, so it was a blackbuck. Perhaps this could have been the "white antelope" mentioned at the beginning, as the species has a tendency toward albinism.

The bird collection was less extensive but nevertheless contained some interesting and impressive species, including a "great sea eagle" (probably the white-tailed species then still breeding on Scottish sea cliffs, but could, in view of the "great," have been the larger and rarer Steller's sea eagle from the Kamchatka area), a golden eagle, a griffon vulture, and what was described as a bearded griffin — the most noteworthy of all as it must have been the Lammergeier or bearded vulture. The bird collection also included a hyacinthine macaw, said to be four feet in length, as indeed it might well have been. There were some sulfur-crested cockatoos fed mainly on hemp seed, although they also liked "sweetmeats and pastry." Hemp, an oily seed, was formerly much used in bird feeding but its excessive use had a tendency to darken the plumage. For example, it was commonly said that a bullfinch (*Pyrrhula pyrrhula*) fed solely on hemp seed for two years turned completely black, so it would have been interesting to see what sort of appearance these normally white birds had. The emus in the collection were bred at Windsor, and there were pelicans that made attempts to breed at the Tower in this direction but without success, even though the male had faithfully fed the female as she had tended their three eggs. These birds had been imported from Hungary, so they may have been the crested or dalmatian pelican (*Pelecanus crispus*), one of the heaviest birds to possess the power of flight. The great horned owl from North America was also represented in the collection.

Rather surprisingly, there was quite a good reptile collection, but regrettably not even the vaguest details have been recorded about how they were housed. Members of this group are not, on the whole, as easy to keep and accommodate as mammals and birds, yet here, in what was probably a damp and drafty room, they were so well cared for and warmed so efficiently that the Indian rock pythons attempted to breed. In fact, Bennett listed them as "Indian boas" and said they were 14 to 15 feet in length, which suggests they were indeed the former species or, less likely, reticulated pythons. After two years in the collection the female laid 14 eggs and, in characteristic fashion, coiled around them — proof they could not have been boas, which produce living young — but by the time *The Tower Menagerie* went to press they had not hatched.

Bennett describes one reptile as an anaconda, but whatever it was, it certainly was not this species. To quote:

> [T]he anaconda has been applied to all the larger and more powerful snakes. It appears to be of Ceylonese origin, and may therefore be of right as well as usage to the present Indian species. The serpent which presides under this title at the Tower ... seems to differ in no essential respects from the Indian boa, the only appreciable distinctions between them consisting of the lighter color, the greater comparative size of the head and the acuteness of the tail of that which at present engages our attention.

No doubt this mysterious creature was a light-phase Indian rock python, which is restricted to western India, grows to a length of 10 or 12 feet and generally has pale shades of brown and pinkish fawn. The anaconda (*Eunectes murinus*), or water boa, is probably the second largest snake in the world, having reached an authenticated length of 28 feet in its native South America, and is distinctively marked with round black blotches on an olive gray background. The reptile collection also included an American alligator, which was fed once a week on raw beef. And somewhere, somehow, room was found for over 100 rattlesnakes. It was noted they varied between four and six feet in length and differed considerably in color and markings, which is to be expected in a family consisting of 30 species and subspecies.

Thus, the Tower menagerie at the end of the third decade of the nineteenth century was rapidly progressing and reawakening after 600 years of near stagnation. The Tower had seen the disappearance of the little princes, the murder of inoffensive King Henry VI, and the executions of the Earl of Essex, Sir Walter Raleigh, Lady Jane Grey, and Anne Boleyn. Within yards of its walls the intrepid Colonel Blood made his bungling attempt to steal the Crown Jewels and doubtless these walls reflected the fiery glow as the great conflagration of 1666 destroyed half of London. Perhaps the lions raised a bass chorus in competition with the garrison's guns that were fired during the plague to dispel the humid air then thought to be responsible for the malady. And past the very gate of the menagerie the heavily pregnant Lady Nithsdale smuggled away to freedom her feminine-garbed husband who was under sentence of death for his part in the 1715 rebellion. If the Tower itself has more than its fair share of ghosts, why should the menagerie not have one as well? Military records relate that during the nineteenth century a sentry died of heart failure after lunging with his bayonet at what appeared to be a spectral bear at the Byward Tower, a matter of literally a stone's throw from the Lion Tower.

Then, apparently suddenly, in late 1831 the collection was presented in its entirety to the Zoological Society of London, with the proviso that any unwanted specimens be sent to the equally embryonic Dublin Zoological Garden, while Mr. Cross at the Exeter 'Change also accepted some. The Tower animals were not all moved away until the spring of the following year. The sole survivors left for Cops to care for were two bears, a Bengal sheep, a leopardess, two emus, and a "cinerous eagle" (which must have been a cinerous vulture). It is not recorded how Cops accepted the loss of his charges and life's work, but one report suggests he was offered the post of head keeper at Regent's Park, where the scope of his vocation would have been vastly increased. Here, one encounters another minor mystery.

In 1828, the London Zoological Garden, which had been open for less than a year, dismissed its head keeper for "several acts of misconduct"; his name was James Cops. As this is hardly the most common of surnames, it would be the ultimate in coincidences if Alfred and James were not related. In fact, they were probably brothers, but no more was heard of, or from, them after the early 1830s, at least not on the European side of the Atlantic. Most interestingly, however, about the middle of that decade an Englishman "from London" turned up in the United States with what must have been the world's first mobile reptile show. His name? Cops! Clearly, there is some fascinating detective work to be done in this direction. The fact remains, however, that Alfred Cops was clearly a first-rate animal man who, unfortunately, has become completely forgotten.

Just why the menagerie's end came so suddenly when real success seemed to be in sight has never been satisfactorily explained. Perhaps it was felt in the highest quarters that the Zoological Society of London was an institution worthy of support, or perhaps that the animals would be better off under its care. It has recently been suggested that the destroyer of the Tower menagerie was also the destroyer of the French Army at the Battle of Waterloo — Arthur Wellesley, Duke of Wellington. At the beginning of the 1830s he was constable (governor) of the Tower of London, which, although it served a variety of purposes, was primarily a military depot and garrison. Wellesley was probably becoming concerned about the way the menagerie was expanding, especially since it was taking up space, and no doubt obstructing the only landward entrance to the Tower. With the establishment of a zoological garden, it was probably felt there was no longer a need for this inconvenience at the Tower. Whatever the reason, the fact remains it came to the end of its long road through history just a decade after taking on its new and vigorous lease on life.

Today, as one looks upon the Lion Tower's foundations, it is possible to wonder just where and how everything was arranged and fitted in, for the dimensions of what was the Lion Tower are modest indeed. It is annoying, too, to bear in mind that it is the only part of the wonderful collection of buildings making up the Tower of London to be demolished in recent times. Almost certainly, no emus, deer, or reptiles could have been accommodated here, so one can but deduce there must have been other adjacent structures that were swept away at the same time and forgotten. The yard casually mentioned in the previous account of the "zebra of the plains" could well be a clue here. And where was the site of England's first elephant house, built with an eye to future unspecified uses when Henry III was king? If at the Tower, stand a moment and think of Alfred Cops — wild animal husbandry today owes much to him.

2.3 Estate Collections and the Windsor Great Park

A brief return to earlier times to consider the kind of collections (ca. 1400s–1800s) that eventually led up to those such as the Windsor Great Park (ca. 1820–1905/06) is in order. From the fifteenth century, private collections in the castles and on the estates of the rich and powerful proliferated. At Brewood in the Midlands there stands an obelisk marking where a leopard escaped from just such a place and was killed with a crossbow in 1513. But, as these collections were just the whims of their owners, mere repositories of strange and wondrous beasts, it was not considered expedient, or worthy of the owner's time, to record any details concerning how the stock was obtained or how long it lived; how it was fed, housed; or whether there were any breeding successes.

For instance, in the eighteenth century at Goodwood House, a few miles north of Chichester on the south coast, there was quite an extensive collection which included two moose from Canada. It was arranged that these treasures should be painted by the famous animal artist George Stubbs, but before he could arrive and set up his easel the moose inconsiderately died. Nothing daunted their owner, the Duke of Richmond, who arranged for them to be strung up in such a way as to appear to be walking, but "the smell was so awful that the artist felt his measurements might have suffered in accuracy." Listed in the duke's collection during the 1740s were "5 wolves, 2 tygers, 1 lyon, 2 lepers [leopards or kangaroos?], 2 tyger cats, 2 musc cats [civets?], 2 volters [vultures?], 3 foxes, 2 Greenland dogs [huskies], 6 eagles, 1 leopard [so the lepers must have been something else], 2 martins, 5 bears, 1 white bear [could have been a polar bear, or the Kermode's form of the American black bear], 1 armadillo, 2 raccoons, 1 manligo [?], 1 jackal, 1 kite, 2 owls, 1 large monkey, 1 sivet cat [so were the musc cats something different?], 1 woman tyger, 3 small monkeys, 1 pecaverre [peccary?], 1 caseawarris, 1 wild boare, 2 hoghs, 1 sow, 1 sow with piggs."[11] This is typical of scores of such species lists; that is, two words were not penned if one would suffice.

At another place, Horton in Northamptonshire, one is informed there were "hogs with their navels on their backs," which could only have been alarmed peccaries exposing their round scent glands situated above the area of the pelvic girdle. At Theobalds Park, to the north of London, no less a personage than King James I set up a collection in the early seventeenth century, but apart from records of an elephant (presumably Asiatic), some dromedary camels that were gifts from the King of Spain, bears, elks (European moose), and a fawn reared with the help of a human wet-nurse, nothing is known about what it looked like and what was done there. More laudably, an attempt was made to establish an English silkworm industry, but apparently the wrong sort of mulberry tree was imported and planted, so the whole enterprise came to naught.

Much later, in 1820, King George IV established the collection already mentioned on several occasions at the Sand Pit Gate, Windsor. Here were wapiti, sambur, chevrotains, zebus, gnus, quaggas, other zebras, "corine" antelopes (presumably korin), llamas, wild swine, emus, ostriches, parrots, and waterfowl. There was also an "enormous tortoise," which arrived from Mile End, London on July 22, 1829, but whether it was one of the Galapagos species, a leopard tortoise, a greaved tortoise from Africa, or merely a larger-than-usual Mediterranean tortoise must remain a mystery, as the Royal Archives say precisely nothing about it.[12]

The *piece de resistance* at Windsor, however, was the first giraffe to be seen in England. This specimen, a young female of the Nubian race, was one Mohammed Ali presented to the king, arriving in London on August 11, 1827, along with several cows that had provided it with milk on the voyage. It was said to be about 18 months old and stood ten and a half feet in height. Three interesting points concern this animal. The first point is the almost unbelievable way in which she was conveyed on the first leg of her journey from Senaar to Cairo. R. B. Davis, the artist commissioned to paint her portrait, noticed the lower limbs seemed to be deformed as the result of recent injuries, whereupon investigations brought to light the astonishing fact that for much of the way she had been carried on the back of a camel, the wounds being caused when the legs were lashed together under the carrier's body.

The second strange thing concerns the unfortunate beast's diet, as somehow George IV firmly got it into his head that this species would take no other nourishment but milk, so he ordered her to be fed on this fluid alone. And on it she contrived to live for over two years! Obviously, toward the end she became very debilitated and her exercise became more and more of a problem until some bright soul hit upon the notion of making an extraordinary device, rather like a gigantic triangle on wheels, in which the creature was somehow secured each day and trundled round her paddock, the hooves just touching the ground. As a point of interest, an old account sheet states, "30th May 1828, expenses to Mr. Bittlestone for steel support for giraffe and adjusting same — £6.6.0." She died on October 14, 1829, having grown 18 inches in height during her sojourn at Windsor.[12]

The other talking point concerns that bane of the zoological historian, conflicting evidence, as although most authorities agree all this took place at Windsor, there are some who assert that the animal was kept at Chiswick House, nearer London. Here the Duke of Devonshire kept a sizable collection, which included an elephant. Some evidence for this position is that until comparatively recently the words "to the giraffe" were barely discernible on a wall in nearby Burlington Lane.

In 1830, William IV came to the throne, and the promptitude with which he sent nearly everything of note to the London Zoological Garden suggests his interest in animals was not great. During the first year of his reign, the society received 13 kangaroos, 14 wapiti, 3 axis and 2 sambar deer, 3 gnus, 2 nylghai, various foreign sheep and goats, 2 llamas, 7 zebus, a pair each of mountain and Burchell's zebras, 2 zebra × donkey hybrids, 1 peccary, 1 European wild swine, 42 peafowl, 11 emus, 4 macaws, 2 eagle owls, 2 sea eagles, 1 peregrine falcon, 2 cockatoos, 1 scarlet lory, 2 golden parakeets (possibly the now rare Queen of Bavaria's conure), 1 rosehill parakeet (later known as the rosehiller, and now rosella), 5 widow birds (whydahs), 1 curassow, 5 kavirondo cranes, 1 scarlet ibis, and 7 cereopsis geese. There is a persistent legend that a bird of paradise was kept at Windsor, but if so it must have died by 1830 as it is not among the list of species sent to London.

A few years later the reign of Queen Victoria began (1837), and here was someone who really was fond of and interested in animals, with the result that in a short time the collection began to take on something of its former glory. Records are scanty, but certainly such impressive species as Impeyan pheasants, bison, and Grevy's zebras were kept. And, as the latter species was not officially discovered until 1882, it is clear the place was still going strong as the twentieth century drew near. King Edward VII, although a generous donor to the Zoological Society of London, did not really share his mother's enthusiasm for natural history, so again the collection was dispersed, this time permanently around 1905/06.

2.4 The Exeter 'Change

The Exeter Exchange, or Exeter 'Change as it was always known during its long lifetime (1800–1828), was a large building situated on London's Strand, where the Strand Palace Hotel now stands. It was in the nature of what would now be called an arcade, with shops and stalls of various kinds. About 1800, Gilbert Pidcock, who owned a traveling menagerie, rented at least two stories, possibly for winter quarters and a base, or more probably because he intended to take his show off the road and was seeking a more permanent venue. Here he exhibited a collection that included a lynx, kangaroos, a secretary bird, an "African ram," a "wolf from Algiers" (perhaps a jackal or even a large and unfriendly dog he had found in the street outside), and a few monkeys. On his death 10 years later Mr. S. Polito bought the place at auction, and added llamas, gnus, sea lions, emus, and large snakes. He also gave it the grandiose title of Royal Menagerie, but after only seven years the whole outfit was sold to Edward Cross, who immediately dubbed it with the even more splendiferous name of the Royal Grand National Menagerie, and set about assembling one of the best collections seen in Europe, and one of the accepted sights of London.

Here, in a multitude of small cages that most today would regard with disgust and horror (that is, those who have forgotten that most species confined in them usually lived long and

Exeter 'Change (1800–1828), London. Courtesy of C. H. Keeling.

healthy lives), the interested visitor could admire an incredible range of species: the first clouded leopard (or clouded tiger, as it was always called then) to be seen alive in Europe, all the big cats (including a lioness that gave birth to 11 cubs in five years and a cape lion that an African woman had suckled when it was a cub), a cheetah (with perfectly sound information about it in the surprisingly good Exeter 'Change guidebook, a great innovation for the early nineteenth century), caracals, spotted and striped hyenas (one of the former, Billy, lived for 26 years, which is not only the species' longevity record in confinement, but speaks volumes for this method of keeping many kinds of animals), polar bears, nylghai, a quagga, capybaras, beavers, civets, jackals, coatis, raccoons, porcupines, zebus, llamas, baboons, and kangaroos. There was also Chunee, the magnificent bull Asiatic elephant who, maddened with toothache, ran amok in 1826 and was butchered when a completely inept firing squad used a total of 152 musket balls to kill him.

Among the birds were an emu (said to have been caught in Van Dieman's Land, so it must have been an example of the now extinct Tasmanian subspecies), pelicans, an adjutant stork, various cranes, parrots in profusion, a "beautiful nondescript eagle from the coast of Guinea" (according to the Exeter 'Change guidebook), spoonbills, storks, owls, many species of ducks and geese, and Egyptian vultures. Among the reptiles was a "boa constrictor, the giant serpent of Tara," whatever that might have been.

The Exeter 'Change Menagerie was open daily from 9 A.M. until 9 P.M. and the admission fee was two shillings, which was decidedly steep for the time. There is an oft-quoted story that a stray cat or dog would be accepted in lieu of payment, as food for the carnivores, but as far as can be ascertained this assertion must be relegated to the realm of folklore or legend. Despite the high admission charge, it proved a very popular display indeed, and its end came in 1828, not because of dwindling support, but because the building was in the way of a major road-widening scheme. Nothing daunted, Cross sought about for another site and eventually found a suitable one in the leafy suburbs of South London where, in 1831, he opened the Surrey Zoological Garden on approximately 13 acres, roughly where Kennington Underground Station now stands.

2.5   Three Nineteenth-Century Private Collections

Many animals are named after Lord Stanley, later the 13th Earl of Derby, for example, the Derbyan eland; the Derbyan parakeet; Lord Derby's zonure; the Derbyan screamer, a species of African water tortoise; the Stanley crane; the Stanley parakeet; etc. This fortunate gentleman lived at Knowsley Hall near Liverpool, where from approximately 1834 until his death in 1851 he kept a large, impressive, and well-cared-for collection.

At Knowsley Hall, harnessed and addax antelopes successfully bred (he bought the breeding female of the latter, rather unpredictable species after discovering it living as a pet in an old woman's cottage garden). These events were not to be emulated elsewhere for a century to come. And for the first and only time, as far as is known, the mandarin and Carolina (wood) ducks hybridized. Other antelope species kept there included saiga, four-horned, oribi, sing-sing, kob, and oryx. Experiments were made to domesticate the eland (it was discovered they were extremely susceptible to tuberculosis) and an unsuccessful attempt was made to introduce, for commercial purposes, a hundred alpacas onto the nearby moors. The rough terrain and fare would in all probability have been much to their liking. The trouble lay in their very poor condition after a particularly long sea voyage. About this time it was recorded that a Burchell's zebra and an anoa arrived from the Jardin des Plantes in Paris in exchange for an eland and a pair of alpacas. There was a thriving herd of American bison, while chevrotains, wapiti, sambar, and muntjac did well there, too.

It is assumed the creature listed in the estate records as the "pita" was a brocket, and there were "wood-eating antelope" (as this species is said to have a curious taste for eating rotten wood) and the Texan deer, the black-tailed species. There were also banded mongooses, Jalarang squirrels, lemurs, and other small primates in an extensive range of small mammals. Birds included Hawaiian geese (which were experimentally bred with a view to possible domestication), cassowaries, rheas, guans, Stanley and wattled cranes, an excellent collection of waterfowl, crowned and Nicobar pigeons, and passenger pigeons. The passenger pigeons bred so freely that periodically they were released from their overcrowded aviaries, which probably accounts for the surprisingly large numbers of this latter species recorded in England during the mid-nineteenth century. The first living lungfish seen in England was kept there, as well. In short, 94 species of mammals were kept there, along with 318 species of birds (of which 84 were bred). Detailed daily occurrence books were kept, including an entry for October 16, 1844 recording what seems to be the first human observation of the birth of a kangaroo.

At Stubton Hall, in the eastern county of Lincolnshire, Sir Robert Heron bred rheas for the first time in England by hatching the eggs under turkey hens, which is all the more surprising when it is borne in mind what unreliable sitters these birds normally are. He too realized the value of keeping daily records, and it is Sir Robert who is responsible for detailed and important accounts of some of the first successful attempts to breed goldfish in the Western world. Sir Robert also may be the first person to warn what nuisances toads can be in the spring as they seek to clasp large fish in amplexus.

Also represented in Sir Robert's collection were kangaroos (which bred constantly), llamas and alpacas, porcupines, ring-tailed lemurs, mountain hares (*Lepus timidus*), axis, roe and hog deer, flying squirrels, nylghai, tapirs, guanacoes, blackbuck, agoutis, jerboas (by no means easy to keep even now), and armadillos. Birds included cassowaries, guans, curassows, "Polish cranes" (probably the common crane, *Grus grus*), choughs, black swans, and secretary birds. The collection also included a few reptiles, such as tortoises and some particularly well-chronicled chameleons. Exchanges between Stubton Hall and Knowsley Hall (both the Earl of Derby and Sir Robert were council members of the Zoological Society of London) were common, and to a certain extent with the Wentworth collection, as well.

This latter collection, owned by Earl Fitzwilliam, was situated near Rotherham in Yorkshire. He also, at that early juncture (the mid-nineteenth century), appreciated the value of note taking. Thus, one reads that on November 14, 1850, £1.4.6 was paid to Mr. Stone for "medicines and attendances," while on the January 24 that year John Bisley was paid 9/6 (nine shillings and six pence, or about $1.00 in today's U.S. currency) for "gathering ants," which is slightly surprising as there should not have been any about at that time of the year. On June 24, four stones of fish, three of grapes, two of raisins, and quantities of oranges, figs, and apples arrived, along with a "bottle of port wine for the emu."

A species list for 1856 details an American bear (presumably black), llamas, alpacas (Knowsley again?), axis deer, ring-tailed lemurs, armadillos, Chinese monkeys, antelope (species not stated, but specified ex Knowsley), golden and sea eagles, a "burnelly" eagle (Bonelli's?), griffon vulture, eagle owls, macaws, purple-capped lories (purchased from the then-flourishing Liverpool Zoological Garden for the comparatively large sum of £8), capercallie (a large species of grouse), king vulture, cereopsis goose, scarlet ibis, bobwhite quail, great-crested grebe, cockatoos, passenger pigeons, and several tortoises (species not specified). Interestingly, some mounted specimens that had once lived at Wentworth are currently kept, albeit behind the scenes, at the Rotherham Museum, while on the estate itself are still some of the foundations of animal enclosures and aviaries built well before the 1860s.[13]

2.6 Zoological Gardens at Surrey and Liverpool

Like its urban and entirely enclosed parent, the Surrey Zoological Garden (1831–1856) published an unexpectedly informative guidebook, which states, for example, that the country's first Tasmanian devil and wedge-tailed eagle were exhibited there. Its most notable structure was a huge circular glass building in which the carnivores were housed, including "an emasculated lion. This is a most singular-looking animal, and at first sight is generally supposed to be a lioness, from the entire absence of all mane, which, together with the organs of voice, are totally wanting in this animal." Also housed within this glass building were normal lions, Asiatic lions, a tiger, leopards in profusion, ocelot, caracal, African civet, spotted and striped hyenas, palm civet, red and brown coatis, and a rhesus monkey. So one can deduce that in high summer the heat and aroma in this glorified greenhouse must have been in a category all its own.

There was a series of enclosures for a variety of curiously dubbed domestic canines, such as the Alpine mastiff, Italian wolf dog, Cuban mastiff, Russian greyhound, and the "wild dog from the Himalaya Mountains" (which was probably a Tibetan mastiff, a large beast still kept for guarding flocks and noted for its truculence and general lack of a sense of humor). On Eagle Rock could be seen both golden and white-tailed sea species, and there was a well-stocked Monkey House that held "coaita," which as far as can be deduced was one of the spider monkey species. The Bird House, home of a profusion of parrots, also contained pythons and boas. There were paddocks for deer, antelopes, and zebras. There appears to be some mystery concerning an Indian rhinoceros, which according to the 1834 guidebook occupied the Rhino Pavilion. The next year's publication does not mention the animal, while its erstwhile home became the repository and abode of a pair of young Asiatic elephants (it must be kept in mind that for another century to come some zoological gardens included in their publications species they *hoped* to obtain). It is quite possible there was an Indian rhinoceros there for a short time, and if so it is one of the few that has slipped through the carefully cast nets of Richard J. Reynolds III of Atlanta, perhaps the world's leading authority on this family in confinement.[14]

In May 1839, there was a bush cow; "hitherto the only evidence of its existence was a portion of a skull brought into the country by Captain Clapperton." Clapperton was quite a noted African explorer. The bush cow is a small race of the Cape Buffalo, rich reddish-brown in color and found in West and Central Africa. It must have been one of the first to be seen in Europe. There were also "naked buffaloes," with the somewhat suspect subspecific name of *seminudus*, although the information that they "appear to be a variety of the domestic buffalo of Egypt with no hair in summer" sounds quite reasonable. The collection included polar bears as well, which, with the Exeter 'Change specimens and another very early one at Regent's Park, makes utter nonsense of Carl Hagenbeck's claim that his father was responsible for bringing the first one to Europe in 1852. In any case, the species is found in northern Europe.

It is now 1842, and curious indeed are the snakes to be seen here, as in addition to an anaconda there is also a water boa (an alternate name for the former) and a "great boa from Ceylon" (despite there being no such species). So one is tempted to muse back to some of the odd information purveyed about snakes that emanated from the Tower of London. This guide also stated that a pelican currently in the collection, which the Emperor Maximillian had once owned, was known to be over 80 years old, a statement whose veracity is doubtful.

From about 1840 there were lavish fireworks displays, which seem gradually to have taken priority over everything else. On Cross' retirement in 1844 his secretary, William Tyler, took over the place, whereupon it seemed to plummet rapidly. Around 1850, an American (no less than Nathaniel Hawthorne, America's consul at Liverpool) who visited the Surrey Zoological Garden, remarked on the general dilapidation and run-down air of the place, besides mentioning camels, lions, a giraffe, polar and black bears, an elephant, monkeys, parrots, waterfowl, an ostrich, stuffed snakes, and an aquarium. He also mentioned the wealth of trees. In the last guidebook no fewer than 202 species of trees are listed.

On a dull November morning in 1856 the collection came under the auctioneer's hammer. At a time when £1 was equivalent to U.S.$5 (and both were worth more), the elephant was sold to a circus for £320, a lion went for £150, and a tigress £79. Two camels were sold for £112 and a giraffe for £250. The giraffe was destined for some collection on the continent, but as it was being lowered onto the deck of the ship in a tackle the ropes broke and it fell to the deck, breaking its back; the skin, however, fetched £25. The ostrich fetched £27, a nylghai £9, and a zebra × donkey hybrid £8. Several species of monkey raised no more than 10 shillings each, 24 shillings changed hands for a pair of jackals, a pair of porcupines (regarded as a good show) realized £8.15.0, and a red deer went to a new home for £2.10.0. Sums ranging from 13 shillings to £2 were paid for two golden eagles, two Chilean sea eagles, a wedge-tailed eagle, and a sea eagle. A pelican (again a showy creature) sold for £20. Four pythons and a boa went for £2 to £5, a pair of flying squirrels £2, an armadillo 30 shillings, a cockatoo £2, a blue and yellow macaw £3.5.0, a common mynah for 9 shillings, and a red-eared waxbill and two cutthroat finches for 8 shillings. A leading hairdresser bought several bears for a few pounds each, so it is clear they ended up in pots for the then popular bear's grease. Perhaps it was as well that Cross was not there to see the end of it all.

Liverpool Zoological Garden (1832–1863), an excellent institution, offered the residents and visitors of Liverpool the opportunity to see a wide range of living animals, the like of which they probably did not realize existed. An ex-showman named Atkins established the zoo. His previous claim to fame had been breeding the first ligers (lion–tigers crosses), at least as far as England was concerned. By all accounts, the zoological garden was attractively laid out and planted, and its collection was superb. For sheer information and accuracy its guidebook was years ahead of its time. The main draw of the collection was what is generally accepted to have been the first Javan rhinoceros to reach Europe alive, although in recent years Kees Rookmaaker (another authority on these animals) has stated his belief that it was, in fact, a small example of the Indian species. The zoo also exhibited two Asiatic elephants; the male in 1848 had grown into a very fine animal, but began to cause serious trouble. Eventually his dangerous behavior became such a worry and a problem that a detachment of soldiers from the 52nd Regiment had to shoot the elephant.

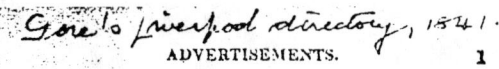

Liverpool Zoological Garden (1832–1863) advertisement of 1834 (left) and 1841 (right). Courtesy of C. H. Keeling.

A characteristic of the garden was the number of flooded pits, which were occupied by waterfowl, waders, pelicans, and storks in far too great profusion to be enumerated here. The bear collection included an example of the spectacled species, which must have been the only one in England, if not in Europe. In a large, circular "menagerie" (at that time the term was not infrequently used specifically to designate a large carnivore house), one could see lions, tigers, leopards, a jaguar, a jungle cat, pumas, a ratel, several species of mongoose, a dasyure (which is interesting as this so-called native cat of Australia has never been really common in confinement), and two ligers which had been born there on July 19, 1833. It would have been interesting to have seen their male parent, as he was the result of a mating between the Barbary and Senegal subspecies — the largest and smallest lions, respectively, with the fullest and the slightest manes. The Monkey House was home

Bear Castle in Leeds, which is the last remaining relic of that city's short-lived Victorian zoological garden. A century and more ago it was normal practice to house bears in these bear castles. © Photograph by Christine Grant.

to species ranging from the entellus (a langur) to Barbary "apes," patas monkeys to drills, and something called the marmondia (which, to judge by its scientific name *Ateles,* was one of the spider monkeys, perhaps the same species as the coaita mentioned earlier). On a hillock above this house was a house for birds of prey, such as the "African black eagle" (which might have been either the bateleur or the Verraux's species), the white-tailed sea eagle, the golden eagle, the "common eagle" (whatever that might have been, but it was a present from the "Gentlemen of the Royal Liverpool Institution"), the sociable vulture from South Africa, a pair of griffon vultures, and a pair of turkey vultures (which no less a personage than Audubon, the great artist of birds, had presented, having spent a considerable time in Liverpool), and great horned owls.

In a smaller "menagerie" visitors could see a fascinating collection that included a striped hyena (born there, but the mother had abandoned it so a bulldog bitch reared it), a leopardess, a pair of jackals, a Virginia opossum, a "Brazilian fox" (probably the Azara's dog, which is not a true dog at all), the European fox, coypus, a gray mongoose, a "zibet" (which must have been the big Indian civet, to go by its scientific name *Viverra zibetha*), and agoutis. Thrown in for good measure was a domestic Persian cat.

An ungulate house contained a wide range of deer and antelopes, including a nylghai, a gnu, a pair of dorcas gazelles, zebus, axis and sambar deer, red kangaroos with young, and a pair of llamas. Other mammals included wolves, a quagga, peccaries, Arctic foxes, "Java hares" (an old name for the agouti), civets, genets, and water buffaloes.

The aviary housed a good collection of Psittacines ranging from macaws to hanging parrots, and the buff-fronted parakeet (a species that defies identification, as today no bird is known by that name, and this is the one species in the entire guidebook whose scientific name is not given). Also in the aviary were marmosets, squirrels, opossums, squirrel monkeys, and reptiles, such as boas, rattlesnakes, iguanas, and an American alligator. There was a well-stocked range of pheasantries and an outdoor enclosure for tortoises.

A very sound dictum is the German one of "the master's foot is the land's best fertilizer," and it is most significant that after Atkins' death in the 1850s the Liverpool zoo rapidly went downhill. In the early days only well-behaved and "fashionable" people were admitted to it, and people conducting themselves in disorderly fashion would quickly find themselves outside the gate. But with the advent of a drinking booth in about 1857, an entirely different type of clientele found its way there, and within a very few years the place had gained such an unfortunate reputation that respectable people chose not to support it. In 1863, the zoo closed its gates for good. Presumably there was a sale of stock, but no details can be located.

2.7 Manchester Zoological Gardens and the Belle Vue Zoo

It is a great pity that the Manchester Zoo operated only from 1838 until 1842, when it was forced to close for economic reasons. This institution is not to be confused with that city's great Belle Vue Zoo (1836–1977). Manchester Zoological Gardens not only had the highest scientific aims, but was also very successful on the practical side of things. During its few years of life its breeding record was quite impressive — lions, various monkeys, kangaroos, zebus, leopards, and marmosets. The breeding of the last two were quite notable successes for the period. The head keeper, Harry Richardson, had many years experience at the Exeter 'Change. In addition, the grounds were under the care of a Mr. Mearns, a leading horticulturist who had come from the Duke of Portland's estates, and Edward Stevens of Bridge Street, Manchester, was the veterinary surgeon. The gardens, situated at Higher Broughton in the northwest corner of the city, covered some 15 acres.

The superb guidebook (which made but one mistake — the last British wolf was killed in the Upper Findhorn Valley in 1743 and not in Lochaber in 1670) contains a wealth of information and lists an excellent collection, of which the star attraction was probably a ground parrot (*Pezoporus wallicus*), a slenderly built, mainly green species from Australia and Tasmania that rarely flies and is nocturnal.

In the elephant house was a five-year-old female Asiatic elephant and, surprisingly, another Indian rhinoceros not listed in chronicles on the rhinoceros. There was also an "arnee ox" (presumably an Indian water buffalo, as arnee is a little-used alternate name) and both Bactrian and one-humped camels. Besides a wealth of parrots in the aviary, there were such Passeriformes as finches, bunting, and weaver birds, along with red and flying squirrels, an alligator, and what was called an Indian boa (which must have been the Indian rock python). Large lakes were home to swans, ducks, geese, pelicans, cormorants, gulls, and storks. A range of flight cages contained the white-tailed sea eagle, tawny and snowy owls, ravens, and a seriema. Small mammals included European and silver foxes, coatis, raccoons, wallabies, crested porcupines, and dingos. Among the larger stock were llamas, guanacos, a vicuna, fat-tailed sheep, deer (red, axis, sambar, and Virginia), and bear (brown and American black).

There was a small but interesting collection in the Monkey House, which included drills and other baboons, crab-eating macaques, and the "smoke-colored monkey," which must have been the sooty mangabey. A "menagerie," which was long and cigar-shaped in plan, contained seven fair-sized cages, as well as six small ones. As far as can be ascertained from the plan, there were no outdoor cages. This menagerie was occupied by two lions (born on November 19, 1836 in Wombwell's Traveling Menagerie), a pair of tigers (one of which came from the island of Saugar), a striped hyena, an African civet, a gray mongoose, and a Virginia opossum. What a tragedy this paradise for naturalists, which also immeasurably broadened the horizons of the more casual visitor, was forced to die such a premature death.

Manchester Zoological Gardens, pursuing a largely scientific policy in the English provinces during the 1830s, was far ahead of its time, while the Belle Vue Zoo — which had opened only three years earlier — more shrewdly set out to attract a larger audience. John Jennison, who established the latter institution, knew full well the value of living wild animals as a money-making public attraction. To make the zoo a place even more attractive to those with more catholic taste, he added all manner of other amenities, such as dance halls, race tracks, fireworks displays, balloon ascents, wrestling, boxing, concerts, and flower shows. Surprisingly, these attractions did not detract from the animal collection; in fact, several polls taken over the many decades consistently showed the overwhelming majority of people considered the place to be primarily a zoological garden. Jennison also pursued a largely successful policy of seeking self-sufficiency, growing vast quantities of fruit and vegetables (for both the animals and the visitors), making his own gas and electricity, brewing his own beer, making lemonade, doing his own building, and printing his own guidebooks and tickets. Particularly resourceful was his millions of bricks baked from the clay dug out to make the boating lake and used to build a 15-foot-high perimeter wall.

Most importantly, the Belle Vue animal collection was excellent and well managed. For example, by 1880, jaguars, lions, and tigers had successfully bred there. Unfortunately, the remaining descendants of the Jennisons lacked their forebear's interest, drive, and enthusiasm. They sold Belle Vue Zoo in 1926 to the first of a whole succession of business enterprises. The last of these, in 1977, realized the zoo was occupying no less than 70 acres of prime building land less than three miles from the center of Manchester, so the zoo was closed and the property sold.[15–18]

2.8 Zoological Gardens at Edinburgh, Bristol, Hull, and Preston

In existence for just 16 years after its 1839 opening, the zoological gardens at Edinburgh (1839–1855) was — to judge by the solitary guidebook and a few press cuttings that have survived — a well-designed six acres, with beautiful and classically planned houses and enclosures, almost in the center of the city. (This zoo should not be confused with the present and excellent institution in that city, the Royal Zoological Society of Scotland's Edinburgh Zoological Park, opened in 1913.)

In addition to a good general collection of mammals and birds then readily available, there was an interesting exhibit that showed unusual perceptiveness and imagination on someone's part — the skeleton of a blue whale, which was afforded its own house and which broadened the educational scope of the living element. After the corpse had been washed up on the nearby coast, the skeleton had been prepared by the notorious Robert Knox, who did appalling business with the "resurrectionists" William Burke and William Hare in 1828 (the pair who suffocated victims and sold them to anatomical schools). The skeleton was set up in a way that fascinated and edified the Edinburgh citizenry, as indeed it still does in the Royal Scottish Museum on Chambers Street. Thus, in a small way, part of the old Edinburgh Zoological Garden lives on. For unknown reasons, as the zoo did not appear to be in financial trouble, the place was sold to a commercial firm of amusement caterers in 1855. Attractions more suited to a carnival were added, the zoo's previous good name began to suffer, and within two years it was all over.[19–26]

Almost alone among the zoos set up during the mid-nineteenth century, the Bristol Zoological Garden, opened in 1835, still survives and flourishes. This is a splendid institution with a good reputation and a most interesting story behind it, so it is a pity that in a way it contrives to keep itself rather to itself. Certainly comparatively little has been written about it, with the exceptions of two slim and rather disappointing books in 1964 and 1985. Unfortunately, both were more concerned with founders and other past personalities, rather than the place itself and its inhabitants.[27–29]

Bristol Zoo, also a beautiful botanical garden with extensive nurseries and plant houses, holds an impressive number of British "firsts," including the first successful breeding of American black bears (in the 1840s), Cape hunting dogs, black rhinoceros, chimpanzee, and okapi. Other first are, the first observation beehive, nocturnal house, and cinema (showing wildlife films). It had the first natural history exhibition to accompany the living collection (in the 1880s) and was the first to vary the big cats' meat diet with whole animals (recognizing the nutritional needs of these predators). The people of Bristol, a once important port in the southwest of England, have always loyally supported their zoological garden, and it is to be hoped they will continue to do so now that the zoo has lost its local character and has moved into line with other British collections, homogenizing its collection and facility.

Other comparatively short-lived nineteenth-century public collections were set up in Hull, Preston, and elsewhere. Hull Zoological Garden, which was based on the old Liverpool Zoological Garden (in fact, some of their buildings were identical), was also rather akin to the original Edinburgh Zoological Garden (adopting carnival policies and offering raucous attractions). It only lasted from 1840 to 1860 and the reasons for this do not seem to have been recorded. Hull Zoo had a good monkey-breeding record and a mainstream collection, with new arrivals steadily arriving via the many ship captains who regularly came to this once-thriving port.

Arguably, the zoo's most interesting exhibit was almost its last, a young hippopotamus that lived at the zoo for a few weeks in the summer of 1860. This, the first of its species to be seen in the country outside of London, was Bucheet (Arabic for "fortunate"), purchased in London from Seth B. Howes, who was touring with his Great United States Circus. After being on deposit at Hull Zoo, Bucheet was shipped to the United States where Phineas T. Barnum exhibited the hippo widely before its death in Canada in July 1867.

Preston Zoological Garden began life in 1865 as an additional attraction in an impressive series of large greenhouses in a public park. These lofty structures, densely packed with a veritable riot of palms, cycads, ferns, vines, creepers, and flowering plants, set around pools and running water, were highly popular in Victorian times. It was often felt their attractions would be greatly enhanced with the introduction of a wide range of animal species. In view of the high degree of warmth and the paucity of ventilation, these structures left a great deal to be desired when it came to good and successful wild animal husbandry. Perhaps this accounts for the remnants of the collection coming under the auctioneer's hammer in 1870, just six years after the opening of the zoo.[19-26]

2.9 Zoological Gardens at Paignton, Chester, and Dudley

During the early 1920s, an eccentric multimillionaire named Herbert Whitley bought Primley House and the estate at Paignton in Devonshire in England's extreme southwest. Here he was able to indulge his very real passion for natural history, with the result that in time his animal and plant houses were spoken of in awe and wonder by those allowed to see them. At the time, his domain was not open to the public, but fellow collectors and enthusiasts were welcome.

Whitley kept very much to himself, and did not seem interested in publishing notes or records. Those notes and records he did keep were destroyed upon his death, as he requested. However, it is widely and genuinely believed that during the 1930s, unrecorded breeding of species ranging from polar bears to kagus, as well as parrots and conures, took place under his care. Shortly before World War II this collection, which came to be known as the Paignton Zoological Garden, opened to the public, and has been in operation ever since.

In its heyday during the 1950s and 1960s, when it was the place to visit if one wanted to see some real gems — true collectors' species, say, tamanduas, epauletted fruit bats, Mount Roraima aracaris, hairy frogs — this was where they could be seen. Eventually, the hand of time removed Whitley, and the trust named after him took over. As a result, the collection has been reduced and is no longer the naturalists' paradise that it used to be, but it has a very active conservation program.

The zoological garden at Chester, in the northwest of England, was founded in 1931. In time, a large and impressive collection, a veritable Noah's Ark, had been amassed. It was, however, purely and simply a place where people came to admire the animals, with no attempt made to establish educational and scientific programs. This orientation would change, but for many years the collection was very popular and experienced a great deal of building and expansion.

In time, the founder, George S. Mottershead, semiretired and the other younger members of the family took over with the result that it became a mainstream institution with a massive and varied collection. An education department and lecture room became part of the facility in 1973. Currently, under the aegis of the North of England Zoological Society, the Chester Zoo has become very conservation oriented. As a result of these efforts, less attention has been paid to its collection, which is no longer as rich and representative as was once the case. However, it is still the largest and most varied in England, except for the London Zoological Garden.

Dudley Zoological Garden, almost in the middle of England, is an interesting yet somewhat unfortunate entity. It opened, to much fanfare and acclaim, in 1937, when it was considered to be the ultimate in modernity and impressive design. Much of it, the entrance, restaurant, penguin enclosure, sea lion enclosure, cat enclosure, elephant house, and birdhouse, was the design of Berthold Lubetkin, who also designed a few London Zoo buildings. However modern visually, his designs were impracticable and unsuitable structures for the animals. Due to preservation status, which protects the buildings for their architectural interest, they must be retained and cannot be replaced with more suitable exhibits. Preservation status for the trees within and adjacent to the zoo property has also prevented the zoo from building new exhibits in these areas.

There is a good collection at the Dudley Zoo, but the place has always suffered from inadequate funding and has had a rapid succession of directors, more so than any other British zoological garden. It is also one of the very few zoological gardens that had to be "rescued" by another (in peacetime); in the 1970s, when facing the very real possibility of closure, the Bristol Zoological Garden bailed out the zoo with money, animals, and other help for several years.[19–26,30]

2.10 London Zoological Garden

The London Zoological Garden (1828) has played an enormous part in the life of many generations of Britons. Over the last two decades, however, the future existence of this great institution has been problematic. If the zoo dies, and it might, its demise will be a bereavement to many. London Zoological Garden pioneered; it learned lessons over the past century and a half that other institutions coming later benefited from for free; it once boasted the finest collection of living animals, not only in the world, but that the world had ever seen; its "firsts," both in breeding and housing, are the most impressive of any zoo; for a long time it led the way in zoological and veterinary research; and it produced more publications than any other such place in the world.[31–36]

From 1859 until 1897, the London Zoological Garden was under the experienced and skillful superintendentship of the almost legendary Abraham D. Bartlett, one of the greatest of all those who have conducted the daily affairs of a living animal collection. Many of the species he received were the first of their kind to be seen in confinement (the aardwolf and the aye aye, to give but two examples). Some were even hitherto unknown and unsuspected type specimens. He knew full well the value of copious note taking, and today his books still make valuable

Raven Cage, preserved on the Fellows Lawn at the London Zoological Garden. Dating from 1828, it is regarded as the Garden's oldest structure, originally housing macaws, then a pair of king vultures, before being devoted to ravens for well over a century. © Photograph by Pamela Keeling.

LONDON ZOOLOGICAL GARDEN FIRSTS

London Zoological Garden, established in 1828, may be considered the first modern zoological garden. An establishment of the Zoological Society of London, the zoo implemented education, research, and conservation programs that ushered other contemporary menageries and zoos into the zoological garden arena. It is certainly within the bounds of possibility that the London Zoological Garden has uncrated more individual animals that were the type specimens of their species than any other similar institution. The number of species exhibited for the first time in Europe (or not seen since the Roman era) or bred for the first time is beyond enumeration.

Three major exhibitions were created for the first time. In 1849, a reptile house was converted from the Cat House, which became empty on the completion of the Carnivore Terrace. Within five years all three British snake species (common viper, grass snake, and smooth snake) bred successfully there. And in 1860, the painted terrapin bred here for the first, and perhaps last, time in England. These are events that would be considered achievements today.

continued

> ## London Zoological Garden
>
> The world's first public aquarium, or Fish House as it was known for most of its innovative life, was opened in May 1853. Throughout its existence, which lasted until the early decades of the twentieth century when its large tanks were converted into the Diving Birds House, its staff had to grapple with temperature problems. For half the year the exhibits were too hot, and too cold the other half. This problem was caused because the house was constructed from that finest of all conductors — glass — one of the advantages, or disadvantages, of early pioneering and experimentation.
>
> The early 1850s coincided with an enormous public interest in marine biology. Within this new exhibition, in tanks placed on stoutly constructed tables and aerated by means that now seem primitive, was an astonishing collection of marine fauna, most of which was collected from inshore waters. In all, 58 species of fish, 75 of mollusk, 41 of crustaceans, 27 of coelenterates, 15 of echinoderms, and 14 of annelids in the first year alone. There were freshwater species, too, one of which, a young pike, inadvertently earned glory as the first living fish to be photographed. Count Montizon, whose father was one of the pretenders to the Spanish throne, was the photographer.
>
> By the turn of the century all the original, and rather temporary-looking, tanks were replaced with far larger fixtures. A contemporary visitor noted not only many fish species, both freshwater and marine, but various amphibians such as amphiumas, and invertebrates like anemones and corals. There were also birds of many kinds — plovers, redshanks, knots, godwits, and other waders, together with penguins, darters, cormorants, and shags, which obligingly dived for living fish in huge glass-fronted tanks at feeding time.
>
> What would today be called an insect or invertebrate house was opened in 1881 when a steel-and-glass structure usurped what had previously served as a refreshment room. The display cases were arranged on stands around the walls, while others were placed on tables in the center. The exhibits were well labeled (another first for the zoo) and use was made of mounted material as well. There was a heavy emphasis on silk moths and their commercial culture, and there were tanks for aquatic species. This insect house even offered its own very informative guidebook. Although the insect house is long since gone, its site still contains a mulberry tree.
>
> *Sources:* Based on Mitchell, P. Chalmers, *Centenary History of the Zoological Society of London,* Zoological Society of London, London, 1929; Lord Zuckerman, S., Ed., *The Zoological Society of London 1826–1976 and Beyond,* Academic Press, London, 1976; and Blunt, Wilfrid, *The Ark in the Park,* Hamish Hamilton, London, 1976.

reading for those engaged in wild animal husbandry.[37,38] As just one example, he noted that whole maize (Indian corn) should not be fed to rhinoceroses as it sometimes started to germinate inside them. And while on the subject of these animals, it is interesting to note that he is still the only person to have had at one time the Indian, Sumatran, Javan, and black rhinoceros species under his care.

Current thinking, or probably more correctly misapprehension, asserts that during the last century most wild species did little more than merely exist in confinement for short periods, and that such breeding as there was could be attributed to pure luck. But, just peruse this list: spotted hyena, crested porcupine, Cuming's octodon, golden agouti, mouflon, Barbary sheep, eland, pygmy hog, scimitar-horned oryx, sika, hog and Mexican deer, red kangaroo, black-faced kangaroo, giraffe, vulpine phalanger, short-headed phalanger, rat kangaroo, Fournier's capromys (a hutia), viscacha, nylghai, collared fruit bat, etc. The common denominator here is that all these mammal species bred successfully in the London Zoological Garden a century and more ago with regularity. Their births were not occasional events or the result of the importation of pregnant females. The births were due to good husbandry on the part of dedicated keepers. The record with birds was even better, with such achievements as brush turkey, sun bittern, and black kite. And the record for reptiles, in which the cyclodus, puff adder, and rattlesnakes predominated, was much better than expected. In 1843 on the completion of the once-famous Carnivore Terrace, the Cat House became empty, and in 1849 it was converted into a Reptile House — the first such in the world.

London Zoological Garden also had other exhibition "firsts," including the world's first public aquarium in 1853, in which the first-ever photograph of a living fish was taken. Other kinds of exhibit buildings, species in the collection, and breeding successes were firsts. In time, this paradise for naturalists, this oasis of pleasure for much of the population, this must-see destination for visitors from a score of other lands became something of a legend in its own

ABRAHAM D. BARTLETT

Arguably one of the greatest of all zoological garden superintendents (directors), Abraham Dee Bartlett was born October 27, 1812. His father, a barber, had a shop situated near the Strand and Bartlett, Sr. was a friend of Mr. Cross. So A. D. became accustomed from infancy to playing with young wild animals behind the scenes at the Exeter 'Change. As A. D. points out, he had not "the remotest recollection of seeing for the first time lions, tigers, elephants or any other wild beasts, simply because I was almost from my birth among them."*

Noting the lad's obvious interest from a very early age, Cross often gave him the cadavers of birds that had died in the collection, and these A. D. learned skillfully to mount without any formal training. This was the first step to becoming a highly successful taxidermist at a time when preserving animals in this way was very much in demand, if not at its peak. Nevertheless, A. D. yearned for contact with living animals and consequently made the acquaintance of Yarrell, Ogilby, Gould, Owen, and other luminaries of the Zoological Society of London. In particular, he impressed D. W. Mitchell who was destined to become the society's secretary. This paved the way for A. D. to be offered the superintendent position at the Garden in 1859, a post he accepted with the greatest of alacrity and never regretted.

Literally thousands of species, a great many of them completely unknown to wild animal husbandry, came under his care during a period when the world was being "opened-up" in every way. So, not unnaturally, much of his work was pioneering. Fortunately, A. D. appreciated the value of exhaustive note taking, which forms the basis for two extremely valuable books, *Wild Animals in Captivity* and *Life Among Wild Beasts in the Zoo*.**

Abraham Dee Bartlett died after a long and painful illness on May 7, 1897, aged 84. It is good to record that, since for A. D. thoughts of retirement were abhorrent, he died "in harness" as Superintendent of the London Zoological Garden. That is the way for an enthusiast to go.

* Bartlett, Abraham D., *Wild Animals in Captivity*, Chapman and Hall, London, 1898.
** Bartlett, Abraham D., *Life among Wild Beasts in the Zoo*, Chapman and Hall, London, 1900.

lifetime, primarily because of the range and richness of its unique collection, which was possibly rivaled only by that of the Berlin Zoo.

Then, during the 1960s, under the leadership of Professor (later Lord) Zuckerman with his emphasis on science, the disposal of many animals, including entire taxonomic orders and families, began: the hippopotamuses, the bears (the family had been there since 1829), the sea lions, the orangutans, the bison, the crocodilians, the giant tortoises, most of the once excellent bird-of-prey collection, most of the parrots, the owls, etc. These animals are all good show species people once would pay to come and admire. Three quarters of the original buildings, cages, and aviaries were demolished and replaced with soul-less modern counterparts. What they emphatically did not replace was the garden's quite unique and slightly eccentric charm, which people loved and would happily support.

These actions precipitated a financial crisis in 1991/92. In addition, the Zoological Society of London (or more precisely, its 15-member council) has long maintained that a major cause of its financial troubles has been the overwhelming increase in the number of other attractions, all competing for the citizen's and tourist's money. Once a huge port and trading center, London has, over the last two decades or so, become very largely a tourism center, with all manner of new tourist attractions. Another point often made is that, unlike so many other zoological gardens in capital cities, London Zoo receives not one cent of government funding, but then, with the exception of Blackpool, neither does any other British zoological garden.

Twice before in its history the society has faced financial crisis, in the 1840s and again at the turn of the century, but at these times the society did not seek to remedy flagging support and coffers via the strange logic of drastically reducing the collection, and subsequently charging the visitor more to see the remnants. Rather, they knew full well that then, as now, people pay to go into such places to see animals, and so, even during the black periods, there was always a large and varied collection to admire and study. How these rather autocratic Victorians would have fared in carrying out this policy today, however, is a matter of interesting — if pointless — conjecture.

Nevertheless, the fact remains that the London Zoological Garden, one of the first great modern zoological gardens, was on the brink of closure in 1991/92. With generous financial support from wealthy friends and with some pruning of its top-heavy administration, the zoo and the society have recovered. Attendance has fluctuated, with 1997 being good but 1998 less so. It remains to be seen how this great institution fares.

2.11 Zoological Gardens at Whipsnade and Jersey

The London Zoological Garden was the first institution of its kind to realize the importance of opening a "country branch." As early as 1831 the London Zoological Society established "The Farm," Kingston Farm at Richmond Hill in Surrey, as a breeding station. However, the society was overambitious with this attempt, and within three years it was closed. The idea was not completely abandoned, as it was an attempt to fulfill part of the society's objectives — captive breeding, including an effort to develop new varieties of domestic stock — and periodically was brought up at council meetings.

In the early 1900s the idea of a country breeding center surfaced again, and plans were ambitiously drawn up, but the advent of World War I soon put them to rest. Again in the 1920s searches were actively made for suitable sites, not too far from London and the garden. Eventually, a derelict estate of some 500 acres was purchased near the village of Whipsnade, in Bedfordshire, on the Chiltern Hills. It opened on May 23, 1931.

Seeing animals in huge, open enclosures was a novelty and it became a popular attraction. Unfortunately, this popularity has dwindled because of its remote rural location and the lack of suitable public transportation. Most of its visitors (95%) come by car, which is an expensive proposition. So to attract the public, what was initially a breeding center is now a public zoo, but one that is in a location more favorable to its original intent. To cope with its new role, the animal collection has been made more varied.

For a long time the animal population at Whipsnade was not particularly varied, as there was a heavy emphasis on ungulates. However, during the 1980s and 1990s the collection has been enormously enriched: the primate population (previously limited to chimpanzees and rhesus monkeys) was increased, a bird garden that enables a wide range of smaller species to be seen at close quarters was built, certain massive paddocks were divided into smaller areas where the hoofed stock can be properly seen, and the Discovery Centre was developed. Within the Discovery Centre dwell dwarf crocodiles, basilisks, leafcutter ants, giant millipedes, toucans, pythons, tamarins, mongooses, Kleinmann's tortoises, red-sided skinks, electric eels, seahorses, tropical marine fish, and piranhas.[39,40]

Jersey Zoological Park was founded March 14, 1959 and opened soon after on March 26. It was renamed the Jersey Wildlife Preservation Trust on July 6, 1963. Gerald Durrell founded the Park at Les Augrés Manor on Jersey Island as a home for his own collection of animals. The emphasis of the trust and the park is on breeding endangered species and conservation of these species. The trust and park have worked cooperatively with several governments on joint captive-field conservation projects. In 1979, the trust acquired the neighboring Les Noyers estate and transformed it into a school to train keepers and representatives from other countries in wildlife conservation and endangered species management. There are now four graduates of this training program with posts in British zoological gardens.[41,42]

Because of the Jersey Zoological Park's emphasis on conservation and endangered species, it is a collection of rarities. And because of this work, it is a highly regarded place. Perhaps it was the first *conservation park*, a term popular in the United States, and the future ideal for zoological gardens. If it is, essential features of zoological gardens have been lost. Although conservation is important, it is only one aspect of a zoological garden. Without an equal effort to educate and interest the public or naturalists, the zoological garden will be a gallant failure. Is the age of magnificent animal collections over, and will the future age rely on an abstruse concept of them?

References

1. Allen, David E., *The Naturalist in Britain: A Social History*, Allen Lane, London, 1976.
2. Thomas, Keith, *Man and the Natural World: Changing Attitudes in England 1500–1800*, Allen Lane, London, 1983.
3. Public Record Office, London.
4. Paris, Matthew, *Chronicles of Matthew Paris [Chronica majora]*, 1200s.
5. Flower, Stanley S., "Contributions to our knowledge of the duration of life in vertebrate animals. 5. Mammals," in *Proceedings of the General Meetings for Scientific Business of the Zoological Society of London*, 1931, 145.
6. *Library of Entertaining Knowledge*, Francis S. Wiggins, New York, 1830–1834.
7. Bennett, Edward T., *The Tower Menagerie*, R. Jennings, London, 1829.
8. Bennett, 1829, xvi.
9. Bennett, 1829, xv.
10. Bennett, 1829, x.
11. Goodwood Collection, Sussex County Record Office, Chichester (Sussex).
12. Royal Archives, Windsor.
13. Fitzwilliam Archives, City Reference Library, Sheffield (Yorkshire).
14. Reynolds, III, Richard J., personal communications, 1980s–1990s.

15. Keeling, C. H., *The Life and Death of Belle Vue*, Clam Publications, Shalford, 1983.
16. Keeling, C. H., *Belle Vue Bygones*, Clam Publications, Shalford, 1990.
17. Keeling, C. H., *The Fragments That Remain (Belle Vue)*, Clam Publications, Shalford, 1992.
18. Keeling, C. H., *Remember Belle Vue*, Clam Publications, Shalford, 1997.
19. Keeling, C. H., *Where the Lion Trod*, Clam Publications, Shalford, 1984.
20. Keeling, C. H., *Where the Crane Danced*, Clam Publications, Shalford, 1985.
21. Keeling, C. H., *Where the Zebu Grazed*, Clam Publications, Shalford, 1989.
22. Keeling, C. H., *A Brief History of British Reptile Keeping*, Clam Publications, Shalford, 1990.
23. Keeling, C. H., *Where the Elephant Walked*, Clam Publications, Shalford, 1991.
24. Keeling, C. H., *Where the Macaw Preened*, Clam Publications, Shalford, 1995.
25. Keeling, C. H., *Where the Penguin Plunged*, Clam Publications, Shalford, 1996.
26. Schomberg, Geoffrey, *British Zoos: A Study of Animals in Captivity*, Allan Wingate, London, 1957.
27. Green-Armytage, A. H. N., *The Story of Bristol Zoo*, J. W. Arrowsmith, Bristol, 1964.
28. Warin, Robert and Warin, Ann, *Portrait of a Zoo*, Redcliffe Press, Bristol, 1985.
29. Keeling, C. H., *The Bristol Book*, Clam Publications, Shalford, 1998.
30. Keeling, C. H., *They Live at the Castle: The Story of the Dudley Zoological Garden*, Clam Publications, Shalford, 1989.
31. Sherren, Henry, *The Zoological Society of London*, Cassell, London, 1905.
32. Mitchell, P. Chalmers, *Centenary History of the Zoological Society of London*, Zoological Society of London, London, 1929.
33. Lord Zuckerman, S., Ed., *The Zoological Society of London 1826–1976 and Beyond*, Academic Press, London, 1976.
34. Blunt, Wilfrid, *The Ark in the Park*, Hamish Hamilton, London, 1976.
35. Edwards, John, *London Zoo from Old Photographs 1852–1914*, John Edwards, London, 1997.
36. Keeling, C. H., *They All Came into the Ark: A Record of the Zoological Society of London in Two World Wars*, Clam Publications, Shalford, 1988.
37. Bartlett, Abraham D., *Wild Animals in Captivity*, Chapman and Hall, London, 1898.
38. Bartlett, Abraham D., *Life among Wild Beasts in the Zoo*, Chapman and Hall, London, 1900.
39. Huxley, Elspeth, *Whipsnade: Captive Breeding for Survival*, Collins, London, 1981.
40. Keeling, C. H., *Whipsnade's War*, Clam Publications, Shalford, 1990.
41. Jersey Zoological Park, *The First Twenty Five Years: The Jersey Zoo*, Jersey Wildlife Preservation Trust, Jersey, 1984.
42. Durrell, Gerald, *The Ark's Anniversary*, Little, Brown, New York, 1990.

3
Zoological Gardens of Western Europe

Harro Strehlow

3.1 Introduction

The sixteenth century in Europe was a dynamic period. Gradually, the first signs of a modern European society appeared. The dogmatic Church was losing its influence, and new thoughts developed in the sciences, arts, and philosophy. Europe began to prepare for the great changes of the coming centuries. There were strong bonds with the Arabic countries, and the rich and civilized Orient. Cities and states that had a monopoly on trade with the Orient became increasingly important, and as the monopoly shifted so did the influence of the states. At the time science and medicine were much more advanced in the Orient than they were in Europe. Therefore, after the crusades, trade was not only an exchange of goods, but a stimulus for the cultural development of Europe.

The end of the sixteenth century brought the most important changes to Europe. Political and scientific centers of Europe shifted from the Italian and German city-states to the Hispanic empires. Portuguese and Spanish voyagers discovered more of the African and Asian coasts, as well as the American coasts. Later, the influence of the Portuguese and Spanish declined as France, England, and Holland became the leading Western European countries with extensive colonies. Cultural and political development within each of the European countries was quite different. In most of the countries, a central power with a king or emperor governing from a capital developed. In some places, such as Germany, the feudalistic society continued without creating a central power. Germany was a loose confederation of dozens, at times hundreds, of principalities of different sizes, each with its own currency, administration, law, and army. Only in 1871 did Prussia unite most of the German states, with Berlin as the capital.

3.2 Post-Medieval Collections to Modern Zoos

The early years of the sixteenth century saw the development of post-Medieval animal collections. These animal collections were as varied as the sociocultural groups and the political regions. Aristocrats and private individuals owned menageries, and there were menageries of only one or a few animals traveling from town to town where owners showed their "beasts" for money. Towns built deer moats and bear pits. Often, the kings or sovereigns established game parks with deer and boar. There were also falconries and, later, pheasantries. In most cases, little is known about these animal collections. Only the names of places, parks, or streets remind us there was any kind of collection. Very often the collections, dependent on one person, either a private owner or a sovereign, were sold or given away after the person retired or died.

It is very difficult to get information on all the different collections, or even just to name and number all of them. In 1912, Gustave Loisel published his outstanding work, *Histoire des ménageries de l'antiquité à nos jours* (*History of Menageries from Antiquity to the Present*).[1] It consists of three volumes and deals with the most prominent menageries of the world. Leopold Joseph Fitzinger wrote a very detailed description of the menageries of the Habsburg monarchy in 1853, *Versuch einer Geschichte der Menagerien* (*Outline of the History of Menageries*), which is full of information about menageries of the Austrian court, including all of the data available about the animals.[2] Another important source is *Hendrik Engel's Alphabetical List of Dutch Zoological Cabinets and Menageries*.[3] This work lists the zoological cabinets, collections, and menageries, as well as short biographies of the collectors, owners, and other persons important to the collections. Altogether the list contains more than 1,700 collections and collectors. Other less comprehensive histories and regional lists have also been written.[4–12]

Because there were so many collections, it is possible only to mention some of the more important and better-known ones, especially those still in existence today. The outstanding number of animal collections in Europe, because of the interest of the aristocracy in native and exotic wild animals, clearly shows that throughout these centuries collections of wild animals were an important part of the culture of European societies.

Interest in wild animals is well documented in animal books printed as early as the sixteenth century. The best known are those of Ulisse Aldrovandi and Conrad Gesner. These books went into numerous editions and were well circulated among educated people. Less well known is the collection of aquarelles and paintings of the Nuremberg physician Michael Rotenbeck from 1615. He inherited many paintings from his uncle Lazarus Rotenbeck and 15 other artists. All of these pictures were bound into the book, *Theatrum Naturae*. It shows the widespread importance of wild animals to the people, and especially to the artists. In many paintings from the sixteenth century onward, artists used wild animals as ornamental decorations in their works. Even more than from the books, one can see from these paintings which animals were known in Europe at that time. Besides books and paintings, there were leaflets with printings of animals like the well-known rhino illustration of Albrecht Dürer from 1515.

3.2.1 Game Parks, Falconries, and Pheasantries

Game parks, falconries, and pheasantries existed in many places. The aristocrats loved game, but it was quite uncomfortable to make expeditions into the wild forests to hunt deer or boar. Sometimes they would go off into the wilderness, but not often as individual hunters with bows, spears, or guns. Rather, they arranged hunting parties with large groups of beaters who beat the bushes to flush the animals, and these events were more like parties where large numbers of animals were shot. The discomfort of wilderness travel and the demand of the court for wild animal meat were reasons for the creation of game parks, which were large areas of forest with some meadows and other landscaping. Often the parks were very large areas; for example, the Berlin Tiergarten, which still exists as a public park, stretched over 500 acres.

Generally, the border of the game park was a wooden lattice fence or a stone wall. A forester, with some helpers, took care of the animals, shot what was needed to supply meat for the court's kitchen, and made arrangements for the sovereign's hunting. The sovereign's family and friends used the park for relaxation, for recreational riding, for walking, and for watching the animals. Use of the game park as a pleasure ground helped transform the game park into a menagerie. Maximilian, oldest son of emperor Ferdinand I, founded the first and oldest menagerie of the Habsburg monarchy in 1552 as part of the game park at Ebersdorf.

The elector (sovereign), Johann Cicero, founded the game park, Tiergarten, in 1490 near Berlin. In the Tiergarten (a common term for game parks, but also the specific name for the Berlin park) there were not only deer and boar, but sometimes aurochs as well. Little is known about these aurochs (the extirpated ancestors of today's cattle) or about the possibility that there were European bison. In the early literature it is difficult to distinguish between the two.

A special kind of animal show developed from the game parks and pleasure grounds. It was the Hetzgarten (hunting garden). These hunting gardens imitated the Roman animal fights, but on a much smaller scale. The elector Joachim II erected the first hunting arena in Berlin in the first half of the sixteenth century. An account has been handed down of a fight between aurochs, a wolf, and a bear in 1543. The bear was so excited it broke through the barrier near Joachim II and only his hunters, hurrying with their spears, saved the elector from the bear's attack.

In 1693, the architect Arnold Nehring erected a baroque-style hunting arena in Berlin similar to a Roman arena. The first king of Prussia, Frederick I (1657–1713), used it to keep a noteworthy collection of animals: lions, tigers, black bears, a polar bear, European bison, deer, and boar. This collection was for public amusement and did not have the cruelty of the Roman-style animal combats because the animals were too valuable and it was too difficult to acquire them. As far as is known, these faux animal fights were considered entertainment and the people enjoyed them. But this kind of entertainment was only a small part of the interest people had in exotic animals and native wild animals. Fredrick's successor Frederick William I (1688–1740) closed the arena because of the high costs of keeping the animals.

Falconry has been popular from ancient times up to the present day. In every epoch, sovereigns engaged falconers for the maintenance of their birds of prey. For their sport they used falcons, buzzards, hawks, and eagles. The falconers tamed and trained the birds of prey, as well as fed and cared for them. The aristocracy especially liked to hunt with falcons for grouse, partridges, small birds, and rabbits. Since the birds of prey being used were plentiful up until the middle of the nineteenth century, it was cheaper and easier to get new birds from the wild than it was to breed them. An exchange of animals also existed among the falconries. Most birds of prey were quite common, but the larger and stronger falcons, like the laggar falcon or the gyrfalcon, were rare and had to be imported. Hawking was such a favorite sport that Marcus Welser, in Augsburg, published a reprint of Emperor Frederick II's magnificent book, *Über die Kunst, mit Vögeln zu Jagen* (*On the Art of Hunting with Birds*), in 1569. The original copy of this book was lost in 1248, but a handwritten duplicate still existed at the time with all of the illustrations.

Pheasantries developed first in Asia where Indian maharajas and Chinese emperors were the first to keep and breed pheasants and peafowl. In Europe, pheasantries have a tradition extending from Greek and Roman times to modern times. In Europe also, the primary reason for having pheasantries was the demand of the court's kitchen. The pheasant master and his helpers fed and cared for the birds, and they had to breed enough birds for the court's demand during the year, catching and delivering the pheasants for this purpose. In the middle of the eighteenth century, the pheasantry at Potsdam supplied the king's kitchen with over 600 pheasants each year.

There were two kinds of pheasantries. In the tame pheasantry, eggs were collected so certain hens could hatch them and then rear the chicks, or sometimes the pheasant master reared the chicks. These pheasants were fed throughout the year and were kept in cages. In the wild pheasantry, the birds lived freely, the hens breeding and rearing the chicks on their own, and they were only given food in the winter. Both kinds of pheasantries covered large areas. Near Berlin the pheasantry was about 120 acres with wooded areas, sandy places, and meadows where the pheasants roamed freely. There were feeding places, breeding houses, cages for rearing the chicks, and a house for the pheasant master.

Pheasantries were sometimes situated near the sovereign's palace or else they were part of a game park. The beauty of the peacocks and the magnificence of the pheasants were a pleasure to the eye and the palate alike. In Brandenburg, pheasants were first imported in the seventeenth century, and the oldest known pheasantry was erected at the royal residence of Brandenburg-Prussia, near Potsdam, before 1671. At the same time, it was more than just a pheasantry. Frederick I kept exotic animals, such as cassowary and deer, together with the pheasants. In 1707, only 45 pheasants were kept there, but some years later, after the closing of another pheasantry near Oranienburg, the number of birds swelled to more than 1,200. Later, Frederick William I shot the pheasants to save the cost of keeping them. Only 800 birds were allowed to survive this shooting spree. At the beginning of the nineteenth century this pheasantry was still in existence, but it is not known when the breeding of the pheasants ended.

Frederick William I closed most of the pheasantries in Brandenburg except the one near Wusterhausen. His son reestablished old pheasantries, as well as establishing new ones. In 1740, the king of Prussia ordered new pheasantries at three places: near Berlin at the border of Charlottenburg, near Ruppin, and near Rheinsberg. However, only the pheasantry near Berlin in the southern part of the former Tiergarten game park was actually established. The Tiergarten was then no longer a game park, but a pleasure ground for the people of Berlin. The king acquired pheasants for his new pheasantry from Halle, which was the most important pheasantry in Germany at that time. In 1744, he received 30 pheasants from the Prince of Cöthen, a well-known pheasant breeder. The new pheasantry existed until 1842, when it was liquidated. The birds were brought to Charlottenhof at Potsdam. The king designated the southern part of the pheasantry as the location for the new Zoologischer Garten bei Berlin.

As in Prussia, pheasantries existed in other areas of Europe. One of the early pheasantries was located in the game park, Neugebäu (Austria), a pleasure park for Maximilian II. Neugebäu was built mostly between 1564 and 1576, but was not completely finished until 1587, including the game park and pheasantry. The birds kept there were common pheasants, and sometimes peafowl.

The common silver pheasant has been well represented in European pheasantries since the eighteenth century, and was often bred in England and France. It is not known when the first golden pheasant was imported, but the first mention of this beautiful bird comes from England in 1740 and it can be assumed that these two beautiful species were kept and bred in Europe from this time forward. Other species of the pheasant family did not arrive in Europe until the nineteenth century. These birds were kept in zoos at first; for example, at London and Antwerp, and later in the pheasantries of private breeders. Private pheasantries became important in the twentieth century. The best known private breeder was Jean Delacour, who imported and bred some of the rarest species of pheasants.

3.2.2 Deer Moats, Bear Pits, and Lion Cages

These animal collections, like those mentioned in the previous section, were privileges of the aristocracy and, to a lesser degree, of the wealthy class. They had three functions: to supply the demand of the court's kitchen, to provide the pleasure of game hunting, and to provide the private enjoyment of watching animals. Many were not open to the public, but some deer moats and bear pits were public animal exhibits. Anybody could visit these public places where the animals were kept. Municipal governments and the citizens themselves established these collections, not the sovereign or the king. It was part of the civic pride of the towns and their citizens. The people were interested in the animals shown, and appreciated the deer moats and bear pits as part of their culture and recreation.

After the Middle Ages, strong fortifications and large moats around the towns lost their strategic importance, so people used the moats for purposes other than their original intention. One of these uses was the deer moat. In part of the moat they erected fences and built an enclosure for the deer. As early as 1410, in Augsburg, the citizens had six deer in their moat for the enjoyment of the people. These deer bred well and Augsburg would give moat-bred deer to other towns.

Frankfurt-on-Main had a deer moat around 1430. There were also roe deer and geese in this moat. The keeper of the gate, which was near the moat, fed these animals. If a deer escaped from the moat, it was shot with a crossbow. Bern, in addition to its bears, kept deer in its moat. In the town of Rhodes, Greece, the old moat fortification still exists with two enclosures for fallow deer, so the tradition of using deer moats continues.

Stags are dangerous when rutting, but the deer moat fences were enough to contain them. Other dangerous animals needed stronger enclosures, such as the bears sometimes kept in the moats. There were also bear trainers who performed shows with their dancing bears at fairs and royal courts. They caught young bears after killing their mothers and then tamed them. The towns, however, kept wild bears in cages or bear pits, especially those towns with a bear in their name or coat of arms. The bear pit was dug several meters into the ground and fortified with a strong stone wall, in the middle of which would be a climbing tree and a small pool of water. A building with cages was attached to the pits. These bear pits may not now seem to be a proper way of keeping bears, but if it was a good pit, the bears often lived for a long time and bred well. Even today, at the few bear pits still in existence, this is true.

Although towns and citizens usually built and maintained bear pits, the sovereigns did so sometimes as well. Bear pits were known to exist at the Stuttgart and Böblingen (southwest of Stuttgart) castles. The sovereign at Dresden also maintained a famous bear pit. It was not so much a pit as a large area surrounded with walls that was part of the Jägerhof on the right embankment of the Elbe River. On maps and paintings it is called the "Bear-garden." A circular pool and four high tree trunks with a platform were part of this bear garden. Other bear pits existed at St. Gallen as part of the monastery's animal collection, and at the island of Reichenau, in the Lake of Constance, where the bishop of Constance ordered a bear pit dug as part of his collection in 1591.

The best known of all the bear pits is the one at Bern. Bern has kept and exhibited bears continuously, from medieval times right up to the present day. The artist, Diebold Schilling, painted his 1485 "Spiezer Chronik" with a scene from 1339 showing one of the bears sitting chained under the stairs of the town hall. Later, in 1441, bears are mentioned in the records of Bern for the first time. In 1549, the building of a bear enclosure was first recorded; two other pits followed in 1764 and 1825. This last pit was said to be very humid and shady, and therefore not appropriate for the bears. In 1857, a new bear pit was built, one that still exists today. This pit is circular with a wall dividing it into two enclosures. In the adjacent building there are cages, a food storage area, and a keeper's room. In 1925, a third pit was added at the back of this building. Another change was made in 1974/75 when modern cages and other equipment were installed inside the building to improve the bear's comfort.

The bear pit is not only popular with tourists, but also with the people of Bern. Despite modernization in 1974/75, the bear pit could not comply with the expectations of people at the end of the twentieth century. In the 1980s the question of closing the bear pit, or changing it again, arose. In 1991, the citizens of Bern and the administrators of Tierpark Dählhölzi decided to save the bear pit (Bärengraben) as part of Bern's cultural identity and to change it into a modern enclosure. In 1996, the new and improved Bärengraben opened, while simultaneously, the Tierpark Dählhölzi built a new bear enclosure that is a large natural wooded area with meadows, rocks, and a pool.[13,14]

It should be mentioned that bears were not always kept in pits. In some courts and towns bears lived in stable cages, bear houses, or animal houses. Often, however, even in menageries and zoos, bears were kept in bear pits. The Jardin des Plantes in Paris and the London Zoological Garden exhibited bears in pits. Even some zoos founded in the 1930s and 1940s built bear pits because they were inexpensive and this was considered a suitable way to keep bears, even at that time. Berlin still maintains its bear pit, which opened in 1939.

In 1937, when Berlin celebrated the 700th anniversary of its founding, the town proposed the construction of a bear pit in Köllnischer Park in addition to the one at the Berlin Zoo. This bear pit opened on August 14, 1939 with a central building and two circular enclosures. Heini Hediger, then Director of Tierpark Dählhölzli at Bern, contributed the first bears for this Berlin exhibit. The bears did well until Allied bombing killed them in 1945. This exhibit opened again on November 30, 1949 with two more bears. When Tierpark Berlin-Friedrichsfelde opened in 1955, its zoo staff took care of the bears in this exhibit. But what in 1939 was a modern bear pit was unsuitable in 1990. Bernhard Blaszkiewitz, the new director of Tierpark Berlin-Friedrichsfelde, tried to close the bear pit, but the public did not want it closed. As a continued result of this disagreement, the bear pit was improved and the government of the Berlin borough of Mitte takes care of the bears. The bears continued to do well and breeding still occurred, although this has now been discontinued through the sterilization of the male. Offspring from 1994 were sent to the zoos in Buenos Aires and Santander.

Keeping lions, the "king of the animals" and a heraldic figure for sovereigns, also has a long tradition in Europe. Lions were often kept in stable cages near the sovereign's castle or at the gate of the town. Indian and Arab sovereigns often gave a lion to a European sovereign or town as a sign of appreciation, and nearly every sovereign and every important town kept lions at one time or another. People believed the lion to be the strongest and most honored animal; therefore, more than any other animal, lions were kept as a sign of the owner's power and dignity. Tame lions were a powerful expression of their owner's sovereignty over nature and people. There is also a long tradition of breeding successes. As early as November 3, 1592, three cubs were born on the island of Reichenau.

3.2.3 Menageries

Menageries were places where a variety of animals were kept for the pleasure of the owner or the people. Menagerie animals were not used for utilitarian purposes, but rather for the improvement of people's knowledge about nature and foreign countries. After the Middle Ages, the three most powerful groups within European society possessed menageries: popes and bishops, sovereigns, and wealthy patricians. In addition to these important stationary collections, traveling menageries also existed, often with one important show animal, such as a rhino or elephant. Later these traveling menageries kept many different animals and sometimes developed into a stationary menagerie.

The earliest important Western European menageries after the Middle Ages began in the Italian, Portuguese, and Spanish courts. The oldest was the menagerie at Palermo in Sicily. Sarazen emirs kept wild exotic animals in this collection. After Sicily was handed over to the Hohenstaufen, Emperor Henry VI (1165–1197) continued this tradition. His son, Emperor Frederick II (1194–1250), exchanged a white bear for a giraffe in the beginning of the thirteenth century. In the fifteenth century, sovereigns of many Italian towns created menageries. The Medici in Florence, the Visconti in Milan, the Este in Ferrara, and the Aragonese in Naples were noted for their animal collections. The Sultan of the Mameluks sent to the menagerie of Lorenzo Magnifico of Florence (1449–1492) a giraffe, which joined another giraffe already living there.

Lorenzo's son, Pope Leo X (1475–1521), created a menagerie in the Vatican. King Emanuel I (1469–1521) of Portugal contributed to this collection by sending an elephant and a rhino from his own menagerie. At the Toscana in San Rossore, near Pisa, Grand Duke Ferdinand II (1610–1670) bred dromedary camels in 1622. This dromedary farm existed for more than 200 years and most of the young were sold to other menageries.

In the Netherlands, the history of menageries goes back to the fourteenth century. At that time the sovereigns of Henegouw established a menagerie near their castle at The Hague. A poultry house, a house for falcons, a dog kennel, a donkey house, and a lion house are known to have existed. In the lion house there were other carnivores in addition to the lions. Each house, and each part of the menagerie, had its own keeper.

The Duke of Gelder owned three collections. Perhaps his first impulse to keep animals came in 1351 when the Duke received a young lion from the sovereign of Henegouw. The first of his menageries was at Nijmegen, the second at Grave, and the third at Rozendaal. At Rozendaal the duke kept falcons and parrots, as well as lions. Records indicate that from October 1398 until July 1399, some 260 sheep fed the lions.

Menageries in Spain and Portugal provided lions for the lion gate at Kampen, the Netherlands in 1477 and 1483. In Kampen the citizens of the town paid for the lion's food. It seems that these lions bred, as some years later the council at Kampen gave five or six lions to Lübeck, the key town of the Hanseatic League. After the lions died, they were stuffed and mounted in the town hall where the town's people admired them for several hundred years.

The Princes of Orange maintained menageries throughout their reign. In 1614, Prince Moritz (1567–1625) received the first cassowary ever brought to Western Europe. And some years later, in 1640, traders of the East India Company presented the first chimpanzee to Prince Frederik Hendrik. This ape lived in his palace at Naadlwijk. In the seventeenth century, a strong trade developed between the Netherlands and the African and Asian regions, a result primarily of the activities of the Dutch East India Company. Very often, their captains or the sailors would bring back animals. Traders and farmers living in foreign countries and colonies would also send home exotic animals. This kind of activity kept Western European menageries well stocked.

Het Loo menagerie near Apeldoorn existed for about 100 years. William III (1650–1702) and Queen Maria Stuart established this collection at the end of the seventeenth century. William IV (1711–1751) purchased Het Kleine Loo, at Voorburg near The Hague, in 1748, and one year later started his collection with a pheasantry where he bred silver and golden pheasants. In exchange for pheasants he received other animals. After his death in 1751 his widow Anna enlarged the collection. In 1756 she bought the collection of Arnout Vosmaer. Vosmaer became the director of the Natural History Collection at The Hague and, in 1771, of the menagerie at Het Kleine Loo. Anna's son William V (1748–1806) was enthusiastic about natural history as well and worked at the menagerie. In 1786 he left The Hague, moved to Apeldoorn, and sold the farm Het Kleine Loo. Most of the animals were taken to the menagerie Het Loo.

Het Kleine Loo menagerie was well known through the descriptions of travelers, and from more than a hundred drawings, aquarelles, and oil paintings by the famous artist Aart Schouman. From these sources we know that in the 37 years the menagerie existed, more than 50 species of exotic mammals and more than 30 species of exotic birds were kept there. These animals came from eastern Asia, southern Africa, and South America. The most prominent of these was the first Bornean orangutan in Western Europe. Also in the collection was the now-rare Cape mountain zebra and the now-extinct quagga. It closed after French troops invaded in 1795. The French Commission of Arts and Science visited the Netherlands, organizing and transporting valuable objects of art and the sciences back to Paris. In 1796, most of the animals of the Het Loo menagerie were sent to Paris as well. Only two Indian elephants remained at Het Loo, until they too were marched to the Jardin des Plantes at Paris in 1798.

Besides these aristocratic menageries, there were those of wealthy patricians, such as Arnoldus Ameshoff near Amsterdam, Jacob Temminck in Amsterdam, and George Clifford at his farm De Hartekampf near Heemstede. The most prominent of these private menageries was the one at Blauwe Jan. Blauwe Jan was the name of a restaurant with a menagerie in its yard. The name goes back to 1675 and the founder of the restaurant, Jan Barentze Westerhof. The menagerie existed until 1874. In the middle of the eighteenth century the owner, in addition to managing the restaurant and menagerie, became involved with the exotic animal trade. For example, he delivered beasts and birds to the menageries of Versailles (Paris) and Schönbrunn (Vienna). Catalogs of the animals were sent to interested aristocrats, as well as to others. Unfortunately, all of these catalogs have been lost; however, artists, such as Jan Velten around 1700, painted the animals of Blauwe Jan.

A birdhouse was established in the center of the yard at Blauwe Jan, and around the birdhouse were mammals and reptiles. It is not known how many species lived at the menagerie, but it is possible to name more than 40 species of mammals. Noteworthy were a hippopotamus, babyrusa, and warthog. A common seal lived there in 1751 as well. The birds shown in the Blauwe Jan menagerie outnumbered the mammals, and included ostriches, pelicans, flamingos, herons, and curassows from South and Central America, as well as parrots, hornbills, and a large number of warblers.

Unlike the menageries of aristocrats and the wealthy, the menagerie of Blauwe Jan was open to the public, although it is not known if they paid admission. The menagerie was also the first well-known business in Europe that specialized in animal trading, although trade in exotic animals had existed as long as trade in other goods. Even in the sixteenth century in Antwerp, traders and trading companies dealt with exotic animals in addition to their regular business. Blauwe Jan was a predecessor of the animal dealers Hagenbeck, Ruhe, and Reiche in Germany during the nineteenth and twentieth centuries.

In France, the menagerie of Versailles was the most important. Similar to the menagerie at Schönbrunn, discussed later, it was best known for its architecture. The prominent feature of classic menageries was the circular layout, in the middle of which stood a beautiful pavilion. Around this pavilion was a walking path and outside this path were the enclosures and cages. Each enclosure had a house or stable at the far end for the animals and was bounded on three sides with walls. There were bars only in the direction of the pavilion. Monkeys and other mammals lived in cages, while birds lived in aviaries near the entrance or along the side of an enclosure. Visitors could sit in the pavilion and watch all of the animals just by looking around. After the French revolution of 1789, the animals at the Versailles menagerie were transferred to the Muséum d'Histoire Naturelle at the Jardin des Plants, where their addition helped form what may be considered the first modern zoo.

Many menageries existed in Germany during the period between 1500 and 1800, most of which were the property of the sovereigns. Nearly every small country, that together formed Germany, had its own menagerie. The menagerie of Karlsruhe was founded in 1714 as a pheasantry.[15] Although the menagerie was closed and reopened, the pheasantry continued to exist. At the beginning of the nineteenth century, more than 3,000 pheasants were in the collection. In 1816 and 1817, Duke Karl sent his forester, von Holzing, to Stuttgart to buy animals from the menagerie of King Frederick, which was closed after his death. Two zebras were the most expensive animals von Holzing bought. He paid 224 Louisdor for the zebras, as much as one and a half kilo of gold. The Dutch animal dealer, and later owner of a traveling menagerie, van Aken, sold monkeys, lorikeets, and other animals for 591 Louisdor to the Karlsruhe menagerie in 1816. The menagerie existed until the opening of the new Zoological Garden in 1865.

Augsburg was an important trading center and the home of the Fugger family who sent their ships to America, Asia, and Africa. In the sixteenth century they abandoned the American trade after losing their American fleet in wrecks. Among the goods they imported were living animals; however, the profits of the animal trade declined over time. The Fuggers' office in Antwerp wrote a letter to Hans Fugger saying they wanted to abandon the trade in monkeys because the business was too small and involved too much work. In 1570, Hans Fugger's collection included mammals and birds from Africa and Central America. His animal collection may have been open to the public. The most important visitor was the Emperor Maximilian II. When the exhibition closed is not known.[16] Others, besides aristocrats and tradesmen, also established menageries. In 1572, Leonhard Thurneisser, physician of the court, established a small menagerie in Berlin. He not only kept monkeys and parrots, but he also had a moose. He even sent a moose as a gift to his hometown, Basel. But the people there were superstitious and believed, for whatever reason, the moose was a creature from hell. An old woman killed the poor animal when she fed it an apple stuffed with nails.

Besides the local, stationary menageries of churchmen, aristocrats, and the wealthy, traveling menageries went from town to town. In the early centuries they often had only one animal, such as an elephant or a rhino. Later they developed larger collections with a variety of animals. The traveling menageries were also involved in the animal trade. Owners of stationary menageries often bought animals from the traveling menageries. Traveling menageries, rather than the stationary menageries, were part of the public's education since it has only been during the last two centuries that stationary menageries have been opened to the public. Before this, only the ruling class and the patricians could visit the exotic animals kept in stationary menageries. The general public only knew these animals from books and leaflets. Therefore, the traveling menageries gave them an excellent opportunity to see living animals they had read or heard so much about. Despite the poor care given the animals, the terrible shows the owners put together, and the misinformation provided, the traveling menageries gave the general public opportunities to learn about the animals and about the world outside their village, town, and country. They, and later the stationary menageries, made an enormous contribution to helping the general public learn about the natural sciences.

3.2.4 Transition from Menagerie to Modern Zoo — Schönbrunn and Berlin

At the turn of the eighteenth century, European society was quite different from what it was during the Middle Ages through the seventeenth century. The bourgeoisie became stronger and more influential. Industrial development started in some countries of Europe. Democratic ideas influenced intellectuals. New developments in science and philosophy affected the educated. Scientific expeditions left for the Americas, Africa, and Asia, multiplying the knowledge about the rest of the world. And the French Revolution was an important step toward the creation of modern Europe.

These new developments and ideas spread among the general public. Progressive and liberal sovereigns accelerated socioeconomic development among their societies. The animal collections, as part of the culture of these societies, were affected by all of these changes as well. The Karlsruhe menagerie was open to the public in the nineteenth century, and other sovereigns did the same with their menageries. The most important transitions to the modern zoo took place at Vienna, Berlin, and Paris.

The oldest menagerie of the Habsburgs, the ruling family of Germany at the time, was founded in 1552 at Ebersdorf as part of a game park. Maximilian acquired an elephant for the collection, which was brought from Spain in 1551, but it died on December 18, 1553. In 1552 he sent this elephant to Vienna for a while, where the exhibition impressed the people so much they named many houses, hotels, and restaurants after the elephant. A life-sized relief was sunk into the wall of a house, documenting this event for more than 165 years.

Other Habsburg menageries existed in Neugebäu and at Belvedere, a palace at Vienna. The former lasted about 200 years from about 1570 until about 1781. Prince Eugene of Savoy (1663–1736) founded the latter in 1716, and it was closed in 1752 when the menagerie at Schönbrunn was established. All the animals were transferred to Schönbrunn except an old vulture and a golden eagle. The vulture died in 1824 after it had been kept, supposedly, for 117 years at Belvedere. The artist Salomon Kleiner made copper engravings of most of the mammals and birds at the Belvedere menagerie, some 97 plates. However, neither a copper engraving nor any other illustration exists of the chimpanzee that lived there in 1734. The menagerie at Belvedere was designed with the classic circular arrangement. In the center of a many-sided yard was a fountain and a pavilion. The enclosures stretched between this central area and the outer wall. Houses with cages were built for the dangerous animals, such as the bears and lions. These were situated near the entrance or alongside the enclosures. There was also a two-story house for the parrots, delicate birds, and small mammals.

The tradition of animal collections at Schönbrunn goes back as far as 1573. In 1569, the Holy Roman Emperor Maximilian II (1527–1576) bought the manor Katerburg on the site where the palace of Schönbrunn is today. He restored and enlarged the Katerburg in 1573, and this included a chicken yard, fishpond, pheasantry, and game park. The Katerburg and its animal collections were destroyed in 1683 during the Turkish siege of Vienna. In 1695, Leopold I (1640–1705) built a summer resort for his son, Josef I, at Schönbrunn. The architect, Johann Bernhard Fischer von Erlach, not only created the palace, but a new garden in the French style. Maria Theresa (1717–1780) and her husband Franz I (1708–1765) enlarged the palace and created its splendid collections. Franz I collected paintings and natural history specimens. He founded the present menagerie at Schönbrunn in 1752, and the botanical garden a year later.[17–20]

The menagerie at Schönbrunn was similar in design to those at Versailles and Belvedere, but it was larger. The Dutch gardener Adrian van Steckhoven designed the menagerie and later became its supervisor. The menagerie's circular center area extended about 465 feet in circumference. Three avenues led from the pleasure garden, and later the botanical garden, into the center. Between the avenues were 13 enclosures for the animals, each with a lawn and a pool with a fountain in its center. The stables resembled small houses with a door and windows, and

Central pavilion in the heart of the Schönbrunn Zoo, Vienna. During the eighteenth century, Maria Theresa enjoyed eating breakfast in the pavilion while watching the animals. The pavilion is still part of the zoo. © Photograph by Harro Strehlow.

all but one were one-story buildings. The house with two stories contained a room for the supervisor of the menagerie, a room for a servant, and on the upper floor there were birdcages. An enclosure near the house, used as an area for walking, contained aviaries. Cages for birds of prey were placed on both sides near the border wall. A gate opened from this enclosure to the chicken yard and duck pond outside the walls of the menagerie. A building for monkeys and delicate birds stood behind the duck pond. Other buildings and stables were used for surplus animals. In 1759, the central octagonal pavilion was built.

Franz I was very much involved with the operation of the menagerie. In 1756, he sent the botanist Nicolaus Jacquin to the Caribbean and South America to collect plants and animals for the natural history collections. The gardener Richard van der Schot and two bird collectors accompanied Jacquin. They returned with many rare plants and animals in 1759. Two further expeditions were sent to Africa and America during the eighteenth century. Other sources of animals for the menagerie were the traveling menageries. In 1794, Franz II bought animals from the menagerie of the Italian Albi, who exhibited his menagerie in Vienna during that year. In 1824 and 1837, the Austrian Emperor Ferdinand I (1793–1875) acquired numerous animals from the menagerie of van Aken, and in 1840 he acquired the menagerie of Polito. In 1851, Franz Josef bought the menagerie of Hartmann. These many new animals required changes to be made in the buildings. New houses for bears and big cats were built instead of using ones that had previously housed the bears and wolves. In 1828, a new stable was built for a giraffe, expected as a gift from Muhammad Ali, Viceroy of Egypt.

Like Muhammad Ali, many kings and emperors of other countries contributed to the menagerie. When the emperors of Austria received gifts from these sovereigns they placed the animals in the menagerie. Members of the royal family, as well as other Austrian aristocrats and wealthy individuals, also sent animals to Schönbrunn. The diplomatic corps of Austria was another important source of exotic animals. Dispersed throughout most of the countries on each of the continents, the diplomats tried to please the emperor by sending him rare animals for his collection. It was made known, via the emperor's administration, that particular animals were wanted and the demand for these animals was conveyed to diplomats located in countries where these animals were native. Often, within a short time, the desired animal would arrive at Schönbrunn.

In its long existence, the menagerie at Schönbrunn exhibited many remarkable animals. The first elephant arrived in 1770. It was a male Asian elephant, and it lived until 1784. In 1799 a pair of Asian elephants arrived. The male, bought from the menagerie of Albi, died in 1810. The female lived until 1845 and reached an age of 53. Matings of the animals took place, but they were not successful. Another pair of Asian elephants is more prominent in European zoo history. On May 18, 1896, the male Pepi and the female Mitzi arrived. Pepi was said to be the largest elephant in Europe at the time. Mitzi became the mother of Madi, the first elephant born in Europe, on July 14, 1906.

The first African elephant arrived in 1870, having been bought from Hagenbeck. He was the first of five elephants to be killed at Schönbrunn because they were too dangerous. The first giraffe at Schönbrunn was a gift from the Egyptian Viceroy, Muhammad Ali. It was the third in Europe since the thirteenth century, following the arrival of giraffe in France (1826) and England (1827). It was caught in 1827 at Darfur and shipped from Alexandria to Venice. From April 27 to July 7 it lived at Venice in quarantine and then left for Schönbrunn. The giraffe traveled in a car and by foot, arriving at Schönbrunn on August 7, 1827.

The first ape at Schönbrunn was an orangutan in 1878, but this animal only lived for two months. Other orangutans arrived in 1899, 1907, and 1913. Their lifetime in Schönbrunn was short as well. Between the two world wars, the orangutan Emil was a favorite of the people. The history of the monkey house at Schönbrunn is typical of many houses in the old European zoos. Schönbrunn did not have a building just for monkeys and apes until 1907. This new house consisted of many indoor cages, but only one outdoor cage. It was a representative systematic collection but was, according to Ludwig Heck, Director of the Berlin Zoo, an outstanding example of unsuitableness. The house remained so until it was modernized in 1930.

Originally, Schönbrunn menagerie had its expenses paid out of Franz I's own income. Succeeding emperors were very interested in the animals as well, and it is said that Franz Josef (1830–1916), the last emperor of Austria, visited the menagerie every day. For the Emperor the menagerie was a place of interest, not just enjoyment. Franz Josef was familiar with each animal in the collection and spent time with new animals when they arrived.

Unlike many other menageries in Europe, Schönbrunn menagerie was open free of charge (until 1918) to the public. Few changes were made to the menagerie during its first 130 years; the number of animals remained about 700 during this period. In the second half of the nineteenth century some groups tried to establish a zoo, an aquarium, or a combination of both in the city of Vienna, but they failed. Only the aquarium from the world's fair in Vienna (1873) existed for a time. The menagerie at Schönbrunn was so popular people preferred to go there instead of visiting a zoo within the city of Vienna. However, competition from these efforts to establish other institutions and a change in curatorship of the menagerie initiated new developments at Schönbrunn.

Alexander Schön retired in 1879, and his vice curator Alois Krause became curator in 1884. Like his predecessors, Krause was not a biologist or a veterinarian, but he was familiar with animals having been a member of two expeditions around the world. On the first expedition he worked on the frigate *Novara* from 1857 to 1859, and on the second he worked on the frigate *Donau* from 1869 to 1871. His job was to care for the animals that were caught. After his return to Vienna he worked at the menagerie beginning in 1871, and continued to do so as vice curator from 1875. During his 33 years as curator he changed Schönbrunn into a modern zoo. The first year after he became curator, 1885, was the year in which the most important changes took place. He removed walls from between the enclosures and replaced them with bars, the menagerie was enlarged, and new enclosures were built for mountain animals and cattle.

Changes continued with the enlargement of the elephant house from 1894 to 1896, the building of a new monkey house in 1906, the building of a hippopotamus house in 1910, and a new aquarium. These changes were part of the transformation of the menagerie into a modern zoo. The new houses and new grounds allowed for an increase in the number of animals that could be exhibited, which now includes about 3,000 specimens.

After the establishment of the Austrian republic in 1918, the menagerie changed ownership from the emperor to the newly installed government. World War I had its impact on the collection. Of more than 3,400 animals on June 30, 1914, only 1,128 were still alive on June 30, 1918. The main reasons for these animal losses were the shortages of food and medicine (unlike World War II when fighting would be the primary cause of animal losses). Numerous animals were sold and some were slaughtered to feed the rare cats or bears. Many of the animals died as a result of the low quality and shortage of food and because the weakened animals succumbed to illness. Like many other zoos in Western Europe, the problems did not end with the armistice. The economic crisis after World War I did not stop at the gates of the zoos. In 1921 there were only a few animals at Schönbrunn. All of the big cats, monkeys, antelope, giraffes, rodents, and many other animals had died. Most of the enclosures and houses were left empty. A question then arose whether or not the menagerie should be closed. A committee, Hilfswerk für den Schönbrunner Tiergarten (Relief Fund for Schönbrunn), was founded. In the autumn of 1921 the press publicized the plight of the menagerie, and the committee's appeal for support was a success.

In 1919, the menagerie hired its first scientific director, the veterinarian Karl Müller, who had worked at the menagerie since 1901. Otto Antonius succeeded Müller in 1924. During the same year the old name, Menagerie Schönbrunn, was changed to Tiergarten Schönbrunn (Tiergarten had by then become synonymous with zoological garden). The community of European zoo directors began to recognize Schönbrunn as a scientific zoo, as the zoo, under Antonius' directorship, regained its former reputation. New buildings, rare animals, and Antonius's scientific work turned it into an important zoo. In particular, Antonius modernized the monkey house in 1930, adding 21 outdoor cages. After World War II chimpanzees finally successfully bred in this building. The southern part of the house was maintained until it was modernized again in the 1980s, giving comfortable indoor accommodation to the chimpanzees and orangutans. In the 1990s, new improved accommodations were provided, but for fewer species. The barbary apes were moved to the barbary sheep exhibit and the mandrills moved into the former elephant house, which was adapted to the needs of these large forest baboons. Other species were moved to a house with outdoor islands, and the last chimpanzee died. The orangutans and other remaining species now have more room and more suitable enclosures, made possible by enlarging the outdoor exhibits and by combining several indoor cages into fewer large exhibits.

World War II destroyed most of Antonius' work. He died in 1945 after having seen the destruction at the zoo that the heavy bombardments had caused. After the war, the new director, Julius Brachetka, began reconstruction of the zoo. New enclosures and buildings were built in a new entrance area of the zoo. His successor, Walter Fiedler, followed suit with rare animals and excellent breeding results. Successes, such as the first zoo-bred white-tailed sea eagles in 1963, underline the importance of the park as a modern zoo.

Until 1991, the Austrian government owned and governed the Schönbrunn Zoo. In 1992, control was transferred to a nonprofit corporation. The new director, Helmut Pechlaner, former director of the Innsbruck Zoo, intends to change the zoo into an even more attractive one, while historically preserving the old menagerie. Within the old menagerie, protected now as a memorial, Pechlaner revived the earlier Schönbrunn methods for keeping large cats. The public now walks through the former cages and looks out on large outdoor enclosures enclosed with a net and landscaped as a natural habitat. Mandrills still occupy the former elephant house, while the elephants and smaller monkey species have new houses. The zoo has now incorporated the adjacent Tyroler Garden, providing additional room for expansion. Along with the garden, the zoo acquired the Tyroler House, which is used as a nature education center, and a farm for rare domestic breeds. At the turn of the millennium, the former menagerie at Schönbrunn is a very good example of a transition from a classic "ivy league zoo" to a modern zoological garden.

The other collection with a successful transition from menagerie to modern zoo is in Berlin. The original menagerie was not in Berlin itself, but on an island between Berlin and Potsdam. Only a few miles from Potsdam, the island of Pfaueninsel (Peacock Island) is situated in the Havel River. At the end of the eighteenth century, the Pfaueninsel was a pleasure ground for the Prussian court. It is not known when the first exotic animals were brought to the island, but at the turn of the century cages existed for the birds of prey and for monkeys. In 1802, a house and enclosure were built for buffalo and deer. In the following years more animals arrived. Cages and small houses were built at random near a romantic castle constructed as a ruin (a popular landscaping technique at that time). The king, Frederick William III (1770–1840), was engaged in collecting animals and his gardener, Gustav Adolph Fintelmann, showed the same interest. In 1822, the famous landscape designer Peter Joseph Lenné redesigned the Pfaueninsel as a new garden.[21–32]

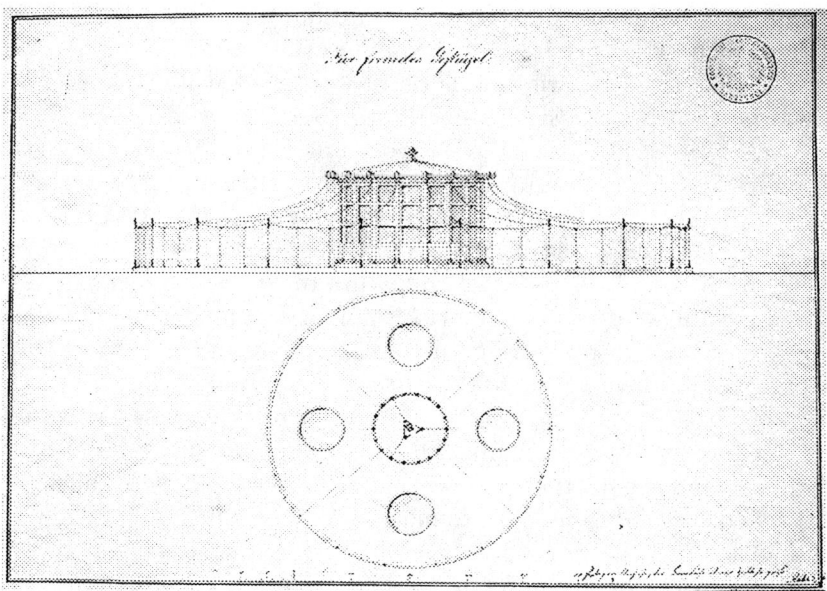

Peacock Island menagerie near Berlin. Drawing of the large aviary, the only building of the menagerie still in existence. © Photograph from Harro Strehlow collection.

In this scheme, the menagerie received a new design as well. Old cages and small houses were removed from the castle and the animals received new ones more centrally located as part of the new garden. Unlike the classic menageries, this one was an integral part of the landscape where the royal family and its guests could visit the animals in natural settings. The menagerie was opened to the public on certain days and was quite popular. After the railway opened from Berlin to Potsdam, the Pfaueninsel got its own station. The king owned the menagerie and a member of the court, the Count of Maltzahn, was supervisor. In the early years, the gardener Fintelmann, and later August Sieber, worked as professors, curators, and animal keepers. Martin Hinrich Lichtenstein, professor of zoology in Berlin, had scientific control over the menagerie. Professor Hertwig worked as veterinarian for the animals. Scientific publications on the animals in the menagerie are not known, although scientific work was being done. In a letter dated April 26, 1837, Lichtenstein wrote about a pair of condors that had recovered from an illness with the help of Hertwig. He instructed the staff to take all carcasses to Hertwig for necropsy. From this research he hoped to get information about improving the maintenance of the animals.

In the 1820s and 1830s the menagerie grew. A bear pit, pheasant aviary, waterfowl aviary, monkey house, and house for the birds of prey provided space for a growing number of animals. Kangaroos roamed their open enclosure, as well as antelope and deer. The wolf house also held lions. The animal inventory of January 1, 1832 listed 847 specimens of 96 species. That was more specimens and species than the Berlin Zoo had in the years after its opening in 1844. Lichtenstein tried to establish a citizen's zoo in Berlin, as well, after his journey to England in 1833. He wanted to create a scientific, yet popular zoo, similar to the London Zoo.

At this time, Lenné was working on a new design for the Tiergarten, the former game park, and now pleasure ground, beyond the Brandenburg Gate of Berlin. He planned the new zoo on an island within the Tiergarten. However, the king did not want a zoo in addition to his menagerie, so he did not agree with the plans. The plans of Lichtenstein and Lenné were therefore abandoned. After Frederick William III died in 1840, Lichtenstein was finally able to carry out his ideas. The new king, Frederick William IV (1795–1861), was not interested in keeping exotic animals, so he gave away most of the deer and other common animals in the Pfaueninsel menagerie. However, he did allow the creation of the Berlin Zoo and donated animals still at the Pfaueninsel menagerie, along with some of its buildings.

3.2.5 Transition from Menagerie to Modern Zoo — Jardin des Plantes

The French Revolution of 1789 was the strongest shock European societies had experienced for several hundred years. The end of the feudalistic society was at hand, and, for the first time, modern democratic society was on the horizon. Among all the events and turmoil the French Revolution engendered was the establishment of what may be considered the first modern zoo, the menagerie at the Jardin des Plantes in Paris.

The Jardin des Plantes menagerie had roots going back as far as the seventeenth century. In 1626, two physicians of the king, Herouard and Gui de La Brosse, asked Louis XIII (1601–1643) to establish a garden for medicinal herbs. The king agreed and the garden was established in what was then a suburb of Paris. In 1640, this garden opened to the public for the demonstration of medicinal herbs under the name Jardin Royal des Plantes. Until 1728, the premier physician of the king also served as the director of the Jardin Royal. After that year, the garden director became an independent position. In the eighteenth century a small museum was added to the plant collections.[33,34]

Buffon (1707–1788) was the most prominent director. He enlarged the small museum and the terrain used for the garden. He engaged other scientists and changed the Jardin Royal des Plantes into a center for science. The museum had a number of chemists, anatomists, zoologists, and botanists at work. After the French Revolution, the Muséum d'Histoire Naturelle was founded on June 23, 1793 and the Jardin Royal was transferred to a new site adjacent to this Muséum as the Jardin des Plantes. Since then, the French government has administered the garden and the museum.

The other origin of the menagerie at the Jardin des Plantes is almost as old. This source was the members of the Academy of Sciences of France who appealed to Louis XIV (1638–1715) to establish a national menagerie. The king agreed and the menagerie at Versailles was opened in 1662. It had three entrances: the first for deliveries, the second for the public, and the third for members of the court. The most remarkable animal was an elephant, a gift of the Portuguese King Pedri II, which arrived in 1668 and died in 1681. During the French Revolution the menagerie, like other royal property, was expropriated. It was not destroyed during the turmoil at Versailles, and was later moved by Bernardin de St. Pierre to the Jardin des Plantes. This was done because the animals could not be cared for at Versailles, and it was thought they would be better off at the Jardin des Plantes as a living part of the Muséum d'Histoire Naturelle. At the museum they were kept in provisional cages and stables; fortunately, the French nation saw the museum, with its menagerie and garden, as a symbol of the new free nation and the new scientific consciousness.

Many people contributed animals to the menagerie. Among the foreign sovereigns, Muhammad Ali of Egypt should be mentioned for sending a giraffe in 1826, and later for sending antelope and an African elephant. The French army brought back rare and valuable animals for the menagerie, including the first two elephants sent to the Muséum d'Histoire Naturelle (which by this time was the official name of the parent organization that included both the museum and the menagerie). The French army occupied the Netherlands in 1795, where they took two elephants from the menagerie at Het Loo in 1798. These elephants originally came from Ceylon and the event of their arrival was so important a small booklet of 16 pages was printed, *Histoire naturelle de l'elephant precedee d'une notice sur l'arivee des deux elephants, male et femelle, au muséum* (*Natural History of the Elephant, Presented on the Occasion of the Arrival of two Elephants, Male and Female, at the Museum*). During the occupation of Vienna, however, Napoleon himself protected the menagerie at Schönbrunn and only a few animals were removed to the museum's menagerie.

Etienne Geoffroy Saint-Hilaire (1772–1844) was the first director of the menagerie. He was a colleague of Georges Cuvier (1769–1832) at the Muséum d'Histoire Naturelle. Cuvier studied the animals at the menagerie as well as those at the Muséum and wrote *Histoire naturelle des mammifères* (*Natural History of the Mammals*) based on these studies. Because of this kind of scientific activity, the menagerie became an important source of scientific information early in the development of the modern natural sciences. After Etienne Geoffroy Saint-Hilaire retired in 1837, his son Isidore Geoffroy Saint-Hilaire (1805–1861) followed him as director, a post he held until his death. For a long time it was the most important animal collection in Europe, along with the London Zoo. The menagerie was used as a living museum, and the directors tried to keep as many species as possible. This menagerie philosophy worked in tandem with the study of animals. The nineteenth century was a time of systematic studies, with the description of new species and their geographical distribution important topics of scientific inquiry.

The Jardin des Plantes menagerie has changed little over the years. Its grounds were enlarged to 15 acres. Many of the old buildings still exist, although conditions for the animals are much better today than they were over the past 200 years. Like the breeding group of bharals (blue sheep), a number of rare species now live in the menagerie.

It is interesting to note that Isidore Geoffroy Saint-Hilaire established another zoo in Paris some 50 years after the founding of the menagerie. In 1854, he established the Societé Zoologique d'Acclimatation (Society for Zoological Acclimatization). Throughout Europe at this time, similar societies for the acclimatization of plants and animals were being developed. Industrialization and the population explosion then occurring in Europe required rapid improvements in agriculture. The societies for acclimatization tried to establish new or improved breeds of animals and plants favorable to the climate of Europe, as well as those that could be used in the European colonies. Since the Muséum d'Histoire Naturelle was not ready for this kind of applied science, and since its grounds were too small for the additional demands these kinds of studies required, the director supported the establishment of a new society and a new zoo.

The Jardin Zoologique d'Acclimatation opened on October 9, 1860 as the second zoo in Paris. Albert Geoffroy Saint-Hilaire, the son of Isidore, was the director from 1865 until 1893. Today, the Bois de Boulogne, where the Jardin Zoologique d'Acclimatation is located, is a pleasure ground and the zoo now keeps only a few domestic breeds.

A third zoological garden, the Parc Zoologique de Paris at Vincennes, opened in Paris in 1934. It was built with artificial rockwork in the Hagenbeck style. It is well known for its rare breeding groups, like the brow-antlered deer, elephants, rare monkeys, and lemurs. Together with the menagerie at the Jardin des Plantes, the Parc Zoologique is an important part of the European zoo world.

3.3 Early Modern Zoos

After the establishment of the menagerie at the Muséum d'Histoire Naturelle, it was 35 years before the next major European zoo opened. This was the garden of the Zoological Society of London at Regent's Park. The next three zoos opened on the British Isles as well. This situation was a sign of England's rapid development as an industrial nation, thanks in part to its constitutional monarchy. This industrial development concentrated the masses in the cities, isolating them from nature. It also concentrated the wealth of the citizens and made colonization of other continents possible. For these reasons, knowledge about exotic animals and the collecting of these animals occurred more readily in England during this time. Industrialization also supported a rapid development of education and culture, which produced scientific societies and animal collections. Similar activities also occurred in other countries, such as Germany, but the earliest zoological societies and collections were in England.

On continental Europe socioeconomic development was slower. It took 45 years from the founding of the menagerie of the Muséum d'Histoire Naturelle before the next continental zoo opened. New zoos developed in the great harbor towns, which were the centers of trade. Zoos opened at Amsterdam in 1838, at Antwerp in 1843, at Marseille in 1855, and at Rotterdam in 1858. The only town far from the sea in which a zoo opened was Berlin in 1844. Of these early zoos, only the one at Marseille no longer exists. These zoos were important cultural institutions at the time they opened, and they have remained important to this day. At the time, the Netherlands were an important colonial power, and trade coming into Europe from many colonies and foreign countries came through Dutch and Belgian harbors. Trade in animals was one of these many commodities. Wealthy citizens and interested scientists founded zoological societies following the standard of the Zoological Society of London.

The first of these societies was established at Amsterdam in 1838. It was the Zoologisch Genootschap Natura Artis Magistra (Zoological Society Natura Artis Magistra), which received the supplement "Koninglijk" (Royal) in 1852. Coenraad Jacob Temminck made the first attempt to establish a zoo in Amsterdam in the early 1830s, but was not successful. However, the book dealer Gerardus Frederick Westerman did succeed. On April 29, 1836, Westerman wrote to King William I (1772–1843) asking for his permission to establish a zoological garden. The king was interested and asked for a detailed plan, which was also sent to the council of Amsterdam. However, the mayor of Amsterdam rejected the idea of keeping wild animals in town as being too dangerous, for he feared an escape, so the king changed his mind. Westerman pursued his interest, and in 1837 another chance arose. The taxidermist Draak lost the rooms where he kept his collection. When Westerman was consulted, he arranged new housing for the collection. This became the first museum of zoology in Amsterdam and it was very successful, proving that the people of Amsterdam were interested in animals. In 1838, along with two friends, Westerman bought some land nearby in Middelaan. This became the first part of Artis, as the Zoo is popularly known. In 1838, an appeal for the formation of a zoological society was published under the title, *Natura Artis Magistra (Nature is the Master of Arts)* and about 120 people joined the new society. In the garden they built cages for parrots, other birds, monkeys, and a leopard. It opened on May 1, 1838. The summer was a popular time for members of the society to visit the gardens, which they did frequently, bringing guests as well. In this way membership increased quickly, and by that first summer it had already reached 400 members.[35,36]

Rijndert Draak was the first director, from 1838 until 1840. Westerman followed, officially becoming the director in 1849, remaining so until his death in 1890. Contrary to Westerman's initial intentions, the zoological garden was not open to the public. Only members of the society and their guests were allowed to visit the garden. This did not change until 1852 when the garden was opened on two Saturdays in September to help raise money for the garden. This opening proved to be a great success, but some of the members were annoyed. The character of the zoological garden as a pleasure ground for the wealthy was lost when it was opened to the poorer masses. Despite these misgivings, the garden continued to be open each year for two days in September to the general public. In 1889, the society agreed to open the garden the entire month of September to the public. After 1922, the public was also allowed to attend concerts the society held, and after World War II the gardens were finally open all year to the public.

The most important event after the opening of the society's garden was the acquisition of the Cornelis van Aken menagerie. Cornelis van Aken was one of seven children of Anthony van Aken, who traveled with a menagerie throughout Europe from 1791 onward. Five of his daughters and sons were engaged in this traveling menagerie, and even had menageries of their own. Cornelis van Aken had his own menagerie from 1828 until 1839, when it was sold to the society. It was late autumn when van Aken arrived in Amsterdam with his elephant, lions, tigers, bears, antelope, and other exotic animals. The zoological garden was not ready to house so many animals because the Amsterdam council had not allowed the society to construct new buildings at the garden. The animals were in danger of dying from the cold, so Westerman discussed the problem with the Amsterdam council and with the Netherlands government. He was finally able to get permission to use stables at the barracks of Orange-Nassau. The animals remained there until the summer of 1840, when they were transferred to the zoological garden.

The zoological society and its garden continued to progress. In 1859, a Museum of Ethnology opened. By 1869, the grounds were 16 times larger than they had been in 1838. In 1882, an important step was taken when the society opened an aquarium, which was open to the public throughout the year. Membership grew, and in 1883 the society included about 6,250 members. Throughout its existence the society has been more than just a zoological society; it has been a society for the sciences in general, as well as culture. Beginning in 1849, concerts became part of the society's life. A large library and scientific lectures for the members, and later for the public, supported the society's educational aims. Other educational activities included a magnificent book by Schlegel about the zoological garden and popular books and guided tours by Anton F. J. Portielje.

Under Director Dr. Coenraad Kerber's management (1890–1927) difficulties arose. Inflation and World War I caused financial problems, which along with a decline in society membership threatened the future of Artis. At the end of Kerber's administration, his successor, A. L. J. Sunier, found Artis in a state of neglect. In 1939, a committee for saving Artis was founded; however, World War II and German occupation brought further difficulties. Not until the war ended did the difficulties begin to resolve. When Sunier retired in 1953, he had partially succeeded in saving Artis, for he had kept it from closing during the war and the occupation, overcoming the shortage of food and fuel that existed at the time.

Development of Artis after 1953, under the Directors E. F. Jacobi and B. M. Lensink (from 1973 on), was one of steady progress. Artis today has many modern buildings and enclosures, in addition to the historically valuable ones that have been adapted to modern animal management. Its most recent enlargement was some years ago, and plans are under way to use this new area for an African plain. The periodical *Artis* has been published since 1955. During 1984 to 1995, together with the Rotterdam Blijdorp Zoo, it published *Dieren (Animals)*. As part of the society's development as a scientific society, a planetarium was opened in 1988, the 150th anniversary of Artis.

Five years after the establishment of the zoological society at Amsterdam, the citizens of Antwerp established their zoological society. Conditions were similar; like Amsterdam, Antwerp was an important harbor town and trading center. Sailors often brought exotic animals to Antwerp, where trade in exotic animals had existed since the sixteenth century. Belgium gained its independence from the Netherlands in 1831, but during the following years it was in turmoil after invasions by the Dutch army and then the French army. This extended the time needed for development of cultural institutions and societies. Perhaps the opening of Artis was a strong influence. On July 19, 1841, a few citizens established a committee for the formation of a zoological society, and on July 21, 1843 the Societé de Zoologie d'Anvers (Zoological Society of Antwerp) was founded as a private society. In 1844, it received recognition as a Royal Society.[37–40]

At first, the society only owned a small area with a few cages. The first director, Jacques Kets, owned a large collection of ethnology and natural history specimens. He became director with permission to build a museum that would house his own collection. The grounds of the small zoo and museum were situated just outside Antwerp, and these grounds were enlarged as quickly as possible. The growing city rapidly enclosed the zoo and built a railway station near the zoo entrance. Today the zoo at Antwerp, with its current 25 acres, is situated near the center of the city. The museum was used until 1897, after which its site was used to build a large concert hall with rooms for the museum on the upper floor. Similarities between the societies of Antwerp and Amsterdam are interesting. Both societies developed from small zoos and important museums, only members of the societies could initially visit the zoos, important libraries were established, and numerous cultural functions developed, including concerts, theater, cinema, lectures, and exhibitions.

Berlin Zoo's Egyptian temple used to house ostriches. This second Egyptian temple in a European zoo was an extraordinary example of the exotic style of zoo architecture. © Photograph from Harro Strehlow collection.

The most important building of the Antwerp Zoo during the nineteenth century was the Egyptian temple for elephants, giraffes, and zebras, built in 1856. Its importance does not lie in its appropriateness for the animals, but rather in its early exemplification of what is known as the "exotic style" of zoo building. It is the most prominent example, and an excellent one. It is interesting to note that the society and its architect, Charles Servais, used the Egyptian style for all kinds of African animals. One reason could be that animals from all over Africa came to Europe via Egypt. Unable to simulate the ecological environment, the architect chose to use a cultural theme. He, no doubt, thought of Egypt as representing Africa, particularly at a time when Egyptomania gripped Europe. Servais had studied the Egyptian Court at the Crystal Palace, which opened in 1854 when he was in London. Egyptology was then of interest to the educated classes, and was popular both in England and France.

Servais built the Egyptian temple with as much scientific accuracy as possible. At the time of its opening, the building was already very impressive. But Servais was not satisfied with just having a building in the Egyptian style. In 1860, the council of the society agreed to his idea of decorating the building with paintings and hieroglyphs. The orientalist, Lodewijk H. Delgeur, supervised this undertaking. The Egyptian temple was unique at the time and it impressed the directors of other zoological gardens who came to Antwerp to buy animals. Many of them imitated the Antwerp style in constructing exotic buildings at their own zoological gardens.

After the Congo became a Belgian colony, Antwerp Zoo exhibited such rare animals from Congo rain forests as the okapi and mountain gorilla, as well as the northern white rhinoceros. Antwerp was the first zoo to open a barless outdoor gorilla enclosure. In 1952, the society established the nature park at Zegge for the protection of its environment. In 1956, the Zoological Park of Plackendael was opened as a breeding center for rare animals. The old zoo in the center of Antwerp with its 25 acres was too small for large groups of animals, whereas Plackendael was four times larger, making it better suited for this purpose. It became the first rural branch of an urban zoo on continental Europe. Today, the Societé Royal de Zoologie d'Anvers consists of a complex of five museums, two zoos, a botanical garden, an aquarium, a library, laboratories, and a planetarium. It has rooms and halls for conferences, concerts, and lectures. It publishes two periodicals, *Acta Zoologica et Pathologica Antverpiensis* (*Antwerp Proceedings for Zoology and Pathology*) and *Zoo Antwerpen*.

The next nineteenth-century zoo to open on continental Europe was quite different from those in Amsterdam and Antwerp. It was not at the seaside, but well inland. The developer was not a zoological society, but a nonprofit stockholding company. The Zoologischer Garten bei Berlin opened in 1844. Unlike the zoos in Amsterdam and Antwerp, the Berlin Zoo was open to the public from the very beginning. Developments leading to the opening of the Berlin Zoo were also different from the other zoos. As has been mentioned, the menagerie on Pfaueninsel existed near Berlin from the early nineteenth century until King Frederick William IV, who was not interested in exotic animals, decided to close the menagerie. Beginning in 1841 a committee was formed to establish a zoological garden. The king gave part of the former pheasantry in the Tiergarten as a hereditary tenure, along with some buildings and most of the animals in the menagerie. On May 21, 1842, the committee accepted the grounds offered by the king and the construction of buildings began. Peter Joseph Lenné created the design and the architect Heinrich Strack created the buildings. The zoo opened to the public on August 1, 1844, while the stockholding company, which had some difficulties to overcome, was established in 1845. The zoo was about one and a half miles west of Berlin. It was popular with the citizens of Berlin and the attendance was good. For a long time the grounds of the zoo were the largest of all the zoos in Europe, but the number of species was small. In 1844 there were fewer than 100 species, and 20 years later there were still only 166 species of birds and mammals.[21,25,26,28,29,31]

The zoo was in a convenient location in Berlin and people liked to go there, but in some respects it was too soon to open the zoo. As of 1868, only 191 of the 500 shares of stock had been sold. Thus, urgently needed money that the shares would have provided was not forthcoming. During the first 25 years, the lack of money was a major problem for the zoo. If not for gifts from the king, the zoo probably would not have been able to continue. With money from Frederick William IV, the company was able to build important new houses like the elephant house and two carnivore houses.

In the 1840s, Germany was primarily an agrarian country and industrialization had only just begun. Wealth had not yet accumulated. There were a few scientific societies like the Gesellschaft Naturforschender Freunde Berlins (Berlin Society of Friends of Nature Research) and the Senckenbergische Naturforschende Gesellschaft zu Frankfurt am Main (Senckenbergian Society of Nature Research at Frankfurt-on-Main), but in the 1840s other societies developed. A Berlin Zoo company was formed with the aim of establishing a zoological society; however, nothing is known about this society except that it existed for a short time.

By the 1850s, times had changed, with the towns and the wealth of citizens growing substantially. The industrialization that had started expanding wealth in England earlier was finally occurring in Germany. A hunger for knowledge and education accompanied this creation of wealth. Many German towns in the 1850s and 1860s established zoological societies and zoos.[41] These trends affected the Berlin Zoo, which after 25 years finally began to take off. In 1869, the company with its 191 shareholders agreed to create a new series of shares. Unlike the first edition of shares, the new shares were sold within a short time. A time of wild speculation overtook Berlin and the citizens had money to support it. Heinrich Bodinus was engaged as the first scientific director of the Berlin Zoo. The old design of the zoo, a forestlike park with enclosures and houses as part of the landscape, was changed into a new design with large ponds and many new large houses.

Within three years the Berlin Zoo developed into the most important zoo in Germany. Not only did the size of the zoo and its collection change, the original ideas regarding the exhibition of animals had changed, too. During the first 20 years, the zoo had been a random exhibition of animals. Then, the zoo's director, Bodinus, established a living museum. He built a systematic zoo with large collections of deer, cattle, cats, seals, monkeys, and birds. The systematic approach remained the main aim of the zoological gardens until about 1914. Berlin started later than other zoos with this systematic collection, but it built an important collection faster than other zoos. Successors of Bodinus, Maximilian Schmidt and Ludwig Heck, followed suit, and by 1914 the zenith was reached with 1,474 species of birds and mammals in the collection.

Before he came to Berlin, Bodinus was Director of the Zoological Garden of Cologne, and prior to that he traveled to zoos in Europe to study their buildings and collections.[42,43] He was most impressed with the Egyptian temple at the Antwerp Zoo and with other exotic buildings. At Cologne he built its first exotic houses, and at Berlin all of its new houses were built in the exotic style. Ludwig Heck continued this pattern. In 1914, the Berlin Zoo was a collection not only of exotic animals, but of fantastic and exotic buildings as well. The last exotic building was the Affenpalmenhaus (Palm House for monkeys), which was built in the mid-1920s.

Because of its vast collection, it is not surprising that the Berlin Zoo, for a while, kept the rarest animals. Well known among these was the gorilla Bobby, one of the first males who reached adulthood in captivity, and who died in 1935. Most interesting were the three milus, or Père David's deer, which arrived in 1876. Existing milus are descendants of these three animals, so this breeding group at the Berlin Zoo was responsible for saving the species. The zoo did not save them purposely, however. A successful captive breeding program happened later, beginning in 1893 to 1895, when zoos gave their animals to the Duke of Bedford, who maintained them at Woburn Abbey (England). His effort was one of the first examples of saving a species from extinction through the management of captive propagation.

It was not until the end of the 1920s that Ludwig Heck accepted barless enclosures at the Berlin Zoo. His son, Lutz Heck, changed the design of the zoo again. Wherever possible, he replaced bars with moats and enlarged the zoo. A "German Zoo" was thus created as a forest with large enclosures and aviaries for European animals. World War II suspended the progressive activities at the zoo. Heavy bombings and the fighting in Berlin destroyed most of the buildings. All but 91 of the animals were killed. Difficulties arose with shortages of food, fuel, and building material. Later, the special situation in Berlin with regard to the communist blockade impeded development of the zoo at a time when other zoos were rebuilding and improving.

Katharina Heinroth, the first director after the war, saved the zoo from closing. New animals arrived and the first repairs to the houses were made. By 1949 the deer house was finished, the first new building after the war. Before she retired in 1956, Heinroth had changed the Berlin Zoo into a modern zoo with the first modern buildings and a remarkable collection of animals, including breeding groups of rare species. Her successor, Heinz-Georg Klös, continued the modernization of the zoo. He renovated some of the old buildings and built many new houses. Berlin Zoo gained a rich collection of species and specimens, but the numbers Ludwig Heck and Lutz Heck had kept could not be reached again. Each species now has more room and most of the species are breeding well. When Klös retired in 1991 his successor, Hans Frädrich, followed along the lines of his predecessor. He gradually reduced the number of species and improved the accommodations for the remaining animals. Today, the Berlin Zoo is an important zoo with many rare breeding groups of mammals and birds, and it still boasts the most varied collection in the world.

Carnivore House, Leipzig Zoo. © Photograph by Harro Strehlow.

OSKAR HEINROTH — ZOO MAN AND FATHER OF ETHOLOGY

The connection between science and animal collections has been strong since the Middle Ages. Important scientists, such as Cuvier, Alfred Brehm, and Paul Matschie, used zoos for their studies. Others like Bernhard Grzimek used the zoo and the zoological society to popularize the idea of nature conservation, and Heini Hediger developed zoo biology as a special branch of biology. Zoo biology translates the ideas and perceptions of other sciences into the practice of zoological garden management and gives stimulus to the use of zoo research in other sciences. Unique among these endeavors was a new branch of biology founded by a zoo man. When Oskar Heinroth (1871–1945) spoke about "Beiträge zur Biologie der Anatiden" ("Contributions to the biology of the Anatidae") at the International Ornithological Congress at Berlin in 1910, it was the hour of birth for comparative ethology.

Oskar Heinroth was born on March 1, 1871 at Kastel near Mainz, Germany. As a schoolboy he reared different species of birds — starlings, redstarts and flycatchers — and he bred canaries and bluebirds. He also studied the behavior of ducks. From 1890 to 1895 he studied medicine, and from 1896 to 1899 biology. After 1894 he worked on the color of feathers, a problem often discussed in the mid-nineteenth century, and which is a mystery still unsolved today. Starting in 1898 he worked as a voluntary assistant at the Berlin Zoo where Ludwig Heck was the director. When Bruno Mencke organized an

3.3.1 The Exotic Style and the Systematic Zoo

After the formation and success of these early zoos, other towns became interested in having their own zoos. In Belgium, seven years following the opening of Antwerp Zoo, the Societé Royale de Zoologie d'Horticulture et d'Agrément de la Ville de Bruxelles (Royal Society for Zoology, Horticulture and Pleasure of Brussels) opened its zoo at what was to become the Parc Leopold in Brussels.[44] In the Netherlands, some 16 years after Artis opened, two railway workers opened a small animal collection in Rotterdam. This first step led to the establishment of the Vereenigung Rotterdamse Diergaarde (Society of the Rotterdam Animal Park), known today as Stichting Koninglijke Rotterdamse Diergaarde (Royal Foundation of the Rotterdam Animal Park). This society was established in 1857 and broke ground for a new zoological garden, which opened in 1858.[45–47] In Germany, some 14 years after the opening of the Berlin Zoo, the Zoological Society of Frankfurt-on-Main opened its zoo.[48–51] Every few years, zoos opened throughout Western and Central Europe. For example, zoos were established at Frankfurt in 1858 (as part of a holding company and a zoological society), at Copenhagen in 1859, at Hanover in 1865 (as part of a zoological society), at Stuttgart in 1871 (privately owned by Johannes Nill, 1871 to 1920s), at Basel in 1874 (the first zoo in Switzerland), and at Leipzig in 1878 (privately owned by Ernst Pinkert).[52–61]

Between 1858 and 1914, zoos opened in more than 40 towns throughout Western and Central Europe. Some only existed for a short time, but most are still extant. In some cities two or more zoos existed simultaneously, or successively. Some collections were specialized, like the

---

expedition to the Pacific islands in 1900, Oskar Heinroth took part as a physician and ornithologist. The expedition was a success at first, but in March 1901 a raid by indigenous people destroyed the camp at Mussau-Islands (now St. Matthias Islands). Some members of the expedition were killed, including Bruno Mencke. Oskar Heinroth was injured when a spear struck his leg, but he survived and saved the collection of bird skins. He returned to Nassau with the surviving members of the expedition and then continued the expedition as its leader.

Oskar Heinroth. © Zoologischer Garten Berlin.

After two months, the expedition returned to Europe. Oskar Heinroth brought back bird skins and about 250 living animals. The latter he sold to the Berlin Zoo and the animal dealer Carl Hagenbeck. Back in Berlin he worked again as a voluntary assistant at the Berlin Zoo. He published the results of the expedition and continued his studies of the Anatidae. On July 1, 1904, Ludwig Heck changed the voluntary assistant status of Oskar Heinroth into the first paid curator of the Berlin Zoo. A month later, on August 10, 1904, Oskar Heinroth married Magdalena Wiebe. Oskar's scientific interest at that time continued to be the behavior of the Anatidae and his specialty was working with hybrids. The Berlin Zoo, with its large collection of ducks, gave him the opportunity to breed and crossbreed the different species. Magdalena helped with the rearing of the birds. Their home had a special room just for this purpose, and when Oskar was at the zoo Magdalena fed and watched the birds.

continued

> ## OSKAR HEINROTH
>
> In 1908, a pair of hand-reared nightjars bred twice in Magdalena and Oskar Heinroth's home; this is believed to be the first breeding of nightjars in captivity. Magdalena published the development of the nightjars in the periodical *Die Gefiederte Welt* (*The Feathered World*) and Oskar published the results in the *Journal für Ornithology*.
>
> In 1910, the International Ornithological Congress was held at Berlin. The Heinroths were very involved in organizing the congress and both presented lectures, Oskar about "Biology of the Anatidae" and Magdalena about "In-House Observations on Some European Birds," a summary of her work in rearing birds.
>
> In 1910, the old Berlin Aquarium closed and Ludwig Heck decided a new aquarium should be built at the Berlin Zoo. Oskar Heinroth was commissioned to plan the new aquarium. The new building was to be much more than just an aquarium. On the first floor was an aquarium, on the second a terrarium, and on the third an insectarium. Oskar created the first walkthrough enclosure in the center of the aquarium. This large hall for crocodiles extended up through the three floors. From the aquarium floor you could see the crocodiles swimming underwater through large windows. On the second floor a bridge led through a tropical river exhibit with sandbanks where the crocodiles lived. And from the third floor you could see into the glasshouse from above. This outstanding design of the aquarium still exists today.
>
> Oskar Heinroth became director of the aquarium and worked there until the aquarium was destroyed in 1943 during World War II. Throughout his directorship his scientific work concerned birds. Together with Magdalena he reared nearly every Central European bird species and documented their development. As director, living in the zoo, he had the logistical background for their research. They could use the photographic laboratory of the zoo, had enough space for their birds, and could make use of the richness of the food at the zoo.
>
> The result of their research was the unique book *Die Vögel Mitteleuropas* (*The Birds of Middle Europe*), published in four volumes between 1924 and 1933. It covered all the species

Tierpark Peter and Paul at St. Gallen, Switzerland. Founded in 1892, it kept only European animals. In its first year there were 12 deer, 8 fallow deer, 4 roe deer, 5 chamois, 8 marmots, and 2 hares. The most important animal at St. Gallen was its breeding group of Alpine ibex, kept for the first time in 1906. In 1911, the collection started reintroducing the alpine ibex back into Switzerland. The Zurich artist Urs Eggenschwyler built the artificial rocks for the ibexes in 1902. Today, the collection also contains boar, sika deer, and lynx.[62]

Most of the collections that opened prior to World War I were small when they began, but developed into important collections. Some were connected with a museum or botanical garden, or had botanical exhibits with greenhouses. Almost every country of Western Europe had one or more zoos at the beginning of World War I. In addition to those mentioned earlier, the zoo in Lisbon was founded in 1884, Helsinki in 1889, Barcelona in 1894, and Rome in 1911. In Germany at the time, zoos ranged from Königsberg and Breslau in the east to Aachen and Mülhausen in the west, from Munich and Stuttgart in the south to Hamburg and Stettin in the north. Some of these zoos, once found in Germany, have changed nationality. Today, Mulhouse (formerly Mülhausen) is in France, Wroclaw (formerly Breslau) is in Poland, and Kaliningrad (formerly Königsberg) is in Russia. In addition, Germans founded the zoo at St. Petersburg in Russia.[63–71]

Each of these European zoos has its own interesting history. Stories about their different directors, financial problems and their solutions, prominent animals, collections as a whole, they reared and included information on their breeding biology in text and photographs. Even today, *Die Vögel Mitteleuropas* is a valuable reference for ornithologists.

Magdalena Heinroth died in 1932. Oskar Heinroth married again in 1933. His second wife, Katharina Berger, a biologist, is well known not only for her scientific work, but much more so for being the first director of the Berlin Zoo after World War II. It is not an exaggeration to say she saved the zoo after the end of the war. Oskar's health had suffered under the deprivations of the war. Two years after the aquarium was destroyed in the war, despite the care of Katharina, he died on May 31, 1945. His grave is in the zoo near the aquarium.

During his life Oskar Heinroth published nine books and more than 300 scientific papers. Of these articles, 146 were written for popular journals, newspapers, or as reviews. He supported other scientists when possible, the most prominent of whom was Konrad Lorenz, who not only studied animals from the Berlin Zoo with the help of Oskar Heinroth, but also learned much from the discussions the two had over many years. The letters of both men show the thorough discussions that took place between them, including the development of many ideas on ethology.

With his studies on the behavior of the Anatidae, Oskar Heinroth laid down the principles of comparative ethology. He was the first to discover the heredity of behavioral elements and to use this as a measure of relatedness between different species. Oskar Heinroth influenced the development of ethology through many discussions in scientific societies and in many meetings and letters with his scientific friends. His accurate criticism and support of young scientists helped to encourage new ideas in ethology and to spread this new science throughout Europe. His other important achievement, the book *Die Vögel Mitteleuropas*, was the intensive and most expansive work about the development of birds ever published. It not only documents the knowledge of the time, but it remains the foundation for the study of European bird development and a husbandry guide to their rearing.

Source: Based on Heinroth, K., *Oskar Heinroth*, Wissenschaftliche Verlagsgesellschaft, Stuttgart, 1971.

---

and buildings could fill books. Books have been written on the history of these zoos, especially for important zoo anniversaries. These zoos have many differences, but there are also interesting similarities. No matter whether scientific societies, stockholding companies, or private individuals administered the zoos, the idea for founding them has been similar — a common interest in animals.

The committee for the Frankfurt Zoo wrote in a paper in 1857:

One turns away from the political and social strife of life to the contemplation of nature with a feeling of contentment, and one is revived and strengthened by the recognition and perception of nature's creativity and her eternal laws. Only very few are in a position to fulfill this desire through studies while in the midst of everyday stress. This has resulted in efforts being made nowadays to arrange leisure programs for people in such a way that as much as possible of the outcome of such studies are demonstrated to them. Polytechnic exhibitions, geological pictures, natural history specimens, etc. belong in this category. These arrangements, however, require an effort on the part of the viewer and also deprive his leisure hours of walks and the fresh air of the outdoors. Consequently, the establishment of zoological gardens meets with complete approval on all sides, and how can it be otherwise! What a pleasant way to relax, walking in a lovely garden, looking at living creatures from all areas and regions — each with its individual characteristics — and admiring their diversity![48]

Education and recreation have been primary aims of the zoos from the beginning, and together with research and conservation, they are still the main aims today. As early as 1859 the Frankfurt Zoo published the journal *Der Zoologische Garten* (*The Zoological Garden*) as a public forum for scientific research and experience at zoos. The zoo community has also worked on exchanging ideas and experiences from the beginning. Before a new zoo was built, the director, or members of the society or company, would visit other zoos and examine the suitability of the facilities, resulting in a strong similarity among European zoo buildings.

Monkey houses, bear castles, deer houses, and aviaries were similar at all the zoos. The model for monkey houses was created at Paris. It was a semicircle around an outdoor cage. Inside were cages for the monkeys and at the back was the public area. In the outdoor cage various kinds of monkeys were kept together. At Berlin, this kind of monkey house also had cages for reptiles and other animals. This kind of monkey exhibit did not change until the 1860s when, to protect the monkeys from infections and cold weather, monkey houses were built with indoor cages only and the cages were built with glass between the bars and the public. In the 1870s, experiments with glass were tried at the Berlin Aquarium, which kept monkeys and other small mammals in addition to the fish, and at the Berlin Zoo several years later.

In 1869, new designs at the Berlin Zoo not only included new buildings, but the existing buildings were also used for different purposes. The former monkey house was enlarged and used as a birdhouse, and the former carnivore house was changed to a monkey house. This latter house retained its outdoor cages and was a house for hardy monkeys. It was the only one at that time with both indoor and outdoor cages. Delicate monkeys and apes were kept in the new birdhouse until 1884, when the new monkey house was built. This monkey house was designed at first without outdoor cages, as that was common in zoos at the time, but the design was changed before it opened to include two outdoor cages in the rear.

As was mentioned earlier, the first exotic design for a zoo building was the Egyptian temple at the Antwerp Zoo. This famous building impressed directors of zoos and their architects who visited Antwerp. Before Bodinus became the director of the Cologne Zoo in 1859 he visited existing zoos, and together with the architect Benz, he published the book *Der Zoologische Garten zu Coeln* (*The Cologne Zoological Garden*). This is a sketchbook of exotic styles that became popular at continental zoos for decades. It shows the fascination, together with the weakness, of this style. The designs were used as exotic settings for exotic animals housed in a European environment. At least the Antwerp Egyptian temple was authentic in its design and accuracy, but this cannot be said for most cases of the exotic style. At many zoos the exotic design was only ornamental. Architects planned animal houses as castles, mosques, Indian palaces, or Arabic houses. Northern animals lived in blockhouses. Another good example is the Minangkabause house at Artis; the Minangkabause is an ethnic group from Sumatra. Today it houses small ungulates like pudu and roe deer. The most remarkable was the Indian temple at Berlin, or Elephantenpagode, a house with sublime design, but one that was unsuitable as a stable. Nevertheless, it was used for about 70 years until bombings during World War II destroyed it. Other houses Bodinus built were more suitable, but were of the fantastic style. While the exotic house designs showed the cultural environment of the animals, the fantastic styles were without any geographic or ethnographic equivalent.

Ludwig Heck, Director of the Berlin Zoo from 1888 until 1931, built some great exotic houses, including the Japanese house for cranes and storks and the Siamese house for Indian buffalo. Both of these were also destroyed during World War II. Two other houses, the Russian blockhouse for European bison and the Indian blockhouse for American bison, are still extant. Philipp Leopold Martin was another outstanding proponent of the exotic style, which he presented in his book, *Die Praxis der Naturgeschichte* (*The Practice of Natural History*), in 1878.[72] His basic idea was to show how animals should be exhibited in houses that mirror the cultural

Minangkabause house at the Artis, Amsterdam. The Minangkabause is an ethnic group from Sumatra, and this house is another example of the exotic style of zoo architecture. © Photograph by Harro Strehlow.

characteristics of the areas from which they came. Sometimes, however, he wrote about the natural environment of the animals, which could be demonstrated using greenhouse exhibits.

The combination of exotic and fantastic architectural styles, together with the gardens and the animals, created a special atmosphere. Zoos were a total experience of another world completely different from the normal urban life. However, biology was developing quickly, as were all the sciences. An important branch was taxonomy, with its description of species. The new colonial regions of Germany, Belgium, and Italy provided their scientists and zoos with new possibilities for collecting animals. Zoos developed as living museums of taxonomy, with houses that had long rows of cages, each with a different taxonomic group. Sometimes two or

Berlin Zoo's Japanese house used to exhibit cranes and storks. It is another example of the exotic style of zoo architecture. It is also an example of the systematic zoo because this house was planned for the exhibit of all crane and stork species. © Photograph from Harro Strehlow collection.

more specimens were housed in the same cage, but zoos did not necessarily try to breed them. They were museums, but were showing them alive. Nevertheless, the standards for maintaining the animals were often high. Longevity was often good and breeding occurred. A proverb about the Berlin Zoo at the time was "Bei Heck heckt alles" ("Everything is breeding at Heck's"). Some births, like that of an echidna in 1901, were delightful events.

Just a few examples of the Berlin Zoo's systematic style in 1914 include 11 species of feral swine and peccaries, 29 species of antelope, 35 species and subspecies of deer, 10 species and subspecies of equids, 15 species of domestic and wild bovids, and 61 species of monkeys. Similar collections of other mammals and birds were also exhibited. The aim of Ludwig Heck was to keep 500 species/subspecies of mammals and more than 1,000 species/subspecies of birds. In 1914, he was not far from this goal as the Berlin Zoo became an outstanding example of the systematic zoo. Other zoos in continental Europe maintained similar kinds of systematic collections. At the time, species richness of a collection was an indicator of its importance.

The most outstanding building within the systematic zoo was the birdhouse. At the Rotterdam Zoo the new birdhouse opened in 1884. Birds were housed in three rows of small cages, one row on top of the other. Larger birds lived in the lower cages, which were the biggest. The upper two rows housed smaller birds in systematic order. Other zoos followed this example and the systematic zoo reached its zenith at the beginning of the twentieth century. To continue this style would have been possible, but public interest changed. And science changed its focus from taxonomy to other disciplines, particularly ethology and ecology. Books, pictures, and the media increased the public's knowledge about foreign countries and animals. The public wanted more than just large numbers of species in long rows of similar cages. Neither the exotic buildings, which gave an interesting frame of reference to the systematic collection, nor the newer systematic buildings could fulfill the new expectations of the visitors.

3.3.2 The Hagenbeck Revolution

The next major change in zoos did not develop at one of the well-established zoos. It was a private individual, an animal dealer, who developed and realized the new ideas. Carl Hagenbeck was born in 1844. His father, also named Carl, was a fish dealer at Hamburg. In 1848 Hagenbeck, Sr. exhibited six common seals, which some fishermen had brought to him, and in the following years more animals began to arrive at his store. The business of trading animals became more important as time passed. Carl Hagenbeck learned the animal trade from his father and in 1866 he took over the animal business. In 1874, Hagenbeck, Jr. opened a second zoo in Hamburg, called Carl Hagenbeck's Thierpark, at Neuer Pferdemarkt. During that year he imported 35 giraffes and had expeditions going to all the continents except Antarctica to bring animals back to Hamburg. His trade extended to all the European zoos and private collections, as well as to the United States. In addition to the animal trade, he became engaged in ethnological shows and circuses. For his own and other circuses he trained animal groups, thus developing a business in animal training and acts.[73–80]

Carl Hagenbeck. © Hagenbeck's Tierpark.

African Panorama exhibit at Hagenbeck's Tierpark, Stellingen (Hamburg). These Panorama exhibits revolutionized the exhibition of animals during the twentieth century. © Photograph by Harro Strehlow.

In the 1890s he developed the Panorama, for which he received a patent in 1896. The Panorama was an exhibit of different species from the same ecological environment. For example, the Northern Panorama exhibited seals and walruses in a pool in the foreground, with reindeer behind them, and polar bears behind the reindeer. The different enclosures were divided with moats not visible to the public, and the successive enclosures were higher than the one in front. The exhibits were landscaped with plants and artificial rocks. This gave the public the impression they were seeing the animals together in one natural habitat. Similar Panoramas existed for other regions. The Panoramas were used at world exhibitions and fairs, as well as at zoos. All of Hagenbeck's experiences with the animal trade, circuses, ethnological shows, and Panoramas influenced his thinking on the development of a new zoo concept.

When Carl Hagenbeck's Tierpark (now without the "h" in Tierpark) opened at Stellingen (near Hamburg) in 1907, it was a shock to the zoo community. Hagenbeck had broken with a strong tradition and created a new style of exhibition that had never before existed. The significant aspect of its design was the use of the Panorama as the most important segment of the zoo landscape. The artist for the artificial rocks was Urs Eggenschwyler, who had built the artificial rocks at St. Gallen some years before. For a change, the individual animal or species was not the focus of the exhibit, but rather it was the composition of the natural landscape with large groups of animals. In the African Panorama, the foreground had ducks and flamingos; behind them were large plains with zebras, antelope, and ostrich; behind them were lions and vultures at the foot of an artificial mountain, on which were ibex or barbary sheep. Enclosures for other animals from his animal business, ethnographic landscapes (such as a Japanese island and Burmese temple ruins), and a large hall for training animals completed the animal park.

The real challenge was turning away from the systematic zoo concept. In Hagenbeck's environments, totally different taxonomic groups were exhibited as if they lived together — hoofed stock, carnivores, and birds. He changed the systematic exhibit into an ecological or geographical one. At first directors of other zoological gardens largely refused to consider Hagenbeck's ideas. In the front rank of opponents stood Ludwig Heck. It took about 20 years before Heck agreed to the first experiments with Hagenbeck's ideas at the Berlin Zoo.

Other zoos and zoological societies reacted faster, for Hagenbeck's exhibit style met the expectations of the public. Soon after the opening of Hagenbeck's Tierpark, other zoos built their first barless enclosures. In Italy, when the new Rome Zoo was opened in 1911, Hagenbeck was its designer. In 1934, the new Zoological Park at Paris-Vincennes used artificial rocks and barless enclosures, but not the Panorama landscape of Hagenbeck. Hagenbeck erected the first monkey-rock exhibit in 1913. It was a large rock of artificial schist, a copy of an Ethiopian cliff taken from photographs. Instead of bars, it had a moat 16 feet wide to contain a large group of some 200 hamadryas baboons that lived in the exhibit. One year later the zoo at Cologne opened its monkey-rock. In many other zoos barless enclosures, monkey-rock exhibits, and pool exhibits without bars for seals came into existence if money and space were available. The ideas of Carl Hagenbeck had started to change the design and animal maintenance systems of the zoo world.

3.4 Modern Zoos

3.4.1 Zoos of the Early Twentieth Century

German society was prospering when Hagenbeck opened his zoo. Some years later, however, political crisis led to World War I, to economic crisis, and to changes in government in many European countries. These difficult times stopped the development of zoos in Europe for about 10 years. Some zoos, like the one in Munich which had opened in 1911, closed in the postwar period. The Berlin Zoo was closed for some months and Hagenbeck's Tierpark closed for two years. The history of nearly every central European zoo indicates that the beginning of the 1920s was a very difficult period when the zoos had to struggle to survive. This struggle did not abate until the mid-1920s when the financial problems began to be resolved and zoos could develop anew.

Despite the problems facing the European world and its zoos, a very important development had begun. World War I and Russian Revolution battles had resulted in the extinction of free-ranging European bison at Bialowieza. The director of the Frankfurt Zoo, Kurt Priemel, collected the available data on European bison still living in zoos. Only 56 specimens were in captivity on October 15, 1922. Of these, six were unable to breed and others were not pure-bred European bison. With the support of Ludwig Heck, and many other scientists and zoo directors, the International Society for Saving the European Bison was founded in August 1923 in Berlin. To maintain successful breeding, the first studbook for a wild animal was established. The aim of the society was not only to save the European bison from extinction, but also to reintroduce it into suitable European forest habitats. Today, more than 3,000 bison live in zoos, on reservations, and in the Bialowieza National Park where they once again roam freely. After the reintroduction of alpine ibex into the Alps and the propagation of Père David's deer, saving the European bison was the most elaborate conservation action of the European zoos to save a species from extinction.

During the period between the wars, most zoos replaced barred cages with moated exhibits where the animals had more space and lived in larger groups. In Leipzig, Director Johannes Gebbing renewed and enlarged the zoo with modern buildings in clinker style (brickwork). He refused to use the fantastic style still being used in some zoos. Many of his modern houses are still in existence.

In Rotterdam the old zoo was closed in 1940 and was replaced with a new zoo built in Blijdorp. Sybold van Ravensteyn served as the architect for the new zoo. This new zoo opened on July 7, but World War II wreaked havoc on the old Rotterdam Zoo just before the animals were delivered to the new zoo. The German military bombed the city of Rotterdam, hitting the zoo, and many of the animals died. Those that survived were brought to Blijdorp. Blijdorp Zoo is known, not only for its collection of animals, with such rare species as okapi, Szechuan

Elefantenpagode (pachyderm house) at the Berlin Zoo after Allied forces bombed it the evening of November 22, 1943. The Elefantenpagode, which resembled an Indian temple, was another example of the exotic style in zoo architecture. Also shown are two of the seven elephants killed during the bombing. © Zoologischer Garten Berlin.

takin, white-lipped deer, and others, but also for its architecture. Originally, it was one of the few zoos designed in its entirety by one architect, van Ravensteyn. Although many changes have been made to van Ravensteyn's design, much of his work, which was then considered the "New Functionalism" style, remains.

New zoos also developed in other cities. In Munich, the zoo at Hellabrunn was closed during the economic crisis of the 1920s. Tierpark Hellabrunn did not open again until 1928, after a new stockholding society had been founded. Some of the old buildings, such as the Ethiopian house (for elephants, giraffe, and rhinos) and the lion house, were renovated or rebuilt in the old style. Heinz Heck, the son of Ludwig Heck and the first director at Hellabrunn, took the zoo in new directions. He used the Panorama idea of Hagenbeck and improved upon it. In Munich-Hellabrunn, the animals lived in large groups in a geographic arrangement. The animals of Europe lived in one part, those of America in another, those of Asia in a third part, and so forth. This was not possible for all the animals, so the large cats from every continent lived in the lion house, elephants from both continents in the Ethiopian house, and apes of both continents lived in the ape station. For the most part, however, Heinz Heck at Munich-Hellabrunn established the zoogeographic concept, which many modern zoos have replicated. The new developments at Blijdorp are also an example of this use of the zoogeographic concept.[81–85]

At the beginning of the 1930s, the Düsseldorf Zoo built a new monkey house.[86] For the first time a barless outdoor enclosure for anthropoid primates was used. The restraint was an electric fence at the top of a small moat used as a corridor for the keepers. The second outdoor enclosure for anthropoid primates without bars was built at the Antwerp Zoo more than 20 years later.

The development of zoos between the two world wars went well. New buildings and the breeding of rare animals represented the high quality of animal husbandry at the zoos. New zoos were established, such as Saarbrücken and those at Zurich and Bern.[87–90] In 1928, the Tiergrotten (Animal Grottos) opened at Bremerhaven. This still remains one of the smallest zoos in the world with only 75,600 square feet, and it specializes in Nordic animals.

The well-known zoo director and author of popular books on zoo biology Heini Hediger became the first director of the Bern Zoo in 1939. He later worked at the Basel Zoo from 1944

Wartime Experiences of Western European Zoos

Events that affect societies often also affect zoos. Prosperous societies have the resources to develop their animal-keeping systems. In contrast, economic crises stress zoos and sometimes destroy them. The worst crisis society can experience is war. It is impossible to know how often menageries have been destroyed during wars, have been plundered by soldiers during or after wars, or how many animals have died from starvation during these times. Just a few historical examples should be mentioned. When the French revolutionaries occupied Versailles they did not think to bring food for the animals in the menagerie, so all the animals faced starvation and some died because of it. It was Bernhardin de Saint-Pierre who saved many of the Versailles animals when he transferred them to the menagerie at the Jardin des Plantes in Paris.

When French troops occupied the Netherlands during 1795, animals in the Dutch menageries were treated as war trophies. As was the case with the elephants at the menagerie Het Loo, they were brought to the national menagerie at the Jardin des Plantes. How often this happened during the campaigns of Napoleon is not known; however, it is known that in Vienna Napoleon protected the menagerie at Schönbrunn. Only a few animals were sent to Paris, while the others were allowed to remain at Schönbrunn.

No matter what has occurred in the past, it is the twentieth century that has seen the most destructive wars. World War I presented difficult problems for the zoos, but the armistice in 1918 prevented serious destruction at the zoos. The economic crisis throughout Europe following that war forced many zoos to close for some time and halted the development of zoos in most European countries up through 1924.

The effects of the World War I were felt very strongly by those zoos that remained open. Problems arose due to lack of staff, lack of food and its bad quality, and lack of fuel. Restricted trade made it impossible to replace animals that died, so the number of specimens and species declined during the war. For example, the Berlin Zoo showed 1,474 species/subspecies of mammals and birds in 1914. At the end of the war only about 700 species/subspecies survived. Similar declines occurred in most European zoos.

A special, and perhaps unique, event was the use of an elephant in the German army. Carl Hagenbeck's Tierpark at Stellingen suffered because so many young keepers were drafted into the army. Hagenbeck negotiated an agreement with the German government to send an elephant instead of more keepers into the army. The elephant Jenny worked in the army felling trees and moving heavy loads, and then returned after the end of the war. Her keeper, on the other hand, was wounded in 1918 and could no longer work as a keeper.

World War I was mainly a stabilized warfare situation. On both the Eastern and Western fronts the German army could not occupy large regions, and the fronts did not change much (notorious places like Verdun or Ypern with their battles and the deaths of hundreds of thousands of young soldiers are well known). The armistice occurred before Allied troops entered Germany, so those European towns suffered relatively little destruction.

The situation during World War II was much more difficult for European zoos. During this war the German army occupied large parts of Europe. Then the Allied troops invaded

Germany and heavy fighting occurred on all fronts until the troops of the former U.S.S.R. occupied Berlin and met other Allied troops near the Elbe River. As they left these areas, the German army used the strategy of "Verbrannte Erde" ("Burned Earth"), destroying as much as possible during their retreats. All of this fighting, together with the bombings, destroyed not only the cities and towns, but also many cultural institutions, including zoos.

When the German army occupied other countries, the Nazi regime destroyed the cultural life of the countries. High schools and universities were closed, museums and libraries were plundered, and treasures of the cultural centers were brought to Germany as war trophies. Similarly, rare animals from the zoos were confiscated as war trophies; however, no German documents are known or published about this.

Zoos suffered again from the lack of manpower, since most keepers were drafted. Some zoos engaged prisoners of war for their work. In contrast to the work situation of prisoners of war in industry, especially the war industry, the work at the zoos was humane. A unique document is Albert Bourez's book, *Zootiere: Freunde in der Fremde* (*Zoo Animals: Friends in a Foreign Country*).[*] Bourez, a prisoner of war who worked in the Zoologischer Garten Münster, reminisces about his time there.

Heavy bombings had the most devastating effect on the zoos. The Berlin Zoo is an excellent example of the fate most zoos suffered. Heinz-Georg Klös gives a detailed list of the bombings of the Berlin Zoo during World War II in his book, *Von der Menagerie zum Tierparadies: 125 Jahre Zoo Berlin* (*From the Menagerie to the Animal Paradise: 125 Years of the Berlin Zoo*).[**] Bombs fell on the zoo 12 times between September 8, 1941 and February 24, 1945. The nights of November 22–23, 1943 and January 29–30, 1944 were the worst of all. In the November bombing alone more than 1,000 incendiary and phosphorus bombs, demolition bombs, and aerial mines damaged or destroyed the zoo buildings. During the two nights in January, a third of the animals were killed, including 7 elephants, 1 rhino, 2 giraffes, 17 antelopes, 11 bovines, 25 deer, many carnivores, 15 monkeys, 1 chimpanzee, and 1 orangutan. The aquarium was hit in its center. All the glass of the crocodile hall, aquariums, and terrariums was broken, the water running down the stairs into the garden and the pools. Those few animals that survived the bombings died within a short time from the cold nights. The second heaviest bombing in January 1944 destroyed nearly all the buildings the first bombing had spared.

In 1945, much fighting took place around the zoo since it was located near the center of Berlin. The fighting between the German army and the Red army of the former Soviet Union did not stop at the walls of the zoo. Trenches ran throughout the zoo, shells rocketed through the zoo, tanks churned the grounds, and the fighting brought further shooting and killing. Keepers and their wives still living at the zoo had to maintain and feed the animals still alive, while citizens expelled from destroyed houses around the zoo searched in the zoo for a place to live. After the war ended only 91 animals were still alive. None of the houses was intact and most were totally destroyed. Water pipes were broken and the electric supply was damaged. The bombing hit the administration building and the director's house, destroying much of the zoo's archives.

The fate of many European zoos was similar. Julius Brachetka wrote in his book *Schönbrunn und sein Tiergarten* (*Schönbrunn and Its Zoo*), "I am indebted to pure chance that I found under the ruins of the hippo house, which was totally destroyed from bombs, documents or rather remainders of documents which obviously were stapled there for heating and which came into daylight in a really uncommon manner by a bomb."[****]

continued

> ## WARTIME EXPERIENCES
>
> Thus, together with the loss of animals and buildings, a portion of the history of many zoos was also lost. For example, when the archives of the Berlin Zoo burned, not only did the documents of the zoo vanish, but also the documents of the Berlin Aquarium. The Berlin Zoo Aquarium was the successor of the old Aquarium and had preserved its archives. Directors of the Berlin Zoo have tried to get duplicates of as many lost publications as they could, but most of the annual reports, guides, and documents pertaining to the Berlin Zoo and Aquarium that are a century or more old are lost. The same thing happened to many of the archives of other European zoos. It has been hard work for their directors, librarians, and archivists to replace these lost documents; many are lost forever.
> * Regensberg, Münster, 1989.
> ** Haude & Spener, Berlin, 1969.
> *** Brachetka, Vienna, 1947.

to 1953, and then the Zurich Zoo beginning in 1954. Hediger developed his ideas of zoo biology during these years, but it took many more years for other zoo administrators to understand his ideas. Even today, modern zoo buildings sometimes do not demonstrate an understanding of his zoo biology principles.[91–95]

At the end of World War II most zoos in Western Europe, and especially in Germany, showed severe damage. In Berlin just 91 animals were alive (out of 3,715 mammals and birds in 1939), in Frankfurt-on-Main about 50, in Cologne 22, in Vienna 100 (out of 2,200 in 1939), in Karlsruhe about 12, and in Hanover about 50. Most zoos had many of their buildings destroyed and the difficulties did not end with the war. Shortages of food, materials, fuel, and money prevented the recovery of most zoos, while some, such as the Düsseldorf Zoo (established in 1874), closed.

3.4.2 Zoos of the Late Twentieth Century

As with the period after World War I, the renovation and improvement of zoos after World War II began slowly, but then quickly accelerated. Until the 1960s, zoos were a primary recreation for people and the number of visitors was high. The influx of money was good and the zoos were able to repair their war-damaged facilities, as well as build new ones. This reconstruction also allowed for the creation of more space for the zoos. The new building designs were as simple as possible without any ornamental accessories. Concrete and glazed tiles were the primary building materials. This kind of design also favored hygienic husbandry, and was necessary because disinfection was an essential tool in the veterinary care of the animals.

In addition to the zoos that were rebuilt, new zoos began. In Gelsenkirchen, the animal dealer Hermann Ruhe opened the Ruhr-Zoo in 1949.[96] In Stuttgart, the Wilhelma Botanical Garden evolved into a botanical garden-zoo combination.[97] Animal buildings and enclosures were built among the large nineteenth-century greenhouses. Stuttgart erected the first building for nocturnal animals to be found in continental Europe. In Münster, the old zoo was abandoned and a new one built outside the city. Its houses were connected with a roofed path, and so became known as the Allwetterzoo Münster (Münster All-Weather-Zoo). Together with a museum of natural history and a planetarium, the zoo is the focus of natural history education in Münster.

In the former German Democratic Republic (GDR) a number of zoos developed, at Rostock, Magdeburg, Schwerin, Eberswalde, Cottbus, and many other cities. A 1987 guide to the animal collections of the GDR indicates there were nine large zoos and about 120 small collections.[9] The most important is the Tierpark Berlin-Friedrichsfelde. A ground of 395 acres provides enough space for breeding large groups of rare animals. Because of its connections and animal dealings with Eastern European zoos, it contains some of the rarest species, such as takins and white-lipped deer. Its former connection with the Academy of Science of the GDR supported research at Tierpark Berlin-Friedrichsfelde. The first director, Heinrich Dathe, was a well-known scientist. He edited five scientific journals and wrote more than 1,000 scientific and popular papers and books.[98] Under his guidance, the Tierpark Berlin-Friedrichsfelde became one of the leading zoos of the world with many rare breeding groups of birds and mammals. In 1963, the Tierpark Berlin-Friedrichsfelde opened the Alfred Brehm House for carnivores and tropical birds. This structure was for a long time the largest animal house in the world, and the only one with indoor barless enclosures for lions and tigers. Other important buildings are the pachyderm house, the house for hummingbirds and crocodiles, and the terrarium.

After the reunification of Germany in 1990 and the death of Heinrich Dathe, the new director, Bernhard Blaszkiewitz, continued the path of his predecessor. For the first time in the history of the Berlin zoos, there is now a close connection between the two zoos in their breeding and education programs. This cooperation has helped alleviate concerns about the zoos in the former GDR. The economic crisis had been severe in the eastern part of Germany and there was the danger that many of the animal collections would close. Even large zoos with a long history were in danger. Today, it seems most of the collections are safe and there are signs of progress. There are new master plans, new enclosures, and new animals. The number of visitors declined in the first years after reunification, but seems to be stable now. Most impressive is the development of the Dresden Zoological Garden, where new orangutan enclosures, an underground zoo, and a large free-flight aviary and breeding station for parrots have been built. A new elephant house was finished in 1999.

In Austria, the Alpenzoo Innsbruck opened in 1962.[99–101] Hans Psenner was the founder and first director. Alpenzoo Innsbruck specializes in the exhibition of animals that live, or used to live, in the Alps. It is known for its breeding of bearded vultures, black (or cinereous) vultures, the waldrapp, and passeriformes (the perching birds).

In addition to this kind of regional specialization, a second kind of specialization has been the bird park. The most prominent is the Vogelpark Walsrode. Private breeder Wolf Brehm established it in 1958, after which it developed into a well-known collection of birds. The Brehm Foundation, established some years later, is a foundation for the research and protection of birds. The Vogelpark Walsrode works with international organizations concerned with the protection and captive propagation of birds. The breeding of rare storks and cranes is its focus. Although there are large breeding facilities not open to the public at Walsrode, the Vogelpark established additional private breeding facilities on the island of Mallorca in the Mediterranean Sea. This facility, the Ornis Mallorca, has large aviaries for hornbills, rare parrots, and other rare birds. Another particularly remarkable bird park is Loro Parque (Parrot Park) on the island of Tenerife, where some of the rarest parrots are kept and bred. Loro Parque is also host to the International Parrot Congress and participates in parrot protection and research activities.

Yet another unique form of specialization exists at the Apenheul in Apeldoorn.[102] Only monkeys and apes are kept here. The smaller species live semifreely, including a troupe of squirrel monkeys. There are also rare South American monkeys that roam freely and gorillas that live in two large groups on islands. Visitors walk through the forest and can see the free-roaming monkeys overhead in the trees or can wait at a feeding place for the monkeys to come to eat. At the Affenberg Salem there is also a group of about 100 Barbary macaques roaming freely in a forest, and the visitor can see these monkeys from outside their exhibit. Similar exhibits exist at the Tierpark Rheine and the Burger Park at Arnheim for chimpanzees and gorillas.

In the 1960s, recreation facilities became commonplace and the competition for people's leisure time reduced the visitation at zoos. Two kinds of facilities became popular, one of which was the safari park, where visitors could view the animals from their cars as they drove through the park. Most safari parks were established for a quick profit and then failed, although a few succeeded, such as the ones at Hodenhagen in Germany, Knuthenborg in Denmark, and Beekse-Bergen in the Netherlands.

The other kind of park that became popular was the dolphinarium. Some dolphin exhibitions traveled through Europe commercially, whereas others were just another attraction at leisure parks. Some zoos established dolphinariums with more educational entertainment, such as at Antwerp, at Hamburg (Hagenbeck's), at Nuremberg, and at Münster. The most prominent one is at Duisburg where the former Director, Wolfgang Gewalt, went on expeditions to bring back inias, toninas, and belugas.[103–105] Research on whales and dolphins is also an important part of the programs at Duisburg. Because of better animal husbandry and improved maintenance of the aquatic systems, European dolphinariums have had good success at keeping and breeding their whales and dolphins, and these mammals live longer than they would in the wild.

Competition for leisure time forced zoos to think about other, new kinds of exhibits during the 1960s. Many ideas, like walk-through aviaries and glass-fronted cages rather than barred cages, were implemented at this time. Many zoos were also enlarged and others created separate facilities for breeding and conservation purposes, as was done at Stuttgart and Basel. In Sweden the very large zoo, Kolmarden, was established in 1965. The primary change was to provide more space with larger groups of individual species for improved propagation and conservation. Other developments in zoos during this period were concerned with protecting animals in the wild, dealing with the difficulty of importing new animals, and satisfying a public desire to see animals in natural settings.

In the 1960s, Bernhard Grzimek, Director of the Frankfurt-on-Main Zoo, began his nature conservation work at the Serengeti in East Africa. Grzimek established his zoological society as a leader in field conservation, produced his monthly television show "Kein Platz für Wilde Tiere" ("No Place for Wild Animals"), and wrote many popular books. He raised a great deal of money for nature protection and deepened the consciousness of many people with regard to wildlife conservation. One of his most significant published works was the 13-volume *Grzimeks Tierleben* (*Grzimek's Animal Life Encyclopedia*), which was an effort to translate the popular *Brehms Thierleben* of 1864 into a modern edition. Grzimek was also successful in his zoo's captive breeding efforts, with many important species bred, such as okapis, rhinos, gerenuks, apes, and many others. In 1978, the small mammal house was named the Grzimek House in his honor.

Zoo training programs, or schools, were established during the 1960s and 1970s in Germany. The school districts, rather than the zoos, financed these schools and the teachers were from the school districts. These schools developed their own programs, but, eventually, the zoos contributed to the development of the programs.

Long-term breeding programs developed in the 1970s and 1980s as Europe's population increased, the habitat of native animals dwindled, and legislation restricted animal trading and acquisitions. Breeding programs, like that developed earlier to save the European bison, were established to save other endangered species. In 1985, the European Endangered Species Program (EEP) was established. That same year, the first EEPs were established for 17 species. Today, more than 100 EEPs and more than 30 European studbooks exist. In 1998, the European Association of Zoos and Aquaria (EAZA) was founded and now has more than 260 members.

New developments can be found in zoos all over Europe. New zoos have been established in Spain, France, Belgium, and elsewhere. Other zoos have enlarged their areas, such as at Tierpark Rheine, Artis (Amsterdam), Rotterdam Zoo, and Zoo Zurich. New masterplans will change other zoos, such as Hanover Zoo. The new Hanover masterplan will create a different zoo where the animals will roam in apparent freedom and their "edutainment" program, so-called by zoo Director Heiner Engel, will give the public an expedition into the world of nature.

New exhibits are also appearing in many zoos. Some are returning to the exotic style of former years. Loro Parque's entrance resembles a Siamese village, while the Berlin Zoo's wild cattle exhibit is in the form of a Siamese house. Cologne Zoo has erected the ruin of a monastery to provide an impressive environment for its owls. Rotterdam Zoo has copied an old Indian temple to house hornbills, Malayan tapirs, Indian rhinoceroses, Indian elephants, monkeys, rodents, and other Asian rain forest animals. Other zoos have developed ecological exhibits based on ecosystems rather than geographical areas, such as the Burgers' Zoo (Arnhem) jungle and desert halls, and its ocean world. The new Rotterdam Zoo masterplan creates a whole new zoo based on a geographical arrangement using natural habitats. Since the older existing buildings are protected, this will be done by changing their function and the landscaping around them.

Many other zoos have had less comprehensive changes, changing individual buildings or exhibits to reflect the new naturalistic style. The old bird house at Wuppertal Zoo has been enlarged and turned into a greenhouse-style free-flight aviary. New chainless elephant exhibits and facilities have been established at many zoos, including those at Vienna, Rotterdam, Münster, Hamburg, Berlin, Munich, and Wuppertal. The new hippopotamus house at the Berlin Zoo is an artificial riverbed with an island, a large grassy outdoor enclosure, and underwater viewing.

For some time the trend was to reduce the number of species and exhibit social groups of fewer species. This reduced the diversity and uniqueness of many zoos. Now this trend is changing, and even though there is still an emphasis on certain species in cooperation with other zoos and captive breeding programs, there is an effort to increase the variety of animals exhibited. In the last few years, a number of new, rare species have been brought into Europe for exhibition, such as the Szechuan takin, white-lipped deer, bearded pig, Baird's tapir, Philippine spotted deer, tufted deer, serow, and red goral, to name only a few.

Hippopotamus House, Berlin Zoo. © Photograph by Harro Strehlow.

Western European zoos are developing, each in its own way, new emphases on public education, research support, and conservation. Conservation involves not only captive breeding, but field programs as well, such as the Endangered Primate Rescue Center in Vietnam, which the zoos of Leipzig, Münster, and Frankfurt-on-Main support; the Prince-Alfred-Hirsch, which the zoos of Mulhouse and Berlin support; and Project Bonobo, which the zoo at Plackendael (a branch of the Antwerp Zoo) supports. Many zoos participate in reintroduction programs for endangered and extinct (in the wild) animals, for example, alpine ibex, European bison, Przewalski's horse, Père David's deer, addax, scimitar-horned oryx, mhorr gazelle, and bearded vulture.

With new ideas, new exhibits and a more sophisticated organization for the welfare of animals, both in the zoos and in the wild, Western European zoos and aquariums will fulfill their mission in the next century. And the public will honor these efforts.

References

1. Loisel, G., *Histoire des ménageries de l'antiquité à nos jours*, O. Doins et Fils, Paris, 1912 [*History of Menageries from Antiquity to the Present*].
2. Fitzinger, L. J., *Versuch einer Geschichte der Menagerien des österreichisch-kaiserlichen Hofes*, Kaiserlich-Königliche Hof- und Staatsdruckerei, Vienna, 1853 [*Outline of the History of Menageries of the Imperial Austrian Court*].
3. Smit, P., Ed., *Hendrik Engel's Alphabetical List of Dutch Zoological Cabinets and Menageries*, Rodopi, Amsterdam, 1986.
4. Hediger, H., *Zoologische Gärten. Gestern–Heute–Morgen*, Hallwag Verlag, Bern, 1977 [*Zoological Gardens: Yesterday, Today, Tomorrow*].
5. Historische Vereniging Holland, *Dierentuinen in Holland*, Regionaal-historische Tijdschrift 20/4–5, 1988 [*Zoological Gardens in Holland*].
6. Hoage, R. J. and Deiss, W. A., Eds., *New Worlds, New Animals. From Menagerie to Zoological Park in the Nineteenth Century*, Johns Hopkins University Press, Baltimore, 1996.
7. Kirchshofer, R., *Zoologische Gärten der Welt. Die Welt des Zoo*, Umschau Verlag, Frankfurt-on-Main, 1966 [*Zoological Gardens of the World: The World of Zoos*].
8. Kourist, W., *400 Jahre Zoo*, Rheinland-Verlag GmbH, Cologne, 1976 [*400 Years of Zoos*].
9. Lemke, K., *Tourist-Führer Tiergärten, Zoos, Aquarien, Wildgehege*, VEB Tourist Verlag, Berlin, 1985 [*A Tourist Guide to Zoos, Aquariums, and Game Farms* (in East Germany)].
10. Schlegel, F., *Die Zoologischen Gärten Europas*, Max Mälzer, Breslau, 1866 [*The Zoological Gardens of Europe*].
11. Stricker, W., *Geschichte der Menagerien und der zoologischen Gärten*, Carl Habel, Berlin, 1880 [*History of Menageries and Zoological Gardens*].
12. Willems, J., *Moderne Dierentuinen in Nederland en Belgie*, Moussault, Wormer, n. d. [1981?] [*Modern Zoological Gardens in the Netherlands and Belgium*].
13. Meyer-Holzapfel, M., *Tierpark Dählhölzli*, Verlag Paul Haupt, Bern, 1962 [*Dählhölzli Zoological Park*].
14. Sägesser, H. and Robin, K., *Das Dählhölzli im Spiegel seiner Tiere*, Stämpfli & Cie AG, Bern, 1987 [*The Dählhölzli Zoo as Mirrored by Its Animals*].
15. Fischer, I., *Tiere bei uns. 125 Jahre Karlsruher Zoo*, Ilse Fischer, Karlsruhe, 1990 [*Animals Among Us: 125 Years of Karlsruhe Zoo*].
16. Gorgas, M. and Schweinberger, W., *Tiere–Kaiser–Anekdoten. Von Fuggers Menagerie zum Großstadtzoo*, Vindelica, Gersthofen, 1986 [*Animals, Emperors, Anecdotes: From Fugger's Menagerie to Urban Zoo*].
17. Brachetka, J., *Schönbrunn und sein Tiergarten*, Brachetka, Vienna, 1947 [*Schönbrunn and Its Zoological Garden*].
18. Fiedler, W., Ed., *Tiergarten Schönbrunn — Geschichte und Aufgabe*, Verband der wissenschaftlichen Gesellschaften, Vienna, 1976 [*Schönbrunn Zoological Garden: History and Purpose*].

19. Kunze, G., *Tiergarten Schönbrunn. Bilder — Geschichte — Geschichten*, Ueberreuter, Vienna, 1993 [*Zoo Schönbrunn: Sketches — History — Stories*].
20. Pechlaner, H., *Meine Schönbrunner Tiergeschichten*, Holzhausen, Vienna, 1997 [*My Animal Stories from Schönbrunn*].
21. Beringuier, R., *Geschichte des Zoologischen Gartens in Berlin*, Alfred Weile, Berlin, 1877 [*History of the Berlin Zoological Garden*].
22. Blaszkiewitz, B., *Untersuchungen zur Entwicklung des Säugetierbestandes im Berliner Zoo für den Zeitraum vom 31.5.1945 bis zum 31.12.1979 — unter besonderer Berücksichtigung der Artenvielfalt und -repräsentanz in Zoologischen Gärten*, Dissertation, Gesamthochschule Kassel, 1987 [*Investigations of the Mammal Stock of the Berlin Zoo for the Period from May 31, 1945 until December 31, 1979 — with Special Regard to the Species Diversity and Representation in Zoological Gardens*].
23. Fintelmann, G. A., *Wegweiser auf der Pfaueninsel*, Henssel, Berlin, 1837 [*Guide to the Peacock Island*; reprinted by M. Seiler, 1986].
24. Heck, Ludw., *Heiter-ernste Lebensbeichte*, Deutscher Verlag, Berlin, 1938 [*Serene–Serious Confession of Life*].
25. Heck, L. and Matschie, P., *Andenken an den Zoologischen Garten zu Berlin*, Zoologischer Garten, Berlin, 1901 [*Reminiscences of the Zoological Garden of Berlin*].
26. Heilborn, A., *Zoo Berlin 1841–1929*, Wilhelm Raue, Berlin, 1929.
27. Klös, H.-G., *Von der Menagerie zum Tierparadies. 125 Jahre Zoo Berlin*, Haude & Spener, Berlin, 1969 [*From the Menagerie to the Animal Paradise: 125 Years of the Berlin Zoo*].
28. Klös, H.-G., Frädrich, H., and Klös, U., *Die Arche Noah an der Spree*, FAB Verlag, Berlin, 1994 [*Noah's Ark at the River Spree*].
29. Klös, H.-G. and Klös, U., Eds., *Der Berliner Zoo im Spiegel seiner Bauten 1841–1989*, Heenemann, Berlin, 1990 [*The Berlin Zoo as Mirrored in Its Architecture, 1841–1989*].
30. Schlawe, L., *Unbekannter Zoologischer Garten bei Berlin. 1844–1869*, AGT, Berlin, 1963 [*The Unknown Zoological Garden near Berlin, 1844–1969*].
31. Schlawe, L., *Die für die Zeit vom 1. August 1844 bis 31. Mai 1888 nachweisbaren Thiere im Zoologischen Garten zu Berlin*, Schlawe, Berlin, 1969 [*Animals in the Berlin Zoological Garden Known to Be There between August 1, 1844 and May 31, 1888*].
32. Stichel, W., *Die Pfaueninsel. Ein Führer durch Geschichte und Natur*, Verlag naturwissenschaftlicher Publikationen, Berlin, 1927 [*The Peacock Island: A Guide through Its History and Nature*].
33. Bernard, P., Couailhac, L., Lemaout, G., and Lemaout, E., *Le Jardin des Plantes*, L. Curmer, Paris, 1842.
34. Boitard, M. and Janin, M. J., *Le Jardin des Plantes*, Gustave, Paris, 1851.
35. Schlegel, F., *De Dierentuin van het Koninklijk Zoologisch Genootschap Natura Artis Magistra te Amsterdam*, Gebr. van Es, Amsterdam, 1872 [*The Zoological Garden of the Royal Zoological Society Natura Artis Magistra of Amsterdam*].
36. Smit, P., *ARTIS. Een Amsterdamse Dierentuin*, Rodopoi, Amsterdam, 1988 [*Artis: An Amsterdam Zoological Garden*].
37. Baetens, R., *De Roep van het Paradijs. 150 jaar Antwerpse Zoo*, Lannoo, Antwerp, 1993 [*The Call of Paradise: 150 Years of Antwerp Zoo*].
38. Cnodder, R. de, *125 Jaar Zoo Antwerpen*, Vlamse Toeristenbond, Antwerp, 1968 [*125 Years of Zoo Antwerp*].
39. Frechkop, S., *Les mammiferes du zoo national d'anvers*, Bibliothek Marabout, Verviers, 1963 [*The Mammals of the National Zoo of Antwerp*].
40. Kruyhooft, C., Ed., *Zoom op Zoo. Antwerp Zoo Focusing on Arts and Sciences*, Royal Zoological Society, Antwerp, 1985.
41. Rieke-Müller, A. and Dittrich, L., *Der Löwe Brüllt Nebenan*, Böhlau, Cologne, 1998 [*The Lion Roars Next Door*].
42. Häßlin, J. J., *Der Zoologische Garten zu Köln*, Zoologischer Garten, Cologne, 1960 [*The Cologne Zoological Garden*].

43. Häßlin, J. J. and Nogge, G., *Der Kölner Zoo*, Greven, Cologne, 1985 [*The Cologne Zoo*].
44. Brauman, A. and Demanat, M., *Le Parc Leopold 1850–1950. Le zoo, la cite scientific et la ville*, Archives d'Architecture Moderne, Brussels, 1985 [*Leopold Park 1850–1950: The Zoo, the Science Center and the Community*].
45. Bakker, J., *Rotterdamsche Diergaarde van 1857 tot 1940 en de overgang naar Blijdorp*, Ridderkerk, Rotterdam, 1985 [*Rotterdam Zoo from 1857 to 1940 and the Transition to Blijdorp*].
46. Vries, J. de, *Ir. S. van Ravenstein — Diergarde Blijdorp*, De Hef, Rotterdam, 1986 [*Ir. S. van Ravenstein: The Blijdorp Zoological Garden* (Rotterdam)].
47. Zwieten, K. van, *125 Jaar Diergaarde 1857 Mei 1982*, Stichting Koninklijke Rotterdamse Diergaarde, Rotterdam, 1982 [*125 Years of the* (Rotterdam) *Zoological Garden, 1857 to 1982*].
48. Anonymous, *Über die Gründung eines Zoologischen Gartens in Frankfurt a. M.*, Frankfort, 1857 [*Establishment of a Zoological Garden at Frankfurt-on-Main*].
49. Goering, V., *Der Zoologische Garten zu Frankfurt a.M. 1858–1908*, Franz Benjamin Auffarth, Frankfurt-on-Main, 1908 [*The Frankfurt-on-Main Zoological Garden, 1858–1908*].
50. Grzimek, B. and Backhaus, D., *Hundertjähriger Zoo in Frankfurt am Main*, Zoologischer Garten, Frankfurt, 1958 [*One-Hundred-Year-Old Zoo at Franfurt-on-Main*].
51. Scherpner, C., *Von Bürgern für Bürger — 125 Jahre Zoologischer Garten Frankfurt am Main*, Zoologischer Garten, Frankfurt, 1983 [*By Citizens for Citizens: 125 Years of the Frankfurt-on-Main Zoological Garden*].
52. Anonymous, *Der Hannover Zoo als EXPOnat. Realisierungskonzept bis zum Jahr 2000*, Zoological Garden, Hanover, 1996 [*The Hanover Zoo as EXPOnat: Concept of Realization until the Year 2000*].
53. Dathe, H., Ed., *Der Zoologische Garten Leipzig. Eine Stätte der Wissenschaft*, Akademische Verlagsgesellschaft Geest & Portig K.-G., Leipzig, 1961 [*The Leipzig Zoological Garden: A Center of Science*].
54. Dittrich, L. and Riecke-Müller, A., *Ein Garten für Menschen und Tiere. 125 Jahre Zoo Hannover*, Grütter, Hanover, 1990 [*A Garden for People and Animals: 125 Years of the Hanover Zoo*].
55. Gebbing, J., *50 Jahre Leipziger Zoo*, Zoologischer Garten, Leipzig, 1928 [*50 Years of the Leipzig Zoo*].
56. Haarhaus, J. R., *Unter Kunden, Komödianten und wilden Tieren. Lebenserinnerungen von Robert Thomas, Wärter im Zoologischen Garten Leipzig*, Fr. Wilh. Grimm, Leipzig, 1905 [*Clients, Comedians, and Wild Animals: Reminiscences of Robert Thomas, Keeper at the Zoological Garden Leipzig*].
57. Knauer, F., *Der Zoologische Garten*, Deutsche Naturwissenschaftliche Gesellschaft, Leipzig, 1914 [*The Zoological Garden*].
58. Knottnerus-Meyer, T., *Tiere im Zoo*, Werner Klinkhardt, Leipzig, 1925 [*Animals in the Zoo*].
59. Mundhenke, H., Ed., *1865–1965. Hundert Jahre Zoo Hannover*, Hannoversche Geschichtsblätter N. F. 19, 1–306, 1965 [*1865–1965: One Hundred Years of the Hanover Zoo*].
60. Schneider, K. M., Ed., *Vom Leipziger Zoo. Aus der Entwicklung einer Volksbildungsstätte*, Akademische Verlagsgesellschaft Geest & Portig K.-G., Leipzig, 1953 [*On the Leipzig Zoo: The History of a Center of Public Education*].
61. Seifert, S., Krische, G., and Puschmann, W., *90 Jahre Leipziger Zoo 1878–1968*, Zoologischer Garten, Leipzig, 1968 [*90 Years of the Leipzig Zoo, 1878–1968*].
62. Maeder, H. and Bächler, H., *Der Sankt Galler Wildpark Peter und Paul*, Verlagsgemeinschaft St. Gallen, St. Gallen, 1974 [*The Peter and Paul Game Park at St. Gallen*].
63. Brehm, R., *Bilder und Skizzen aus der Thierwelt im zoologischen Garten zu Hamburg*, H. Krumbhaar, Liegnitz, 1865 [*Pictures and Sketches from the Zoological Garden at Hamburg*].
64. Cereja, P.-L., Fischbach, B., Lernould, J.-M., Reduron, J.-P., and Thouvenin, C., *Mulhouse Parc Zoologique & Botanique*, Alsatia, Mulhouse, 1991 [*The Botanical and Zoological Garden of Mulhouse*].
65. Geigy, R., Lang, E. M., Wackernagel, H., Studer, P., and Brägger, K., *100 Jahre Zoologischer Garten Basel 1874–1974*, Helbing & Lichtenhahn, Basel, 1974 [*100 Years of the Basel Zoological Garden, 1874–1974*].
66. Jørgensen, B., *Zoo — en historie om dyr og mennesker gennem 125 ar*, Rhodos, Copenhagen, 1984 [*Zoo: A 125-Year History of Animals and People*].

67. Kourist, W., *Aus dem Tierbestand des Zoologischen Gartens Hamburg*, Selbstverlag, Berlin, 1969 [*From the Animal Stock of the Zoological Garden of Hamburg*].
68. Lang, E. M., *Mit Tieren unterwegs. Aus dem Reisetagebuch eines Zoodirektors*, Buchvertrieb Basler Zeitung, Basel, 1995 [*With Animals en Route: The Travel-Diary of a Zoo Director*].
69. Poulsen, H. and Parbst, E., *Dyrene og os*, København Zoo, Copenhagen, n. d. [*The World of Animals and Us*].
70. Staehelin, B., *Völkerschauen im Zoologischen Garten Basel 1879–1935*, Basler Afrika-Bibliographien, Basel, 1993 [*Ethnographic Exhibitions at the Basel Zoological Garden 1879–1935*].
71. Stettner, F., *Der Aachener Tierpark*, Meyer & Meyer, Aachen, 1990 [*The Zoological Garden of Aachen*].
72. Martin, P. L., *Die Praxis der Naturgeschichte*, Bernhard Friedrich, Weimar, 1878 [*The Practice of Natural History*].
73. Dittrich, L. and Rieke-Müller, A., *Carl Hagenbeck (1844–1913)*, Peter Lang, Frankfurt-on-Main, 1998.
74. Gretzschel, M. and Pelc, O., *Hagenbeck: Tiere, Menschen, Illusionen*, Hamburger Abendblatt (Hamburg Museum), Hamburg, 1998 [*Hagenbeck: Animals, People, Illusions*].
75. Hagenbeck, C., *Von Tieren und Menschen*, Vita, Berlin, 1909 [*Beasts and Men*].
76. Niemeyer, G. H. W., *Hagenbeck*, Hans Christians Verlag, Hamburg, 1972.
77. Rothfels, N. T., *Bring 'em Back Alive: Carl Hagenbeck and Exotic Animal and People Trades in Germany*, Bell & Howlett, Ann Arbor, MI, 1994.
78. Thode-Arora, H., *Für fünfzig Pfennig um die Welt. Die Hagenbeckschen Völkerschauen*, Campus, Frankfurt, 1989 [*For 50 Cents Around the World: The Ethnographic Exhibitions of Hagenbeck*].
79. Wiese, E., *Das Hagenbeck-Buch*, Historika-Verlag, Hamburg, 1995 [*The Hagenbeck Book*].
80. Zukowsky, L., *Carl Hagenbecks Reich*, Wegweiser-Verlag, Berlin, 1929 [*The Reign of Carl Hagenbeck*].
81. Baumgärtner, G. A., *Denkschrift zum Wieder-Aufbau des Münchener Tierparks Hellabrunn*, Tierpark-Ausschuß, Munich, 1927 [*Memorandum for the Reconstruction of the Munich Animal Park in Hellabrunn*].
82. Hirsch, F. and Wiesner, H., *75 Jahre Münchener Tierpark Hellabrunn*, Tierpark Hellabrunn AG, Munich, 1986 [*75 Years of the Hellabrunn Zoological Park of Munich*].
83. Hirsch, F. and Wünschmann, A., Eds., *Hellabrunn — Gestern und Heute ein Tierpark für Morgen*, Tierpark Hellabrunn, Munich, 1979 [*Hellabrunn: Yesterday and Today — A Zoological Park for Tomorrow*].
84. Roth, H., *Die Errichtung eines Zoologischen Gartens in München*, Verein Zoologischer Garten, Munich, 1907 [*The Establishment of a Zoological Garden at Munich*].
85. Wiesner, H. and Hirsch, F., *Hellabrunn. Der Münchener Tierpark*, Bayerland, Dachau, 1984 [*Hellabrunn: The Munich Zoological Park*].
86. Müller, O. and Wache, C., Eds., *Unser Zoo*, Verlag Neuzeitliche Architektur, A. Kosmala, Düsseldorf, 1927 [*Our Zoo*].
87. Gerlach, R., *Mein Zoo-Buch*, Albert Müller Verlag, Zurich, 1959 [*My Zoo Book*].
88. Meyer-Holzapfel, M., *Tiere, meine täglichen Gefährten*, Benteli Verlag, Bern, 1966 [*Animals, My Daily Companions*].
89. Meyer-Holzapfel, M., *Tierpark kleine Heimat*, Benteli Verlag, Bern, 1968 [*Animal Park, the Little Homeland*].
90. Peterhans, T., *Zürcher Zoogeschichten aus einem halben Jahrhundert*, Pendo, Zürich, 1979 [*A Half Century of Zoo Stories at Zurich*].
91. Hediger, H., *Wildtiere in Gefangenschaft*, Benno Schwabe & Co. Verlag, Basel, 1942 [*Wild Animals in Captivity*].
92. Hediger, H., *Der Zoologische Garten als Asyl und Forschungsstätte*, Gute Schriften, Basel, 1948 [*The Zoological Garden, Asylum and Place of Research*].
93. Hediger, H., *Skizzen zu einer Tierspychologie im Zoo und im Zirkus*, Büchergilde Gutenberg, Zurich, 1954 [*Sketches of an Animal Psychology in Zoo and Circus*].

94. Hediger, H., *Mensch und Tier im Zoo: Tiergartenbiologie*, Albert Müller Verlag, Zurich, 1965 [*Man and Animal in the Zoo: The Biology of Zoological Gardens*].
95. Hediger, H., *Ein Leben mit Tieren im Zoo und in aller Welt*, WerdVerlag, Zurich, 1990 [*A Life with Animals in the Zoo and All over the World*].
96. Ruhe, H., *Wilde Tiere frei Haus*, Copress Verlag, Munich, 1960 [*Wild Animals Free of Freight Charges*].
97. Neugebauer, W., *Die Wilhelma*, Konrad Theiss, Stuttgart, 1993.
98. Spitzer, G., *Heinrich Dathe. Ein Leben für die Tierwelt*, Staatsbibliothek zu Berlin, Preussischer Kulturbesitz, Berlin, 1995 [*Henry Dathe: A Life for the Animals*].
99. Pechlaner, H., *Alpenzoo Innsbruck–Tirol–Austria. 25 Jahre Forschung und Naturschutz, Erholung und Bildung*, Alpenzoo, Innsbruck, 1987 [*The Alpine Zoo at Innsbruck, Tyrol, Austria: 25 Years of Science and Conservation, Recreation, and Education*].
100. Psenner, H., *Der Alpenzoo Innsbruck 1962–1972*, Alpenzoo, Innsbruck, 1972 [*The Alpine Zoo Innsbruck, 1962–1972*].
101. Psenner, H., *Der Alpenzoo — Mein Leben*, Perlinger, Wörgl, 1982 [*The Alpine Zoo: My Life*].
102. Mager, W. and Vermeer, J., *Apen in Apeldoorn: Ambassadeurs van het regenwoud*, La Rivière & Voorhoeve, Apeldoorn, 1993 [*Monkeys at Apeldoorn: Ambassadors of the Rain Forest*].
103. Anonymous, *60 Jahre jung*, Zoo Duisburg AG, Duisburg, 1994 [*60 Years Young*].
104. Gewalt, W., *50 Jahre Zoo Duisburg - 50 Jahre und noch kein bißchen leise!*, Zoologischer Garten, Duisburg, 1984 [*50 Years of the Duisburg Zoo: 50 Years and Not One Bit Quiet!*].
105. Heubach, G., *Der Duisburger Tierpark besteht 25 Jahre*, Duisburger Tierpark AG, Duisburg, 1959 [*Duisburg's Animal Park Exists 25 Years*].

Additional Sources

1. Bourez, A., *Zootiere. Freunde in der Fremde*, Regensberg, Münster, 1989 [*Zoo Animals: Friends in a Foreign Country*].
2. Heinroth, K., *Oskar Heinroth*, Wissenschaftliche Verlagsgesellschaft, Stuttgart, 1971.

4
Zoological Gardens of Central-Eastern Europe and Russia

Leszek Solski

4.1 Introduction

Europe! Everyone knows very well what it is, but not everyone is sure what is meant when speaking about the central-eastern portion of the European continent. From a modern geographical point of view it is also difficult to distinguish the parts of the Old Continent. Historically and geopolitically, and for our purpose, the shifting configuration of political borders has placed several Central and Eastern European zoos in different countries at different points in their histories. The most recent configuration is considered here, one resulting from endless changes that have taken place since the Middle Ages. The zoo and aquariums discussed here, while less well known than the zoos and aquariums of Western continental Europe, have had an equally long history during which they have made significant cultural and scientific contributions.

4.2 Poland

4.2.1 Wild Animal Keeping in Poland through the Nineteenth Century

In Poland during the Middle Ages, wild animal keeping was the privilege of royalty and wealthy noble families. Menageries were maintained for hunting and entertainment purposes. The early days of the Renaissance did not bring any significant changes. Horses and packs of hunting dogs were still kept, as were falcons, and sometimes wolves, deer, lynx, foxes, and bears. Very often these wild animals were the participants in "spectacles," fights between different kinds of animals.

Very slowly, some truly exotic animals were introduced. The Polish and Lithuanian King Wladyslaw Jagiello (1386–1434) received the very first lions to be seen in Poland. A Polish delegation, which had been part of a special mission to the Pope in Rome, brought a pair of lions back with them to Krakow Castle (Krakow was then the Polish capital) in 1406. On their way back from Rome, the delegation stopped in Florence, where — probably by special order of the King — they acquired a pair of lions from the famous Medici breeding center (as it would be called today). We do not know exactly what happened to these animals, but there is evidence that lions were kept at Krakow, almost without interruption, up to 1570.

The first Indian elephant was seen in Poland in 1569 during the wedding ceremony of Lord Chancellor Zamoyski and the daughter of King Batory's brother. The ceremony took place at Zamosc, a town that the Zamojski family built and owned, and this elephant was probably borrowed for show purposes. Unfortunately, there is no written information identifying the true owner of this animal.

A few years later, another Polish magnate, Mikolaj Radziwill, brought back many exotic animals when he returned in 1584 from his pilgrimage to the Holy Land. As noted in his diary, he brought back two leopards, two ichneumons (mongoose), a family of patas monkeys, and a few species of "sea cats" (an old Polish name for monkeys of *Cercopithecus* sp.). He also had a few Arabic gazelles, but they died during the sea voyage. And, again, we do not know what exactly happened to all these animals.

The last decades of the sixteenth century were the last years of the Jagiellon family of Polish kings, and the years during which the Polish capital was transferred from Krakow to Warsaw. The last Jagiellons were not true animal lovers and it was probably during their reigns that the old menagerie at Krakow Castle was closed. But the last two Polish kings of the Jagiellon dynasty were both connoisseurs of the arts. In the years 1548 to 1567 they ordered and bought 350 arras (wall tapestries). These arras were made in factories at Brussels and depicted paintings of the then-famous artists M. van Coxcie and W. Tons. Most of them depicted different religious subjects, but there were also arras with landscapes and animal stories. Even on arras with religious scenes (such as the Flood), many species of animals are shown. We can find not only common European animals, but also well-known representatives of the African and Asian faunas. The big surprise is that these artists also immortalized such animals as llamas, coatis, uistitis (marmosets), macaws, amazonas (South American parrots) and turkeys. These paintings therefore provide an idea of which American animals were also known in Europe at the beginning of the sixteenth century.

Although the menageries of the Polish kings were not as impressive as those of the Habsburg kings of Central Europe, it was Poland, starting in the sixteenth century, that became the main supplier of wild European animals for menageries. Throughout Western Europe such animals as wolves, lynx, wisents, bears, and even foxes became relatively dispersed and uncommon, while the vast Polish forests were still filled with these species.

Poland was the home of the last true aurochs (*Bos primigenius*), and it was the first animal species to be officially protected in Poland. Its protection began in the sixteenth century, but the king's orders came too late. In the mid-1500s there were 38 aurochs, and by 1594 there were only 24. These last aurochs were gathered together in the strictly protected Jaktorowska Forest, but unfortunately with no success. The poor breeding results and continued poaching reduced this last herd of aurochs to four specimens by 1604. In 1627, the last aurochs cow died, the species gone forever from Earth.

Another example of Poland's role as an animal supplier concerns the non-native camel. For centuries Poland was in conflict with the Turks, Mongols, and Tartars. In battles that the Polish armies won prisoners were taken along with their property. In this way Bactrian camels became very common animals in Poland during the fifteenth century. These camels were used for hard work at the king's palaces and, as far as is known, also reproduced very well. Many of the camels were sent out as special gifts to other kings and emperors all over Europe in the fifteenth to seventeenth centuries.

During the seventeenth century private hunting menageries blossomed in Central and Eastern Europe, and Polish nobles owned most of them. In 1629, the major landowner in the Poznan district, Jan Ostrorog, printed his own publication about establishing hunting menageries. Unfortunately, only pieces of this work have survived. Surely it was a useful guide for the beginner on how to keep some wild animal species safe and healthy in captivity. In those days hunting with falcons was very popular. Again, Poland supplied other areas of Europe with some very rare species of falcons, the kind that were especially valuable for hunting.

Karol Radziwill, a Polish noble, owned an important private menagerie that was established in 1758 at the small village of Alba in central Poland. He was by no means an animal lover, but rather a rich eccentric. In Alba he settled some 180 families — lesser nobles, as well as relatives and friends. Along with their barns and houses he ordered the building of special cages and enclosures for animals. There were also areas for cranes, thrushes, rabbits, and swans. In addition, Radziwill kept large animals on about 490 acres of old forest and vast meadows. Between 1758 and 1785, some 96 wisents, 120 moose, 56 reindeer, 140 roe deer, 66 fallow deer, 130 red deer, and numerous wild boar and domestic breeds of goats and sheep were kept there.

The most eccentric story about Radziwill concerns his famed collection of tamed and trained bears. These animals were trained specifically to serve as draft animals for a special kind of coach. During the visit of the last Polish king, Stanislaw August Poniatowski, to the properties of Radziwill in 1784, he went out for a drive in this coach, which was pulled by eight brown bears. This story attained international fame and a Turk sultan asked Radziwill to sell him these trained animals. Instead, Radziwill sent him 32 trained bears as gifts. In return, he received 60 pairs of Bactrian camels, which had become well-known animals throughout Poland by the end of the eighteenth century.

Poniatowski was not particularly interested in animals; yet in 1791 he imported three African ostriches from Naples, which were kept near his summer palace in Warsaw. Unfortunately, in 1795 Poland lost its independence and this small royal menagerie was destroyed when the Russian tsar's army killed the animals for food.

Toward the end of the eighteenth century, and during the nineteenth century, traveling menageries criss-crossed Poland and the annexed Polish territories. There is written evidence that this era of traveling menageries began in 1757 when the first one (Italian perhaps) came to Warsaw, and an Indian rhino was shown for the first time in Poland.

Finally, we could not come to the end of this period without a few words about the large animal park of the Polish lord J. Potocki. Potocki established his animal park in 1901 at Pilawin on Wolyn (now the Ukraine) on an area of about 17,290 acres, encircled with a nine-foot-high fence. In different types of enclosures he kept beavers, bears, seven different species of deer, and even some antelopes (such as eland and saiga). These antelope were probably acquired from the collection at Askaniya Nova. In addition to mammals, the park also contained tamed birds. The literature provides very little written information about this collection, so it is not possible to say anything about its role in the acclimatization of animals. Nor is it known what finally happened to this collection.

4.2.2 Origin of Modern Polish Zoos to 1939

Poland disappeared from European maps for almost 125 years. An independent Polish state came into existence again in 1918, just after the last volleys of World War I died away. Therefore, it should be no surprise that the main task of such a young state was to build a good economic base and provide political stability. Many important structures of state, as well as many cultural and educational institutions, did not exist during these early years of freedom. Who could think about zoos?

Poland had lost its independence in 1795 when three neighboring countries annexed its territory: Russia, Prussia, and the Habsburg Empire (the Austro-Hungarian monarchy). The reborn Poland was created from pieces of these three countries, but in a very new shape. No zoos existed in the territory Poland eventually regained from Russia and the Habsburg Empire. Only one zoo existed at Poznan, in the territory regained from Prussia. The Poznan Zoo originated in the 1870s. Technically, the early years of the Poznan Zoo should be included in a general history of German zoos, but for many different reasons this zoo is often omitted. On the other hand, the creation of the Poznan Zoo was the result of Polish efforts.

A historical problem is the date that should be regarded as the official beginning of the Poznan Zoo. In the commemorative book published for the 100th anniversary of the zoo,[1] the authors expressed the opinion that the zoo originated in 1871 from a small animal garden kept by a restaurant owner near the old Poznan railway station. The authors wrote, as well, that there is no evidence the first meeting of the Zoological Society, held on February 24, 1874, was successful. Nonetheless, the 100th anniversary ceremonies were held in 1974. All agree, however, that the name "Zoological Gardens" in connection with this small Poznan menagerie was used for the first time in 1875.[1]

The first period of prosperity for the Poznan Zoo began in 1883, when Robert Jaeckel became the director. The zoo, which until then had occupied a leased area, obtained its own property, and the total area was enlarged to 13 acres, still its current size. There were many new improvements to the facilities, not only for the animals (dens, aviaries, and ponds), but for the visitors as well (such as a restaurant). When Jaeckel died in 1907, the zoo owned about 900 animals of 400 species and had an annual attendance of more than 250,000.

The Poznan Zoo then had some bad years, especially under the direction of M. Meissner from 1907 to 1913, when the zoo stagnanted. The next director, H. Laackmann, strived to regain the zoo's good reputation. He acquired a second elephant for the zoo (an adult male Indian elephant from a circus) and probably the most curious animal in the zoo's history — an echidna. Information on this echidna is based only on a short note that appeared in the Poznan daily newspaper in 1913. Laackmann was also the author of the first guidebook to the Poznan Zoo, written in German and published in 1914, only a few months before the outbreak of the war.

World War I totally ruined the zoo and it was left with only 243 animals of 75 species. The most valuable animals that survived were a pair of Indian elephants. Polish authorities took over its operation in June 1919, but the first three years after the war were very difficult, and the zoo faced closure several times. Fortunately, the zoo survived these recession years. More surprising, at exactly the same time, in 1921, the well-known and important zoo in Breslau (Wroclaw) closed and did not reopen until 1927.

K. W. Szczerkowski became the director of Poznan Zoo in 1922 and held this position until 1940. He modernized the zoo buildings, mostly by giving them new electrical and heating systems. During 1923/24, the Poznan Zoo received its first new animals from abroad. In 1923, the International Society for the Protection of European Bison came into existence, and the Poznan Zoo immediately became a member. Within the next few years a pair of pure-blooded European bison (wisent) were on exhibit at the Poznan Zoo. During the 1920s, the zoo's collection was enriched primarily through exchange of animals with other zoos. In a few cases, such Polish travelers as A. Fiedler and F. Ossendowski donated animals from the wild to the zoo (for example, coatis, capuchin monkeys, and the first chimp in 1926).

In 1924, the zoo produced its first guidebook, which was the first such guidebook written in Polish. The same year saw the construction of a huge aviary for sea and wading birds; roughly 42 feet high, 66 feet long, and 60 feet wide, it exists today. In 1926, for the first time, there were discussions on transferring the zoo to a new location. This was again discussed in 1933, but the idea never developed into any serious projects until 1967. In the meantime, the zoo published the second edition of its guidebook and modernized most of its buildings.

The second Polish zoo was established in the capital when the members of the municipal council of Warsaw voted, almost unanimously, for a zoo in June 1927. The site they chose for the future zoo was Praga Park on the Vistula River (the eastern part of Warsaw on the right bank of the Vistula River is called Praga, which in the Polish language is the same name as the capital city of the Czech Republic). It was not the first attempt to establish a zoo in Warsaw. During the nineteenth century, as well as the first years of the twentieth century, there were many attempts to establish a zoo in Warsaw. The more serious attempts were undertaken in 1874, 1884, and 1926, but each of these menageries survived only a few years, with no chance of becoming a zoological garden. This was because they were private collections and each ended with the death of its owner.[2]

Finally, all went well. By September 1927, ground was broken and the idea of a zoological garden started to become reality. In November, Wenanty Burdzinski was appointed the first director of the Warsaw Zoo. He was the appropriate person for this position, as he had organized the zoo at Kiev in 1908, and was its director until 1923. A year later, with problems at Kiev, he came to the newly independent Poland.

Warsaw Zoo was opened for visitors on March 11, 1928. On opening day, the zoo only had a few animals, but they were important ones, such as lions, tigers, and a female Indian elephant named Kasia (Katie). Unfortunately, Burdzinski died before the end of 1928 from acute pneumonia. His successor was appointed in 1929, the young zoologist Jan Zabinski, who kept the director's position until 1939. In 1929, the first guidebook to the Warsaw Zoo was produced, the only one to be published prior to World War II.

Before World War II several buildings and enclosures were built, such as the monkey house, elephant house, separate antelope houses, a special pond for seals, and the giraffe barn. In the pre-war years over 200 species of animals were exhibited. Before the war, the Warsaw Zoo had already received international fame thanks to the successful breeding of two species. First, in 1937 the female Indian elephant Kasia gave birth to a healthy young calf. It was the first elephant born in any Polish zoo. The young female was named Tuzinka (the feminine form of Tuzin), which in Polish means "dozen"; zoo officials were convinced at the time, based on the zoo literature and information available to them, this baby was the twelfth elephant to be born at a zoo. Many years later the scientific staff learned they were wrong, and in fact Tuzinka was the 27th Indian elephant born in captivity. In the following year, 1938, the zoo recorded two litters of the African hunting dog, which two of the females reared together. This event is still cited in the zoological literature as an outstanding breeding success. Warsaw Zoo, during those pre-war years, also successfully bred such difficult animals as roe deer, moose, lynx, and Przewalski's horse.

Indian elephant born at the Warsaw Zoo in 1937, as pictured on a pre-World War II postcard. © Warsaw Zoo Archives.

The third Polish zoo was established at Krakow. The organizing committee of the Zoo Society had its first meeting in 1927, and the zoo opened to the public on July 6, 1929. The zoo was set in the heart of Las Wolski (Wolski Forest), which had (and fortunately still has) a total area of about 1,200 acres on the outskirts of the town. But the area of the zoo itself was only 10 acres. One of the small group of men who started this zoo was a young actor with a keen interest in zoology, Karol Lukaszewicz. He soon became the scientific adviser to the zoo and wrote its first guidebook.

In the beginning, Krakow Zoo exhibited mainly native birds and mammals, which were housed in a wooded site having a few small aviaries and several enclosures, stables, and barns. Because of its location in the forest, the zoo was under the strict supervision and control of the forest guards. Since these officials did not like exotic and strange creatures in the forest that might be dangerous to their beloved trees, only one exotic species was on exhibit at the zoo — porcupines, which bred there every year.

The fourth Polish zoo was established at Lodz in 1938, only one year before the start of World War II. The zoo had only enough time to collect some 50 different species of animals, mostly native fauna. There were no buildings, only rather small temporary enclosures.

To complete the pre-war history of Polish zoos, one must mention a very small zoo at Zamosc in the eastern part of the country. This very small zoo, only one acre, was set up as a school zoo in 1917 by a lycée (secondary school) professor, Stefan Miller. This zoo lived its independent life until 1980, when it closed. Two years later a new zoo at Zamosc opened to the public.

Small collections existed in many other towns of Poland during pre-war times. These could be found in the municipal parks of such cities as Katowice, Leszno, Czestochowa, Vilnius, Kaunas, and Grodno. Those in the two latter cities (then located in the Soviet Union) became true zoological gardens after World War II.

Polish zoos before the war were characterized by rather poor financial resources and animal collections. The Warsaw Zoo made the best effort to overcome these constraints, but World War II interrupted all efforts. During the war years Polish zoos stagnated and the few breeding successes were more accidental than the result of planned breeding programs. Nevertheless, Polish zoos were represented in international forums by two members during this time. K. W. Szczerkowski, of the Poznan Zoo, was one of the founders of the International Union of Directors of Zoological Gardens (IUDZG) and a very active member. The second Polish member of IUDZG was the young director of the Warsaw Zoo, Jan Zabinski.

4.2.3 Polish Zoos during and after World War II

For the Polish nation World War II started the early morning of September 1, 1939, when predominately German forces suddenly attacked. On September 17, another large army moved into Poland, this time from the Soviet Union. The next period in Poland's history began, this time with the partitioning of Poland between fascist Germany and communist Soviet Union. Polish zoological gardens found themselves under German administration.

Fascist policy was not only to destroy Polish cities, towns, villages, farms, and factories, but also — or perhaps even first of all — to destroy Polish culture and science. The Germans considered Warsaw a small provincial town, and as such they felt it should not have a zoo. The Praga Park Zoo was emptied (and would have no animals until the end of 1946). The animals that survived the September state of siege were transferred to German zoos, and even now, after half a century, there is no evidence documenting this. The information may well be buried deep in the archives in some German zoos, but likely no one has ever tried to find it or (as Strehlow points out in Chapter 3 on Western Europe) the archives were destroyed during the war.

Poznan Zoo lost about 50 animals during the first days of the war. The German administration of the town appointed a new director, Richard Müller, who previously was a director of Königsberg Zoo in East Prussia. The most difficult time for the zoo came in 1944 when the Soviet Army heavily destroyed it during the liberation. The Germans had already shot many of the dangerous animals, such as tigers, lions, bears, and the big ungulates, while others died of starvation and cold. Soviet bombs added to the animal death toll. At the Poznan Zoo, only 176 specimens survived out of the more than 1,200 that had been there during the pre-war years.

Krakow Zoo, situated outside the city, was not destroyed at all. But the Germans removed some of the animals; for example, six pure-blooded European bisons were transferred to the Berlin Zoo, where they all died in 1945 during the Allied bombings.

The small zoo in Lodz found itself in the best situation. For unknown reasons, the Germans transferred many animals from circuses to this zoo, and evacuated many animals from their own German zoos to the Lodz Zoo during the last months of the war. Under the German administration, the zoo was enlarged and some other work was done. As has been mentioned, the Lodz Zoo at the very beginning had only 50 species. At the end of the war this zoo had over 600 animals of 117 species. The most valuable animals were a pair of Indian elephants (the female lived in Lodz until 1960).

After the fall of Germany in 1945, Poland found itself with totally new borders and under a new, and unfamiliar, communist government. Its cities were ruined, factories were destroyed, and the soil was littered with shells and mines. In addition, Poland lost over six million of its citizens during the war. Zoological gardens are institutions of peace; for them, every war means destruction. And so it was this time. Two of the most important pre-war collections, those at Poznan and Warsaw, were totally destroyed. The other three — at Lodz, Krakow, and Zamosc — were little changed (these zoos had not become fully developed zoological gardens before the war).

At the Poznan Zoo the cleaning and rebuilding effort started in February 1945, before the official end of the war. First, the carcasses of animals that had been shot, such as the lions, tigers, and antelopes, were dug out and their skeletons given to zoological museums. Then, after some necessary repairs to the buildings and enclosures, the zoo opened to the first visitors in April 1945. And, once again, the Poznan Zoo was the only major zoo in Poland.

In Warsaw the old zoo grounds were refurbished and some new enclosures were built on newly acquired land. The first visitors came to the zoo in September 1948. Then, the zoo only had 300 animals, all representatives of native fauna. These animals, without exception, were given to the zoo as gifts from the citizens of Warsaw.

During these difficult postwar days, the Warsaw Zoo hired its previous director, Jan Zabinski, who remained at this position from 1947 to 1951. He retired from the zoo when he was only 54 years old, probably because of misunderstandings with members of the municipal council. Nevertheless, the rest of his life was devoted to zoos and animals. He was a professional zoologist and his knowledge about animals and zoos was substantial. When he retired from the zoo he immediately started writing articles and books, as well as doing broadcasts on animal subjects for Polish radio. He wrote many popular books for general readers, while his wife wrote entertaining stories about young animals for children. One of his first books was *Przekroj Przez Zoo* (*A Walk through the Zoo*), published in 1953. From 1964 to 1969 he published a six-volume set of his radio stories, which involved 144 different stories about animals. Jan Zabinski (1897–1974) was not only a good zoo director, but also an excellent zoo education officer.[3]

Wroclaw (formerly Breslau) was the largest city in the territories returned to Poland in 1945. In pre-war times it was one of the largest cities in all of Germany, and also the location of the leading European zoological garden. Breslauer Zoologischer Garten was opened on July 16, 1865 and its history before 1945 was a very important part of zoo history in Germany. This zoo was also important to the entire European zoo community and many professionals considered it to be one of the best animal collections in the nineteenth and early twentieth centuries.

This opinion has also been expressed by such notable authors as Peel[4] and Loisel.[5] Breslau Zoo achieved fame about 1885 when O. Anschütz used its animal collection as subject for some of the first photographs of live zoo animals.[6] Some 32 of these photographs were used in the Breslau Zoo's 25th anniversary guidebook, published in 1890, which may be the first photo-illustrated zoo guidebook.[7]

The history of the Wroclaw (Breslau) Zoo as a Polish institution starts in May 1945. Some 75% of the city of Wroclaw, including the zoo, was destroyed. All the buildings were heavily damaged. There was no heat, water, or electricity. Here again, German soldiers had shot the dangerous animals, such as the big cats, bears, and elephants. Many of the other rare animals such as the Amazon manatee, three adult chimps, and a pigmy hippo died from cold and starvation. Only about 300 animals survived, roughly one sixth of the pre-war population.

Wroclaw Zoo orangutan cage as it looked in 1956. At the time, these orangutans were the only ones on exhibit in Eastern Europe. © Photograph by S. Poradowski.

Life in Wroclaw was still unstable and sometimes dangerous. The German inhabitants were still there and so was the Soviet army. Even so, the first Polish immigrants from the eastern territories that the Soviet Union annexed started to come back into the city. There was no food for the people, so the temporary military government of Wroclaw decided to close the zoo for good so they would not have to feed the animals. At the beginning of 1946 the animals that survived were transported to zoos in Poznan and Lodz. The situation changed in 1948 when Wroclaw finally came under a Polish administration; there were organized Polish schools, academies, theaters, hospitals, and other social and cultural institutions. The most important was Wroclaw University, which was formed mostly by professors from Lvov University (Lvov was the largest Polish city in the eastern section, but after the war became part of Soviet Ukraine). Although Lvov had never had a zoo, the professors formed a group to support the rebuilding of the zoo in Wroclaw. This was not a formal group, but it was soon transformed into a legal advisory board, and this body appointed Karol Lukaszewicz (the organizer of the pre-war zoo in Krakow) as the zoo director.

Drawing of a modern zoo exhibit. Drawn by T. Zipser for the Wroclaw Zoo in the late 1950s. © Photograph by S. Poradowski.

Karol Lukaszewicz — A Zoo Man Story

Dr. Karol Lukaszewicz (1901–1973) was instrumental in the origin of Krakow Zoo and the postwar period of Wroclaw Zoo. His contributions to the development of Polish zoos and to the education of people about conservation, wildlife, and zoos were so significant that he deserves recognition, especially because his name is being lost even among Polish zoo people.

Lukaszewicz was born at Krakow in 1901. In his early life he was interested not only in animals and nature, but also in the arts, music, and theater. He received his art academy diploma in 1924 and went on to become a qualified actor. But in the very same year, instead of undertaking a well-paid job in the theater, he accepted a scholarship to study in France. It was a turning point in his life, because he chose zoology as his first love. During that time he visited scientific institutions, museums, and zoos in Paris, Dijon, Lyon, Arles, Marseille, and Nice. On his way back to Poland, he also visited the animal collections at Vienna and Berlin. Back in Poland he organized a nature exhibition at the Katowice District Museum. In 1928, the municipality of Krakow charged him with organizing a zoo in the city. After its opening in 1929 he was, for the next 10 years, its scientific adviser and wrote its first guidebook, which was printed in the early part of 1939.

Before World War II, he was also busy writing a popular monograph on the animal kingdom under the ambitious title (in English), *From Amoeba to Gorilla*. The book contained over 600 of the author's original drawings, and 10,000 copies were printed just after the outbreak of the war. Unfortunately, the Nazis confiscated and destroyed all of the copies. Lukaszewicz survived the war years in his native town of Krakow giving

continued

> ### Karol Lukaszewicz
>
> language lessons. He had studied Latin and ancient Greek in school, but he also had knowledge in speaking, writing, and reading German, English, Russian, French, Spanish, and Italian. This knowledge of languages helped him later in making private contacts with other zoo directors from all over Europe.
>
> After the war, in 1947, he was invited to reorganize and rebuild the totally destroyed Wroclaw (Breslau) Zoo. This he accomplished, starting with one rhesus monkey and an old black kite. His personal knowledge about the good pre-war zoos in different countries helped him turn the old, ruined German zoo into one of the more important zoological collections in Europe. As director of the Wroclaw Zoo, he spent a great deal of time traveling and visiting other zoos and animal dealers all over Europe, including those in Germany, Holland, Belgium, France, Denmark, and Italy. He also attempted to establish a Zoological Gardens Section within the Polish Zoological Society. He edited a special zoo bulletin, which the department of zoos of the Ministry of Communal Affairs published irregularly. He was one of the lecturers and organizers of a postgraduate course for scientific staff at Polish zoos. He was coauthor of the very first Polish handbook for zookeepers published in 1958 (perhaps the first at any zoo). One of his main ideas was to establish a special faculty for wild animal husbandry at the Agricultural Academy, but this idea still has not been realized. He wrote his doctoral dissertation in 1952 about aurochs (*Bos primigenius*), but unfortunately (very probably because of political reasons) he did not receive confirmation of this degree from Wroclaw University until 1968. He retired from Wroclaw Zoo in the same year and immediately started to write his book about the world's zoos. He died suddenly of a heart attack in July 1973.
>
> Altogether, since the war, he wrote more than 100 articles and books, some for *International Zoo News* and *Zoo Revue* (Antwerp). One of his most important contributions is his

Rebuilding the Wroclaw Zoo was an uphill task, not only for the new director, but for the entire inexperienced staff. Still, the zoo opened (for the third time in German/Polish history) for visitors in 1948, with a fairly good collection. Most of the animals were gifts and were representatives of native fauna. Lukaszewicz was very fortunate, mostly because of his own respected authority, to get back from the zoos in Poznan and Lodz almost all of the original Wroclaw animals that had been sent to these zoos in 1946. Wroclaw Zoo therefore had such animals as tapirs, hippopotamuses, mandrills, several species of antelopes, and many rare birds.

However, the status of zoos in Poland had changed because of new legislation by the socialist government. Zoos became national institutions under the direct management of their municipal governments. They were put into a department for zoological gardens within the Ministry of Public (Communal) Affairs, together with other municipal parks, cemeteries, sanitation facilities, etc. This organization was typical of central planning in a socialist society. The zoos still had some degree of independence, even though their budgets were allocated centrally in Warsaw. Sometimes, however, the amount of money given to each zoo depended upon the private connections their directors had with the clerks at the ministerial offices.

In general, the times were very difficult politically and economically under Stalin's dictatorship. But in a society that survived six dark years of war cruelty, the demand to have its own zoos was very strong. At the beginning of the 1950s there were still only six zoos in Poland: those at Warsaw, Lodz, Poznan, Wroclaw, Krakow, and Zamosc. With central planning it was very difficult to make plans for new zoos. Therefore, new collections in Poland were started from the initiative of small local societies within a given area or city. In such a way, four new zoos were established in Poland during the 1950s: at Plock (1951), Opole (1953), Gdansk first book, *Zwierzeta wytepione* (*Extinct Animals*) published in 1958. In this book he described in detail 10 different species of extinct animals, along with short notes about another 20. Many of his articles straddled the border between nature and the arts. His most comprehensive works described the world of animals on the arras tapestries.

His main work, *Ogrody Zoologiczne-Wczoraj-Dzis-Jutro* (*Zoological Gardens — Yesterday, Today, Tomorrow*),* was published two years after his death. In this book he describes the history and development of zoos from earliest times, covering in more or less detail over 150 zoos from around the world, including chapters on the social and cultural role of the zoo in modern societies. He also covers the future of zoos and their role in a global conservation strategy. Bearing this breadth of coverage in mind, and the fact that this book was written in the early 1970s, it is an amazing accomplishment. However, it did not receive worldwide recognition, probably because it was published only in Polish.

Lukaszewicz was convinced that the zoo environment is something different from the two concepts associated with zoos — freedom and captivity. He agreed that zoos could not be regarded as a substitute for freedom, but he did not like the word *captivity*. He wrote, "One thing is for sure — as long as we are looking for elusive gold, freedom in nature itself, and perceive the situation of animals in zoos as inexplicable captivity, all of the discussions on these two subjects should be treated as some kind of argle-bargle [misunderstanding]." Lukaszewicz saw the zoo environment as a third state. "It is not freedom, nor is it captivity," he wrote. His proposal was to use a new word, *Vivariology*, a word formed from the Latin *vivarius* and *vivarium*, which were used in ancient Rome. His idea is still fresh and by making it better known perhaps it will become more popular and and widely used.

* Wiedza Powszechna, Warsaw, 1975.

Wroclaw Zoo's staff at an exotic animal anatomy lesson at the city's university during the late 1950s. Dr. Karol Lukaszewicz, then director of Wroclaw Zoo, is in the first row, fourth from the right. © Photograph by S. Poradowski.

The Flood and International Zoo Solidarity

It is not only war that is a great threat to zoos. Another serious danger is a natural disaster, such as a flood. In July 1997, the biggest flood of the twentieth century inundated vast areas of southern Poland. Many cities, towns, and villages, including their schools, theaters, hospitals, and homes, were flooded. Among the institutions that suffered from the high water were two zoological gardens.

In Opole the zoo is located on a flat, small island in the Odra River. For several days and nights the zoo staff and many volunteers worked to evacuate animals. Many of these animals were transferred to zoos at Katowice, Lodz, and Poznan. However, the main surge, a 15-foot wave, came through the town 12 hours earlier than expected. The entire zoo was flooded and many animals, some still in transport cages, were lost, for example, hippos, antelope, goats — altogether over 40 large hoofed mammals. Most of the buildings and enclosures were destroyed or heavily damaged. The Opole Zoo nearly came to a tragic end.

In Wroclaw, the surge of water affected its zoo as well. The grounds of the Wroclaw Zoo are such that the river's high water and strong current found a new bed just under the wall of the oldest part of the zoo. This situation threatened the zoo wall and a nearby bridge into town for three days and nights. Because of the historic buildings and the many large, old, valuable trees in this part of the zoo, it was not possible to use heavy equipment. So military helicopters had to dump much of the material needed to bolster the wall. Some animals near the threatened part of the zoo were moved. In the end, a large portion of the zoo wall was destroyed, as were hundred-year-old oaks. Several animals were lost due to the panic and stress created by the work and the helicopters. The most tragic loss was three snow leopards, which died a few days after the work.

Help for both zoos came just a few weeks after the flood. Many zoos from other European countries, mainly Germany and Holland, offered help, food for the animals, and money for needed repairs. In Opole, the town council decided the zoo would be rebuilt in the same location. And thanks to international contributions, work on both zoos began as soon as the flood waters receded to their normal level. A small part of Opole Zoo opened to the public in June 1998, but there is still much to be done. And it will be done.

(1954), and Katowice (1958). Each of these zoos was established with very limited funds, collected by local citizens. Members of the small local societies worked voluntarily on afternoons and weekends to build their zoos. After the initial work had been done, some additional help came from local municipal treasuries.

Plock Zoo, opened in 1951, is located on the high bank of the Vistula River and is still not completed. The original area was only 17 acres, which has now been doubled. The uniqueness of the zoo lies in its successful breeding of South American animals. The emphasis on South American animals was due to the zoo director, Tadeusz Taworski, who was born in Argentina and had very good connections there. In the early 1960s he came to Poland with the idea of collecting books for the Polish communities in Argentina and Brazil. This activity was very successful and thousands of books were sent. In return, Plock Zoo received many interesting

South American animals, mostly reptiles. Even today, the Plock Zoo has the second largest reptile collection in Poland, with very good breeding results.[8]

In Opole, a small zoo of only six acres, opened in 1953 on an island in the Odra River. This place was also the site of a small German zoo between 1936 and 1944. Although Opole Zoo still has no significant breeding achievements, it was substantially enlarged in 1980 and there are now very spacious enclosures for ungulates.

Gdansk Zoo opened in 1954. It is the largest zoo in Poland, encompassing 336 acres. Set in a beautiful mixed forest with hills, small valleys, and streams, Gdansk Zoo is also the most charming zoo in Poland. In the beginning, the small town of Oliwa owned this zoo, but the zoo is now an administrative part of Gdansk. For this reason, it is still often called the Oliwa Zoo. This zoo has not achieved its potential since, from its very early days, the zoo has always been in some kind of financial trouble. There were also many unexpected changes in the zoo directorship, which did not permit directors enough time to put their ideas for improvements into practice. In addition, what would normally be an advantage — the city's seaport — was not beneficial to the zoo. Since Poland had very strict veterinary regulations, the importation of exotic animals through the seaport was a difficult and time-consuming task. Therefore, many private individuals and animal dealers sold their animals in Hamburg and other European cities, rather than in Gdansk.[9]

Katowice Zoo, established in 1958, was the last zoo established in the 1950s. Katowice is the capital city of the Silesia region in central-southern Poland, which is known for its coal mines and steelworks. This zoo is also known as the Silesian Zoological Park. It is part of a very large (about 1,480 acres) District Park used for leisure and recreation. As the park is located between the cities of Katowice and Chorzow, this zoo is also called the Chorzow Zoo. For a long time this zoo was the only one in the country to exhibit African elephants and white rhinos. Unfortunately, after the zoo opened there were no funds for further development and the zoo has spent many years in stagnation.

As mentioned earlier, there were many attempts to build a new zoo in Poznan. These efforts began to succeed in 1967 when an independent public committee for a new zoo was formed. The site for the new zoo in Poznan was chosen in a municipal forest in the eastern part of the city, about six miles from Poznan's center and in close proximity to the main road connecting Berlin, Poznan, and Warsaw. And again, as in all previously described cases, much of the work on the new zoo was done by volunteers. However, this time it was done with a lot of support from local factories and with a subsidy from the municipal government. The new zoo, on 280 acres, opened in September 1974. The plans were (and still are) to build it gradually into a very modern zoo and to close the old one. Reality has been different. Because of a lack of funds, there has not been any progress in building new enclosures and winter buildings. There are also problems with keeping the buildings at the old zoo in good condition, which are now regarded as monuments of nineteenth-century architecture that should not be destroyed or closed. Because of these circumstances, Poznan still has two zoos and a very big problem to solve. Attendance at the old zoo is four times greater than at the new zoo. This difference in popularity is primarily a result of the lack of good public transportation to the new zoo. With such a situation, no one can decide what to do, or which — if either — zoo should be closed. One can hope both will survive.

Volunteer workers also built another zoo at Bydgoszcz (in central Poland). It is a relatively small zoo, and the only one in Poland totally devoted to European fauna. This zoo was opened in 1978 and is part of the very large Forest Park of Leisure and Recreation. Gradually, some exotic animals have been introduced, which may help to increase attendance, but this is contrary to the previous policy of keeping only European fauna. The last zoo in Poland was opened in 1982 at Zamosc, and is the only one in the southeastern region.[10]

4.2.4 General Characteristics and Comments on Polish Zoos

The history of Polish zoos has affected the development of programs in recreation, education, scientific investigation, and conservation.[11] Recreation is important; however, total annual attendance at Polish zoos is not very high compared with that of other countries. For example, in the former Czechloslovakia attendance has reached more than one fourth the total population, while in the former East Germany it has reached one third of the country's population. In Poland the total number of visitors is well under 5,000,000, which makes this attendance only one seventh to one eighth of the total population. Zoos in Poland do not attract as many visitors as do zoos in some of the other Eastern European countries.

To have good attendance, it is not enough to exhibit rare and popular animals; a zoo should have appropriate facilities for the public. From this point of view the situation in Polish zoos could be regarded as unsatisfactory. Restaurants, bars, fast-food services, playgrounds for children, souvenir shops, and restrooms are lacking in Polish zoos. However, this situation is getting better and new improvements are being made at the zoos, which should help increase the number of visitors.

Education is also important; however, the labeling system in Polish zoos is often poor and rather old-fashioned. The labels are mostly metal or wood, easy to destroy, and very often are not readable. Only a few zoos (Poznan, Warsaw, and Wroclaw) use properly designed labels with graphics and sufficient information about the given animal or group of animals.

The situation is even worse with regard to printed materials. In 1998, the Warsaw Zoo published its new guidebook — the very first guide to any Polish zoo published during the past 20 years. In recent years, 1989 to 1990, only a few zoos (Plock, Poznan, Wroclaw, Opole, and Lodz) have published small colorful leaflets with zoo maps. Annual reports and animal inventories are published irregularly at the Poznan, Wroclaw, and Lodz zoos. Staff of the Lodz Zoo publish an informative book about all of the Polish zoos and their animals once a year.

Poland's first zoo-school, Dom Mlodego Biologa (House of the Young Biologist), existed at the Wroclaw Zoo between 1959 and 1967. This education center, located in the old section of the zoo, had its own building with classrooms, aquariums, terrariums, and greenhouses. Children attending the school were divided by age into classes for lessons appropriate to their age groups. Lessons were concerned with practical biology, such as the use of microscopes, how to keep a home aquarium, how to transplant plants, and even how to milk venomous snakes. All the children had free access to the zoo and volunteered to help do simple zoo-keeping chores. This school, for whatever reason, closed, but now a similar school will open at the Poznan Zoo. The post of education officer only exists at two Polish zoos: Poznan and Warsaw.

Publicity about Polish zoos is poor and there is no zoo-oriented magazine published in Poland. The only magazine that publishes scientific zoo articles is the *Przeglad Zoologiczny* (*Zoological Review*), which the Polish Zoological Society publishes quarterly. Brief news items about, or from, the zoos are printed occasionally in local daily newspapers. In the other mass-media publications, zoo information is even rarer. On the radio, the late Jan Zabinski once narrated short essays, but that was in the 1960s. Since then, there have been no programs on zoo subjects.

Polish television (which remains a function of the national government) has shown only three different series related to zoos. One was a German production called "Zoos of the World — The World of Zoos," which was produced in the late 1960s. Recently, Polish television aired the well-known J. Cherfas production, "Zoo 2000," and the American series "Zoo Life" with Jack Hanna. For over 20 years there has been a program entitled "With a Camera among the Animals," which began in 1969. Its creators are Antoni Gucwinski, director of Wroclaw Zoo, and his wife Hanna Gucwinski. In September 1991, the Gucwinskis celebrated the release of their 500th program. "With a Camera among the Animals" has become a regular program on television each Saturday evening. It is now only 15 minutes long, although the show was 30 minutes when it started.

"With a Camera among the Animals" has provided very good information about zoological gardens in general, but emphasizes the Wroclaw Zoo. This show's programs have featured many different zoos from all over the world, such as at Berlin, Prague, Amsterdam, Tokyo, São Paulo, as well as throughout Canada and the United States. Unfortunately, during its many years of existence it has shown only one program about another Polish zoo. Such a situation had, and still has, measurable benefits for the Wroclaw Zoo. Whenever someone finds an injured or orphaned animal (such as a roe deer, storks, hedgehogs, rabbits, or sometimes otters), even if it is found far from Wroclaw, the animal is brought to the Wroclaw Zoo. The program has also helped increase attendance at the zoo. In January 1999, the president of Poland honored Dr. and Mrs. Gucwinski with the Cross Medal, the country's highest honor for civil service, for their work on behalf of animals, nature education, and nature conservation.

Dr. Antoni Gucwinski (center) and his wife Hanna Gucwinski (left) with their friend Mr. Szymon Poradowski (right) in 1976. The popular Polish educators were honored with Poland's highest award for their work on behalf of animals, nature education, and nature conservation. © Photograph by S. Poradowski.

Last, the role of science and conservation in Polish zoos should be discussed. Scientific investigations are conducted at all the Polish zoos. Each zoo works in close collaboration with local scientific institutions. This contact includes cooperation with universities and academies in many fields of study: zoology, animal diseases and zoonoses, anatomy and physiology, and reproduction. Many students have written their master theses or doctoral dissertations on animals at the zoo, or have used the data and materials collected at the zoo from its dead animals.[12]

Polish zoos twice hosted international symposia on zoo and wild animal diseases, at Wroclaw in 1967 and at Poznan in 1976. There is also a branch within the Polish Veterinary Scientific Society that deals specifically with zoo animal medicine. Each Polish zoo has its own veterinarian. However, since veterinary services are now privatized, some of these services undoubtedly will change.

Scientific staff at the Poznan and Warsaw Zoos were involved in the protection of the wisent (European bison), including the publication of the first studbook for a wild animal. In 1923, the International Society for the Protection of Wisent held its first meeting and a few months later the Polish branch of the society was established. One outcome of the effort to save the wisent was the studbook, which von Groeben of the Frankfurt-on-Main Zoo's scientific staff and Erna Mohr compiled initially. Their first three editions were published between 1932 and 1937, and covered 270 animals starting with the male, named Planet, born in 1899. In 1946, Jan Zabinski of Warsaw Zoo took over the studbook, and the first Polish edition was published in 1947.[13] The Department of National Parks at the Ministry of Forests and Agriculture has since published this studbook independently of the zoos. Now the future of European bison is safe and the future need for the studbook is doubtful. This species has become so abundant in zoological gardens around the world that it is difficult to get all the data required for the studbook. The second studbook kept in Poland was established in 1991 at the Poznan Zoo for the pygmy lori (*Nycticebus pygmaeus*). Poznan Zoo was chosen because of its good breeding results with this species.

Poland signed the Convention on International Trade in Endangered Species of Wild Fauna and Flora (CITES) in 1990. The full text of this convention was published in a government bulletin in April 1991, and it came into force in Polish territories six months later. In October 1991, the government also published new regulations on the protection of nature and animals. These replaced the previous regulations, which had existed in Poland since 1929. The new regulations are intended to control importations and exportations, and to promote cooperation among zoos. However, this cooperation has not been good, so there has not been an effort to establish a needed conservation center for native animals. At each meeting of the Polish zoo directors there have been proposals to set up such a center for otter, sea eagles, and other native animals, but none of these projects has ever been realized. Endangered native animals are still dispersed among the 11 Polish zoos, with no chance for any significant breeding success. Perhaps the new government regulations will encourage the establishment of this center.

4.3 Russia

4.3.1 Wild Animal Keeping in Russia through the Nineteenth Century

The history of wild animal keeping in Russia dates to the eleventh century. There is evidence that the state first kept wild animals beginning in 1061 at what was then known as the Great Duchy of Novgorod. Hunting with cheetahs and falcons was popular among the great princes of the Russian states and their noble class (boyars). In Eastern Europe the custom of hunting with cheetahs and falcons arose as a result of direct contact with the lands of central Asia that the Mongols, Turks, Persians, and Tartars inhabited. Among these peoples this hunting style was very popular because, on the vast, flat, and endless steppe, it was the most effective way to hunt.

Keeping and training cheetahs, as well as different species of falcons, was done in special areas of the princes' palaces and the boyars' manors. In most cases, these areas bore the name of falconries. Gradually, these places began to be designed, not only for falcons and cheetahs, but for keeping hunting dogs and the hunters' horses. In Russian literature these larger places were called "hunting manors" and were found in close proximity to the princes' and boyars' residences. As time went by, wild animals were introduced to the hunting manors, mostly young animals that were captured during the hunts. And very soon, the most popular wild animal to be kept in the hunting manor was the brown bear.

In Russia, since earliest times, the bear has been the symbol of strength and courage. Tsar Ivan Groznoj (Ivan the Terrible), who reigned from 1530 to 1584, erected the first well-known bear pits and palisades in Moscow near the tsar's palace (very close to the present Kremlin). He liked bears, and the animals sometimes played an important role in his internal politics. For example, once the Novgorod bishop was unfortunate enough to fall into disgrace with Ivan. The tsar ordered him to prison and after a couple of days had him sewn into a bear skin. In this guise, the poor bishop was torn to pieces by a pack of hunting dogs that were used in bear hunts. Similarly, seven monks who once opposed Tsar Ivan were captured and imprisoned, but in this case Tsar Ivan was generous. Each monk was given a long wooden dart to defend himself against a hungry and angry bear. Of course, their chances were poor and very quickly all the monks, one by one, were killed by the bear, to the applause of spectators.

Ivan prepared different kinds of spectacles for noble guests from neighboring countries and for diplomatic envoys serving other kings and emperors. These were fights between bears and peasants armed only with wooden hay forks. It sometimes happened that a peasant would kill the bear and would become relatively rich because the tsar, or one of the noble spectators, would give the victorious peasant a wealth of prizes. This kind of activity changed the hunting manors into "entertainment manors" featuring amusements for the tsar and his boyars.

In 1556, Ivan received a special present from the royal English family — the first lions (a female, a male, and a cub) to be seen in Russia. The next pair of lions, also from the royal English family, was donated to Tsar Fiedor in 1586. And around 1625/26, the Persian shah donated the first elephants to the tsar's collection. These elephants were the first exotic animals to be kept at the tsar's entertainment manor.

During the time of Tsar Aleksei Michailovitch (in the seventeenth century), the largest and most famous menagerie in Russia was established. It was located in a small village east of Moscow called Izmailovo where the tsar had maintained a summer mansion since the early years of the seventeenth century. In 1663, by order of the tsar, it became a center for horticulture, beekeeping, and cattle breeding. After this came the introduction of wild native species, and later some exotic species. By the turn of the century there were animals such as bears, moose, wild boars, deer, lynx, foxes, lions, tigers, snow leopards, a valuable colony of sable, swans, and many species of pheasants and ducks. In 1664, Aleksei Michailovitch received his first polar bears, and this pair was probably kept at his entertainment manor outside of Moscow. In 1714, the Izmailovo collection received its first Indian elephant, which again was a present from the Persian shah to the tsar. Unfortunately, the royal family lost interest in the Izmailovo collection in the second half of the eighteenth century. The total destruction and final end of this large zoological collection came in 1812 when Napoleon's army surrounded Moscow.

By the eighteenth century there were animal collections in northern Russia in the proximity of St. Petersburg. The entertainment manors of Letnij Sad and Carskoje Siolo were typical mixed collections of domestic and wild native animals. Not typical, yet very famous, was the "elephant manor" of St. Petersburg. The exact date is not known, but in the 1780s there were special barns built for elephants at the manor, including heating systems for the very cold winters. At times, as many as 15 Indian elephants were kept for the tsar's amusement. During the short, hot summers, the elephants were taken to the Neva River. The route used was first known as Elephant Road, then as Elephant Prospect, but since Soviet times only as Suvorov Prospekt. Who knows, now that Leningrad is again St. Petersburg, Suvorov Prospekt may once again be known as Elephant Road.

4.3.2 Origin of Modern Zoos in Russia to 1917

The first true menageries were established for entertainment purposes in Moscow during the second half of the nineteenth century. The very first, started in 1855, belonged to two Frenchmen, whose names are not known. The Kreuzberg family owned the next private menagerie, which first opened to the public in 1862. These menageries featured animals such as lions, parrots, tigers, a few monkeys, and constrictors. Later these collections were to enrich a new zoological garden, the Moscow Zoo, by forming the nucleus of its collection.

In the 1850s, professors at the Moscow University established the first Society of Acclimatization. The idea to establish a zoological garden in Moscow came in 1857. In 1862, the society was reorganized and given the name Tsar's Society for the Acclimatization of Plants and Animals. In 1863, this society was eventually able to buy property for its future zoo, where it still exists. The very first visitors came to the zoo unofficially in the spring of 1863; however, the opening ceremony took place on February 12, 1864. On opening day, the Moscow Zoological Gardens exhibited 134 specimens of domestic animals and 153 specimens of wild exotic animals, including seven reptile specimens. In those days there were no elephants, hippos, or monkeys, but visitors could see two tigers, two lions, a jaguar, a leopard, and a rhino.[14–16]

A second Russian zoological garden was opened only one year later. Organized in St. Petersburg, it was a private enterprise of Sophia and Jules Gebhardt. The zoo opened to the public at its present site on August 1, 1865. From the very beginning this zoo had very few animals, but it was proud of its theater and two restaurants. The zoo was close to the Petropavlosk Citadel, and therefore the officers of the tsar's army visited in great numbers. Their main interests were to drink and dance, rather than to look at the animals, and the zoo's facilities accommodated these activities. The first period of prosperity for the St. Petersburg Zoo came under the direction of E. A. Rost (1873–1897). At the end of his directorship the number of animals exhibited at the zoo reached 1,100 specimens, including for the first time in Russia giraffes, orangutans, and African elephants. Director Rost also produced several ethnographic exhibits, mostly with central Asian tribes, to demonstrate the role of animals in the daily life of the given tribe. After the death of Rost, the zoo went into many years of stagnancy and decline, and faced closure several times. The new director, S. N. Novikov, started to rebuild the zoo in 1910, acquiring new animals with the idea of restoring the zoo's good reputation. But World War I and then the Revolution interrupted this effort.

Neither the St. Petersburg Zoo nor the Moscow Zoo developed successfully. The Moscow Acclimatization Society was in constant financial trouble, and the zoo, with a large debt, went into private hands for the first time in 1874. After four years, in 1878, the Moscow Zoo found itself again under the direction of the society, which successfully managed it until 1903. And once more, this time with an even larger debt, the zoo passed into private hands. Later, during the December Revolution of 1905, the zoo grounds were demolished. Afterward, the Society took over the zoo for the third time and began revitalizing it yet again. In 1909, a new aquarium building was erected and in 1913 new water and sewage systems were completed. During all of these difficult years, from 1878 to 1912, the zoo grounds were used as the traditional site for many different expositions, including those dealing with hunting, new breeds of domestic animals, ornamental birds, houseplants, and others.

Any discussion of Russian, and later Soviet, zoo history would not be complete without at least a few words about Askaniya Nova. One might question whether Askaniya Nova was really a true zoo. But even if it was not, the role of this collection in the history of keeping wild animals in captivity and its influence on other zoos are so important that it merits some attention.[17]

The area of Askaniya Nova is an untouched steppe, spreading along the northern shore of the Black Sea, with the region's typical continental climate of strong winters and hot, dry summers. This property, with its harsh environment, was one the tsar government donated to foreign newcomers in the nineteenth century. Many of the new landowners in this region were of German origin. In 1828, a German Duke from Anhalt-Köthen bought the area of the present reserve and zoological park. In 1841, he built a farm there and gave it a name — Askaniya

Nova (New Ascania) — reminiscent of Ascania, his German homeland, which is itself derived from ancient Greek mythology. The farm belonged to his family for the next 25 years and an effort was made to breed merino sheep and Arabian horses.

The next owner of this estate was also German, Friedrich Falz-Fein. His keen interest in zoology, botany, and traveling started the zoological and botanical gardens at Askaniya Nova. The botanical gardens were planned in 1885 on an area of 170 acres, including a 67-acre arboretum park. In a very short time more than 220 kinds of trees and shrubs were planted. Many different kinds of animals were slowly introduced onto the estate, such as marmots, saiga antelopes, Przewalski's horses, pheasants, ostrich, nilgai, yak, eland, wisent, and deer. The status of animals in 1887 was 45 species of birds and 4 species of mammals; in 1894, there were 78 species of birds and 13 species of mammals.

The first work with domestication and the hybridization of some wild species (for example, bison and eland) started in 1904. In 1910, the first scientific institute for animal physiology was established as a branch of the veterinary department of the Ministry of Agriculture. Animals bred at Askaniya Nova were exhibited at several All Russian Exhibition of Domestic Animals (1908, 1910, 1911), where they won many first prizes and gold medals. In this way Askaniya Nova became famous even before World War I. In 1917, because of the Great October Socialist Revolution, the owner of Askaniya Nova, Friedrich von Falz-Fein, escaped with his family back to Germany.

The three institutions already mentioned, in Moscow, St. Petersburg, and Askaniya Nova, were the leading zoological collections in Russia before 1917. But there were three other zoos and one aquarium in the Russia of the tsars. The history of the Charkov Zoo dates back to 1895. This zoo was started as a small bee-keeping station and museum, then on an area of seven acres. The next was organized at Nikolaev in the center of town at a small corner of the municipal park and was opened in 1901. The last one in Russia was opened at Kiev in 1908, with only a few cages in the University Botanical Gardens. This zoo was then transferred to a new location in 1913 with only enough time to erect several wooden barns and some fenced enclosures. The zoo director, from the zoo's origin to 1928, was W. Burdzinski, a Pole, who later established a zoo at the Polish capital, Warsaw.

Last, the only aquarium should be mentioned. The Marine Science Institute was organized at Sevastopol in 1871, with a building erected in the typical neoclassical style (very similar to the marine station buildings at Naples, Varna, and Monaco). The public aquarium itself opened in 1897, and except for a few changes, it looks the same today. It has 12 wall tanks, and one round tank in the center of the hall. Only marine fauna representative of the Black Sea has been exhibited at the aquarium.[18]

4.4 Soviet Union

4.4.1 Soviet Zoos — Political and Economic Realities

In 1914, with the outbreak of World War I, very significant changes began throughout Europe. Only three years later in 1917, these changes climaxed with the start of the Great October Socialist Revolution. The Soviet Union replaced the Russia of the tsars. The new nation remained in a civil war, as well as being involved with other local wars, in 1921. Then, in 1922, the state was officially proclaimed as the Union of Soviet Socialist Republics. It soon became clear that the everyday practice of socialism was a little different from the idealistic theories of Marx and Engels. It is impossible to explain in a few words all these differences, but several main rules of practical socialism will be mentioned that affected and influenced the existence and work of the zoological gardens.

First, nationalization needs to be understood, since, as a general rule, zoological gardens found themselves nationalized. Zoos, together with other institutions, such as theaters, opera houses, cinemas, philharmonic companies, museums, libraries, etc., were declared cultural establishments, and as such were regarded as the property of the entire nation. They were designated to teach and entertain all the people of the Soviet Union. Among the previously Russian zoos, the very first one nationalized was the Moscow Zoo. The edict of nationalization of the Moscow Zoo was issued on March 27, 1919 and was signed by Vladimir I. Lenin himself. Similarly, other zoos were nationalized: St. Petersburg, Askaniya Nova, Charkov, Nikolaev, and Kiev, as well as the Sevastopol Aquarium.

Nationalization has, of course, a very simple meaning — turning private property into national property. Thus, zoos came under the direct supervision of so-called municipal (but nationally controlled) governments. As scientific and cultural institutions, zoos had a reduction in the taxes (such as property tax) that supported them, and in the salaries of their workers. The money for each zoo was distributed centrally and was included in the general budget of the given city that owned the zoo. However, it did not mean the director of a given zoo could use the money according to the zoo's needs. The money was already divided into certain categories; for example, money for salaries, money for animal food, money for modernization and renovation of buildings, money for animal acquisition, and so forth. No other use could be made of these monies. So, very typically, at the end of the year there might still be money to buy animals, but none could be bought because there might not be any money left to buy food for them.

Second, in addition to nationalization, another important aspect of a socialistic economy that has affected zoos is the concept of central planning. In socialism there is no place for a free or independent market, so everything must be planned. This is as true for zoos as it is for other cultural institutions, all of which must plan in writing for each coming year. In each annual report written for the government and municipal authorities, the zoo director must document not only what had happened in the zoo during the past year, but also the plans for the next year. So one can read in such fortunetelling plans how many visitors will come, how many new animals will be acquired, how many births will occur, and how much money will be needed for food, salaries, animals, renovations, heating, electricity, etc. Surprisingly, the plans sometimes came true.

Third, science had its rightful place in the socialistic state, except that socialistic ideology did not officially accept all scientific theories. Philosophies contrary to Marxist theory were rejected. The theories of Darwin were accepted, but with some modifications by I. Michurin and T. Lysenko, who mistakenly held that theories concerning the inheritance of only some acquired features were valid. Many scientific institutions in the Soviet Union attempted to prove Michurin and Lysenko's limited theories, for example, Askaniya Nova. Unfortunately, these restraints influenced all aspects of scientific work undertaken at the zoos.

Fourth, collectivization followed the cancellation of private ownership of land, which then became the property of the Soviet nation, that is, the property of everyone in the Soviet Union. Through the process of collectivization, enormous agricultural factories (later known as kholchoz or sovchoz) were developed from many pieces of small, previously privately owned farms. These consisted of tens of thousands of acres of land and many thousands of farm animals. Former peasants became land workers or farmworkers. With such large accumulations of animals, there was a need for knowledgeable staff. And for this purpose the Soviet Union established the science of zoo-technics, a discipline somewhere between zoology and veterinary science, or more properly, practical zoology. In the big kholchoz those with a degree in zoo-technics were responsible for animal management, including breeding, genetics, monitoring diseases, hybridization, and milking. Zoo-technicians also became members of zoo staffs. For example, in the mid-1950s, of the 200-member Moscow Zoo staff, 70 had this degree.

In the Soviet Union, and later in other socialistic countries, a very important development was the "voluntary job." People devoted their free time to work voluntarily. In this way the Soviet Union was able to build many cultural institutions, such as parks, playgrounds, sports facilities, and, of course, zoos.

One should also remember that politics and ideology dominated everything, including science, culture, and the economy, permeating everyday life and every institution, including zoos. To become a director, or to keep any other kind of important position in a zoo, required not only appropriate qualifications (and the zoo-technics degree was considered sufficient), but one also had to be a member of the Communist Party. Each zoo had its own party organization, which had an important advisory voice in zoo matters. How ideology influenced the daily aims of a zoo can be found in the writings of I. P. Sosnowskij, who contrasted zoos elsewhere, "Abroad, in many zoological gardens there is not any anti-religious propaganda! No proper materialistic explanations are given to visitors to teach them about the true origin of animal species. Collections are maintained only for commercial purposes, by exhibiting only those animals which attract more visitors."[19]

Finally, one should bear in mind that the Soviet Union started its new economic politics in 1922, and that the next war began in 1941. These 19 years were too brief for the Soviet Union policies to develop fully, and too brief for these policies to have a permanent effect on institutions like zoos.

4.4.2 Russian Zoos in the Soviet Union

The Moscow Zoo was designated national property in 1919, but its reconstruction and modernization did not begin until 1921. Construction work began on the new zoo grounds, which were added to the old zoo on the opposite side of the street. The opening ceremony took place in 1926, with the total area of the zoo increased to 44 acres. In the meantime, many expeditions were sent into Central Asia, the Far East, and the Caucasus to collect wildlife for the zoo. Moscow Zoo changed its name in 1923 from Zoological Gardens to Zoological Park, and since then it has familiarly been called the Zoopark. This shortened name has become popular and has been used in the daily Russian language for all zoo collections in the country. In 1924, the Young Zoologist Club was organized, which has survived without interruption to the present day. Many young girls and boys have started their adventure with zoology and wild animals in this club. In 1926, the zoo began its scientific work, and one of the first tasks was to set up a breeding program for sable. In 1928, the first successful breeding of this species in captivity took place, which resulted in the establishment of large sable farms in Siberia.

The zoo in St. Petersburg was nationalized in 1919, the same year as the one in Moscow. A year later the total area was extended to nine acres and new enclosures were added. In 1924, St. Petersburg (which had been changed to Petrograd during and after World War I) got a new name, Leningrad — and so did the zoo. The zoo's Young Biologist Club was formed in 1929, and its educational work with children started. Leningrad Zoo had its biggest pre-war achievement in 1930 when it bred polar bears. In 1940, for its 75th anniversary, the zoo received government funds that allowed it to buy 420 acres of forestland outside the city on which to build a new zoo. But this was only one year before World War II came to the Soviet Union, and these plans still remain unfulfilled.

As for Askaniya Nova, its territories as of the end of 1920 had been abandoned because of the Revolution. During this time, many of its animals were killed and many plants were destroyed. In February 1921, Askaniya Nova was declared the property of the Soviet nation and a national reserve was established there. In the following years many scientific institutions were located at this facility, such as a zoo-technical experiment station for cattle breeding and a phyto-technical station. In 1932, a Scientific Research Institution for Hybridization and Acclimatization of Animals was also created. Of course, as has been mentioned, many of the so-called scientific efforts and investigations done at these institutions were under the strong influence of the Michurin and Lysenko theories. Despite this, many good works were accomplished during this time, especially on hybridization, crossbreeding, domestication, and the acclimatization of a variety of wild species of animals.

World War II continued on Soviet territory for four years (1941–1944). The German army came very close to Moscow and Leningrad, and annexed the Askaniya Nova reserve. Most of the animals in Askaniya Nova were killed either during the German occupation or during the Soviet liberation efforts. By far the greatest loss was the scientific library at Askaniya. Many books and scientific papers were destroyed, and others stolen. Surprisingly, the Leningrad Zoo survived. During the German army's long state of siege, the people of this city, mainly the elderly and children, died from cold and starvation. So it is incredible that the zoo animals survived, especially such culinary animals as hippos and antelopes.

A unique situation existed at the Moscow Zoo during the war. In the first two years, 1941–1942, the Germans bombed the area of the zoo several times. Some buildings were destroyed, however, there was no significant loss of animals. The danger came instead from elsewhere: the Moscow municipality, which discontinued financial help for the zoo. Fortunately, during these difficult years attendance of the zoo was very high and provided the primary source of funds for the zoo. In addition, the zoo had a working domestic animal farm and all its surplus eggs, chickens, milk, vegetables, etc. were sold to the people, providing a second source of income for the zoo.

After the war, the zoos were rebuilt and reorganized, starting with the leading institutions. The Askaniya Nova Reserve and Zoo were reorganized once again in 1956, with all of their scientific institutions under the umbrella of the Ukrainian Academy of Science. Today, Askaniya Nova Reserve is a complex of scientific institutions, including various research departments and laboratories, as well as the zoological garden, botanical park, and steppe reserve. Animals at the zoo and reserve are kept in large herds with good breeding results, keeping prices for the offspring low. For this reason many zoos choose to buy animals from Askaniya Nova. Such animals as eland, Przewalski's horses, saiga, Grant's zebras, ostrich, and others at present kept in many formerly Soviet zoos were born at Askaniya. Many of these offspring have also been exported to zoos all over the world.

The idea of building a new zoo in Leningrad seems to have been forgotten. For many years after the war the existing zoo was under constant reconstruction, but unfortunately there has been no opportunity to enlarge its territory. After the war the zoo became famous for its breeding record with giraffes, Père David's deer, Przewalski's horses, and hippos. These have become common animals, and not very difficult to breed under normal conditions. But conditions are not normal in Leningrad. Looking at the map one can see that the Leningrad Zoo, together with Helsinki Zoo (Finland), Oslo Zoo (Norway, which no longer exists), and the Anchorage Zoo (Alaska, United States), are the northernmost zoos of the world. Leningrad has long, cold winters and short, hot summers, which surely influence animal physiology; therefore, the breeding results for these species obtained at the Leningrad Zoo are exceptional. This climate is, however, very good for other species, such as Japanese macaques. Leningrad Zoo received a group from the Osaka Zoo, and these animals have bred well there.

Although all the zoological gardens in the Soviet Union were municipal (national) properties, it is clear the Moscow Zoopark had advantages and privileges by virtue of its location in the capital city. For example, it was usually the recipient of animals that official guests and visitors donated to the state. There were such gifts as Indian elephants (from former Ceylon, now Sri Lanka), musk ox (from Canada), pumas (from the United States), and Malayan bears (from Indonesia). However, it was the gifts of the giant panda that provide the most interesting story. The first giant panda exhibited in the Moscow Zoo was the male Ping-Ping. He came to the Moscow Zoo as a special gift from the Chinese people to the people of the Soviet Union. This animal was put on exhibit in May 1957. Ping-Ping was thought to be a female (only after the panda's death was it discovered to be a male), so in 1959 the next giant panda sent from China to the Moscow Zoo was the male An-An, a special gift for all the Soviet children. Ping-Ping died in 1961, leaving only two giant pandas outside of China, one in the Moscow Zoo and one in the London Zoo. By 1965, the London Zoo was sure its Chi-Chi was a female, and the London and Moscow zoos started to negotiate the possibility of putting both animals together. By the mid-1960s, during the days of the "Cold War," it was surprising that the director of the Moscow Zoo allowed such a possibility. Chi-Chi came to Moscow in 1966 and stayed for eight months. The only effect of this visit was that Chi-Chi became not pregnant but very fond of sweet tea. Likewise, the visit of An-An to London for ten months in 1968–1969 was unsuccessful. Although unsuccessful, this was one of the first attempts at international cooperation in the breeding of endangered animals. Both pandas died in 1972.

It is not possible to mention all of the important species kept and bred at the Moscow Zoo, but some are significant. The first Indian elephants were born there in 1948 and 1952. In 1984, four African elephants were exhibited for the first time. The zoo is very proud of its bear collection, which includes all seven species. All of their bear species have bred very well, especially the Malayan and sloth bears. The first gorillas came to the zoo in 1972, but have not yet bred. However, the zoo has been successful in breeding orangutans in 1981, 1985 (two births), and 1986 (with the young mother-reared). The zoo is also proud of its breeding results with two subspecies of tigers, two subspecies of leopards, snow leopards, maned wolves, cheetahs, and many species of birds, mostly waterfowl and pheasants. The breeding results could be better, but the space for new facilities is limited. Plans for a new zoo were finally drawn in 1970, and in 1971 the Moscow municipal authority donated a 370-acre site in the southwestern part of the city. The plans are now ready, although they have become somewhat outdated — but sooner or later, the Moscow Zoo will be transferred to the new site.

4.4.3 Other Zoos in the Soviet Union

Between 1919 and 1941 at least 11 new zoological gardens were established in the Soviet Union. All of these had their origin in small "zoo corners" set up in municipal parks or on school grounds. Many of them have not reached the status of a zoological garden, but all of them are very popular among the inhabitants of their towns. The most important collections are those at Rostov-na-Donu and Alma-Ata.

A general history of Soviet Union zoos should also include those zoos established in what were once independent countries, but which the Soviet Union later annexed. In 1939, the Soviet Union annexed a part of Poland, which had a small zoo in Grodno. In 1940, the Soviet Union annexed three independent Baltic states: Latvia, Lithuania, and Estonia. With this annexation the Soviet administration acquired three more zoos, at Riga, Kaunas, and Tallin. Of all these zoos, the most important collection was Tallin. This zoo began in 1937 when Estonia took part in the World Championship of Shooting in Helsinki and received a live young lynx. This animal was put into a specially built cage on the grounds of a tree nursery near Tallin. The zoo proper began in 1939 on an area of seven acres with no more than 100 animals. Its success can be attributed to the fact that it is the one zoo in the entire Soviet Union that was planned and well designed. The new zoo in Tallin opened to the public on June 26, 1983.

4.4.4 General Characteristics and Comments on Soviet Zoos

Altogether, the Soviet Union had 32 zoological gardens, 2 public aquariums, and 1 oceanarium. All the zoos were designated as cultural institutions under the direct supervision of ZooCentre, an administrative branch of the Ministry of Culture. At least 15 traveling menageries have also been under this same administration. These menageries have traveled to all parts of the region, even to the most distant small towns of the former Soviet Union. These traveling menageries have been popular and have some 15 to 16 million visitors each year.

The Soviet Union was the first socialistic country to adopt the regulations of the IUCN (International Union for the Conservation of Nature) and WWF (World Wildlife Fund). The U.S.S.R. Red Data Book was first published in 1974, with a second edition 10 years later. This book has two appendixes for native wildlife: one (I) for endangered animals, and one (II) for vulnerable animals. In the second edition appendix I lists 17 species of mammals, 21 species of birds, 7 species of reptiles, and 1 species of amphibians. Appendix II lists 17 species of mammals, 24 species of birds, 7 species of reptiles, and 6 species of amphibians. Of all these animals, Soviet zoos have only bred about 12 to 14 species of mammals, 8 to 9 species of birds, 2 to 3 species of reptiles, and 1 species of amphibians. The Soviet Union was also the first socialistic country to sign the CITES Convention (Convention for International Trade in Endangered Species). The Soviet Parliament did this in 1976.

Total area for all the zoos was 1,400 acres, with an average per zoo of only 44 acres. Soviet authorities knew very well that improvements to existing zoos and new zoos were needed. The most easterly zoo is in Novosibirsk, and the entire region to Vladivostok is without a single zoo. But with the dissolution of the Soviet Union, it is now a new political situation, with a new economy, and a new way of thinking. It is difficult to predict the future of zoos in those countries of the former Soviet Union, but it will be interesting to watch.

4.5 Czech and Slovak Republics

During the Middle Ages there were also menageries in the country that is now the Czech Republic. These collections were kept in feudal castles and the nobles' manors. However, it is difficult to find written evidence about these collections, with names, places, and dates. What little evidence there is indicates that these privately owned collections consisted of the most common wild animals and domestic breeds of animals.

There is, however, some written evidence that the Prague Castle has contained a collection of lions since the earliest years of the eleventh century; however, the origin of these first animals is not known. The great love of lions was a common characteristic of most Czech kings, so much so that in the thirteenth century the lion became the official heraldic emblem of the Czech State, and has replaced the eagle as the most common emblem in this part of Europe. In contrast, the eagle remains the national emblem of Germany, Russia, Austria, and Poland.

In 1409, the Bohemian king, Vaclav VI (1361–1419), built a lion pit near his castle in Prague. For many years only lions were kept there, but other exotic animals were introduced into the castle collection during the reign of Emperor Ferdinand I (1503–1564). Ferdinand I was emperor of the Holy Roman Empire, and also had the official title of King of Bohemia. The next emperor, Maximilian II (1527–1576), built a special lion pit and a small manor house, which still exists in the old town of Prague. But the best-known menagerie, and probably the biggest one at that time in Europe, belonged to Emperor Rudolf II (1552–1612). Rudolf II was well known as an eccentric, but also as an animal lover. During his reign the Prague menagerie had an international reputation. As far as is known today, he introduced many new exotic species into his menagerie, even such rare and (in those days) unexpected ones as apes (chimps or perhaps orangutans, we do not know) and macaws. But the rarest inhabitant was the Mauritius dodo, which lived in the collection for a few years. When these animals died they were stuffed and placed in a curiosity room (a natural history cabinet). Unfortunately, Rudolf II's successors did not share his interest in animals and after his death the Prague menagerie declined very quickly. By 1724, there were only two old brown bears.

In the eighteenth century Prague declined in importance as the main scientific, cultural, and political center of the Habsburg Empire. In the second half of the century many members of the Habsburg family preferred to spend their time in Vienna, so it was in Vienna, in 1752, that their menagerie was established in the gardens of Schönbrunn, which some would later regard as the first modern zoo. This menagerie, for the next several decades, dominated attempts to establish an independent zoological collection in the other cities of the Habsburg Empire.

Evidence concerning attempts to establish a zoo in Prague during the nineteenth century is scarce. Unlike other cities, where the birth of a zoo was widely discussed in the local newspapers, both pro and con, the Prague daily press of the second half of the nineteenth century is silent. Information can finally be found in 1891, the year in which the Society for Zoo and Acclimatization was formed. There was even a chosen site for the society's future zoo and Carl Hagenbeck, the well-known German animal dealer, was invited to see the place and express his opinion. Unfortunately, Hagenbeck did not approve of the place. The society's members later proposed two other sites, but these were not accepted by the municipal council. Finally, one member of the municipal council, Dr. Gros, proposed a site in Prague known as Stvanici, but the area was only 10 acres without any possibility of further growth. Gros privately published, at his own expense, the newspaper *Zoologicka Zahrada* (*Zoological Gardens*). But he only had enough money for two issues, the last of which was published in 1909.

Taking all these problems into account, it is not surprising that the first Czech zoo — the Severoceska Zoologicka Zahrada Liberec — was not established in Prague, but rather in Liberec. Liberec became an important city at the end of the nineteenth and beginning of the twentieth centuries. Most of the middle class in Liberec was of German origin, and in a short time a significant textile manufacturing industry developed. Liberec's wealthy citizens built hotels, banks, theaters, and a new town hall. These citizens also created municipal parks and in 1877 they established a botanical garden. They started a Private Society of Nature Lovers and in a short time had their own ornithological club. From the private donations of the club members a large aviary for birds was erected in 1904 (this structure was still in use at the zoo until a few years after World War II). Three years later, in 1907, the club members introduced several species of ornamental waterfowl to the small lake in the park next to the aviary. Gradually, other animals were added and temporary enclosures were built. Then came World War I, which impeded development, but finally in 1919 the zoological garden in Liberec was officially inaugurated.

After the war, the Czech and Slovak Republics became a single independent country. And then it was the time for a zoo in Prague. The municipal authority gave permission for Jiriho Janda to find a place for the zoo and to draw up a set of plans. Janda was a professor of biology at the lycée (high school) and he was also the general advisor to the Ministry of Education, so his position in the new country was very influential. For the first three years he spent his time visiting and planning. He made the first important step in 1924 when, from the 14 different locations the municipal councils proposed, he choose one on the right bank of the Veltava River close to the nineteenth-century Troja Castle. As Janda wrote in his report, "All others [zoo personnel] will envy us because of this place!" In 1926, the Zoological Gardens Government Society was created, whose main aim was to build and manage the zoo. Despite serious financial troubles, the zoological garden in Prague was officially inaugurated on September 28, 1931. At that time the total area was only 20 acres, but it had several good buildings, such as open enclosures for bears, a parrot aviary, a carnivore house, and one of the largest free-flight aviaries for birds of prey. In 1933, the Prague Zoo received its first Indian elephant and had its first success in breeding Przewalski's horses. In 1937, the Prague Zoo received international fame again, this time because of its successful breeding of the Andean condor. In 1938, Janda, the director of the Prague Zoo since its inception, died.[20]

Before 1939 the Czech and Slovak Republics had only two zoos, in Liberec and Prague. Under German occupation the zoo labels were in two languages and Jewish people forbidden to visit the zoos. Work at the zoos went on quietly. During 1942/43, the Prague Zoo once again received international fame, this time because of its first successful breeding of polar bears. In the years after the war 15 Czech Republic and Slovak Republic zoos have been established, each local authority developing its own zoo. However, only one of these will be mentioned here. The zoo in Dvur Kralove was established in 1946 and was a typical small-town zoo for almost 20 years. Josef Vagner became its director in 1965. The former area of 20 acres was enlarged by adding 148 acres. Very simple enclosures and wood barns were built. One of Vagner's ideas was to get permission from some African states (such as Sudan, Kenya, and Tanzania) to permit the capture of wild animals for the zoo's collection. This permission was given to the Czech and Slovak Republics because of their support in building factories in these African countries. Altogether, between 1968 and 1975, some 21 large shipments of wild-caught animals arrived from Africa. It was, perhaps, the largest consignment of African animals to Europe since World War II. Gradually, other species were introduced at the zoo, in addition to those that had come from Africa.

All the zoological gardens in the Czech and Slovak Republics are municipally owned. Of the 15 zoos, 12 are in the Czech Republic and 3 are in the Slovak Republic. These zoos are under the supervision of the Ministry of Culture in both countries. In addition, the Czech and Slovak Republics appear to be the first countries to establish special schools for zookeepers. The schools are similar to a secondary school lasting four years, and, when completed, graduates receive a secondary school certificate (abitury). New students to these schools are recruited every two years. With this certificate a graduate can get a position at a zoo, circus, or any other animal facility.[21]

4.6 Hungary

From the sixteenth century to the twentieth century Hungary was closely associated with the Habsburg (Austrian) realm. During part of this period Hungary and Austria formed, on an equal basis, the Habsburg Empire, which was also sometimes called the Austro-Hungarian Empire. In 1867, the Hungarian parliament in Budapest received some degree of political autonomy, but the true Hungarian state came into existence only after World War I in 1918. After World War II it became the Hungarian People's (socialist) Republic.

The first Hungarian zoo, in Budapest, was established thanks to the efforts of the Hungarian scientific elite. A few of these men are mentioned here, because without them the Budapest Zoo would not have been possible; they are Janos Xantus (a traveler and naturalist), Jozsef Szabo (a university professor), Jozsef Gerenday (director of the University Botanical Gardens), and Agoston Kubinyi (director of the Hungarian National Museum). These men met for the first time in 1864, after which they convinced the local municipal council members of the necessity for a zoo in Budapest. The municipal authorities felt the only possible site for the zoo was in the city park where a botanical garden already existed. The small zoo of 40 acres opened on the grounds of the botanical garden on August 9, 1866, with Janos Xantos as its first director.[22]

The Hungarian Academy of Science helped establish the zoo as a private, nonprofit company. In 1872, this company was transformed into the Society for the Acclimatization of Plants and Animals, a name that mirrored the status of this institution as a zoo-botanical gardens. From the beginning it exhibited not only exotic animals, but also domesticated plants and animals from many parts of the world. In the years of prosperity, between 1866 and 1912, many species were exhibited and bred, including some very rare ones.

Budapest Zoo received its first giraffe from the Schönbrunn Menagerie in 1868. The first African elephant arrived in 1875, while an Indian elephant came in 1883. The zoo became well known for its breeding of native wild animals. The most successful breeding and reintroduction program was that of the great European bustard. The zoo also bred native domestic animals such as the three original breeds of Hungarian dogs (the puli, puni, and vizsla), the great horned Hungarian cattle, and various breeds of Hungarian goats and sheep. Although many interesting species arrived at the zoo throughout its history, the zoo received one of its rarest species in 1894 — the Sumatran rhino. From the long list of animal species that have been bred here, the most important have been hippos, giraffes, and Indian elephants.

Budapest Zoo was in constant financial trouble during the years 1866 to 1897, as it was part of a struggling independent society. But in 1907 the municipal government took over the zoo, which started a period of revival. In 1912, an aquarium was constructed with two independent sections, one for marine fish and one for freshwater fish. Unfortunately, after 1912 the Budapest Zoo went into a long period of stagnancy. The independent Hungarian government had very little money to spend on the zoo, with so many other needs of a young state. But even during these difficult times, it built one of the earliest public terrariums for amphibians and reptiles, as well as a botanical greenhouse, which opened in 1922. Again, as has been mentioned in other cases, fighting during World War II heavily damaged the zoo and it had to be rebuilt and modernized. The zoo was not to recover fully from this war destruction until 1955. A new greenhouse was built, which exhibited many rare species of plants, along with a modern and well-designed aquarium, which was built underground beneath the greenhouse.

Educational and scientific work at the Budapest Zoo started in 1958. Zoo authorities organized special courses for keepers and the scientific staff to establish a higher standard for their work. A few years later the Budapest Zoo staff wrote a significant publication, entitled (in English) *The Fundamentals of Zoological Garden Work*. This publication was some 335 pages in two volumes. Most of the text is now outdated, but at the time it was the only source of information for new staff members, as well as one of the first zoo training manuals.

Four other zoos were established in Hungary at the end of the 1950s and beginning of the 1960s. These zoos were built with the help of local citizens, with much of the work performed by volunteers. It should also be noted that each of the Hungarian zoos kept and bred domestic (but not strictly Hungarian) breeds of animals. In 1989, the latest Hungarian zoo was established at Szeged. This zoo has an area of some 148 acres, but only a small portion of it has been developed. There are three areas of specialization at this zoo: South American monkeys (including seven species of marmosets and tamarins), canids (such as maned wolves, big-eared fox, and pampas fox), and a good breeding program for domestic cattle, sheep, horses, and goats (without which it would not be a good Hungarian zoo).

4.7 Bulgaria

Territories comprising what is now Bulgaria (plus most of southeastern Europe) were part of the Turkish Ottoman Empire for many centuries. Bulgaria received some degree of independence in 1878 after the war between Russia and the Turkish Empire; it became totally independent as a kingdom in 1908. Only one year after Ferdinand Saxe-Coburg-Gotha, a fanatical amateur ornithologist, became head of state in 1888, he built a birdcage in the palace park to hold a black vulture. In a very short time this rather small park around the castle was full of animals, such as game birds and deer. In 1890, this collection was transferred to another park in Sofia where it remained as the Sofia Zoo until 1984. This site had been designed as the prince's botanical garden, but Ferdinand decided to use it for the animals instead. Cages for birds and a very modern (for the time) cow barn were built on this site. Fresh milk from the cows supplied the tsar's kitchen. In 1889, Ernst Hulein was hired as the first zoo manager. Hulein was a sculptor, but perhaps not a particularly good one, as none of his works has survived to the present. However, he was a good taxidermist and he stuffed and displayed all the animals that died. These specimens began Ferdinand's well-known natural history museum. Hulein also caught native animals for this collection, and later exotic animals as well.

In 1892, Hulein bought the zoo's first pair of African lions from the Leipzig Zoo. That same year these lions had their first litter and the cow barn was rebuilt as a temporary carnivore house. During the years 1893 to 1896 the zoo had a new manager, the ornithologist Paul Leverkuhn, but he turned out to be an ineffective administrator. Nevertheless, during his management two aviaries were built. The first aviary was a long one with 24 separate cages designed for pheasants and other gallinaceous birds. The second was built in 1895 and was a large, open aviary for bearded vultures. A pair of these birds was acquired, and for the next 20 years these birds laid eggs and raised young almost every year.

Until 1893 the zoo was never open to the public. Only members of the royal family and special guests could visit this zoo. Since the summer of 1893, however, the zoo has been open to ordinary visitors, at first only three days a week, free of charge. Daily attendance then reached four to five thousand. At the end of the nineteenth century the zoo had 1,384 animals of 266 species, including 30 species of birds of prey and 20 species of pheasants. Surplus stock from the Sofia Zoo was sent to parks in other towns (Varna, Ruse, and Kritchin), the nucleus of future zoos. There is some evidence that Ferdinand decided at one time to close the Sofia Zoo and transfer the animals to the larger park in Varna. Fortunately for Sofia residents, these plans never came about (and after World War I all such ideas of closing the Sofia Zoo were finally dropped).

In 1902 the zoo received its first pair of lynx, in 1905 its first two Tibetan yaks, and in 1918 its first American bison. Sometime before 1914 a terrarium was built, and soon its most important occupant was a Mississippi alligator, which was given to the zoo as a private donation. In 1912, the zoo received its first pair of Indian elephants from Hagenbeck's Hamburg Zoo, although the animals were first put in temporary enclosures at Varna. The authorities at the Sofia Zoo had problems building a proper enclosure for the elephants, so their temporary stay at Varna was extended to the end of 1928. These animals were finally transported to Sofia at the beginning of 1929. Very soon it became clear that the fence around the enclosure was not strong enough for the young male elephant. One day he refused to go inside the barn and started to destroy the strong metal fence systematically. The story ended well, however, with some help from the town's fire engines, but the entire fence had to be rebuilt. This male elephant, Nol, died of anthrax before 1940, while the female, Damiatin, lived until 1953 when she died from an internal parasitic infection.

In 1911, Adolf Schumann was appointed general inspector of the Sofia Zoo. Previously he had worked in zoos and with private collections in Austria and Germany. Schumann was well known as a specialist in breeding birds, reptiles, and fish. He knew less about management and administration, but proved a capable administrator and was able to improve the Sofia Zoo's reputation among the international zoo community. He held this position until 1940.

Sofia Zoo received little damage from the heavy Allied bombing in 1944. During World War II Bulgaria supported Germany, at least politically. Therefore, in 1944 the Soviet Union declared war on Bulgaria, whereupon the Soviet army invaded the country. In 1945, Bulgaria became an independent socialist republic. Sofia Zoo, while being rebuilt, became part of the Bulgarian Academy of Science in 1947. The zoo acquired its first shipment of new animals from the Soviet Union, East Germany, and other socialist countries in the mid-1950s. For example, lions arrived in 1955 from the Leipzig Zoo. In the previous year the zoo received a special gift, a pair of Indian elephants, from Vietnam. The wild-born female was named Saviti and the male was named Sativan. They were shipped to Varna harbor on May 14, 1954 and arrived at the Sofia Zoo two days later by train. Unfortunately, the male elephant died in 1964, but the female was already pregnant and gave birth to a female calf on April 13, 1965.

Since 1960, the Sofia Zoo has been under the management and administration of the municipal government. In 1965, for the first time, it remained open through the winter season (previously it had only been open from April to October). In 1966, efforts were made to develop a new zoo for Sofia since the zoo's six acres had become too small, both for its animals and for the increasing number of visitors. Building a new zoo was an enormous task for the municipal authorities, but was eventually accomplished with significant help from the citizens, who worked voluntarily. During this time, in 1968, the zoo also started its education program with children in a so-called zoo school. The new zoo was built on 60 acres, and another 94 acres remained for future development. The last animal was transferred from the old site in August 1984, and the new zoo opened to the public on the last day of the same month.

As 1888 is accepted as the first year of the Sofia Zoo, the zoo staff only had four years after its new opening to prepare for the 100th anniversary celebrations. Having spent most of its money on the new facility, there was none left over for new animal acquisitions. The zoo authorities publicized this difficulty and received many grants, gifts, presents, legacies, and donations, not only from Bulgarian citizens, but also from many famous and wealthy Bulgarians living abroad. On the day of its 100th anniversary, the Sofia Zoo had almost 2,500 animals of no fewer than 250 species.

In addition to the Sofia Zoo, Bulgaria has smaller zoos in municipal parks in cities such as Varna, Ruse, and Lovetsch. Bulgaria also has one public aquarium. This aquarium is an integral part of the Institute of Aquatic Resources at Varna, which is under the supervision of the Food and Agriculture Ministry. The main building, in the neoclassic style, is very similar to the biological stations at Naples and Sevastopol. It was built in 1906 to 1912, and was open to the public on July 17, 1932 with exhibits featuring animals from the Black Sea.

4.8 General Comments on Central-Eastern European Zoos

To draw some conclusions from the history of Central-Eastern European zoos and aquariums, one should remember that the region has been less well developed economically than its Western counterpart. The political and economic transformations have been difficult, but the zoos and aquariums have survived and none have closed. New political and economic realities have affected the zoos. Many have now established their own nonprofit societies and many have modernized their facilities. New construction is under way, new exhibits are being built, and improvements for visitors are being made. Most importantly, the number of visitors to these zoos has not declined. The citizens still regard the zoos and aquariums as important parts of the community and they support them. Central-Eastern European zoos will not only survive the latest of many historic transitions, but will take their places alongside other European zoos as excellent examples of modern zoos and aquariums of the twenty-first century.

References

1. Taborski, A. and Urbanski, J., Eds., *Ogrod Zoologiczny w Poznaniu*, PWN, Warsaw-Poznan, 1975 [*Poznan Zoological Gardens*; commemorative book published for Poznan Zoo's 100th anniversary].
2. Wolinski, Z., Ed., *30 Lat Warszawskiego Ogrodu Zoologicznego 1928–1958*, MOZ, Warsaw, 1959 [*30 Years of the Warsaw Zoo 1928–1958*].

3. Zabinski, J., *Przekroj Przez Zoo*, Wiedza Powszechna, Warsaw, 1953 [*A Walk through the Zoo*].
4. Peel, C. V. A., *The Zoological Gardens of Europe: Their History and Chief Features*, Robinson & Co., London, 1903.
5. Loisel, Gustave, *Histoire des ménageries de l'antiquité à nos jours*, Octave Doin et Fils and Henri Laurens, Paris, 1912.
6. Kourist, G., *Zivilisation und Wildtierhaltung im Europa*, privately published, Linz am Rhein, 1989 [*Civilization and Wild Animal Keeping in Europe*].
7. Gleiss, H., *Unter Robbe, Gnus und Tigerschlangen: Chronik des Zoologischer Garten Breslau 1865–1965*, Natura et Patria Verlag, Hamburg, 1967 [*The History of Breslau Zoo, 1865–1965*].
8. Poradowski, Sz., "Ogrod Zoologiczny w Plocku," *Przeglad Zoologiczny* (Wroclaw), 2, 53, 1958 ["The zoo in Plock," *Zoological Review*].
9. Poradowski, Sz., "Zoo w Oliwie," *Przeglad Zoologiczny* (Wroclaw), 1, 324, 1957 ["The Oliwa-Gdansk Zoo," *Zoological Review*].
10. Kotarba, I., "Powstawanie ogrodow zoologicznych," *Przeglad Zoologiczny* (Wroclaw), 13, 117, 1969 ["The origin of the new zoos," *Zoological Review*].
11. Xiezopolska, I. and Taborski, A., "Sytuacja obecna i prognozy rozwojowe Polskich ogrodow zoologicznych," *Przeglad Zoologiczny* (Wroclaw), 18, 192, 1974 ["The present status and perspectives for Polish zoos," *Zoological Review*].
12. Lukaszewicz, K., "Znaczenie ogrodow zoologicznych dla nauk zoologicznych," *Biul. Tech. Min. Gosp. Kom.*, No. 1–59, 5, 1959 ["Role of the zoos in zoological science," *Technical Bulletin of the Ministry of Communal Affairs*].
13. Zabinski, J., Ed., *Przedruk Pierwszych Ksiag Rodowodowych Zubra 1932–1937*, PWN, Warsaw, 1966 [reprint of first wisent studbooks, 1932–1937 editions].
14. Ogniev, C. I., Ed., *Moskowskij Zoopark*, Moskowskij Rabotchij, Moscow, 1949 [*Moscow Zoo*].
15. Sosnovskij, I. P., *Pitomcy Moskowskogo Zooparka*, Moskowskij Rabotchij, Moscow, 1974 [*The Inhabitants of the Moscow Zoo*].
16. Sosnovskij, I. P., *Za Kulisami Zooparka*, Agropromizdat, Moscow, 1989 [*Behind the Scenes at the Moscow Zoo*; commemorative book published for Moscow Zoo's 125th anniversary].
17. Salganskij, A. A., *Zoopark Askaniya Nova*, Gosudarstviennoye Isdatielstvo Selskhochozaystviennoy Literatury (The National Publisher of Agricultural Literature), Kiev, 1963 [*Askaniya Nova Zoo*].
18. Sosnovskij, I. P., "Zooparki i zoosady SSSR, Moskowskij Zoopark-sbornik statiej," *Wypusk*, 3, 126, 1961 ["Zoos of the U.S.S.R.," *Newsletter of the Moscow Zoo*].
19. Sosnovskij, I. P., *Pitomcy Moskowskogo Zooparka*, 1974, 86.
20. Veselovsky, Z., *Vsedni den v Prazske Zoo*, Albatros, Prague, 1990 [*A Day at the Prague Zoo*].
21. Dobroruka, L. J., *Zoologicke Zahrady*, Statni Pedagogicke Nakladatelstvi, Prague, 1989 [*Zoological Gardens* (of the Czech and Slovak Republics)].
22. Szidaine, C. A., *A 125 Eves Budapesti Allat-es Novenykert Tortenete*, Athenaeum Nyomda, Budapest, 1991 [*125 Years of the Budapest Zoo*].

Additional Source

1. Lukaszewicz, K., *Ogrody Zoologiczne — Wczoraj–Dzis–Jutro*, Wiedza Powszechna, Warsaw, 1975 [*Zoological Gardens — Yesterday, Today, Tomorrow*].

5
Zoological Gardens of the United States

Vernon N. Kisling, Jr.

5.1 Introduction

Menageries in what was to become the United States emerged just as those in Britain and Europe were becoming modern public zoological gardens. Although very little has been known about the early historical development of American zoos and aquariums, information on this subject has increased significantly during the 1980s and 1990s. The discussion here is only intended as an introduction, since it would not be possible to consider all the varied cultural influences affecting the development of these institutions and because information is still lacking for many of the individual institutions.

Prior to the European colonization of North America, there is no information on primitive animal collections, although the southwest native American tribes kept, and possibly bred, tropical parrots obtained in trade from the Aztecs of Central America and Incas of South America.[1,2] Hard work, frugality, simple pleasures, and the need to establish a new society in an overwhelming and threatening wilderness characterized America's colonial period. Colonists did not look favorably upon frivolous pursuits and amusements, and this viewpoint applied to the itinerant animal acts that appeared early in the eighteenth century. Nevertheless, fascination with wild animals attracted onlookers to these early itinerant acts, which featured bears and other native animals. These animals were brought to a tavern or village square for show and then the owner would pass around the hat to collect enough money to take him and his animals to the next town.[3] The intellectual study of America's indigenous wildlife during the colonial period was the domain of visiting European naturalists.[4,5]

5.2 Eighteenth- and Nineteenth-Century Menageries

Exhibiting native species in itinerant animal shows remained common during the eighteenth century, but occasionally the native animals were replaced with more fascinating exotic animals from distant lands. A lion had the distinction of being the first exotic species exhibited when it was brought to Boston, Massachusetts in 1716. This representative of the "dark continent" was housed at the home of Captain Arthur Savage before it was moved in 1720 to the home of Martha Adams. Adams advertised in the newspaper, welcoming all to visit her lion whenever they wanted. She also had a sign on her house that read, "The Lion King of Beasts is to be seen here." This lion was shipped to the West Indies in 1726, but returned in 1727 to be shown in

Philadelphia, Pennsylvania. In 1728, it appeared in New York and New Jersey, and finally in New London, Connecticut before vanishing from the historian's view. It would be 1791 before another lion made it to American shores.[3,6]

Another exotic species reached America soon after the lion arrived in Boston. This was a camel, which arrived from Africa and was on display for about four weeks in 1721. Other camels arrived in 1739 in Boston and in 1787 in New York. This last occurrence involved a pair of camels that were exhibited all over New England, New York, and Pennsylvania for about 10 years.[3,6] Camels had actually been imported into Virginia in 1701, but for the purpose of domestication rather than exhibition.[7] A "ferocious Greenland bear," the "great white [polar] bear" was brought to Boston in 1733 and kept on its owner's property among the hustle and bustle of the Boston wharves. Another polar bear made its way to Boston in 1798.[3] Several ostriches were shown at Woart's Tavern, Boston in 1794 before leaving for South Carolina, the first documented occurrence of this species.[3]

These exotic animal arrivals were few and far between. Most exotic animal shipments during the eighteenth century went to British and European menageries. However, American sea captains occasionally ventured to transport exotic animals to America, where they hoped to sell these animals for a high profit. It was then up to the itinerant showmen who purchased these animals to keep them alive long enough to recoup their huge investment. To do so, the animals had to travel to as many communities as possible, an arduous undertaking at the time. Exotic animals usually appeared at one of the major commercial ports in the northeast and then traveled to other major cities, such as Boston, New York, Philadelphia, and Baltimore, Maryland. These cities provided the largest audiences, but on the way to these cities the animals were exhibited at smaller invervening towns. To attract customers, the traveling animal shows were advertised in local newspapers and printed flyers. However, these documents were ephemeral and few have survived.

Itinerant animal acts exhibiting several species began to appear in the larger cities during the late 1700s and carried over into the early 1800s when traveling menageries and circus menageries became popular. A menagerie containing reptiles and birds, in addition to the usual mammals, was exhibited in New York in 1781. Another menagerie, consisting of a tiger, orangutan, sloth, baboon, buffalo, crocodile, lizards, snakes, and other creatures was exhibited at New York in 1789. In 1796, a menagerie of birds, a seal, and about 20 other animals appeared in New York, followed the next year by another, which consisted of wolves, monkeys, a mongoose, numerous small mammals, and numerous birds. Although the exact identity of some of these animals is not known, some no doubt represented species that were probably making their first appearances in the United States.[3,8]

The tiger promoted in the 1789 menagerie was part of a collection from Africa and Brazil so it may, or may not, actually have been a tiger since tigers are not from these regions. Two tigers from India did appear at Crombie's Tavern in Salem, Massachusetts in 1806. These tigers appeared later in 1806 in New York, and another tiger arrived in Salem in 1816. Likewise, the orangutan is not from the regions advertised. This orangutan, however, was followed by others shown in New York in 1828, in Salem in 1831, and in Boston in 1836.

The first elephant to arrive in America was a two-year-old female Captain Jacob Crowninshield brought to New York in 1796 aboard his ship *America*. After exhibition at Beaver Street and Broadway in New York, the elephant was sold and taken on tour to Philadelphia (1796), Baltimore (1796), Charleston, South Carolina (1796), Philadelphia (1797), New York (1797), Providence, Rhode Island (1797), Cambridge, Massachusetts (1797), Salem (1797), Philadelphia (1798), Charleston (1798), Boston (1804), Philadelphia (1806), New York (1808), Baltimore (1811), Gettysburg, Pennsylvania (1812), New York (1818), and many points in between. While at Cambridge in July 1797, the elephant (*the* elephant because she was the only elephant

Poster dated 1797 advertising the Boston exhibition of the first elephant brought to the United States. Facsimile of the original poster. Courtesy of Vernon Kisling.

in America and her owner felt no obligation to name her) supposedly attended commencement ceremonies at Harvard University.[3,6]

Captain Crowninshield bought the elephant for $450 and sold it for $10,000, a fortune in 1796. The price of admission to see the elephant was 50 cents, so many paying customers were needed to recover the investment and maintenance costs. At the time (1790s) there were only two cities with more than 25,000 residents and most towns had fewer than 2,500.[9] Fortunately for the owner, the elephant was immensely popular and stayed healthy for two decades. Many other exhibitors were not as fortunate.[3]

After the American Revolution, European naturalists returned home, but since their work had primarily supported European scientific endeavors, there was little effect because there was no American scientific infrastructure to collapse. During the early republic period, American naturalists, after initially relying on European resources, began developing their own education programs, research support, societies, communication networks, and collections. However, financial resources and human efforts had to be devoted to endeavors that would allow the newly independent nation to survive, and therefore the natural sciences emerged only slowly. American knowledge about natural science improved in the early to mid-1800s as colleges were established, state-sponsored survey expeditions increased, journals and publications became more numerous, and the number of naturalists grew.[10–13]

American attitudes during the first half of the 1800s were still a mixture of a practical need to survive, a fear of the unknown wilderness, and a need to conquer and cultivate that wilderness.[14] Itinerant animal acts continued to make the rounds of the big cities, as well as the new frontier towns, traveling the back roads and along the rivers. A lion, elephant, and camel made their way to the remote frontier town of Natchez, Mississippi in 1806 and 1808, and two trained bears appeared at the Mississippi Hotel, Natchez in 1835. These itinerant acts, along with circus acts and other public amusements, were on their way to or from New Orleans, Louisiana.[15]

Other new species were displayed in America for the first time during the antebellum years, including a zebra in 1805, a rhinoceros in 1826, a quagga in 1833, a giraffe in 1837, and a hippopotamus in 1850.[3,16] A one-horned rhinoceros walked onto the stage at Peale's Museum, New York in 1826 and was still being exhibited in New York in 1829. Another rhinoceros appeared at Washington Gardens, Boston in 1830 and was on the road from 1831 to 1835. The quagga, a species (*Equus quagga*) that soon became extinct, was an important part of one traveling menagerie in 1833. Rufus Welch, Zebedee Macomber, and Eisenhart Purdy, the managers of this menagerie, were able to obtain this rare animal because they fielded their own expeditions during 1833/34. Three giraffes from Cape Town, South Africa were brought to America in 1837 and exhibited for several years. By 1842, only one survived and was still exhibited. Another was imported in 1839 and shown at Peale's Museum, New York. George Bailey imported a hippopotamus in 1850.

In 1835, a gathering of traveling menagerie owners representing most of the menageries then in existence met at the Elephant Hotel in Somers, New York to form the Zoological Institute. The institute was formed as a corporation that owned and managed its menageries jointly in a coordinated manner, monopolizing the traveling menagerie business to control expenses. The institute was short-lived, however, since it succumbed to the financial crisis of 1837. From the 1830s until the Civil War the circus was coming of age, and many exhibited their own menageries, often including trained animal acts.[17,18]

5.3 Nineteenth-Century Zoos and Aquariums

Joel R. Poinsett expressed interest in establishing a zoological garden in the United States during his Washington, D.C. address to the National Institution for the Promotion of Science in 1841. Poinsett called for a national institution that would include, among other organizations, a zoological garden.[19] His concern was expressed as part of a broader debate on establishing the Smithsonian Institution. Poinsett thought the Smithsonian should be modeled after the Jardin des Plantes in Paris, which had "an observatory, a museum containing collections of all the productions of nature, a botanic and zoological garden, and the necessary apparatus for illustrating every branch of physical science."[19,20] This suggestion, however, was in the minority and there was little interest in the zoological garden idea. The Smithsonian was eventually established in 1846 with a museum of natural history rather than a zoological garden.[21] The zoological garden idea was not publicly debated again until 1859 in Philadelphia, 1870 in Washington, and 1872/73 in Boston, but no zoological garden was established until 1874.

Until 1800, Philadelphia was the capital of the American republic, as well as its largest city and a leading cultural center. It was a city distinctive in its civic pride and its patronage of science. It was the location of America's first scientific society (the American Philosophical Society, 1743), the first botanical garden (the Bartram Botanical Garden, 1731), and the second natural history museum (the Peale Museum, 1784). The antebellum years found Philadelphia with a number of public parks, gardens, museums, circuses, menageries, concerts, theaters, and other cultural entertainments.[22] And with the determined assistance of some dedicated naturalists and civic leaders, Philadelphia would eventually find itself with America's first zoological garden.

A new kind of natural history society (for the United States), the Zoological Society of Philadelphia, was chartered in 1859. The society's purpose was to establish a living collection of wild animals on a grand scale; it was not to be a handful of animals in cheaply made enclosures. It was to be a zoological garden with several hundred animals, several permanent buildings, a keeper staff, a community-supported society, and an emphasis on education and science. During the 1860s, small urban menageries were established in New York (at Central Park) and Chicago, Illinois (at Lincoln Park), but these could not compare with the kind of facility Philadelphia's civic leaders were planning. Other menageries were no doubt established as well, but these would close, although the menageries in Central Park and Lincoln Park would continue and eventually become modern zoological gardens.

At the request of William Camac, M.D., a meeting was held at his home to discuss the establishment of a zoo in Philadelphia. This meeting of naturalists and civic-minded citizens led to the incorporation of the Zoological Society of Philadelphia on March 21, 1859. According to the society's charter, "The object of this corporation shall be the purchase and collection of living wild and other animals, for the purpose of public exhibition at some suitable place in the City of Philadelphia, for the instruction and recreation of the people."[23] In addition, an early report of the Society Board of Directors stated that it was "the aim of the Managers, not only to afford the public an agreeable resort for rational recreation, but by the extent of their collection, to furnish the greatest facilities for scientific observation."[24]

Camac had traveled widely in Europe and was well aware of the value of a zoo to the community; however, many others in Philadelphia could not understand the objectives of a zoological society or the benefits to be derived from such a large collection of animals. This

Number of American Zoos, Aquariums, and Conservation Centers by Decade Based on the Year They Were Established

Decade	Zoological Parks		Aquariums		Conservation Centers	
	No.	%	No.	%	No.	%
1859–1899	29	16	2	5	0	
1900–1909	15	8	2	5	0	
1910–1919	17	9	0		0	
1920–1929	18	11	1	3	0	
1930–1939	22	12	3	8	0	
1940–1949	5	3	0		0	
1950–1959	17	9	2	5	0	
1960–1969	23	13	5	13	0	
1970–1979	24	13	7	18	4	100
1980–1989	7	4	4	10	0	
1990–1999	4	2	13	33	0	
Totals:	181	100	39	100	4	100

opposition resulted in several failed attempts to raise funds for the new facility. A further difficulty was the site assigned to the society in Fairmount Park. This site was in an unfortunate location between two railroads and the Schuylkill River with no easy access. Because of these difficulties, the society was reluctant to develop the site for the zoo.

The Civil War interrupted the development of newly emerging urban menageries and Philadelphia's zoological garden. After a period of social and economic recovery, the zoological garden idea reemerged during the 1870s. On June 21, 1870, the Washington Zoological Society was established for the purpose of establishing a zoological garden in the nation's capital. However, nothing came of this society or its effort to establish a zoo, even though the members purchased 56 acres of land to use for the zoo site.[25] The Boston Society of Natural History considered establishing a zoological garden in 1872/73, but determined it was not economically feasible and therefore would not work in Boston, where the tradition was to have facilities that paid for themselves without an entrance fee. The society made another attempt in the 1890s to establish a zoo and freshwater aquarium in Franklin Park, as well as a marine aquarium at City Point in South Boston. But this time the society could not get the community support it needed and was not able to raise enough funds.[26,27] In 1874, Philadelphia succeeded where other cities had not.

In March 1872, Dr. Camac, still committed to establishing a new kind of zoo, held a reorganization meeting. As funds were being raised, a different site within Fairmount Park was obtained in 1873, a fence was erected around the site, an engineer was selected to design the zoo, and work began. Frank J. Thompson, an animal collector then in Australia, was authorized to obtain animals for the collection and was appointed the first superintendent of the zoo. On July 1, 1874, the Philadelphia Zoological Garden opened to the public with 282 mammals on exhibit, including antelopes, lions, zebras, kangaroos, an elephant, a rhinoceros, a tiger, some 50 monkeys, and numerous rare species from Asia and Australia. There were also some 674 birds and 8 reptiles. Just two years after its opening, the U.S. Centennial Exhibition was held adjacent to the zoo in Fairmount Park. The centennial exhibit helped draw the zoo's largest crowds until recent times. However, when one contrasts the zoo's 677,630 visitors to the fact that the neighboring centennial exhibit drew 9,910,966 visitors, it is apparent that the zoo was well attended, but was not the kind of attraction that enjoyed widespread popular appeal. The lack of appeal engendered several periods of hard times for the zoo; however, the City of Philadelphia finally recognized the value of the zoo to the city in 1891 with the first of its financial contributions. With this help, and with growing public support, America's first zoological garden succeeded, paving the way for many others to come during the next century.[28,29]

The zoological gardens of Europe exerted a major influence on the establishment of the Philadelphia Zoological Garden, as well as others to come in the nineteenth century. Wealthy and influential American civic leaders like Dr. Camac visited zoological gardens whenever they were in Europe. European science, technology, and culture still impressed American civic leaders during the later part of the nineteenth century. At the same time, however, they had strong nationalistic feelings and wanted to surpass the European endeavors. These civic leaders brought back European knowledge and ideas in an effort to improve and "Americanize" them. American botanical gardens and natural history museums were already established, and it was felt that zoological gardens would be a natural complement to these civic institutions.

The development and popularity of urban parks also influenced the establishment of American zoos. Large, landscaped urban parks became popular in the mid- to late-1800s, as did zoos. Parks offered building-bound city dwellers an opportunity to escape to a nearby pleasant natural environment. Recreational activities compatible with these natural settings became increasingly popular, and institutions compatible with these natural settings became suitable park facilities. A zoo with its animals was one of these suitable facilities. As the name implies,

Philadelphia Zoological Garden's north gate, date unknown. © Zoological Society of Philadelphia.

Philadelphia Zoological Garden's original Aviary, later remodeled to be the Reptile House, 1874. From *The Second Annual Report of the Board of Managers of the Zoological Society of Philadelphia*. © Zoological Society of Philadelphia.

Philadelphia Zoological Garden's original bear pits, 1874. From *The Second Annual Report of the Board of Managers of the Zoological Society of Philadelphia*. © Zoological Society of Philadelphia.

zoological gardens (or zoological parks) were gardens (or parks) with animals in them. In addition, most zoos in the United States began as municipal facilities, operated by parks (or parks and recreation) departments, so placing a municipal zoo in a municipal park was commonplace. Zoos and urban parks grew up together and were symbiotic.[30-32]

When the Philadelphia Zoological Garden opened, a few small urban menageries within parks existed in other cities. Central Park Zoo in New York had an uncertain beginning about 1861/62 when the first animal donations, mostly unwanted pets, were made to Central Park employees. Early animals consisted of a black bear, a pair of cows, deer, monkeys, raccoons, foxes, opossums, ducks, swans, pelicans, eagles, and parrots. Originally the animals were housed in the basement of the Arsenal Building on Fifth Avenue and in the park in small cages. Later, as the collection grew, the animals were moved to the floor of the Arsenal in 1865. The first known formal report to the Superintendent of Parks on the animal collection was made in 1873. This report indicated a much larger collection with more exotic species, such as African elephant, giraffe, Cape buffalo, eland, zebra, Malayan tapir, Brazilian tapir, kangaroo, hyena, sloth bear, and aardvark. Some of these were being temporarily housed for P. T. Barnum and other circus owners.[33-35] William Conklin, a veterinarian, was the zoo's director beginning in either 1862 or 1870 (according to two different accounts) through 1892, and the zoo may have had a library as early as 1877.[36-38]

Lincoln Park Zoological Gardens in Chicago opened in 1868 with two pairs of mute swans sent to it from the Central Park menagerie. Accommodations for the animals were built in 1870 and by 1873 there were 27 mammals and 48 birds in the collection. The first expenditures for animals was made in 1874 and the first director was hired in 1888. By 1908, there were 782 mammals and birds representing 117 species. Each park district in Chicago was independently operated and many of them had their own small menageries. After the park districts were consolidated in the 1930s, these regional menageries were discontinued, leaving Lincoln Park with the only collection.[39]

Roger Williams Park Zoo in Providence, Rhode Island was established in 1872 with a small collection of mostly native animals. A menagerie building was built in 1890 and by 1900 there were 47 species on exhibit.[40]

None of these zoos (or menageries as they were known when they opened) was equal to the Philadelphia Zoological Garden facilities, staff, or collection. The first zoo comparable to the Philadelphia Zoological Garden was a zoo founded in a city considered to be "out west." In 1872, the Society for the Acclimatization of Birds was established in Cincinnati to acquire insect-eating birds from Europe to control a severe outbreak of caterpillars. This task accomplished, they turned their attention to establishing a zoo. With advice from Alfred Brehm (of Berlin, Germany), Andrew Erkenbrecher and others in the society incorporated the Zoological Society of

Cincinnati Zoo aviary exhibits, 1875. One of these exhibits, in which the last passenger pigeon lived, has been preserved as a memorial to all endangered species. © Cincinnati Zoo and Botanical Garden.

Postcard with a composite of various scenes of the Cincinnati Zoo, late 1880s. © Cincinnati Zoo and Botanical Garden.

Cincinnati in 1873. A site was acquired in 1874 and the zoo opened to the public on September 18, 1875. The opening collection consisted of an elephant, tiger, and hyena, but was soon supplemented with many other animals. Some animals were bought from Carl Hagenbeck in Germany, at the auction sale of Phineas T. Barnum's collection, and from General James Brisbin in Omaha, as well as other sources. Despite several setbacks and difficulties, by 1891 the collection was reported to be "the largest and most complete zoological gardens in the country," as well as "the only important institution of its kind that [had] neither state nor municipal assistance."[41] Among its many accomplishments, the zoo is also known for having the last passenger pigeon, which died at the zoo in 1914, and the last Carolina parakeet, which died at the zoo in 1918.[41,42]

Public aquariums in the United States were initially commercial operations that began shortly after the aquarium concept was developed in England during the 1850s. One of the first books on the subject written and published in the United States was *Life beneath the Waters; or, the Aquarium in America*.[43] This book stated that New York stores already, in 1858, had aquatic specimens and aquariums for sale. After an 1855 trip to England, where aquariums were popular, Phineas T. Barnum established an aquarium at his American Museum, New York in 1856. James Cutting, who had been experimenting with aquariums since about 1854, opened the Boston Aquarial Gardens in 1859. In 1860, this facility became the Boston Aquarial and Zoological Gardens, with one floor devoted to opossums, raccoons, muskrats, guinea pigs, monkeys, bats, lions, a leopard, a grizzly bear, black bears, deer, foxes, a kangaroo, birds, an alligator, an anaconda, pythons, and two trained seals. In 1861, three beluga whales, a bottlenose dolphin, and a gray shark were exhibited. In 1862, Barnum bought the gardens and renamed them the Barnum Aquarial Gardens. However, in 1863 this facility closed and the animals were transferred to Barnum's American Museum, which burned to the ground in 1865. In 1876, the Great New York Aquarium, another private endeavor, opened with 42 tanks of fresh and salt water, but soon closed.[44–46]

What was to become the original National Aquarium opened initially as a public facility of the U.S. Commission of Fish and Fisheries at Woods Hole, Massachusetts in 1873. In the 1880s,

Cutting & Butler's Grand Aquaria at the Aquarial Gardens, Boston during the latter nineteenth century. This was one of the earliest public aquariums established in the United States. © The Boston Athenæum.

LAST PASSENGER PIGEON, CINCINNATI ZOO

The passenger pigeon was once common in the eastern United States and Canada. A migratory species, its range extended from Canada to north Florida and from the eastern seaboard to the midwestern states. In some areas, the birds were the avian equivalent of the bison. Early accounts describe flocks so extensive they blocked the sun like a giant feathered cloud.

Like the bison, passenger pigeons were so common they were hunted with little thought given to the possibility of their extinction. A serious decline in the pigeon's population began in the 1850s as a result of commercial hunting for food, which increased as a result of improved rail transportation to urban marketplaces. In the 1850s, the New York City food markets alone consumed about 300,000 birds annually. By 1871, the same amount was shipped East each month, and millions of birds were killed every year. The pigeon's forest habitats were cut down, and the species was considered an agricultural pest.

Attempts to breed the passenger pigeon were few because it was so common in North America and because it was considered a pest. John James Audubon took some pigeons to the London Zoological Gardens in 1832, which later bred. Others began breeding colonies the aquarium was transferred to the Washington headquarters of the commission and in 1888 this Central Station Aquarium opened to the public, while its work on fishery research and its role as a hatchery continued. In 1903, the commission became part of the U.S. Department of Commerce and the aquarium moved into a new Commerce Building in 1932 with a collection of some 57 species and 1,000 specimens. The National Aquarium Society was organized in 1932 as a private support group for the renamed Bureau of Fisheries Aquarium. The Bureau of Fisheries was transferred to the U.S. Department of Interior as part of the U.S. Fish and Wildlife Service in 1939; however, the aquarium remained where it was. As far back as 1871, attempts were made to establish this aquarium as a national aquarium, but it was not until 1961 that funds were finally appropriated to establish a National Fisheries Center and Aquarium; however, this newer facility never became a reality. Instead, in 1979 Congress granted the "National Aquarium" title to the new aquarium in Baltimore. During 1981/82, Congress attempted to close the Washington facility, also known as the National Aquarium. In 1982, the National Aquarium Society reconstituted itself and took over operation of Washington's National Aquarium, which still resides in the Commerce Building.[47]

Until the 1880s, zoos and aquariums were, for the most part, phenomena of the northeastern cities. A few more were established in this region's cities, including Binghamton, New York (1875) and Buffalo, New York, where a local businessman donated a pair of deer he did not know how to care for that were set to graze in a city park in 1875. As the herd multiplied, it was fenced in and joined by a few cows, sheep, and buffalo. In 1894, the City Parks Commission took control of the collection, which became known as the Buffalo Zoological Gardens and included 22 animals by 1895. New facilities were built, but in 1899, with 138 animals, it was reported that what the city needed was "not a menagerie, and a poor one at that, but a Zoological Park — not the penning of a few animals and birds in inadequate fields, houses and cages, but places for them to move about."[48] In the Mid-Atlantic region, the Baltimore Zoo began similarly

---

when the wild population began to decline. A Wisconsin prospector and naturalist, David W. Whittaker, owned one of the largest captive flocks. This flock of about 50 birds was raised from a pair he received from a Native American in 1888. These were sold to Charles O. Whitman, a professor at the University of Chicago. However, Whitman had no luck raising the birds and the last of them died in 1907.

Soon after its opening in 1875, the Cincinnati Zoo acquired some pigeons, which also bred, but not well enough to sustain their captive population. Additional pigeons were bought or captured in 1876 and 1878. By 1881, the zoo had 20 pigeons, but its captive population was declining. From 1899 onward the zoo offered $1,000 — quite a sum of money at the time — for another pair, but the reward was never claimed. Before anyone realized it, the species had vanished.

By 1907, when Whitman's flock died out, Cincinnati Zoo only had three pigeons left. One died in 1909, another in 1910. The last, a female named Martha (named after Martha Washington), was part of a trio obtained from the Whitman flock in 1902. Martha was found dead on September 1, 1914. Martha's body was given to the Smithsonian Institution where it was displayed at the U.S. National Museum. The Cincinnati Zoo has preserved the birdhouse where she lived as a memorial to the passenger pigeon and all endangered species.

Sources: Based on Gale, Oliver M., "The Cincinnati Zoo: one hundred years of trial and triumph," *Cincinnati Historical Society Bulletin*, 33, 87, 1975; Kisling, Vernon N., Jr., "The last passenger pigeon dies," in *Great Events from History II: Ecology and the Environment*, Salem Press, Pasadena, CA, 111, 1995; Schorger, A. W., *The Passenger Pigeon: Its Natural History and Extinction*, University of Oklahoma Press, Norman, 1973.

with a few deer and a flock of sheep in 1876. By the 1890s the zoo contained an extensive collection — camels, monkeys, cats, an alligator, and various birds.[49]

During the 1880s, zoos appeared in what was then perceived as the frontier. Frontier life was shifting from the midwestern region to the western territories, as well as lingering in much of the South. As the population of these regions increased and major cities developed, the cultural amenities of the large, well-established eastern cities began to appear in these western and southern cities. These cultural amenities included zoos. The Cleveland Zoo (Ohio) began in 1882, when a herd of deer and a site for the zoo were donated to the city. Enclosures and a zoo building soon followed, and in 1888 the collection included two black bears, a family of raccoons, a pair of foxes, and a colony of prairie dogs. By 1895, there were 321 animals, including, however, 200 doves.[50] Zoos appeared in Portland, Oregon and Dallas, Texas in 1887 and 1888, respectively. The year 1889 saw the beginnings of two other notable "frontier" zoos, one in the South and one in the Far West. A prominent Atlanta, Georgia businessman purchased the animals of a bankrupt circus — two lions, two cougars, two wild cats, monkeys, a black bear, a raccoon, a hyena, a jaguar, a Mexican hog, a Bactrian camel, a dromedary camel, an elk, and a gazelle. He donated the collection to the city, which accepted and built a menagerie building that housed animals for the next 70 years.[51] The San Francisco Zoo began with the exhibition of a grizzly bear in Golden Gate Park. The zoo moved to its present location in 1928 when it contained two zebras, a Cape buffalo, a Barbary sheep, five rhesus macaques, two spider monkeys, and three elephants.[49]

For some, the late nineteenth-century zoological park was one of the many symbolic achievements of America's greatness, power, and influence, just as European collections symbolized the global achievements, power, and influence of their home countries.[52] Others, however, felt the new zoos were falling short of their stated objectives.[53] Discussions at the time concerned "the real mission of zoological gardens in contradistinction to menageries."[54] This "mission" separated the zoological garden from the menagerie because "a zoological garden ... has within its means to powerfully aid, encourage, and stimulate human progress, education, and science in an infinite variety of ways; and such an institution stands among the very best of investments to be made either on the part of State or city."[55] Others involved in the discussion simply promoted the idea of animal rights, which is ironic, considering the state of human rights at the time in American society.[56]

Although the capitals of Europe, major cities in other countries, and several cities in the United States had zoological gardens as the turn of the century approached, the U.S. capital was still without a national collection of animals. As discussed earlier, there was an attempt in 1870 to establish a zoo in Washington, D.C., but it was unsuccessful. And the federal government, which had a difficult time supporting the Smithsonian Institution, did not feel it should fund scientific or cultural activities.[57]

In 1887, the Smithsonian Institution's Museum of Natural History established a Department of Living Animals to provide its taxidermists with an opportunity to observe realistic habits and positions of the various species they were preparing for exhibition at the Museum. William T. Hornaday, the Smithsonian's chief taxidermist, was appointed curator of this living collection in 1888. Hornaday had, by then, already conducted an expedition to several northwestern states and territories, returning with a large collection of deer, bears, foxes, lynxes, eagles, and other animals. These animals formed the nucleus of this new living collection. A wooden structure was built for the collection south of the Smithsonian building, and on December 31, 1887 it opened to the public. It became a popular attraction, resulting in the donation of many additional animals. By March 1888, the collection had nearly doubled in size and required a full-time keeper to care for the animals.

Previously, animals received on behalf of the federal government had often been sent to existing zoos in other cities. Hornaday and others felt these animals should stay at a national

Bison on view at the Smithsonian castle in 1888, before they were transferred to the National Zoological Park as part of its original collection in 1891. © National Zoological Park, Smithsonian Institution.

collection in Washington because "a nation so far in advance in the march of progress as the United States should [have] some such institution under Government protection." Hornaday felt this national collection would be "one of the leading attractions of Washington," and should be concerned with the "importance of preserving living North American mammals."[58]

Three bills were introduced in Congress during 1888 to establish a national zoo, but they were all defeated. Despite these setbacks, on March 2, 1889 an amendment to the annual appropriations bill authorizing expenditures for Washington included funds to establish a zoo for the advancement of science and the recreation of the people. A 175-acre site was chosen that provided large, open habitat exhibits in a scenic area of the capital and the 185 animals at the Smithsonian's Museum were transferred to the new site. Frederick Law Olmsted, who would become involved with the design of many other zoos, designed the

William Hornaday, chief taxidermist at the U.S. National Museum, Smithsonian Institution, with a bison calf on the Mall near the Smithsonian castle, 1886. © National Zoological Park, Smithsonian Institution.

A public school group looks at parrots in front of the National Zoo's original Lion House, 1899. © National Zoological Park, Smithsonian Institution.

Tasmanian wolf (thylacine) on exhibit at the National Zoo. This female arrived in 1902 with three young in her pouch, two of which survived. This species became extinct in the 1930s when the last wild and captive specimens died. © National Zoological Park, Smithsonian Institution.

grounds. The National Zoological Park opened to the public on April 30, 1891 under the direction of Frank Baker as superintendent (director) and William Blackburn as head keeper. Hornaday was not chosen as the director of this new zoo and he resigned in 1890 before it opened.[21,58–61]

Several new zoos were established during the last decade of the nineteenth century, not only in the East, but in the West and South as well. The Denver Zoo (Colorado) began modestly with the donation of a bear cub to the mayor in 1896. The bear was taken to the city park and tethered to a haystack; other species soon joined the bear — eagles, wolves, deer — and the city zoo was born. In Pittsburgh, Pennsylvania the effort to found a zoo began in 1895, and by the time the

William Blackburne, head keeper, stands in front of the National Zoo's first permanent building, the Lion House. Built in 1892, it was replaced with the Lion/Tiger Exhibit in 1974. © National Zoological Park, Smithsonian Institution.

William Blackburne, National Zoo's first head keeper, gives a young camel a drink, 1893. © National Zoological Park, Smithsonian Institution.

zoo opened in 1898 as the Highland Park Zoo it contained elephants, lions, tigers, antelope, zebras, and a building to house them.[62] In 1899, the Toledo Zoo (Ohio) obtained a woodchuck, a wolf, a pair of foxes, raccoons, rabbits, alligators, a badger, an opossum, and an assortment of birds in anticipation of its opening the next year. Many other zoos opened during the era, including in Springfield, Missouri (1890), St. Louis, Missouri (1890), Grand Rapids, Michigan (1891), Bloomington, Illinois (1891), Milwaukee, Wisconsin (1892), Brooklyn, New York (Prospect Park Zoo, 1893), St. Augustine, Florida (1893), New Bedford, Massachusetts (1894), Rochester, New York (1894), New York (New York Aquarium, 1896), St. Paul, Minnesota (1897), Alamogordo, New Mexico (1898), and Omaha, Nebraska (1898).

New York City at this time had numerous animal attractions, menageries, and aquariums, as well as one of the earliest urban park systems. In 1884, a New York State commission report on public parks stated that "no park system can be regarded as complete without suitable tracts for botanical and zoological gardens" and that "a park system that failed to include a zoological garden would be wanting in one of the most essential requisite."[63] With the help of Theodore Roosevelt and prominent New York citizens, the Boone and Crockett Club (a prestigious sportsmen's group) turned these grand statements into reality.

The New York Zoological Society (now the Wildlife Conservation Society) was incorporated on April 26, 1895. Objectives of the society were "to establish and maintain in [New York] a zoological garden for the purpose of encouraging and advancing the study of zoology, original researches in the same and kindred subjects, and of furnishing instruction and recreation to the people."[64] The first annual report of the society, stated the objects of the society:

> [T]he establishment of a free zoological park containing collections of North American and exotic animals, for the benefit and enjoyment of the general public, the zoologist, the sportsman and every lover of nature;… the systematic encouragement of interest in animal life, or zoology, amongst all classes of the people, and the promotion of zoological science in general;… and … cooperation with other organizations in the preservation of the native animals of North America, and encouragement of the growing sentiment against their wanton destruction.[65]

Elephant House at the New York Zoological Park (now the Wildlife Conservation Park), 1910. © Wildlife Conservation Society/Bronx Zoo.

Bear exhibit at the New York Zoological Park (now the Wildlife Conservation Park), early 1900s. © Wildlife Conservation Society/Bronx Zoo.

The New York Zoological Park (the Bronx Zoo, now the Wildlife Conservation Park) opened to the public on November 8, 1899 with a collection of 843 specimens representing 157 species. After failing to be appointed director of the National Zoo, William T. Hornaday had become thoroughly disappointed with zoos and had spent several years in private business selling real estate. Fortunately, he was persuaded to return to zoo administration and became the first director of the New York Zoological Park. Hornaday was finally able to implement his ideas concerning zoo management, and in the process he set the standard for other American zoos. By 1912, when the last buildings of the original plan were completed, the collection had almost 5,000 animals, and this success was primarily the result of Hornaday's efforts.[33]

Baird Court at the New York Zoological Park (now the Wildlife Conservation Park), 1910. © Wildlife Conservation Society/Bronx Zoo.

In 1896, the New York Aquarium opened to the public in a structure that had been built as a fort in 1807 and subsequently used for a variety of purposes before becoming an aquarium. The City of New York managed the aquarium until it was turned over to the New York Zoological Society in 1902. Its collection consisted of an extensive selection of native freshwater and marine fishes, as well as amphibians and aquatic reptiles. Aquatic mammals such as harbor seals, West Indian seals, and manatees were also exhibited, and in 1897 the aquarium exhibited white whales.[33]

5.4 Twentieth-Century Zoos and Aquariums

As the twentieth century began, major collections existed at Philadelphia, Cincinnati, Washington, and New York, and there were 25 other zoos and two public aquariums (these numbers include institutions that are still extant; others existed, but closed and little is known about them). The first half of the twentieth century was a strong growth period for American zoological gardens, with 77 zoos and 6 aquariums established between 1900 and 1949. It was also a period of professional growth and accomplishment.

Sometime during the years of 1901 to 1905, scientific work began at a Philadelphia Zoological Garden facility that would become the Penrose Research Laboratory, the first research center within an American zoo. Early research included veterinary pathology, animal nutrition, acclimatization of tropical animals to temperate climates, the use of glass between primates and visitors to control diseases, and the treatment of tuberculosis.[29] Work done at this laboratory resulted in what may be the first veterinary book pertaining to captive wildlife, *Disease in Captive Wild Mammals and Birds*, by Herbert Fox.[66]

The New York Zoological Society established the first American zoo-related scientific journal, *Zoologica* (published from 1907 to 1973); the first veterinary clinic in 1916; and the first zoo-based field research program, the Department of Tropical Research, in 1916. Starting with a Tropical Research Station in British Guyana, the society has developed one of the largest field conservation programs in the world. Hornaday and the New York Zoological Park were also instrumental in saving the American bison from extinction.[33]

After Carl Hagenbeck opened his Tierpark at Stellingen, Germany in 1907, showcasing his natural, open, moated panoramic exhibition techniques, the barless exhibit concept received worldwide attention. In the United States, the earliest zoos to use open, moated exhibits were the Denver Zoo (1918), St. Louis Zoo (1919), Detroit Zoo (1920s), Chicago Zoo (1930s), and Cincinnati Zoo (1930s). There were many zoo directors who did not approve of this exhibit design, or who could not afford to replace existing exhibits with this new style. However, these new exhibits gained popularity and were precursors to the naturalistic, immersion exhibits that zoos would eventually adopt in the latter half of the century.[33,41,67–69]

Hagenbeck's barless concept might have influenced American zoos sooner; however, the country's attention had been concentrated on European conflicts that would soon spread. World War I did not immediately affect American zoos, as the United States attempted to remain neutral when the war began. By 1915, however, zoo and aquarium employees began enlisting. The New York Zoological Society allowed its employees to take leaves of absence with full pay and the society set up a cooperative store to help its employees obtain items needed for their families. For the most part, however, zoo and aquarium operations continued as usual until the United States finally declared war in 1917. Afterward, the New York Zoological Park's lion house was turned over to the American Red Cross and a company of soldiers (zoo employees) was formed, armed, uniformed, and drilled at society expense. Zoo employees assisted with a number of war-preparedness activities. When food became scarce, the zoo plowed the elk range and grew its own crops. Animal importation ceased because of problems

Zoological Gardens of the United States 165

San Diego Zoo's original lion enclosure with Miss Scripps, who donated the funds for this enclosure, in the foreground, early 1900s. © Zoological Society of San Diego.

with transportation related to the war effort, but this did not significantly affect the collections. Similar wartime activities occurred at other zoos throughout the country. Although times were difficult, American zoos fared better than did European zoos, and after the war American zoos helped their European counterparts in unique ways. For example, the New York Zoological Park was able to send the Antwerp Zoo 329 animals, along with a large quantity of animal food.[33]

With the end of the war, new zoos and aquariums opened across the United States, among these the San Diego Zoo. A small collection of animals in circuslike cages abandoned after the 1915/16 Panama–California International Exposition in Balboa Park inspired a local physician, Dr. Harry M. Wegeforth, to establish a zoo to benefit his city. He and four other men founded the Zoological Society of San Diego on October 26, 1916, but it was not until after the war that the city granted the present site, that the grounds were laid out, including roads for the bus tours that still run today, and that Dr. Wegeforth, as director of the zoo, embarked on aggressive animal collection ventures.[70]

Following the war, Dr. Wegeforth and several directors of other zoos felt there was a need for organization and cooperation among the growing number of zoos. Attempts at organizing a professional association occurred at Chicago, St. Louis, and San Diego during 1924. For a year or so prior to 1924 there was also discussions within the American Institute of Park Executives (AIPE) on how to accommodate better the institute's zoo executives. Dr. Wegeforth was one who felt there needed to be more cooperation among zoo executives, especially when it came to acquiring surplus animals from each other's facilities or when expeditions were needed to acquire new specimens.

In April 1924, Dr. Wegeforth and his colleagues from St. Louis, Nashville, and Washington met in San Diego to form the National Association of Zoological Executives. However,

Saving the Bison from Extinction, New York Zoological Park

American bison were common, both east and west of the Mississippi River. Eastern herds ranged from New York south to Florida and western herds ranged throughout the Plains states. Eastern herds were smaller and disappeared early, but western herds were so extensive little thought was given to the possibility that they would ever disappear. Nevertheless, the once abundant bison (buffalo) was to become extinct in the wild.

William Hornaday, Director of the New York Zoological Park (now the Wildlife Conservation Park), with one of the first bison from the zoo's collection to be shipped west to be reintroduced into the wild, ca. 1907. © Wildlife Conservation Society/Bronx Zoo.

Eastern bison herds died out by 1830, the southern half of the western herds were exterminated by 1889, and the northern half of the western herds were gone by 1897. At the turn of the century, the only wild herds left were in Canada (the wood bison) and Yellowstone National Park (the plains bison). The American Bison Society was established on December 8, 1905 for the purpose of preserving and increasing the number of bison and other North American big game species. The society coordinated the establishment of wild herds through the use of captive specimens. William T. Hornaday conducted a 1908 census that accounted for 1,116 bison in 15 zoos and private collections (compared with 25 managed wild bison surviving in Yellowstone National Park and none in the wild outside of the park).

According to Hornaday's survey, the largest zoo herds were at the New York Zoological Park (33), Cincinnati Zoo (21), San Francisco Zoo (18), and Denver Zoo (16). In 1907, the New York Zoological Park was the first zoo to help the American Bison Society with its reintroduction project, sending 15 bison to the Wichita Forest Reserve in Oklahoma. Other reservation herds were established in succeeding years using additional zoo-bred animals. By 1933, there were 4,404 bison in the United States and 17,043 in Canada.

Sources: American Bison Society, *Annual Report of the American Bison Society 1905–1907*, American Bison Society, New York, 1908; Garretson, Martin S., *A Short History of the American Bison*, American Bison Society, New York, 1934.

San Diego Zoo's original flight cage, then the world's largest, early 1900s. © Zoological Society of San Diego.

San Diego Zoo's first tour bus, early 1900s. © Zoological Society of San Diego.

Some of the founders of a national zoo association in April 1924. This association was superseded by the American Association of Zoological Parks and Aquariums in October 1924. Top (left to right), A. D. Luchrman (St. Louis Zoo), Harry M. Wegeforth (San Diego Zoo), George P. Vierheller (St. Louis Zoo). Seated (left to right), Fred W. Pape (St. Louis Zoo), Frank Schwartz (St. Louis Zoo). © Zoological Society of San Diego.

before this association could become established, these individuals along with other zoo executives met at the American Institute of Park Executives annual conference in October 1924 to form the American Association of Zoological Parks and Aquariums (AAZPA) as an affiliate of the institute. Its purpose was to promote and advance zoological parks and aquariums; to aid in the exchange and importation of zoological specimens; to provide exhibits for scientific, educational, and recreational purposes; and to aid in the preservation of wildlife.[71–74]

By 1924, aquariums had already come and gone in Miami Beach, Florida and Venice, California, while others would close shortly afterward in Boston (eventually replaced with the New England Aquarium), Chicago (at Lincoln Park), New Orleans (Odenheimer Aquarium, replaced later by the Aquarium of the Americas), and Philadelphia (at Fairmount Park). In addition to the Washington and New York aquariums already mentioned, others had opened at Detroit (Belle Isle Aquarium, 1904), Honolulu, Hawaii (Waikiki Aquarium, 1904), and San Francisco (Steinhart Aquarium, 1923). Aquariums were planned for Cleveland, Madison (Wisconsin), Milwaukee (Wisconsin), San Diego, St. Louis, and Wichita (Kansas).[45]

In October 1925, San Diego Zoo hired a bookkeeper, Belle J. Benchley, who would become the first woman to be a zoo director at a public zoological park (San Diego Zoo, 1927–1953), as well as the first and only woman, to date, to be president of the American Association of Zoological Parks and Aquariums (1949–1950). Women zookeepers were hired, perhaps for the first time, about 1931. And Patricia O'Connor was hired in 1943 as the first woman veterinarian at the Barrett Park Zoo (Staten Island, New York).[74–78] (Much later, from 1972 to 1975, Margaret A. Dankworth would be the first executive director of the newly independent AAZPA.)

Recreation, in the form of wholesome family entertainment, was often viewed as an important function of the early zoological gardens. As one visitor to the New York Zoological Park put it in 1904:

> [L]earning natural history ... is not the greatest good this Zoo does for the multitude. It matters little whether Michael Flynn knows the difference between the caribou and the red deer. It does matter a lot, however, that he has not sat around the flat disconsolate, or in the back room of the saloon, but has taken the little Flynns and Madam Flynn out into the fresh air and sunshine for one mighty good day in which they have forgotten themselves and their perhaps stuffy city rooms.[79]

Learning natural history, however, was an important aspect of a zoo or aquarium visit. Information and signage was necessarily basic, but zoos and aquariums were important educational sources for information about animals and nature. Education programs at the New York Zoological Park, perhaps the best at the time, came under legal scrutiny in 1927 when Anna M. Harkness willed her inheritance to the zoo. Her will was probated in Ohio and the State of Ohio claimed an inheritance tax based on its opinion that the New York Zoological Park was not an educational institution and therefore not exempt from the tax. The courts, however, upheld the zoo's educational status, based on its charter and its demonstrated efforts to educate the public through its publications, tours, and lectures. The zoo hired a docent (so-called by the zoo's director, W. Reid Blair) in 1929 to work with school teachers and their students in a more formal manner as curator of educational activities.[33]

During the decades of the 1930s and 1940s the devastation of the Depression caused budgets to be severely cut, making it difficult to pay employees and feed animals. Programs of the Works Progress Administration (1935 to 1939), and to a lesser extent the Civil Works Administration (1933 to 1934) and Federal Emergency Relief Administration (1933 to 1938), offset these difficulties somewhat with funds and labor to build new or renovated zoo and aquarium exhibits and facilities. Zoos were completely modernized at Buffalo, Belle Isle (Detroit), Toledo, San Diego, and St. Paul. Others received major improvements or new exhibits: a primate building at Fair Park Zoo (Little Rock, Arkansas), a bird and reptile building at Lincoln Zoo (Nebraska), a hippopotamus building and various grounds improvements at Cincinnati Zoo, a tropical birdhouse at Pueblo Zoo (Colorado), a feed-storage barn at Lafayette Zoo (Indiana), bear dens at New Bedford Zoo (Massachusetts), general repairs at Potter Park Zoo (Lansing, Michigan), an elephant building at Duluth Zoo (Minnesota), extensive repairs at Philadelphia Zoo, a hospital and commissary at San Antonio (Texas), and a monkey exhibit and other displays at Cleveland Zoo.[80,81]

This period also marked the first American exhibition of the giant panda, in 1937, at the Brookfield Zoo (Chicago). This was a great media event, a modern version of the excitement rarely seen wildlife have caused throughout history. And in 1938, Marineland (St. Augustine) opened as the first oceanarium, housing marine mammals.

With World War II came a reprise of the problems zoos across the world faced during the previous war, but to a much greater extent. While U.S. zoos shared in these difficulties, they did not suffer the destruction European zoos experienced. Budgets were reduced, food and utilities were in short supply, building materials were difficult to obtain, and employees left for military service. Jobs were filled by women to some extent, and the role of women at zoos and aquariums would continue to grow. Only five zoos, and no aquariums, were opened during the 1940s, the lowest number of any decade in the twentieth century.

Post–World War II years were a relatively prosperous period. While the 1940s saw the establishment of new facilities come to a near standstill, recovery began in the 1950s with the

First Zoo Women — Directors, Veterinarians, and Keepers

Mary and John Elitch, Jr., owned the Elitch Amusement Gardens (open in 1890), one of several private zoos existing in Denver prior to the opening of the Denver Zoo in 1896. A year after opening, John died and Mary took over. She took charge of the animals' care and even descended into the bear pits daily to feed them. Mary received a fair amount of fame, and was asked to establish zoos in other cities. She was even recognized in Europe for her efforts. A London magazine noted her achievements as the only woman who owned and operated a zoo. Contemporary accounts considered the Elitch collection to be "first class." It included elephants, giraffe, camels, lions, a polar bear, bison, seals, kangaroos, and cassowaries.

Belle J. Benchley was the first woman to be director of an American public zoo, the San Diego Zoo. Hired in 1925 as a temporary bookkeeper, Benchley did so well she was retained as a full-time employee. She made public appearances for the zoo, handled the trading and selling of animals, and had other responsibilities. After the zoo's founding director, Harry M. Wegeforth, was succeeded by four unsatisfactory directors in a row, the Board of the Zoological Society of San Diego hired Benchley as its director on Wegeforth's recommendation. The Board was reluctant to hire Benchley, however. As Wegeforth put it, "So far as we knew, no woman had ever filled a position of such importance in a zoo, and the Board felt it would be an experiment with an element of risk." Overcoming this "element of risk," Benchley remained director until she retired in 1953. Benchley also served as the first (and so far only) woman president of the American Association of Zoological Parks and Aquariums during 1949/50." Heini Hediger, Director of the Zurich Zoological Gardens, dedicated one of his widely read books to her. *The Psychology and Behaviour of Animals in Zoos*

Elitch Amusement Gardens bear pit with Mary Elitch feeding the bears, ca. 1893. At the time, she was the only known (and perhaps first) woman zoo director. Photograph by Jones and Lehman. © Western History Collection, Denver Public Library.

opening of 17 new zoos and 2 new aquariums. This post–World War II period was also a time for renewed professionalism, building new exhibits, and developing new programs. The American Association of Zoo Veterinarians (AAZV) was founded in 1946. Although an informal association began with a meeting in 1936, the AAZV was not established until 1946 at the annual conference of the American Veterinary Medical Association. Its aims were to inform *and Circuses* was "Dedicated in respect and gratitude to the perfect Zoo Director Mrs. Belle J. Benchley."****

Patricia O'Connor became the first woman zoo veterinarian at an American zoo when she was hired at the Barrett Park (Staten Island) Zoo, New York in 1943. O'Connor received her veterinary degree from Cornell University in 1939 and worked with several private practices before replacing Daniel Gates at the zoo. She was also the first president of the American Association of Zoo Veterinarians when it formed in 1946 (serving from 1946 to 1957 and again from 1965 to 1966), as well as the Association's program chair (serving from 1946 to 1954 and from 1956 to 1958). O'Connor retired from the Staten Island Zoo in 1970.†

It is difficult to know for certain who were the first women zookeepers at an American zoo or aquarium. The first known instance occurred about 1931 at the Woodland Park Zoological Gardens, Seattle, Washinton. Director Gus Knudson explained his reason for hiring women zookeepers, "Since women can be found who are really interested in animal life, and are able to do the work, I see no reason why they shouldn't be given the opportunity of working in the zoological gardens. The women we have added to our staff are keen observers, interested in their work, neat, humane, and are very attentive, hardly any absenteeism." The first of these women keepers, Margaret Wheeler, was raised on a farm and had worked as a veterinary assistant. She worked with many different animals at the zoo, as well as the education program. In 1943, there were two other women working at the zoo.††

When the American Association of Zoological Parks and Aquariums (AAZPA) became an independent Association in 1972, Margaret A. Dankworth became its first Executive Director. Dankworth had been the Office Manager for the National Recreation Association in New York, as well as its field representative in the Great Lakes District (1948 to 1958). She then became the Assistant Executive Secretary of American Institute of Park Executives (AIPE) and later the National Recreation and Park Association (NRPA), which absorbed AIPE (1958 to 1970). As secretary of these two associations, she handled the affairs of the AAZPA, a branch affiliate. After 1970, she remained with the NRPA at the Oglebay Park office in Wheeling, West Virginia, where she continued to handle AAZPA affairs until AAZPA became independent from NRPA. Dankworth was chosen to continue her AAZPA work as its executive secretary (later this title was changed to executive director) and served from 1972 to 1975.†††

* Etter, Carolyn and Etter, Don, *The Denver Zoo: A Centennial History*, Roberts Rinehart, Denver, 1995.
** Wegeforth, Harry M. and Morgan, Neil, *It Began with a Roar: The Beginning of the World-Famous San Diego Zoo*, Rev. ed., Zoological Society of San Diego, San Diego, 1990.
*** Dover, New York, 1968 [Reprint of 1955 edition].
† Conant, Roger, "First woman zoo veterinarian," *Parks and Recreation*, 26, 340, 1943 see also Fowler, Murray E., *Historical Perspectives of the American Association of Zoo Veterinarians*, American Association of Zoo Veterinarians, Davis, 1977.
†† Knudson, Gus, "Seattle has women zoo keepers," *Parks and Recreation*, 26, 385, 1943.
††† Dankworth, Margaret A., personal communication, 1998.

members of new developments in captive wild animal veterinary medicine, to publish and disseminate information, and to encourage the use of veterinary services at all zoos.[77,82]

Veterinary research, nutritional studies, and disease studies were under way at the Penrose Research Laboratory (Philadelphia Zoological Garden), and the New York Zoological Park set an example with its veterinary clinic. Nutritional research on artificial diets at the Penrose Research Laboratory began in earnest after 1935, helping lead to the development of commercially produced, artificial diets that became popular in the 1960s and continue to improve nutritionally.[83]

Chemical immobilization of wildlife using remote drug delivery systems (rifles and darts) began in 1953 with the immobilization of free-ranging white-tailed deer. This practice led to a number of studies to improve the delivery systems and the chemicals used. Sophisticated systems using a variety of instruments (rifles, pistols, arrows, blowpipes, and a variety of darts), as well as a diversity of new drugs, have since replaced the original crude systems and generic drugs.[84]

Captive animal husbandry has been greatly improved by these many advances in veterinary medicine, including the treatment of exotic animal diseases, reproductive biology, genetics, nutrition, surgery, pathology, and others too numerous to mention. In addition to improved animal health, there have been improvements in the shipment of animals and their survivability after arrival, in animal handling and restraint, in propagation success and control, and in exhibition techniques.

Although most new zoos and aquariums developed along the traditional exhibit and program lines established by their predecessors, the Arizona–Sonora Desert Museum established unique ecological exhibits and programs. The Arizona–Sonora Desert Museum was founded in Tucson, Arizona in 1952 as a living museum of animals and plants native to the Arizona–Sonora Desert region. Buildings and enclosures were integrated into the habitat and were hardly noticeable. Enclosures were miniature natural habitats and education was the zoo's primary purpose. Information was provided on the entire ecosystem, not just the animals and plants. These were ideas that would be adopted by other zoos in later years.[85]

FIRST GIANT PANDA, BROOKFIELD ZOO

Ruth Harkness brought the first giant panda to the United States in 1936, and in 1937 it was purchased for $14,000 by the Brookfield Zoo in Chicago. Harkness wrote about her exploits in China and her great "find" in the book, *The Lady and the Panda*. Based on the recent panda craze, one can imagine the excitement that surrounded this first panda, an animal little known to the public or scientists at the time. The lady and the panda arrived in San Francisco aboard the *President McKinley* on December 18, 1936 to an enthusiastic crowd of reporters looking for a break from Depression-era stories. The panda, Su Lin (originally thought to mean "something very cute," a recent interpretation of the Chinese characters indicates the more appropriate meaning, "joyous gem'"), was an infant being hand-raised, which made it an irresistible "attention getter."

Harkness' adventure in the wilds of China and her panda find may not be exactly as described in her book, however. Ruth's husband, William, died in 1936 while in China on an expedition with Tangier Smith. William Harkness was a wealthy adventurer who had already been on an expedition to Indonesia, where he captured Komodo monitor lizards for the New York Zoological Park in 1934, before traveling to China. Tangier Smith spent many years in China capturing animals for Asian and Western zoos. His dream was to capture the elusive giant panda and he went to a great deal of trouble and expense to do so — establishing contacts with Chinese officials, setting up an extensive system of camps, hiring competent assistants, and making important contacts.

Progress was also made in the publication and dissemination of the professional literature. Books, journals, conference proceedings, and newsletters focusing on professional issues increased significantly during the 1960s and 1970s, coinciding with a growth of professional activities in traditional zoo programs (animal husbandry, exhibition, education, research, conservation) and newer zoo programs (marketing, public relations, historic preservation).[86,87] Professionalization among the zookeepers resulted in the establishment of the American Association of Zoo Keepers (AAZK) in 1967.[88] Canadian zoo and aquarium professionals, formally part of the AAZPA, established their own Canadian Association of Zoological Parks and Aquariums (CAZPA) in 1974.

Conservation efforts have always existed as part of zoo and aquarium programs, but the extent of these efforts varied from one institution to another and were limited due to what was (and was not) known about the biological and conservation needs of the species at risk. As part of a renewed national concern for conservation in the 1960s and 1970s, zoos and aquariums intensified their efforts to conserve endangered species through improvements in programs specifically concerned with conservation. Significant federal legislation also affected these zoo efforts, including the Animal Welfare Act of 1966, the Marine Mammal Protection Act of 1972, the Endangered Species Act of 1973, and adoption of the Convention on International Trade in Endangered Species (CITES) of 1973.

In 1966, the AAZPA became a professional branch affiliate of the newly formed National Recreation and Park Association (which absorbed the AIPE). In the fall of 1971, the AAZPA membership voted to become an independent association and, in January 1972, it was chartered as the AAZPA with its executive office in Wheeling, West Virginia (now located in Silver Spring,

---

When word reached Ruth that her husband had died, she said she may have lost a husband, but she had gained an expedition — and then she promptly set off for China. A cultured New York dress designer, Ruth was not a likely candidate to trek through the wilds of China, dodging both Chinese Nationalist Army and Red Army units at war, looking for a seldom-seen animal. Nevertheless, she claims to have traveled the Yangtze River across China, set up a series of camps, avoided the warring armies, and found an infant panda sitting alone in a huge rotting tree — on one of her first tries and all within a six-month period. She immediately reached Shanghai with the panda, avoided Chinese officials upset with her unauthorized acquisition and headed for San Francisco.

Evidence indicates that she did not get along with Tangier Smith, but in discussions with him she became familiar with his established network. Taking advantage of Smith's absence from the field, Ruth supposedly went straight to one of his camps where an infant panda was being held for Smith, and bought it. Some accounts say she was able to do this because she posed as Smith's associate. Smith was unable to stop her because he was ill in a Shanghai hospital at the time. The press, having a grand time with the little panda, ignored Smith's claims.

Panda Su Lin was in the Brookfield Zoo for less than a year when it died in 1938. Smith recovered from his illness long enough to bring five pandas to London in 1938, but he was hospitalized shortly afterward and died in 1939. Ruth made two more trips to China in 1937 and 1938, returning with one more panda, Mei Mei. This panda also went to the Brookfield Zoo. Ruth died in 1947, alone in a Pittsburgh hotel room.

* Reichenbach, Herman, personal communication, 1999.

Sources: Harkness, Ruth, *The Lady and the Panda,* Carrick and Evans, New York, 1938; Brady, Erika, "First U. S. panda, shanghaied in China, stirred up a ruckus," *Smithsonian,* 14, 145, 1983.

Maryland). Accreditation of institutions was enacted on a voluntary basis in 1972, became mandatory for new members in 1980, and mandatory for existing members in 1985. A Code of Professional Ethics for individual members was adopted in 1976. In January 1994, the shorter name American Zoo and Aquarium Association (AZA) was adopted.[89]

A number of AZA programs were established during the 1980s and 1990s, including Species Survival Plans (SSP) for individual species, Taxon Advisory Groups (TAG) for groups of species, Fauna Interest Groups (FIG) for species within geographic regions, a Reintroduction Advisory Group, an In Situ/Field Conservation Committee, a Wildlife Conservation and Management Committee, and a Small Population Management Advisory Group. Several studbooks for rare or endangered species have been established, and there has been zoo staff participation on U.S. Fish and Wildlife Service species recovery teams. During 1996/97, American zoological parks and aquariums had over 1,200 research projects in progress.[90,91]

Santa Fe Teaching Zoo and its parent, the Zoo Animal Technology Program, began in 1972 at Santa Fe Community College (Gainesville, Florida) to provide academic and practical training in the captive husbandry of wildlife. Although this program built its own zoo, other similar programs usually work with an established zoo to provide students with practical working experience. Other programs include the Exotic Animal Training and Management Program at Moorpark College (Moorpark, California), Zoo Biology and Animal Management Training at the National Zoological Park Conservation and Research Center (Front Royal, Virginia), International Crane Foundation Training and Internship Program at the Foundation (Baraboo, Wisconsin), Zoo Studies Program at Friends University (Wichita, Kansas), Captive Wildlife Management Program at the University of Wisconsin (Stevens Point, Wisconsin), Zoo Animal

PROFESSIONALIZATION OF CAPTIVE WILDLIFE MANAGEMENT

America's zoo profession, professional literature, and zoo libraries developed to a significant degree in tandem during the 1960s and 1970s. Most American zoo libraries were established after 1960 in response to a tremendous increase in professional literature that began to appear during that decade. This increase in literature was, in turn, a reflection of an increase in professional activities at the zoos and aquariums. Prior to this time, professional knowledge was passed along by word of mouth from experienced to new staff and required very little in the way of formal training or outside professional expertise. As this situation changed, there was a need to publish the accumulating information.

The first American zoo-related research journal, *Zoologica*, was a New York Zoological Society publication (published 1907 to 1973), and was primarily concerned with contributions from New York Zoological Park staff. The Zoological Society of London's *International Zoo Yearbook* (1960+) received numerous contributions from American zoo and aquarium personnel. Prior to this time most American zoo-related material was published as articles in the AAZPA magazine or, if possible, in scientific journals. The *Yearbook* was particularly important because it encouraged zoo staff members unfamiliar with writing for scientific journals to publish their material.

Technology Program at Pensacola Junior College (Pensacola, Florida), and Zookeeping Technology Program at Pikes Peak Community College (Colorado Springs, Colorado).

The American Zoo and Aquarium Association has developed formal training programs as well. These programs include the following: Professional Management Development (1975), Applied Zoo and Aquarium Biology (1987), Studbook (1991), Population Management (1992), Science of Zoo and Aquarium Animal Management (1994), Principles of Elephant Management (1995), Conservation Educators Training Program (1997), and Institutional Records Keeping (1998).

An electronic animal record-keeping system, the International Species Information System (ISIS), was initiated in 1973 to gather census and vital statistics on animals in captivity. Its objective has been to collect, analyze, and disseminate information essential for the genetic and demographic management of wild species populations in captivity. Starting with inventory data (originally extracted with a great deal of effort from years-old information filed in obscure filing cabinets), ISIS has expanded into veterinary and other management data. ISIS has also expanded from a national database into an international database.

It might be said that the 1970s to the 1990s were the age of aquariums in the United States. Of all the existing aquariums, 61% were established during these three decades, and over half of these were established just in the last decade. This period has also been a time for large, independent aquariums, whereas in the past aquariums tended to be exhibits or buildings within zoos. These new aquariums have taken advantage of improved technology to provide crystal clear-water, expansive viewing vistas, and elaborate natural habitat exhibits. New aquariums have also taken advantage of the public's intensified interest in aquatic conservation,

---

Professionalism was advanced with the establishment of the American Association of Zoological Parks and Aquariums (AAZPA, now the AZA) in 1924, the American Association of Zoo Veterinarians (AAZV) in 1946, and the American Association of Zoo Keepers (AAZK) in 1967. Association publications added considerably to the professional literature, and have included the AZA *Communiqué* (a monthly newsletter, 1960+), AZA annual conference proceedings (1968+), AAZV annual conference proceedings (1968+), AAZV *Journal of Zoo Animal Medicine* (1970+), AZA regional conference proceedings (1973+), AAZK *Animal Keepers Forum* (a monthly journal, 1974+), and AAZK annual conference proceedings (1974+).

Prior to 1960, pertinent books related to captive wildlife management included *A Handbook of the Management of Animals in Captivity in Lower Bengal* (1892), *Disease in Captive Wild Mammals and Birds* (1923), *Wild Animals in Captivity* (1950), and *The Psychology and Behaviour of Animals in Zoos and Circuses* (1955). Books of significance to the American zoo profession published after 1960 include *The Management of Wild Mammals in Captivity* (1964), *Zoological Park Fundamentals* (1968, revised in 1982 as *Zoological Park and Aquarium Fundamentals*), *Zookeeper Training: A Suggested Guide for Instructors* (1968), *Man and Animal in the Zoo* (1969), *General Principles of Zoo Design* (1972), *Zoo Design: International Symposium on Design and Construction* (1975, 1976, 1980), *Capture and Care of Wild Animals* (1975), *Breeding Endangered Species in Captivity* (1975), *The Chemical Capture of Animals* (1976), *Zoo and Wild Animal Medicine* (1978), *Restraint and Handling of Wild and Domestic Animals* (1978), and *Behavior of Captive Wild Animals* (1978).

Since the 1980s, the number of titles has mushroomed. However, the only zoo-related research journal published since *Zoologica* has been *Zoo Biology* (1982+). In addition to the

continued

providing environmental education and exhibits on freshwater and marine habitats, particularly ones of local interest.

Conservation centers were established for the first time in the United States for the purpose of off-site captive reproduction and conservation at the San Diego Wild Animal Park (Escondido, California, 1972), International Crane Foundation (Baraboo, Wisconsin, 1973), St. Catherines Wildlife Conservation Center (Midway, Georgia, 1974), and National Zoological Park Conservation and Research Center (Front Royal, Virginia, 1975). Zoos manage all these centers except the International Crane Foundation, and all are off exhibit (not open to the public) except the San Diego Wild Animal Park.

Except for one national zoo (the National Zoological Park, part of the Smithsonian Institution), an aquarium designated as a national aquarium (the National Aquarium in Baltimore, which the City of Baltimore owns and the National Aquarium in Baltimore, Inc., manages), an aviary designated as a national aviary (the National Aviary in Pittsburgh, which the City of Pittsburgh owns and the National Aviary in Pittsburgh, Inc., manages), and a couple of state-operated zoos, the majority of zoos and aquariums are municipal institutions. Almost every zoo and aquarium has a "friends of the zoo" society with varying degrees of operational control over the institutions. And almost every society-managed zoo and aquarium receives funding from the local municipality. Few of the twentieth-century zoo societies have been chartered as scientific societies for the sole purpose of establishing a zoo, as was the case in the nineteenth century. A trend in the latter part of the twentieth century has been to incorporate (privatize) the zoo or aquarium. American zoos and aquariums have become businesses with significant programs in public relations, fund-raising, and marketing. Directors at these institutions have become presidents and CEOs, and the AZA conferences are concerned as much with administrative matters as they are with animal matters.

Late twentieth-century trends at many zoological gardens also include projecting themselves as conservation parks, involvement with *in situ* (field) conservation programs, providing dis-

---

CAPTIVE WILDLIFE MANAGEMENT

International Zoo Yearbook, there have been other important international serials, such as the *Dodo* and *International Zoo News,* along with those published by European zoological gardens.

The complexity of professional knowledge and the diversity of expertise needed to manage zoos and aquariums properly has increased tremendously in the post–World War II era. During the 1970s and 1980s, AZA established a code of ethics for individuals and accreditation standards for institutions. Formal training now exists at most zoos and aquariums, and many positions require academic studies. Several zoo technology programs have been established at community colleges and the AZA provides its own training programs. ISIS (the International Species Information System) also provides needed population management and veterinary data. Administrative needs and the level of care given to animal collections can no longer rely on word-of-mouth training provided by experienced staff. And as species become increasingly endangered, these collections and their proper care will become increasingly important.

Sources: Kisling, Vernon N., Jr., "Libraries and archives in the historical and professional development of American zoological parks," *Libraries & Culture,* 28, 247, 1993; Kisling, Vernon N., Jr., "Libraries and archives in American zoological parks and aquariums," in *Encyclopedia of Library and Information Science,* Volume 57, Supplement 20, Marcel Dekker, New York, 1996, 292.

tance-learning educational programs, designing more naturalistic immersion exhibits, and utilizing biotechnology (such as frozen tissues) to supplement propagation programs. American zoos and aquariums have taken advantage of improved electronic technology, biotechnology, and animal husbandry to advance their programs. Architects and commercial firms working with zoo and aquarium professionals also provide expertise in designing new exhibits and developing new programs. Funding has come from a variety of sources (such as taxes, admission fees, concessions, marketing programs, donations, grants, corporate sponsorships, etc.), providing ample opportunity to develop needed exhibits and programs. Of particular note is the AZA Conservation Endowment Fund for funding conservation projects, which exceeded $5 million in 1999.

5.5 Summary

In 1999, the American Zoo and Aquarium Association celebrated its 75th anniversary, and American zoos their 140th anniversary. It was also the 283rd year since that first lion was lowered onto a busy wharf in Boston. There have been many changes in American zoos and aquariums during this time and there will be many more. By making use of technology, taking advantage of a variety of funding sources, and striving to improve their professional knowledge, American zoo and aquarium professionals continue their tradition of providing state-of-the-art facilities and continue to make significant contributions to the conservation of wildlife.

References

1. DiPeso, Charles C., "Macaws…crotals…and trumpet shells," *Early Man*, 2, 4, 1980.
2. Wood, W. Raymond, "Plains trade in prehistoric and protohistoric intertribal relations," in *Anthropology on the Great Plains*, Wood, W. Raymond and Liberty, Margot, Eds., University of Nebraska Press, Lincoln, 1980.
3. Vail, R.W.G., *Random Notes on the History of the Early American Circus*, Barre Gazette, Barre, MA, 1956.
4. Hindle, Brooke, *The Pursuit of Science in Revolutionary America, 1735–1789*, University of North Carolina Press, Chapel Hill, 1956.
5. Stearns, Raymond P., *Science in the British Colonies of America*, University of Illinois Press, Chicago, 1970.
6. Benes, Peter, "To the curious: bird and animal exhibitions in New England, 1716–1825," in *New England's Creatures: 1400–1900*, Benes, Peter, Ed., Boston University, Boston, 1995, 147 [Annual Proceedings of the 1993 Dublin Seminar for New England Folklife, Volume 18].
7. Fleming, Walter L., *Jefferson Davis's Camel Experiment*, Louisiana State University, Baton Rouge, 1909 [University Bulletin Series 7, no. 1, pt. 2].
8. Odell, George, C. D., *Annals of the New York Stage*, Columbia University Press, New York, 1927.
9. U.S. Department of Commerce, *Historical Statistics of the United States: Colonial Times to 1970*, U.S. Government Printing Office, Washington, D.C., 1975.
10. Daniels, George H., *Science in American Society: A Social History*, Knopf, New York, 1971.
11. Greene, John C., *American Science in the Age of Jefferson*, Iowa State University Press, Ames, 1984.
12. Hughes, Arthur F. W., *The American Biologist through Four Centuries*, Charles C Thomas, Springfield, IL, 1982.
13. Smallwood, William M., *Natural History and the American Mind*, AMS Press, New York, 1967.
14. Nash, Roderick, *Wilderness and the American Mind*, 3rd ed., Yale University Press, New Haven, CT, 1982.
15. James, D. Clayton, *Antebellum Natchez*, Louisiana State University Press, Baton Rouge, 1968.

16. Thayer, Stuart, *The Travelling Menagerie in America, 1813–1834*, Privately published by the author, Seattle, WA, 1989.
17. Thayer, Stuart, *Annals of the American Circus, 1830–1847*, Privately published by the author, Seattle, WA, 1986.
18. Joys, Joanne C., *The Wild Animal Trainer in America*, Pruett Publishing, Boulder, CO, 1983.
19. Poinsett, Joel R., *Discourse on the Objects and Importance of the National Institution for the Promotion of Science, Established at Washington, 1840*, National Institution for the Promotion of Science, Washington, D.C., 1841.
20. Goode, George B., "The genesis of the United States National Museum," in *A Memorial of George Brown Goode, Together with a Selection of His Papers on Museums and on the History of Science in America, Annual Report of The Board of Regents of the Smithsonian Institution, Showing the Operations, Expenditures, and Condition of the Institution for the Year Ending June 30, 1897. Report of the U.S. National Museum*, Part II, U.S. Government Printing Office, Washington, D.C., 1901, 103.
21. Rhees, William J., Ed., *The Smithsonian Institution: Documents Relative to Its Origin and History, 1835–1899*, U.S. Government Printing Office, Washington, D.C., 1901.
22. Scharf, J. Thomas and Westcott, Thompson, *History of Philadelphia, 1609–1884*, L. H. Everts, Philadelphia, 1884.
23. Zoological Society of Philadelphia, *An Act to Incorporate the Zoological Society of Philadelphia*, Zoological Society of Philadelphia, Philadelphia, 1859.
24. Scheuermann, Alyssa N., *"Firsts" at the Zoological Society of Philadelphia*, Zoological Society of Philadelphia, Unpublished typescript, 1984.
25. Washington Zoological Society, *Charter, Etc. of Washington Zoological Society*, Washington Zoological Society, Washington, 1872 [Smithsonian Institution Archives, Record Unit 7002, Spencer F. Baird Collection].
26. Kohlstedt, Sally G., "From learned society to public museum: the Boston Society of Natural History," in *The Organization of Knowledge in Modern America, 1860–1920*, Oleson, Alexandra and Voss, John, Eds., Johns Hopkins University Press, Baltimore, MD, 1979, 386.
27. Scudder, Samuel H., *Can We Have a Zoo in Boston?*, Boston Society of Natural History, Boston, 1891 [Remarks made at a meeting of the Thursday Club January 15, 1891].
28. Cadwalader, Williams B., *Bears, Owls, Tigers, and Others! Philadelphia's Zoo, 1874–1949*, Newcomen Society in North America, New York, 1949.
29. Zoological Society of Philadelphia, *An Animal Garden in Fairmount Park*, Zoological Society of Philadelphia, Philadelphia, 1988.
30. McKenzie, Rod C., "American zoological gardens: elements of metropolitan landscapes," *Journal of Cultural Geography*, 6,1, 1986.
31. Hanson, Elizabeth A., *Nature Civilized: A Cultural History of American Zoos, 1870–1940*, Ph.D. dissertation, University of Pennsylvania, Philadelphia, 1996.
32. Wirtz, Patrick, "Zoo city: bourgeois values and scientific culture in the industrial landscape," *Journal of Urban Design*, 2, 61, 1997.
33. Bridges, William, *Gathering of Animals: An Unconventional History of The New York Zoological Society*, Harper & Row, New York, 1974.
34. Smith, John W., "Central Park animals as their keeper knows them," *Outing Magazine*, 42, 248, 1903.
35. Conklin, William A., *Report of the Director of the Central Park Menagerie*, Department of Public Parks of the City of New York-William C. Bryant Co., New York, 1873.
36. Johnson, Steve, archivist, Wildlife Conservation Society, personal communication, 1999.
37. Johnson, Rossiter, Ed., *The Twentieth Century Biographical Dictionary of Notable Americans*, Biographical Society, Boston, 1904.
38. Wilson, James G. and Fiske, John, Eds., *Appleton's Cyclopaedia of American Biography*, Appleton, New York, 1888, 706.
39. Rosenthal, Mark A., "It began with a gift," *The Ark*, 4, 2, 1976/77.

40. Marshall, David, *The Jewel of Providence: An Illustrated History of Roger Williams Park 1871–1971*, Providence Parks Department, Providence, 1987.
41. Gale, Oliver M., "The Cincinnati Zoo: one hundred years of trial and triumph," *Cincinnati Historical Society Bulletin*, 33, 87, 1975.
42. Ehrlinger, David, *The Cincinnati Zoo and Botanical Garden: From Past to Present*, Cincinnati Zoo and Botanical Garden, Cincinnati, 1993.
43. Edwards, Arthur M., *Life beneath the Waters; or, the Aquarium in America*, Bailliere, New York, 1858.
44. Barber, Lynn, *The Heyday of Natural History 1820–1870*, Doubleday, New York, 1980.
45. Mellen, Ida M., "A table of information regarding the large aquariums of the world," in *Twenty-Ninth Annual Report, New York Zoological Society*, New York Zoological Society, New York, 1925, 83.
46. Ryan, Jerry, *The Forgotten Aquariums of Boston*, New England Aquarium, Boston, 1995.
47. Anonymous, *History*, National Aquarium Society, Washington, 1994 [Unpublished typescript].
48. Heap, Mildred F., *The Buffalo Zoo Story*, Buffalo Zoological Gardens, Buffalo, 1982.
49. Kisling, Vernon N., Jr., "History of Zoological and Botanical Collections: A Survey," unpublished, 1985–1989 [American Zoo and Aquarium Association Archives, Smithsonian Institution].
50. Corell, Margaret, "1882 ... Beginning of the zoo: the early years," *Zoo News*, Winter, 1–12, 1982.
51. Reynolds, Richard J., "History of the Atlanta Zoo," in *Atlanta's Zoo*, City of Atlanta, Atlanta, 1969.
52. Stott, R. Jeffrey, *The American Idea of a Zoological Park: An Intellectual History*, Ph.D. dissertation, University of California, Santa Barbara, 1981.
53. Anonymous, "Zoo versus menagerie," *Living Age*, 317, 375, 1923.
54. Link, Theodore, "Zoological gardens: a critical essay," *American Naturalist*, 17, 1225, 1883, 1226.
55. Shufeldt, R. W., "Zoological gardens: their uses and management," *Popular Science Monthly*, 34, 782, 1889, 791.
56. Salt, Henry S., *Animals' Rights: Considered in Relation to Social Progress*, Society for Animals Rights, Clarks Summit, PA, 1980 [Reprint of 1892 edition].
57. Dupree, A. Hunter, *Science in the Federal Government: A History of Policies and Activities*, 2nd ed., Johns Hopkins University Press, Baltimore, 1986.
58. Smithsonian Institution, *Annual Report of the Board of Regents of the Smithsonian Institution, Showing the Operations, Expenditures, and Condition of the Institution to July 1888*, U.S. Government Printing Office, Washington, 1890, 44–45.
59. Smithsonian Institution, *Annual Report of the Board of Regents of the Smithsonian Institution, Showing the Operations, Expenditures, and Condition of the Institution to July 1889*, U.S. Government Printing Office, Washington, 1890.
60. Smithsonian Institution, *Annual Report of the Board of Regents of the Smithsonian Institution, Showing the Operations, Expenditures, and Condition of the Institution to July 1890*, U.S. Government Printing Office, Washington, 1891.
61. True, Webster P., *The Smithsonian Institution*, 2nd ed., Smithsonian Institution Series, New York, 1943.
62. Bamrick, Ray, "The Pittsburgh Zoo: a look back," *Animal Talk*, Spring, 17–19, 1988.
63. Bridges, 1974, 2.
64. New York Zoological Society, *First Annual Report of the New York Zoological Society*, New York Zoological Society, New York, 1897, 52.
65. New York Zoological Society, 1897, 13.
66. Fox, Herbert, *Disease in Captive Wild Mammals and Birds*, Lippincott, Philadelphia, 1923.
67. Ehrlinger, David, "The Hagenbeck legacy," *International Zoo Yearbook*, 29, 6, 1990.
68. Hagenbeck, Carl, *Beasts and Men*, Longmans, Green and Company, London, 1909 [Abridged translation of *Von Tieren und Menschen* by Hugh S. R. Elliot and A. G. Thacker].
69. Reichenbach, Herman, "A tale of two zoos: the Hamburg Zoological Garden and Carl Hagenbeck's Tierpark," in *New Worlds, New Animals: From Menagerie to Zoological Park in the Nineteenth Century*, Hoage, Robert J. and Deiss, William A., Eds., Johns Hopkins University Press, Baltimore, MD, 1996.
70. Gray, Bob, *The San Diego Zoo*, Zoological Society of San Diego, San Diego, 1997.

71. Anonymous, "Rambling thoughts, published as friendly chat with the zoo directors," *Parks and Recreation*, 7, 196, 1923.
72. Anonymous, "National association of zoological executives formed here; Dr. Wegeforth, first president," *San Diego Union*, April 17, 1924.
73. Merkel, Hermann W., "Zoo men organize, association affiliated with institute formed," *Parks and Recreation*, 8, 121, 1924.
74. Wegeforth, Harry M. and Morgan, Neil, *It Began with a Roar: The Beginning of the World-Famous San Diego Zoo*, Rev. ed., Zoological Society of San Diego, San Diego, 1990.
75. Etter, Carolyn and Etter, Don, *The Denver Zoo: A Centennial History*, Roberts Rinehart, Denver, 1995.
76. Conant, Roger, "First woman zoo veterinarian," *Parks and Recreation*, 26, 340, 1943.
77. Fowler, Murray E., *Historical Perspectives of the American Association of Zoo Veterinarians*, American Association of Zoo Veterinarians, Davis, CA, 1977.
78. Knudson, Gus, "Seattle has women zoo keepers," *Parks and Recreation*, 26, 385, 1943.
79. Hubbard, Leonides, Jr., "What a Big Zoo Means to the People," *Outing Magazine*, 44, 678, 1904.
80. Barker, John H. and Tolson, Peter J., *A Short History of the W.P.A. Construction Projects at the Toledo Zoo, 1933–1941*, Toledo Zoo, Toledo, n. d. (198?).
81. Conant, Roger, "WPA helps zoos toward ideal," *Parks and Recreation*, 20, 42, 1936.
82. Conant, Roger, "Zoo veterinarians meet," *Parks and Recreation*, 20, 50, 1936.
83. Crawford, M. A., Ed., *Comparative Nutrition of Wild Animals*, Academic Press, New York, 1968 [Symposia of the Zoological Society of London, Number 21].
84. Bush, Mitchell, "Remote drug delivery systems," *Journal of Zoo and Wildlife Medicine*, 23, 159, 1992.
85. Carr, William H., *Pebbles in Your Shoes*, Arizona–Sonora Desert Museum, Tucson, 1982.
86. Kisling, Vernon N., Jr., "Libraries and archives in the historical and professional development of American zoological parks," *Libraries & Culture*, 28, 247, 1993.
87. Kisling, Vernon N., Jr., "Libraries and archives in American zoological parks and aquariums," in *Encyclopedia of Library and Information Science*, Volume 57, Supplement 20, Marcel Dekker, New York, 292, 1996.
88. Rogers, Rachel W., *Zoo and Aquarium Professionals: The History of AAZK*, American Association of Zoo Keepers, Topeka, KS, 1992.
89. Wagner, Robert O., "The independence of AAZPA," in *Regional Conference Proceedings 1996*, American Zoo and Aquarium Association, Wheeling, WV, 1996, 393.
90. Hodskins, L. G., Ed., *AZA Annual Report on Conservation and Science*, American Zoo and Aquarium Association, Bethesda, MD, 1998.
91. Hutchins, Michael and Conway, William G., "Beyond Noah's Ark: the evolving role of modern zoological parks and aquariums in field conservation," *International Zoo Yearbook*, 34, 117, 1995.

Additional Sources

1. American Bison Society, *Annual Report of the American Bison Society 1905–1907*, American Bison Society, New York, 1908.
2. Brady, Erika, "First U. S. panda, shanghaied in China, stirred up a ruckus," *Smithsonian*, 14, 145, 1983.
3. Garretson, Martin S., *A Short History of the American Bison*, American Bison Society, New York, 1934.
4. Harkness, Ruth, *The Lady and the Panda*, Carrick and Evans, New York, 1938.
5. Hediger, H., *The Psychology and Behaviour of Animals in Zoos and Circuses*, Dover, New York, 1968 [Reprint of 1955 edition].
6. Kisling, Vernon N., Jr., "The last passenger pigeon dies," in *Great Events from History II: Ecology and the Environment*, Salem Press, Pasadena, CA, 111, 1995.
7. Schorger, A. W., *The Passenger Pigeon: Its Natural History and Extinction*, University of Oklahoma Press, Norman, 1973.

6
Zoological Gardens of Australia

Catherine de Courcy

6.1 Introduction

Australia's national zoo network comprises four urban zoos, their associated wildlife parks and sanctuaries, and the Territory Wildlife Park near Darwin. These zoos are all publicly owned and work closely with regional and global authorities to fulfill the four objectives of the international zoo community: recreation, education, conservation, and research. The metropolitan zoos in Melbourne, Adelaide, Sydney, and Perth were established in the nineteenth century along traditional lines. All have developed additional animal parks to exhibit and care for different sections of their collections. Territory Wildlife Park, a regional fauna park in the Northern Territory, was established independently of the other zoos. These facilities now constitute the highly successful Australian zoo network, which has the popular support of their local communities.

Privately owned zoos and small publicly owned fauna parks in Australia have come and gone over the past two centuries. A few, such as the Hobart Municipal Zoo and Lord McAlpine's Pearl Coast Zoological Gardens in Broome, held exotic animals; most, however, exhibited only native animals. These zoos filled a void, particularly in Queensland, where an early attempt to establish a zoo failed, and in Tasmania. However, the ephemeral nature of their organizations and the diverse influence of private ownership on their development make them of no great relevance to the history of mainstream Australian zoos.

There have been three phases in the history of major Australian zoos. The foundation phase covers the period from the establishment of the first zoo in 1857 to about 1920. By then, all the elements of a traditional zoo were in place and the institutions had a solid footing in their respective cities. The second phase is the dull and difficult stage in the mid-twentieth century when the zoos had little standing in their communities and were under threat of closure. The third phase began in the 1960s and is marked by the adoption of clear objectives together with local and international cooperation.

This history of Australia's zoos considers each of these three phases. While the pace of the zoos' development follows an international pattern, local influences also left their mark. Geographic isolation, small urban populations, an interesting local fauna, the demands of a young colony, and other major factors have had an impact on how the zoos evolved. In the earlier years, the general effect on the zoos was to keep them small and simple. In recent years, some of these factors have contributed to the significance of the role they are playing in modern

Australian society. They are now very popular institutions that attract large crowds, and their work in environmental education and animal conservation is widely recognized.

6.2 Foundation and Development, 1857–1920

6.2.1 Origins

The forces behind the foundation of the four urban Australian zoos were similar to those behind many other nineteenth-century zoos. Civic pride, individual enthusiasm, education, recreation and vague scientific interest were important factors. However, one unusual aspect that was common to all Australian zoos was a connection between their founders and the acclimatization movement. The early years of each of the zoos were shaped by the drive to import and acclimatize unusual domestic animals for the benefit of the colonies. Such was the significance of this movement to the history of Australia zoos that some details of it are included here.

The Zoological Society of Victoria was established in 1857 and was the first to develop a major zoo in Australia. The founders, who were lawyers, doctors, and other prominent citizens, were inspired by the success of the London Zoo. Together they formed the Zoological Society of Victoria to emulate this success. The city of Melbourne had undergone significant social and economic changes in the years following the gold rush of the early 1850s. During the decade; the small pastoral town had become a major colonial city as its population rose from 77,345 in 1851 to 410,766 in 1857. Several major cultural institutions were created in that decade, including the University of Melbourne, the Museum of Natural History, and the Public Library and Art Gallery. Learned societies provided forums for discussion on a wide range of issues; the Philosophical Institute, modeled on the Royal Society in Britain, was the most important of these.

However, the demands of the young colony forced the Zoological Society to drop its initial plans for a zoo. Within a year its priorities had shifted from collecting and displaying exotic animals to importing large numbers of unusual domestic animals in the hope that they would become standard farm stock. To reflect this emphasis, the society renamed itself the Acclimatisation Society of Victoria in 1861 and concentrated its efforts on acquiring useful animals.

The animals in question fell into three broad categories: economic, game, and ornamental.[1] The animals whose primary function was perceived as having economic value, such as wool-bearing goats, sheep, silkworm, ostriches, and bees occupied much of the society's agenda. For example, 93 angora goats were imported from Britain at great expense. Their heavy fleeces and the belief that they could survive in harsh conditions similar to those in Victoria suggested they were suited for land that could not be used for sheep. Although these animals seemed to fare well in the zoo, the society decided to sell them in 1869. However, there was no interest in buying them and, eventually, a long-time supporter of the organization offered to care for them on his property.

The second group of animals comprised those seen as a combination of game and food and included various types of deer, partridge, quail, and hares. Shooting and hunting were popular pastimes in Victoria; indeed, the activity was encouraged to keep young men away from cafés and casinos. Bored sportsmen went so far as to import deer, partridge, and other game because Australian animals were not considered exciting enough to sustain the demands of the colonial gentleman hunter. Many of the animals in this category that the Acclimatisation Society imported were poached or died in the bush. Unfortunately, a few survived too well, becoming pests that have since caused great problems to natural habitats of local wildlife.

The Le Souefs — A Family of Zoo Directors

The Le Souef family dominated the early development of Australia's zoos. For 69 years, there was a Le Souef in charge of at least one of the major zoos, and for 20 years, three of the four zoos were run by a Le Souef.

The reign began when Albert Le Souef was appointed honorary secretary to Melbourne Zoo in 1870. At the time he was Usher of the Black Rod, a part-time position in the Legislative Council, and had many influential contacts, which he used to gain support for the floundering zoo. Within 15 years Le Souef had turned the zoo into a thriving institution. Thrift, constant collections of native animals, extensive use of horticulture, and awareness of international zoo developments were characteristics of his management style.

Le Souef lived at the zoo and used his large family to compensate for the lack of staff. His eldest son, Dudley, was groomed to take over Melbourne Zoo. He became a well-known naturalist, a founding member of the Royal Australasian Ornithologists' Union. As expected, on his father's death in 1902, Dudley became director of Melbourne Zoo and served in that capacity until his own death in 1923.

Albert's second son, Ernest, studied veterinarian science before being appointed foundation director of Perth Zoo in 1897. Perth Zoo was established on

Albert LeSouef. © Zoological Parks and Gardens Board of Victoria.

continued

The third group was the one that has brought the nineteenth-century acclimatization movement much ridicule today. This group involved those animals imported to enliven the environment of the colony. Birds, usually referred to as songbirds, were the principal animals in this category. Frederick McCoy, foundation professor of natural science at the University of Melbourne and a member of the society, described them as the source of "those varied, touching, joyous, strains of Heaven-taught melody, which our earliest records show, have always done good to man."[2] English thrushes, blackbirds, larks, starlings, and others were transported from Britain in cages. They were then given time to recover from the journey before being released at selected locations in Melbourne and around the colony. Some of these also adapted quickly and, within a few years, were drawing voluble protests from farmers.

By the mid-1860s, the popular interest in acclimatization in Victoria had diminished. The results produced by the animals in the economic category were poor, sparrows and other song birds were interfering with orchards, and the society was being accused — falsely — of importing

> ## THE LE SOUEFS
>
> the model suggested by Albert Le Souef, and his son further developed the zoo along these same lines until his retirement in 1932. Sherbourne, the fourth son, was appointed director of the Moore Park Zoo in 1903 at the age of 23. He supervised the development of Taronga and continued there as director until 1939.
>
> The three brothers worked closely together, exchanging animals, sharing ideas on enclosures and other buildings, and discussing plans for the development of their respective zoos. They were keen supporters of Hagenbeck's enclosure design and introduced it in various forms to their zoos. In 1910, Dudley approached the Queensland government to encourage it to create a zoo in Brisbane, but after some discussion the plan fell through.
>
> The Le Souefs' reign drew to an end in each of the zoos without the triumph their influence deserved. Dudley died just before the Melbourne Zoo Council was to fire him. He had lost control of the zoo since suffering a stroke caused when a disgruntled former member of staff attacked him. Ernest retired as financial pressure forced a change of management at Perth Zoo. And Sherbourne was dismissed from his position as director at the age of 62 following friction with the new superintendent, a qualified veterinarian.
>
> The long-term influence of the Le Souef family on the development of Australia's zoos is unquestioned. They brought the zoos through times when there was little public enthusiasm for their work, and through times of deep recession. By the end of their reigns, each of the three zoos was firmly established with a strong tradition for beautiful gardens and good native fauna collections.

one of the greatest pests in the country today, the rabbit. In response to these criticisms, the government, the main source of funding, had reduced its support to negligible amounts. By 1870, the Acclimatisation Society was on the point of collapse when Albert Le Souef was appointed to the position of honorary secretary. Le Souef was determined to keep the acclimatization work on trout and ostriches going and, in order to do this, he began to lay the groundwork for a more traditional zoo.

Le Souef lived with his large family on zoo property. By managing the limited finances of the society carefully, he gathered a modest collection of exotic animals and augmented that with a substantial collection of native animals. He also put considerable emphasis on the presentation of the gardens as well as providing low-key amusements and refreshments for visitors. In the 1880s, Melbourne experienced a period of economic boom, which gave Le Souef the opportunity to place the zoo on a solid footing. By 1900, Melbourne Zoo was an established city institution with all of the major elements of a traditional zoo in place. Le Souef's approach to zoo management continued in Melbourne under his son Dudley, who succeeded him as director after his death in 1902. It was also carried to other parts of Australia by two younger sons, Ernest and Sherbourne, who were appointed to manage the zoos in Perth and Sydney, respectively.

The second zoo founded was the Adelaide Zoo in South Australia, which the South Australian Zoological and Acclimatisation Society opened in 1883. This society was founded five years earlier solely for the purposes of animal acclimatization. Despite encouragement from its patron, the governor of South Australia, who wanted it to liberate Indian blackbuck and zebra, the society was soon putting all of its effort into creating a public zoological garden.

The small zoo, situated near the city center, was laid out on 16 acres of the Botanic Gardens. Since then, it has expanded and is the only Australian zoo still controlled by its founding

Early scene of the Melbourne Zoo, 1880s. © Zoological Parks and Gardens Board of Victoria.

Flight aviary at the Melbourne Zoo, 1908. © Zoological Parks and Gardens Board of Victoria.

Front gate of Adelaide Zoo, 1883. © Adelaide Zoo.

organization. Richard Minchin, who had been involved in the inauguration of the society in 1878, was appointed the first director and was largely responsible for shaping the zoo in its early years. He died in 1893 as a result of a virus he had picked up on one of his collection trips in Southeast Asia. His son, Alfred, began acting as assistant director in 1891, having spent some time in Melbourne Zoo to gain experience. Alfred Minchin continued as director of the Adelaide Zoo until his death in 1934 at the age of 77. His son, Ronald, succeeded him and continued the Minchin style until his untimely death six years later.[3]

Sydney's Moore Park Zoo was also opened in 1883 under the auspices of the New South Wales Zoological Society. The society, founded in 1879, had a declared interest in acclimatization rather than in the creation of a zoo. Songbirds, game, and other animals were imported. But then, as facilities in Moore Park were arranged to accommodate them, the society's interests gradually shifted away from the importation and liberation of animals toward the development of an attractive and profitable zoological gardens close to the city.[4]

This was the second attempt to create a public zoo in Australia's oldest city. Earlier plans in 1852 had failed. Later, in 1861, an acclimatization society, modeled on the Victorian society, had been established and animals donated to it had been displayed in the Botanic Gardens. These included many species that were of little use to an acclimatization society. The staff of the Botanic Gardens, already known for its skills in bird management, had been happy to receive the wallabies, kangaroos, deer, raccoons, peccaries, monkeys, the anteater, and other animals that were sent to them. After 22 years, the animals remaining in the Botanic Gardens were transferred to the new Moore Park Zoo.

In 1903, Sherbourne Le Souef was appointed secretary to Moore Park Zoo. Shortly afterwards, preparations began to move the zoo to a location that would accommodate expansion. His older brother, Dudley, traveled from Melbourne to give his opinion on the selection of the site. The current harbor location with its steep slope, southerly aspect, and beautiful views was chosen. In 1916, Sydney's Moore Park Zoo was renamed Taronga Zoo and was opened to the public. The elder Le Souef's subsequent efforts to persuade the Queensland authorities to establish a zoo farther up the coast around the same time failed despite some local interest.

The fourth zoo established was in Perth. In 1897, the newly formed Western Australian Acclimatisation Committee called on Albert Le Souef for advice on creating a zoo in Perth. The city had grown rapidly as a result of a gold rush; over a 13-year period to 1901, its population had risen from 44,000 to 180,000. The Acclimatisation Committee had been established with much the same intentions as the Victorian Acclimatisation Society and it opened the Perth Zoo in 1898 with Ernest Le Souef, then aged 28, as director. Unlike its Melbourne counterpart, the Perth society's acclimatization work continued well into the twentieth century. White swans, Californian quail, and deer were among the species it released into the wild over a 20-year period.

Perth Zoo was not a wealthy institution. Even with the gold rush, the city was small and isolated with few other settlements in the region. The Le Souef management style ensured a careful approach to developing the zoo and its collection, purchasing animals within budget, building simple enclosures, and maintaining beautiful gardens. Such was the importance of the gardens to the young Le Souef that a senior member of the grounds staff from Melbourne Zoo was seconded to Perth for a while to give advice.

6.2.2 Enclosure Design

Australian zoos have always kept abreast of contemporary styles of enclosure design. The early administrators used their personal holidays and collection trips to gather ideas about the latest

Elephant temple at Moore Park Zoo (now Taronga Zoo, Sydney) with the monkey exhibit in the foreground, 1920s. © Zoological Parks Board of New South Wales.

innovations in animal containment and accommodation. The relative poverty of the zoos meant there were no buildings as fine as those in major European and American zoos, but the mild Australian climate was a considerable compensation.

Their nineteenth-century enclosures for exotic or carnivorous animals were traditional cages with bars, often made from wood and purchased from circuses or dealers along with the animals. As wooden floors rotted, the enclosures were replaced with solid brick and bar cages. Adelaide Zoo, for example, had replaced all of its early wooden cages with solid brick and bar cages within 40 years of its foundation. Herbivorous animals, many of them native animals, were kept in paddocks and open enclosures.

One of the earlier attempts at following classic zoo design was the creation of a bear pit in Melbourne Zoo in 1875. The pit was nine feet deep, made of brick and cement, and featured a 16-foot pole in the center. But shortly after it was completed, Le Souef admitted that he had made a mistake "by building one of the old fashioned bear pits which are now, I understand, being altered in the European gardens, and which I find are unsuited to this climate."[5] The pit was replaced some years later by a bear house.

Exotic brick buildings designed to convey a sense of place were rare in the early zoos. The Taronga Zoo (previously the Moore Park Zoo) Asian elephant enclosure was one of the few substantial buildings in this category. It was built before the zoo opened in 1916 and was modeled on a Hindu temple in an attempt to represent the exotic, faraway home of the elephant. It has since been retained because of its historical interest.

The idea of displaying animals in bar and brick cages was not acceptable to all nineteenth-century Australians. A social columnist suggested that the temper of a newly acquired Bengal tiger, held in a traditional cage in Melbourne Zoo, was "anything but good, for he watched the people gazing at him with a savage look, as though their curious stares added the last touch of bitterness to his ignoble bondage."[6] "Vagabond," the byline of a Victorian journalist who often presented an alternative view of life, wanted to know why lions could not be kept in large, sandy enclosures rather than be cooped up behind bars.[7] There was no official response to these or similar comments at the time.

The mild Australian climate and the propensity for outdoor enclosures may have contributed to the speed with which Melbourne and Sydney zoos picked up on Carl Hagenbeck's barless design principles. In 1914 in Melbourne, Dudley Le Souef incorporated these principles into

Lion pit at Moore Park Zoo (now Taronga Zoo, Sydney) on opening day, 1916. © Zoological Parks Board of New South Wales.

enclosures for zebras, wallabies, and goats. The old wooden buildings, he said, were to be replaced "by concrete structures resembling natural rocks." He suggested that the new enclosures would require no painting or upkeep, adding that they "do not in any way overshadow the animals."[8]

In Sydney, however, Sherbourne Le Souef was prepared to take on the full challenge of the barless exhibit by putting carnivorous animals into moated enclosures in Taronga Zoo. For some years prior to the opening of the zoo, Le Souef and the trustees had taken every opportunity to examine the latest developments in animal enclosures when they were in Europe. The geographic arrangement of species in Munich, the dry moats in Nuremburg, and the Mappin Terraces in London were particularly noted, but Hagenbeck's barless enclosures impressed most of all. So, when the zoo was officially opened in 1916, the bears, lions, and monkeys were kept in moated enclosures with a combination of artificial rock and natural rock.

6.2.3 The Animal Collections

While geographic isolation was not an impediment to enclosure design, distance from the major sources of animals and the big international animal traders had a considerable impact on animal acquisition policies of Australian zoos in the early years. In addition, the zoos' income from entrance fees and government benevolence was relatively small in international terms. Consequently, the zoos often found they could not compete against the wealthier European and American zoos for the popular, larger animals.

On the other hand, Australian zoos had a ready supply of native animals that were appealing to international zoos and could therefore be used for sale or exchange. Ferdinand von Mueller brought Melbourne Zoo into the international exchange network in 1859. Using contacts he made through his work as director of the Botanic Gardens, Mueller opened exchange transactions when he sent pairs of black swans to various zoos and gardens around the world, including those in Copenhagen, Cologne, Java, Calcutta, Paris, and, of course, London. Some time later

Polar bear exhibit at Moore Park Zoo (now Taronga Zoo, Sydney), ca. 1919. © Zoological Parks Board of New South Wales.

Melbourne Zoo began to receive animals in return, but many of these had little value as zoo exhibits, being small game animals such as hares, fowl, doves, and waterbirds.

Nevertheless, native animals were an important commodity for the Australian zoos. In the nineteenth century, animals that might have cost £20 to acquire in Australia could be exchanged for stock worth up to £200 from overseas zoos. A yak worth £40, for example, could be exchanged for a pair of kangaroos that may have cost little or nothing to acquire. The thylacine was, by far, the most valuable of the Australian animals on the nineteenth-century market. Zoo officials with good connections could acquire one for £12; they could then exchange this for animals worth £90 in European zoos.[9] Perhaps the most lucrative single exchange in early Australian zoo history was when the King of Siam sent elephants to Melbourne Zoo and Moore Park Zoo in exchange for a collection of native fauna sent to him. At the time, Melbourne Zoo already had an elephant and sold this new arrival to Adelaide Zoo.

Ship captains and circuses were useful and reliable sources of animals. The captains acquired exotic animals in foreign ports and, on arrival in Australia, offered them for sale to the local zoos. Circuses traveling around Australia often chose to sell their animals to the zoos at the end of a season. In both of these cases, the zoo staff could see clearly the condition of the stock they were purchasing. The animals in question were usually the more common large vertebrates such as lions, tigers, and bears, although, in 1894, an exchange between Adelaide Zoo and a circus involved swapping native fauna for a zebra.

The more straightforward method of acquiring animals through dealers and traders did not work well for the Australian zoos in the earlier years. Expensive animals often arrived in poor condition and did not survive long enough to make the investment worthwhile. Distance from sources made it difficult to know whether an animal had left a European or Asian port in bad health or whether it had not traveled well. The Australian zoos relied to some extent on agents in foreign ports but, such were the problems of transporting animals to Australia, it was usually cheaper to send a member of the staff to the marketplace to negotiate the deal and then to accompany the animals back.

The Le Souefs and the Minchins traveled extensively in Europe, North America, Southeast Asia, and South Africa on behalf of their respective zoos. They usually brought a selection of native animals to exchange with the zoos they visited. In 1896, Albert Le Souef, then aged 70, undertook a 4,800-mile train trip around South Africa to collect hunting cheetahs, tortoises, zebra, a chacma baboon, and other animals.

The personal trips were not always successful. In one eventful, three-month trip in 1884, Dudley Le Souef went to Singapore to purchase an Asian rhinoceros, tapirs, and other regional fauna. After a couple of months, he had acquired two tapirs, a black panther, a leopard, a tiger, an orangutan, and other animals, all of which he sent back to Melbourne unaccompanied. But the rhinoceros, the prize of the trip, was proving extremely difficult to acquire. Le Souef spent the next month in a cheap hotel in Singapore waiting for news of one. Ignoring the cockroaches and the colorful life around him, he amused himself by collecting, killing, and preserving snakes for the small museum in Melbourne Zoo. Then, just when he was about to give up hope, he heard that a rhinoceros had been caught in Klang and he went to receive it. On the way back to Melbourne, to his intense dismay, the rhinoceros became ill during an unscheduled transshipment in Sydney and died before it reached the zoo. To compound his misery, the tapirs that he had sent back unaccompanied also died, one on the voyage and the other shortly after its arrival.

All the problems associated with acquiring animals had a great influence on the composition of early Australian zoo collections. The lions, tigers, monkeys, and other exotic vertebrates that were so readily available from the ship captains and the circuses formed the core of the foreign collection. However, to attract visitors, the zoos also needed more expensive exotic animals: elephant, large primates, rhinoceros, hippopotamus, and giraffe. Great care had to be taken over the selection and purchase of these to maintain a degree of novelty for regular visitors and to stay within budget. Albert Le Souef was so conscious of the drawing power of these animals that he deliberately excluded them from the collection until the government allowed him to charge an entrance fee. In the meantime, he relied on the more common foreign animals to draw visitors. He also saw great value in the native collection he was building up, but he was well aware that local visitors paid little heed to it.

As soon as the government permitted him to charge an entrance fee, Le Souef bought an elephant. This was not only because of its curiosity value, as most Melbournians would have seen one when the big circuses visited the city, but because rides on its back promised to be both a popular attraction and a source of income. Asian elephants were relatively easy to acquire. In 1883, he purchased a seven-year-old female elephant from the Calcutta Zoological Gardens. It survived a stormy sea journey by wrapping its trunk around the stanchion of the ship and, within weeks of arrival, was giving rides to children.

The other three zoos also made the acquisition of an elephant a priority. The exchange from the King of Siam in 1883 provided Sydney and Adelaide zoos with elephants within months of their opening. They also began to give rides for a small fee shortly after arriving. From then on, until rides ceased in the 1960s, the absence of an elephant in the collection was treated with urgency; the zoos could not afford to be without one of their most popular animals and, of course, the income they generated.

Larger primates were probably as important as the elephants in early Australian zoo collections. Orangutans and chimpanzees had curiosity value; more important, they were often promoted as individuals with names, habits, and history. Consequently, they became immensely popular with visitors returning regularly to visit these well-known characters. Orangutans were the first of the larger primates to appear in Australian zoos. They were fairly easy to get in the nineteenth century and, in the 1880s, both Adelaide and Melbourne introduced them. After suffering a high mortality rate among the earlier arrivals, Melbourne acquired a young female in 1901 who was to live for over 21 years in the zoo.

This orangutan soon became known as Mollie. When early attempts to teach her to ride a bicycle and put on gloves failed, her ability to drink large quantities of beer without appearing drunk and to light and smoke cigarettes became her public performance. Despite her great popularity, however, she became a target for abuse as objectionable visitors poked her with umbrellas and walking sticks, and teased her by throwing food just out of her reach. Nevertheless, her significance was such that, when she died, she was skinned, mounted, and returned to her old cage. Her enduring status as a Melbourne character was demonstrated 10 years later when a photo of her appeared in a daily newspaper marking the anniversary of her death.[10] This strong emotional response was not confined to early generations either. When, in 1975, the orangutan George died after living for 25 years in the Adelaide Zoo, the public acted spontaneously and set up a fund. The money raised was used to commission a bronze bust based on his death mask.

Elephants and orangutans were easy to acquire when compared with the problems associated with some of the other major exotic animals. After the heartbreaking loss of the rhinoceros that Dudley Le Souef bought in 1884, Melbourne Zoo had to wait seven years before it managed to acquire another one of these great representatives of the exotic. Eventually, its agent in Calcutta purchased one on its behalf despite strong competition from English buyers. He sent it to Melbourne on the *SS Bancoora*, a ship that had carried animals for the Acclimatisation Society and the zoo for many years. This time, however, the ship ran into a storm and was wrecked. The animal survived and, a week later, was removed and sent on to Melbourne. It survived for just two weeks, drawing in massive crowds despite poor weather. Then it succumbed to the injuries sustained on the ship and died. Melbourne did not get another rhinoceros until 1915.

Adelaide Zoo had more success with its first rhinoceros. Richard Minchin purchased it in 1886 on one of his collection trips in Southeast Asia. The animal, initially described as an Indian rhinoceros, lived for 21 years in the Zoo. On its death in 1907, it was transferred to the South Australian Museum where it was skinned and mounted; 41 years later, it was correctly identified as the smaller Javan rhinoceros.

Adelaide Zoo was not so fortunate with its hippopotamus, the first to be displayed in an Australian zoo. A young female was purchased in 1900, partly through public donation, a method of fund-raising the Minchin family used to great effect. The popularity of the animal exceeded all expectations with huge attendance. But within a year, it was dead. The director believed it had been about a year old when captured and too young to survive captivity, transportation, and a cold winter. Its next hippopotamus, which arrived in 1920, died nine years later after swallowing a tennis ball. A public fund sprang up in response to this death and sufficient money was gathered to purchase a young hippopotamus from the Auckland Zoo. This locally bred animal, probably related to the prolific pair that had been living in Melbourne Zoo since 1912, cost less than one third the price of the hippopotamus the zoo purchased in 1900.

While zoo managers concentrated their efforts on buying their few large exotic animals, they filled many of their enclosures with native and smaller, less prominent animals. Native animals had always been a feature of the Australian collections but had seldom been given prominence because they did not attract local visitors. Although kangaroos and koalas might have been popular in European or American zoo collections at the time, tourists and specialists were the only visitors who expressed strong interest in them in Australia.

In the 1870s, Albert Le Souef ignored the public's lack of interest and deliberately maintained a substantial native collection in Melbourne Zoo. Some contemporary commentators supported him. A reporter in the *Age*, a metropolitan daily newspaper, said that when funds were low, a representative collection of native fauna was a more pressing need than the acquisition of "zoological luxuries, such as elephants and tigers."[11] In 1876, "Vagabond" advised a similar approach: "First obtain a complete collection of every Australian animal and bird; fishes and an aquarium you might add in time. But let the gardens be an Australian menagerie, where the rising generation may learn what living things surround them on this continent."[12]

Eight years later "Vagabond" complimented the zoo's council on the collection of native fauna: "Boys who have never seen the joys of nature which make up a juvenile poacher's experiences, or know not the glories of animal life on a Riverina plain, may learn lessons of the highest importance by studying the habits of the marsupials, the birds, and the reptiles of his native land ... our youngsters will learn all this in the Zoological-gardens."[13] As anticipated, however, the majority of local visitors clustered around the few imported animals and paid scant attention to the indigenous collection.

In terms of public interest, the smaller animals did not fare much better. From the earliest days, each zoo had several aviaries. The wide range of beautiful and talkative local birds provided tremendous displays in enclosures of all shapes and sizes. Keen ornithologists were associated with the zoos at various stages. Albert Le Souef's wife, Caroline, was the daughter of John Cotton, one of Australia's early ornithologists, and several members of the next generation of cousins, including Dudley Le Souef, were active members of the Australasian Ornithologists Union.

Aquaria and insect exhibits were not standard features of the early zoos, although they were considered occasionally. In the 1880s, several people proposed building an insect house in Melbourne Zoo similar to the one in London Zoo; however, they dropped the idea in order to build the exotic collection instead. Reptile exhibits, on the other hand, were common. Predictably, the large constrictors were popular; indeed, a governor of Victoria, Sir Charles Darling, even wanted the Acclimatisation Society to import them. He believed they could "be made one of the most interesting drawing-room pets possible." He had seen them "introduced suddenly amongst a party and made to rear their heads over a piano; and although a little alarm was at first created, the creature soon became an object of interest and curiosity."[14]

However, the majority of reptiles in the collections were local and were usually kept in a pit. This style of exhibit caused problems when caring for the health of the animals and protecting them from visitors. In the 1890s, Adelaide Zoo lost many of its snakes after visitors dropped stones on them. The snake pit also precluded any educational value in the display. A commentator asked Melbourne Zoo to display reptiles in such a way as to allow the local people and new settlers to study them to identify those that were dangerous and those that were not. However, this call to draw on the educational value of lesser-known animals was premature and Melbourne Zoo persisted in displaying its snakes in a pit until the mid-twentieth century.

6.2.4 Early Development of the Four Objectives

Education and recreation were the objectives of the early Australian zoo administrators, but it is also worth noting the relationship between the zoos and the other two modern objectives, conservation and research. There was a general acceptance of the educational value of Australian zoos in the early years. The intention of the founders of Melbourne Zoo was "both for the purposes of science and for that of affording the public the advantage of studying the habits of the animal creation in properly arranged zoological gardens."[15]

The intrinsic, if passive, educational role of the zoos was acknowledged in the recorded comments of contemporary politicians, observers, and zoo staff. However, formal education, insofar as it existed, consisted of occasional guidebooks and labels containing basic information such as the common and Latin names of the animals and their approximate area of origin. The arrangement of animals in taxonomic groups, and in cages that permitted a full view of the exhibit, was considered adequate provision for comparative analysis of varieties within species. If the aim of zoology was "to furnish every possible link in the procession of organized life," as "Vagabond" described it, then a zoo was a good place to study it.[16]

Despite the inadequacy of the educational programs in modern terms, the usefulness of the zoos as educational institutions was frequently mentioned in the nineteenth century. Graham Berry, Chief Secretary of Victoria, regarded Melbourne Zoo as being "a most interesting mode of amusement and instruction."[17] The *Argus*, a daily newspaper in Melbourne, said that, "In a collection of this nature, instruction and amusement are so happily combined that one is enabled to imbibe valuable doses of natural history with an ease which could not otherwise be provided."[18] Without the zoo, the paper noted some years later, "Literature must lose some of its intelligibility to [young people]."[19] One writer went as far as to suggest that a zoo "might be properly attached to the Education department, and ... a good deal of money now wasted in over educating or vainly educating youths and maidens for very ordinary work in life would be better spent in support of this great instructive and entertaining institution."[20]

The *Herald*, another daily Melbourne newspaper, summarized the role of a zoo succinctly: "There be some people, doubtless, who are on principle opposed to detaining any animals in captivity, but the large majority agree in thinking that collections of zoology are a valuable and important adjunct to practical education."[21] In 1900, a South Australian newspaper echoed the opinions of the Victorian newspapers: "Nowhere within the neighborhood of a city are the natural powers of observation possessed by the young more successfully brought out with truly instructive effect than in a well-stocked zoological garden like that of Adelaide."[22]

It was not just the children who were seen to benefit from the educational role. Adults, according to various sources, could ponder Darwin's theory of evolution while visiting a zoo. Although Darwin's central thesis was not generally accepted in Australia until the 1890s, it was the subject of debate and speculation.[23] "Vagabond," for example, noted that on a visit to Melbourne Zoo, "The origin of the species may be studied, that progression of human life which ends in man and assumes no higher form." [24] And, an *Age* reporter suggested that, "To the Darwinian, a zoological garden is a history of his ancestors, replete with thrilling genealogical interest."[25]

The inherent value of the zoos as recreational venues was also widely accepted. While exotic animals provided the main focus of trip to the zoo, beautiful grounds, good shade or shelter, reasonable refreshment facilities, and perhaps sideshows or concerts were also added to ensure a pleasant day out to visitors and to encourage them to return repeatedly.

Albert Le Souef was keenly aware of this when he took control of Melbourne Zoo. Although the zoo was initially established as a recreational venue with an educational role, its early diversion into acclimatization brought a halt to efforts to make the place attractive and comfortable for visitors. In 1861, the grounds were laid out in a formal manner, confirming the intention of the founders to attract visitors for a day out much like the London Zoo had done in the 1830s, but none of the basic facilities for visitors was provided. There were no picnic lawns, shelter sheds, refreshment rooms, or even public conveniences. In 1870, Le Souef made the addition of these features his priority. When he went to Europe in 1880, he was very impressed by the restaurants, concert halls, and other recreational facilities of the continental zoos. On his return, he opened a milk bar and invited a local band to play in the zoo.

The other three zoos learned the value of providing visitor facilities from Melbourne's experience and incorporated them into their earliest construction plans. Moore Park Zoo and

> # A Universal Story —
> # Feeding Lions at Melbourne Zoo in the Nineteenth Century
>
> Watching the lions being fed was one of the favorite pastimes of visitors to Melbourne Zoo in the early years. A local journalist captured the excitement surrounding this performance in his article in the *Daily Telegraph* in 1884: "Turning away [from the lion cages], we hear behind us a hum of voices and tramping of feet …. Here coming down the wide paths towards us some two hundred people. Men with chimney pot hats, and men with roguish-looking deer-stalkers … on one side of their heads, men leading a boy and a girl by either hand, men carrying infants; and women and girls … dressed in all the shades of all colors …. On they come, laughing and talking loudly, but still as if on some fixed purpose bent, and we can discover no cause, until the crowd reaches us, parts, and from the middle there emerges a man, pushing before him a hand cart, on which are large joints of beef. Now the truth dawns … the time has arrived for feeding the wild beasts."*
>
> * *Daily Telegraph* (Melbourne), January 2, 1884.

Perth Zoo both had tea rooms when they opened, and Moore Park added a Chinese tea pavilion to its attractions in 1885. By the time Taronga Zoo opened in 1916, the value of refreshment facilities was no longer in question. Food stalls were placed around the zoo, picnic facilities were laid on, a hot-water shop was set up, and 20,000 native shrubs were planted along the paths and around the lawns.

Each of the zoos used its animals in various ways to increase their entertainment value. Elephant rides were the most popular of these, providing many happy memories for visitors who experienced them. Pony and camel rides were also provided on occasion. When Adelaide Zoo's elephant died in 1904, a Bactrian camel was fitted out with a howdah to give rides to children until a new elephant was acquired.

After the elephant ride, watching the lions being fed was perhaps the second most popular attraction. Cecil Rix, in his history of Adelaide Zoo, referred to the bravado and challenge of the public to the big cats behind bars. "So long as the barriers are there, many people will taunt these animals."[26] In Melbourne, a journalist with the byline of "Liber" speculated on the attraction of carnivores in zoos and wondered how many people visiting this and similar exhibitions did so in the hope of seeing a man "torn to pieces"; "exhibitions of a dangerous kind," he observed, "are far more attractive than those where a broken neck is not a likely contingency."[27]

The carnivores were not fed on Sundays at the Melbourne Zoo in front of the non-fee-paying audience, a system which probably had less to do with the animals' health than an awareness on the part of the zoo's administration of the appeal of this exhibition. In the early decades of the twentieth century, the zoos added other entertainments to their programs such as miniature trains and fun-fair amusements. These were usually placed in the vicinity of the elephant rides, bandstands, and refreshment facilities. By this stage, the novelty value of exotic animals was waning; on the other hand, the gardens were maturing and the role of the zoo as a pleasant place to spend a day with the family was taking hold. Elephant rides, simple refreshments, and limited additional amusements seemed to be enough to draw steady crowds.

Work on the conservation and preservation of animals and research behind the scenes were minimal. The association with acclimatization introduced Melbourne Zoo to issues relating to the protection of animals in the wild at an early stage. Ironically, initial concerns were for introduced species. Birds, particularly game birds, were especially vulnerable in bush areas just outside cities. In the 1860s, the Acclimatisation Society of Victoria was responsible for many details of the colony's game laws, advising on seasons, lobbying on the use of the swivel gun, and suggesting additions to the protection lists. It also worked at encouraging the enforcement of protection laws by sending a circular to suburban municipal councils asking them to advertise the provisions of the act to provide for the protection of native and imported game. In addition, the police were advised of the locations of animals being liberated and asked to assist with their protection.

Despite these measures the animals and birds were never safe from destruction in or near the metropolitan areas. An appeal was issued to the women of Victoria to remonstrate with children "on the cruelty, folly, and selfishness of killing creatures," adding that, "If the Ladies would do all this for us, the evil would disappear, and they would have the honor of assisting in a most material degree in accomplishing the great objects of the Acclimatisation Society."[28] The society went so far as to form a bird protection group designed to appeal to young boys who were believed to be responsible for raiding nests. After complaints were received from farmers and fruit growers about the destruction to crops caused by the sparrow, the mynah, and the hare, the Acclimatisation Society reluctantly agreed that these species should be taken off the protected list.

Some years later, as the acclimatization activities waned and the zoological collection became more important, Melbourne Zoo's governing authority, now the Zoological and Acclimatisation Society of Victoria, began to grow concerned about the protection of native animals. In 1891, for example, it recommended a closed season for the bustard and, the following year, a closed season for the kangaroo. In 1895, after much persuasion, the society agreed to support the Field Naturalists' Club in its attempts to have a desolate region in southern Victoria, called Wilson's Promontory, declared a permanent reserve. It is likely that Dudley Le Souef significantly influenced the later concern about native animals because his knowledge on the subject was considerable.

In 1913, Dudley Le Souef received a copy of William Hornaday's book, *Our Vanishing Wild Life*, and saw it as an opportunity to promote methods of protecting Australia's own vanishing wildlife. He suggested that all forest reserves should be made game sanctuaries under the care of wardens, and he supported international moves to protect birds from being slaughtered for their plumage. He was also keenly aware that the thylacine was becoming so difficult to acquire that its survival was in question.

The thylacine, popularly known as the Tasmanian tiger or Tasmanian wolf, was a marsupial resembling a dog that once roamed the Australian continent. Its extinction on the mainland was possibly due to the introduction of the dingo from Asia. By the time of colonial settlement in 1788, it was only to be found on Van Diemen's Land, later renamed Tasmania. There it was considered a threat to sheep and settlers hunted it until its population was decimated. By the end of the nineteenth century, it was confined mostly to the northern regions.

At the beginning of the twentieth century, the mainland zoos were finding it more and more difficult to acquire thylacines for their own collections and for exchange purposes. Adelaide Zoo featured thylacines occasionally from 1886, with the last one appearing in its records in

1903. Melbourne's access to the rare animal continued sporadically until 1929. Seven years later, the last thylacine in captivity died in the Hobart Municipal Zoo. Since then, occasional unsubstantiated sightings of the thylacine have been reported in Tasmania but, officially, the animal is extinct.

Although Melbourne Zoo had several thylacines in its collection, even after Le Souef knew that they were endangered, there is nothing in the zoo's records to suggest there were attempts to breed them. In fairness, it will be remembered that the first studbook for a wild animal was not opened until 1923 when one was created for the European bison. The thylacine has become an emotional symbol for the impact European settlement has had on the wildlife population of the continent.

Early research at the Australian zoos was a lost opportunity rather than a period of activity. Historically, virtually no research was undertaken even though the zoos were ideally situated to become involved in nineteenth-century zoological studies. When the Zoological Society of Victoria was established in 1857, there were several important areas of scientific research concerning Australian fauna, which had not yet been resolved. There was an enormous interest in Australian zoology in Britain, particularly with regard to marsupials and monotremes, both of which confounded the classification schemes already in operation.[29] The problems associated with collecting and preparing animal specimens and amassing data about animals in their natural habitat for transmission to Britain made the study of the zoology of Australia potentially very rewarding.

The creation of the Victorian Zoological Society could have provided an ideal platform from which the colonial scientist might have contributed to the study of zoology. But the objectives of the new society did not include public discussions, lectures, published proceedings, or other features of a traditional specialized scientific society, and no provision was made to promote these studies within the organization.

In his study of societies in colonial Victoria during the 1850s, M. E. Hoare identified some of the difficulties that the founders of learned societies had to face: "Amateurism; a shortage of skilled men; the layman's obstinacy and the utilitarian emphases of a pioneer society."[30] He also points out, with reference to the Philosophical Institute of Victoria, that colonial societies had "to tread a precarious road between the 'useful' and the 'abstract' sides of science."[31] The demands on Australian science and societies to develop economic resources were powerful and the support for a purely intellectual approach were weak. As a result, colonial science in the mid-nineteenth century was localized and utilitarian; in the case of zoos, this manifested itself in their acclimatization activities.

The Victorian Zoological Society sent some indigenous animals abroad for research and study, usually to the Zoological Society of London. Dingoes, native hens, fish, ducks, and magpies were among the animals that followed the earlier batches of black swans and eagles to London in the early 1860s. There is no evidence in the Melbourne Zoo archive to suggest the London Society maintained a scientific correspondence with any member of Melbourne's Zoological Society, or to suggest it attempted to solicit scientific data from this source.

Many years later, the Zoological Society of New South Wales was criticized for its perceived antiscientific attitude and the standard of its management of Moore Park Zoo. In 1910, Count Mörner, the chief critic, a member of the society's council, and a founder of the Wildlife Preservation Society, resigned from the council declaring that the Zoological Society was not good enough to maintain an institution "of use to science, a study place of the disappearing native fauna of this country."[32] The internal squabble became public with popular support on Mörner's side. As a result, the society elected zoologists and ornithologists to join its council. Scientific papers appeared for the first time in the annual report of 1912. A year later, the *Australian Zoologist*, a scientific journal, was produced and has continued ever since.

Ironically, the society lost control of Moore Park Zoo around the time of the controversy. Prior to the transfer of the zoo to its new harbor site in 1912, the government appointed a new seven-member Zoological Gardens Trust to administer Taronga Zoo. Nevertheless, the society, now known as the Royal Zoological Society of New South Wales, has continued to flourish as an important Australian scientific institution. There is no evidence to suggest the trust that replaced the society gave any thought to the role of science and conservation in Taronga Zoo.

6.2.5 Visitors

The early visitors drew whatever educational and recreational value they wanted from these urban centers and were largely indifferent to any work going on behind the scenes. With fairly small regional populations, the zoos did not put restrictions on access. Although societies founded these zoos, few privileges were accorded their members. Indeed, by relying on government funding, Melbourne and Adelaide Zoos were not allowed to charge an entrance fee on some, or all, days of the week.

In the 1860s, few people visited the Melbourne Zoo. After all, with its emphasis on acclimatization, those animals that were in residence were not very exciting to the public. After Le Souef developed the collection in the 1870s, however, the zoo attracted hundreds, sometimes thousands of people every weekend. Many of the visitors came from the surrounding working-class neighborhood and they treated the place as a parkland with some exotic distractions. The standard visit seemed to involve spending time teasing the monkeys and watching the big cats being fed, then retiring to the shade of the trees to relax and doze off.

"Vagabond" described what he referred to as "the usual heterogeneous Victorian crowd" in the gardens in the 1870s: "Working men with their wives and children, tradespeople, clerks, schoolboys, sewing machine and show girls, with a touch of the larrikin element .... I was struck by what I am told is true colonial style — the fact of the husband so often walking holding his children by the hand, and the wife pacing in front, as if she did not belong to them."[33]

As the railway network was developed, Melbourne Zoo became part of a rural visitor's city itinerary. At the time, entrance to the zoo was free and, with virtually no subscribers, the management was largely reliant on government funding. After much persuasion, Le Souef was allowed to charge an entrance fee on all days but Sunday without losing the government grant. He also tried to attract the wealthy families of Melbourne but with little success. For the rest of the century, the visitors remained predominantly working class and occasionally rough. On several occasions in the later 1880s, Le Souef estimated that crowds of 30 to 40,000 people visited the gardens on free days, and he and his staff needed the assistance of the police to keep order. In the early twentieth century, the entrance fee was standardized for all days of the week and the size of the crowds became more manageable.

Across the country, as part of the deal with the South Australian Government, Adelaide Zoo was also free to the public one day a week. For a while, the free day was Monday, but soon it was changed to Saturday. The zoo maintained this arrangement for nearly 30 years before it decided to introduce half rates in 1914, initially for war-related charities, then for itself. The Minchins in Adelaide were perhaps more democratic than Albert Le Souef because, before 1914, they did not hide newly arrived exotic animals from the free Saturday crowds. As new animals arrived, there were put on display without restriction and each time the numbers of visitors on free Saturdays soared. It was estimated that, over those years, about 60% of the zoo's visitors went there on the free days.[34]

Gradually, each of the zoos built up a steady number of visitors, the numbers fluctuating with the economy. In 1904, for example, Perth Zoo received 71,000, about one third of the state's population. Just before World War I, the other zoos were receiving at least three times that number. Adelaide Zoo had 210,000 visitors in 1912, and this was in a city of about 200,000; Melbourne received 240,000 in a city of 590,000; and Sydney's Moore Park Zoo received 250,000 in a city of 630,000. It was estimated that by the time Moore Park Zoo was transferred to its harbor site, it had received about 7.5 million visitors in total.[35]

6.3 Survival, 1920–1960

All four zoos moved into the twentieth century as established institutions with good local support. By about 1920, all the major exotic animals, including the giraffe, bison, zebra, and tapir had been exhibited in at least one of the Australian zoos at some stage or another and, although price was still an impediment to displaying more than one or two of these species, it was no longer a strain or even a great novelty to acquire one. Indeed, when Adelaide Zoo bought its first giraffe in 1929, there was virtually no fuss or widespread interest. However, despite a positive beginning, the middle decades of the twentieth century were to be difficult for the zoos in Melbourne, Adelaide, and Perth because of a lack of focus, the great depression, the war, and a general run-down atmosphere, the last drawing considerable criticism. Only Taronga Zoo maintained its standards, thanks largely to the interest of a wealthy benefactor.

6.3.1 Enclosure Design and the Animal Collection

In terms of enclosure design, both Melbourne and Taronga Zoos were strangely retrogressive in the 1920s. After the challenge and success of the barless enclosures a few years before, they reverted to traditional cages during this period. Melbourne's new buildings, including those for the orangutan and chimpanzees, were made with bars and concrete and left a minimum amount of distance between the visitors and the animals. Sydney Zoo's new enclosures for chimpanzees, giraffe, and hippopotamus, built around the same time, were also old-fashioned. Then, in 1933, moats in the lion enclosure were filled in and a wire mesh barrier erected; most of the other moats were filled in a short while later. This was to allow visitors to move closer to the animals and to ensure that the animals spent more time out of the shadows of their dens.

An aquarium, one of the major exhibits constructed in Taronga Zoo at this time, was also a disappointment. The aquarium was designed to look like an underground grotto with lots of artificial rock and, while it was very popular with the public, it was a disaster from a maintenance point of view, the area behind the tanks being virtually inaccessible. Many of the problems associated with the construction of aquaria such as aeration, water circulation, and filtration had been resolved some time before in other zoos; however, these problems were all inherent in Sydney's aquarium when it was built, costing a great deal in alterations. One of the criticisms leveled at it in 1929 was that the "world's worst aquarium … made no provision for research."[36]

The chimpanzee was probably the most significant addition to the zoos' collections during this period. Melbourne Zoo received its first chimpanzees in 1929. The director, Andrew Wilkie, had been having trouble with the Dutch Colonial Government over permission to purchase orangutans. The Dutch believed that animals they had sold to an Australian zoo had later been sold to American zoos.[37] There is nothing to suggest Melbourne was party to such a transaction, but the Dutch refused to sell any more orangutans to the Australians. Desperate for large primates, Melbourne sought out the more expensive chimpanzees. A pair arrived and, although the male died of flu shortly afterward, the popularity of the remaining female made the investment worthwhile.

Melbourne and Adelaide, which received their first chimpanzees in 1935, bred these popular animals successfully through the middle decades of the century. In both cases, the wives of keepers hand-reared several generations of young chimpanzees. They were featured in publicity,

Aquarium exhibit at Moore Park Zoo (now Taronga Zoo, Sydney) showing an underwater grotto theme, early 1900s. © Zoological Parks Board of New South Wales.

newspaper photographs, and, later, on television. But the zoo's managers gradually learned that, as the young animals matured, they became too difficult to handle and could not be integrated into an existing group. The practice of hand-rearing and excessive human handling of young chimpanzees in both of these zoos ended in the late 1960s.

In the later 1920s and early 1930s, the zoos in Perth, Adelaide, and Melbourne each entered its most difficult decades as the country struggled through the Great Depression. In Perth, the financial problems were so severe that staff were retrenched, buildings deteriorated, rats multiplied, pipes corroded, and fences collapsed. The director was even forced to kill a young buffalo and sell the meat to staff because it could not afford to maintain it. Perth Zoo augmented its income to some extent by trading in native fauna. Finches were particularly popular abroad and in 10 years, to 1943, nearly 40,000 were exported.[38]

In Melbourne, there were calls to close the zoo, some of them from influential scientific quarters. The management was having problems servicing the loans for its building projects and was very dependent on the goodwill of the government to survive. One innovation during this period was the creation of an ambitious native fauna section within the zoo. David Fleay, a scientist with a specialized knowledge of Australian fauna, was appointed to build a representative collection and develop suitable enclosures. A platypussary displaying the delicate and unique monotreme was the central feature of the section, and a large group of koalas also provided a popular exhibit.

Koalas and platypus are particularly difficult species to exhibit. The problem with koalas is finding a constant supply of leaves from the few species of eucalyptus the animals will eat. The platypus, a freshwater animal, provides an even more complex challenge. It is so stressed by capture that, if it is not handled with great care, it dies quickly in captivity. Consequently, for many years, the mortality was exceptionally high. At the Melbourne Zoo, Fleay designed a platypussary and arranged a daily diet of about 600 worms, two dozen tadpoles, and two egg custards for the resident platypus named Binghi. The platypus lived in the Australian section for three years.

Despite the success of the Australian fauna section, it became embroiled in a conflict over the future of Melbourne Zoo. Some supporters of this section wanted the entire zoo to be devoted to the study of native animals under the care of a specialist staff. Although this idea had some influential support among medical scientists working in Melbourne, there was no question of getting rid of the exotic collection at this stage and the idea was dropped. As a result of the conflict, the Zoological and Acclimatisation Society of Victoria lost control of Melbourne Zoo and a government-appointed Zoological Board of Victoria was put in its place. Fleay left shortly afterward and, within months, the platypus and most of the koalas were dead. Fleay went to work in Healesville and, some time later, the first platypus ever to have been bred in captivity was born under his care.

The years of World War II compounded the zoos' problems. The younger staff enlisted, animals were virtually impossible to acquire, and equipment and building materials were not available to repair the enclosures. Melbourne Zoo was so anxious to maintain its collection that it suspended its own ruling on dealing with circuses. In 1924, the zoo's management had decided to end their dealings with circuses in response to a letter received from the Royal Society for the Prevention of Cruelty to Animals concerning the sale of tiger cubs to a Sydney circus. At the end of the war, the ruling was reimposed although other Australian zoos were not quite so reluctant to deal with circuses; as late as 1965, Adelaide Zoo sold an elephant to circus interests.[39]

6.3.2 Hallstrom and Taronga Zoo

In the meantime, a major benefactor had joined Taronga Zoo's trustees and, at the end of the war, this was the only major Australian zoo in a position to rebuild its collection. Indeed, it was about to enter a particularly prosperous phase, due largely to the extraordinary interest and generosity of Edward Hallstrom, a wealthy refrigerator manufacturer. In 1938, Hallstrom donated two black rhinoceros to Taronga Zoo and, in 1941, he was appointed a member of the Zoological Park Trust. From then until the late 1960s, his donations and his involvement in every aspect of the daily operations dominated the development of the zoo.

In 1946, Hallstrom sponsored a collection trip to Africa, led by his son and experienced staff from the zoo. Over 18 months, the expedition sent back three shipments of animals that included African elephants, rhinoceros, chimpanzees, baboons, various big cats, zebra, giraffe, reptiles, and about 1,800 birds. Importations of exotic animals continued, including shipments from Papua and New Guinea, South America, and Africa.

During this time, the other zoos were finding it difficult to deal with the increased stringency of quarantine laws. The laws, in place since 1910, were felt strongly during and after the war. In 1941, Melbourne Zoo was not permitted to import zebra and a tapir from South Africa, one of the few places where it seemed relatively easy to obtain animals at the time. Australian quarantine laws seemed so strict for a period that, at one stage, Melbourne Zoo's director was going to give up importing exotics because he felt it was too difficult to meet government regulations. Adelaide Zoo also found the laws difficult at times. In 1957, Canada offered several bison to the zoo, but they had to be refused because of quarantine. Bison from New Zealand, however, were allowed in after a quarantine period of some weeks spent in Sydney.

Somehow, Hallstrom's work was not hampered by these laws. His generosity spread to the other zoos, whose exotic collections were very meager at this stage. Elephants and hippopotamus seem to have been the only large vertebrate species that were constantly represented. Hallstrom donated, sold, or exchanged leopards, capybaras, eland, giraffe, chimpanzees, and other animals, becoming an important source of foreign animals throughout the continent. He also lent a Père David's deer to Adelaide for four years before bringing it back to Sydney for a breeding program. He even provided funds to build enclosures for his donations.

Under Hallstrom's influence, the reversion to old-fashioned enclosure design continued in Taronga Zoo. In 1949, an elephant slipped into the moat around its enclosure and became trapped; eventually it had to be killed. A few days later another animal fell into it, this time without injury. Hallstrom declared moats a failure and replaced the one in the elephant enclosure with bars. He then developed a fixation for concrete flooring. In 1943, a giraffe, two hippopotamuses, and a rhinoceros died within weeks of each other. Although veterinarians conducting the autopsies provided open findings on the causes of death, Hallstrom was convinced they had died from eating sand and gravel with their food, which came from the floors of their enclosures. He paid for repairs on 16 enclosures that did not pass his personal inspection. By 1956, all of Taronga Zoo's animals were living on concrete floors.

Ironically, Hallstrom's generosity with exotic species led to overcrowding at the Taronga Zoo and, by the time Ronald Strahan took over as director in 1967, there was a disproportionately high number of animals in every species group. Strahan did a survey of the number of individuals per species in zoos around the world. Although the average number of mammals in a species group was between three and five, Sydney had between eight and nine; of the birds, there were usually between two and four in other zoos, but in Taronga there were between 13 and 14. Most were not rare.[40]

6.3.3 Survival of the Four Objectives

Of the four objectives, recreation was the only one that received active attention during this period as the zoos sought ways of augmenting their falling incomes. Perth organized treasure hunts and baby shows in addition to its other amusements. It also boasted the only mineral bath and massage facility in an Australian zoo since it had an artesian bore, which also provided warm water to heat cages and to irrigate the gardens. Melbourne Zoo introduced a merry-go-round, a children's playground, and a small circus, which continued until the early 1960s. Adelaide also introduced a circus, although this only lasted for about two years.[41] Even the managers of Taronga Zoo felt the need to add a miniature train and a merry-go-round near its elephant circuit.

With the upsurge in the fun-fair atmosphere, the educational function of zoos was being questioned in some quarters, and still the zoos made no attempt to put their educational programs on a formal footing. School groups had long been accommodated in one way or another. In Adelaide, school students had always been admitted free to the zoo with their teachers; in 1887, for example, nearly 1,500 children visited the zoo under this arrangement. In Melbourne, the staff often left their other duties to talk to groups of school children. But throughout this second phase of their history, the zoos' attitude toward education remained passive. The labels were no more descriptive than they had been 50 years before. Strahan even remarked in 1967 that the hand-painted enamel labels which Hallstrom paid for were "uninformative and often inaccurate."[42]

Conservation and research were even less prominent than they had been before. There was a hint, at one stage, that Melbourne Zoo might be encouraged to take a more active role in research. Professor Colin MacKenzie, a comparative anatomist, orthopedic surgeon, and member of the zoo's governing authority, declared that zoos were a valuable place in which to research the anatomy and behavior of animals. He saw a need to study Melbourne Zoo's animals to throw light on "the basic problems of midwifery and the cell unrest as in cancer" and suggested that the success of the zoo was a matter of concern "to every patient in our public hospitals."[43] However, the greatest impact of MacKenzie's presence on the governing authority was to increase the competition to get hold of the zoo's dead animals. On one occasion, several scientists in different institutions examined the brain and nervous system of a dead orangutan before the skeleton was finally sent to the museum.

Of course, there was some need for ongoing research, if only to maintain the health of the animals through reasonable feeding practices, standards of hygiene, and veterinary care. The absence of veterinarians on zoo staff was compensated for by the skills of long-serving staff. Keepers who started working in the zoos as youths, often continued working at the zoos until they retired. It was not unusual for staff to retire after 45 or 50 years of service. Andrew Wilkie, for example, began working at Melbourne Zoo in 1867 as a junior gardener; he retired in 1936 as director and then came back to work as a volunteer until his death 10 years later. Keepers learned how to look after the animals from older members of staff, but, until relatively recently, they were poorly paid and there was little acknowledgment of the skills required to work with exotic animals. This lack of recognition of the staff, together with uninspired management, was symptomatic of the miserable state and poor image of most of Australia's zoos in the mid-twentieth century.

6.4 Modernization, 1960 to the Present

Despite the difficulties of the preceding decades, Australian zoos had enough life left in them to participate in the great modernization program that spread throughout the international zoo world in the 1960s. Over the past 35 years, they have each developed at a different pace and are now incorporating the four objectives into their operations in practical ways. Taronga, Melbourne, and Adelaide Zoos have developed wildlife parks while the Perth Zoo has developed a center for conservation and research. In addition, the Zoological Board of Victoria, now called the Zoological Parks and Gardens Board of Victoria, has taken over responsibility for the native fauna sanctuary in Healesville.

In the north of the country near Darwin, the Territory Conservation Commission opened a large native fauna park, the Territory Wildlife Park, in 1989. The park covers over 900 acres of tropical woodland, monsoonal rain forest, and natural wetlands, and is laid out to promote awareness of the local environment among territorians as well as tourists. All these institutions are now working closely with government authorities and, with the effective application of the four objectives, the nineteenth-century zoos have placed themselves in a strong position to play a major role in Australian society for the twenty-first century.

The first challenge for the zoos' managers in this period has been to improve the accommodation for animals and the appearance of the zoos. Melbourne Zoo began its modern period in 1963 amid criticism about the conditions in which the animals were living. A television broadcast further fueled this situation with a tearful and emotional Joy Adamson standing in front of the lion enclosure. Rather than try to defend the zoo, Dr. Alfred Dunbavin Butcher, chairman of the Zoological Board of Victoria, used the publicity to encourage the government to support the modernization program.

Over the next 20 years, Butcher gradually transformed the zoo, drawing on whatever resources were available. Then in 1987, the renowned zoo architect, David Hancocks, prepared a master plan to develop the zoo using bioclimatic zones with landscape-immersion-style enclosures. Several of these enclosures have already been created, covering over a quarter of the zoo. They are immensely popular with visitors and contribute to the exciting atmosphere in the zoo today.

As Melbourne Zoo's enclosures developed, the Victorian Zoological Board seized opportunities to extend its activities to a wildlife park and a native fauna sanctuary. Werribee Zoological Park was created by the Zoological Board in 1975 on the outskirts of the city, and opened to the public in 1983. It is now called Victoria's Open Range Zoo at Werribee. Visitors currently view the animals from a safari bus driven through the vast enclosures. Some species, including the large hippopotamus and the southern white rhinoceros, have been transferred from the urban facility to more suitable accommodation in the wildlife park. Although Werribee Zoological Park was designed to display African animals only, the scope of the collection has been extended to include grassland animals from Australia and abroad under the direction of David Hancocks.

There is no such dilemma with regard to the choice of species displayed in the Zoological Board of Victoria's native fauna sanctuary in the mountains on the other side of the city. Healesville Sanctuary was transferred to the control of the Zoological Board in 1978. It had been established in 1934 on the site of Sir Colin MacKenzie's former research station and was then known as the Sir Colin MacKenzie Sanctuary; an independent committee managed it. Its inaugural curator was Robert Eadie, famous for his work with the platypus in the 1930s.[44] Healesville Sanctuary is now attracting local and international visitors to view native animals in a bush setting. Even though many of the species may also be seen in Melbourne Zoo, visitors who make their way to the sanctuary treat it as something more than a zoo. There is such a feeling of natural bushland in the sanctuary that keen naturalists and families alike spend their time observing and learning, as well as experiencing the pleasure of a day out in the Victorian mountains.

Taronga Zoo began its modernization program with the appointment of Ronald Strahan as director in 1967. Strahan, a professional zoologist, was faced with an overly large collection and heavy concrete enclosures. In response to this, the trustees established a wildlife park in rural New South Wales. Before drawing up plans, senior members of the zoo's management traveled around the world to look at similar large-scale zoos. They found the European wildlife parks of limited value, being designed for cold, even snow-covered winters. The zoos in Singapore and San Diego, on the other hand, contributed valuable information. Strahan noted that the San Diego Wild Animal Park provided some points on what not to do; he considered the animals too far from the visitors and the enclosures "boringly similar and, lacking trees or ground cover, had already become dustbowls."[45]

In 1977, Western Plains Zoo in Dubbo was opened. The species selected for exhibition were those that could be accommodated in moated enclosures. To allow visitors to get fairly close to the animals, the enclosures were long and relatively narrow along the moated sides. A considerable range of species have been exhibited in Dubbo since then; these include hippopotamus, African elephant, cheetah, white rhinocero, Persian onager, and Przewalski's horse.

Back in the urban zoo in Sydney, reduction in the number of animals gave the staff space to create new, up-to-date enclosures. The famous chimpanzee enclosure with a stream, rocks, long grass, and an artificial termite mound opened in 1980 for a group of 23 animals and is now one of its most successful exhibits. At first, the social grouping was difficult to manage in the large moated area because the animals were accustomed to living in much smaller groups. Now, the animals use all of the features of the enclosure and their interactivity almost lends a theatrical effect to the exhibition.

The modernization work at Adelaide Zoo took off slightly later than in the eastern zoos. Two walk-through aviaries were among the first of new enclosures to be built in this period since Adelaide Zoo was known for its bird collections. These aviaries and the ones that have been developed since have proved to be both popular with visitors and positive for breeding programs. The scarlet macaws, for example, have produced a clutch of eggs since they were transferred to the aviary in the Amazon rain forest enclosure where there is flowing water, a rock face, and an automatic sprinkler spray. In 1993, a master plan was produced for the zoo. Designed to be completed over a 20-year period, it is based on the Gondwana continental drift theory. Six of the Southern Hemisphere landmasses that made up the supercontinent of Gondwana will be featured in the zoo. Construction work on the Southeast Asia enclosure began a year later.

One of the keys to the recent turnaround in the appearance of the Adelaide Zoo was the establishment of the Monarto Zoological Park at Monarto in 1993 and the subsequent transfer of some of the larger vertebrates, including the elephant, to the new facility. The 2,470-acre Monarto Zoological Park is about 40 miles outside Adelaide and its natural features include Mallee woodland, rocky outcrops, creeks, and wetlands. It displays antelope, zebra, ostrich, and giraffe in its African plains exhibit, and deer, antelope, and Przewalski's horse in the Asian grasslands enclosure. Like other wildlife parks, it also pays considerable attention to regional animals. In addition, the staff care for regional plants, such as the endangered Monarto mint bush, rare orchids, and acacias.

Perth Zoo also started its modernization program later than the two eastern coast zoos. Plans for a major redevelopment began in the 1970s, and facilities for education, conservation, and research have since featured prominently. Its orangutan enclosure has been designed to encourage a successful breeding program. Perth did not receive its first orangutans until 1968, and since then it has developed a style of caring for these animals that takes their proneness to stress into account. The males are given as much space as possible to allow them to live solitary lives. Females are kept in small groups, and young females are encouraged to mix with adult females who are nursing infants to learn maternal skills. Climbing frames have been built in the enclosures to allow the animals to exercise.

Perth's work with local animals has become very important. The Department of Conservation and Land Management believes that 88 species in the state are endangered. The zoo's center for research and education outside the city will now be part of the statewide conservation and preservation program. It is set near a state forest and a mine rehabilitation project, and both of these features will be incorporated into educational programs to generate interest in the natural world.

The new approach to enclosure design and animal distribution in the Australian zoos was complemented by radically different policies on animal acquisition. In 1967, Strahan had been forced to address urgently the problem of indiscriminate animal acquisition in the Taronga Zoo. He made it clear to the trustees that, from then on, every addition to the collection was to be "a deliberate act, justified by the 'message' that the species could convey to visitors." The staff referred to this as "the principle of thematic exhibition."[46]

This declaration marked a new spirit of cooperation across the country. The export of native fauna was already under control with the introduction of a ban on the general export of animals in January 1960. Permits could be granted for the exportation of animals to approved zoos, a provision that caused a few problems for the Australian zoo officials who were obliged to establish the credentials of people claiming to be agents for foreign zoos. Now, beginning with an interstate meeting in 1968 at the Perth Zoo, the Association of Zoo Directors of Australia and New Zealand (AZDANZ) was established.

Among the earlier agreements of AZDANZ were to set up a price list for Australian species, to establish studbooks for exotic species in the zoos, and to arrange free and indefinite loans between zoos for animals that might be part of a breeding program. Later, AZDANZ accepted the principle that the collection of animals in the public zoos of Australia and New Zealand should be managed as a single collection. There is now cooperation on which exotic species are imported, and where they are exhibited, to ensure that distribution is in the best interests of the species. The development of the Australasian Species Management Plan (ASMP), a program of the Australasian Regional Association of Zoological Parks and Aquaria (ARAZPA), the successor to AZDANZ, is the latest approach to listing, tracking, and coordinating the distribution of species in the zoos. Each zoo pays a fee to support this program. Through these means, old-style rivalry between the zoos has been eliminated.[47]

Modernization programs have been highly effective and all the old Australian zoos have shaken off the nineteenth-century look of bars and bricks. Some exhibits have yet to be fully developed, but now it is a matter of time, money, and imagination rather than of interest as every effort is made to ensure enclosures accommodate the animal to the highest standards. Moreover, great emphasis is placed on incorporating some or all of the objectives in each enclosure to maximize the value of every animal on exhibition.

6.4.1 Modernization of the Four Objectives

Enhancement of the educational value of the zoos was fundamental to much of the modernization work in the 1960s. Various programs were established to encourage the use of zoos as educational centers. Adelaide Zoo invited the state education department to conduct biology and science classes there in 1966. Students were offered free access as long as their visit was connected with the school curriculum. University students were also allowed in free of charge for study-related research. Five years later, a teacher was seconded from the Department of Education to cope with the annual turnout of 10,000 students and to ensure their day would indeed have an educational focus. Student teachers joined the overworked zoo teacher; later, additional professional teachers were added to the staff, and classrooms were constructed.

In Sydney, Ronald Strahan laid the foundations of a zoo education service in 1968. Although his ambitious plans involved 12 teachers and a comprehensive course in animal biology for primary and secondary school children, he had to be content with one teacher initially. Nevertheless, the program was a great success. In 1974/75, the zoo received over 100,000 children through its education program. In 1976, the Education Centre was built, comprising classrooms, a library, and offices. Several teachers were employed to conduct programs aimed variously at school children, undergraduates, and science teachers. Taronga Zoo education staff now work with about 30,000 school children directly, and produce kits and other material for an additional 60,000 students.

From its earliest days, the Melbourne Zoo education service identified itself with environmental education. It saw itself as part of a new network consisting of staff from museums, galleries, botanic gardens, parks, and other bodies concerned with wildlife. The educators from the related institutions met regularly to discuss problems and exchange ideas on various means to encourage widespread interest in the natural world. There was concern that the students might be overwhelmed by the problems facing animals in the wild on a global basis, and feel helpless if they could not see and participate in preservation and conservation programs of immediate interest and impact. Consequently, with statewide cooperation, education programs that empowered the students were developed.

In the early 1970s, Graham Morris, the first officer in charge of the zoo education service, believed it had a valuable role to play in promoting "desirable attitudes towards the natural environment" among children. A majority of children, he wrote, were growing up in cities and had limited contact with the natural world. Recreational activities such as visits to zoos were ideal opportunities to encourage "an appreciation and understanding of the animals on display and [assist] the development of worthwhile attitudes towards wildlife in general."[48] He set out to ensure that the thousands of children who visited the zoo would leave having experienced pleasure and excitement, as well as having picked up some information about wildlife.

Now Melbourne's programs are graded for different levels. Some groups are accommodated in specially designed classrooms that graphically demonstrate environmental problems. Students also have access to small animals, which they may touch under careful supervision. Pop the Pelican, a staff member in costume, gives students guidelines and a kit before sending them off on specified trails around the zoo.

In Perth Zoo, the insect exhibit is a highlight of an educational visit. Microworld introduces children to the world of insects. A member of the staff, dressed as a spider, shows children the effect of erosion and salinity on the soil in a display area the education staff have labeled Devastation Theatre. The children are then led into a giant-size model of a burrow, complete with giant-size models of termites, fly maggots, mites, an earthworm, and a wolf spider. After this experience, they are then invited to look at real insects at work on soil under a microscope.

The informal methods of transmitting information to casual visitors have proved more challenging. The principal media available to zoos to reach these visitors are keeper talks, enclosure designs, labels, and the promotion of celebrity species. Institutions are reached through publicity releases. As with the students' programs, much of the information conveyed through these means relates largely to environmental awareness and the lives of species in their natural habitats.

Keeper talks are used in various ways. In wildlife park safari buses, the driver is usually trained to give fairly detailed information about the animals on exhibition. In other zoos, keepers sometimes wear a microphone and talk while they are feeding their animals. In Sydney, a special theater was built in 1979 to showcase seals, while a keeper explains their behaviors and gives information about their habitats. In 1993, Adelaide Zoo began a series of evening walks around the native animal section. Visitors were offered special barbecued food and bush music, as well as a talk by keepers about the animals. The event was such a success it has been continued and developed. In 1994, Aboriginal guides talked to visitors about the relationship between Aboriginal people and Australian animals. Aboriginal music, dance, and dreaming stories were also part of the evening.

In the Territory Wildlife Park, white-bellied sea eagles and wedge-tailed eagles have been trained by the staff to fly on command in front of an audience. The large birds of prey were usually brought to the zoo after being injured, abandoned, or when sick. While they are being rehabilitated and, in most cases, being prepared for release back into the wild, they take part in an impressive display for the public. The Territory Wildlife Park also organizes keeper talks in various areas including the animal care center, the aquarium during feeding time, the wader lagoon when the pelicans are being fed, and the nocturnal house, where visitors are told about the behavior of the black flying fox.

The Australian mass media, through discussions of popular species, have been used as a direct way to reach to a wide audience. The visit of two giant pandas from China to Taronga Zoo and Melbourne Zoo in 1988 had a huge impact on the public's general knowledge about the plight of animals in the wild. Every detail of the pandas' diet and habits, and the threats they face in their natural habitat, was discussed on television, in the newspapers, and in school programs. By the time the pandas left, there were few people in the country who were not aware of zoos' work in animal conservation generally and with the giant panda in particular. As planned, the zoos were able to use the extra income generated by the pandas' visit to support local conservation programs on less popular animals.

In recent years, carefully controlled publicity in relation to the hidden work of zoos has had a fundamental impact on the attitudes of visitors. This was evident in 1993 when Melbourne Zoo transferred its young male gorilla, named Mzuri, to Jersey Wildlife Park as part of an international breeding program. Sydney first exhibited gorillas in 1959 and, within a few years, Hallstrom had increased the group to seven. In 1973, Melbourne acquired a pair of gorillas and, over the past 20 years, has established an important family group. The species quickly became the most popular animals in the zoo.

When a young male was born as a result of artificial insemination in 1984, he became an instant celebrity. A competition was organized to find a name and, over the years, it was featured regularly in the media. However, Judith Henke, the publicity officer, ensured that the information accompanying the cute photographs explained the threats facing the species in the wild and this animal's role in an international breeding program. So when the time came to send

Mzuri off to Jersey, many people wrote to the zoo and to the newspapers expressing their feelings of loss but, in virtually all cases, acknowledging his role in conservation work.

The use of enclosures to convey messages is much more difficult as it relies to a great extent on visitors reading labels or drawing on information from other sources to interpret the exhibit. Consequently, enclosure design now involves staff from all sections of the zoos. The philosophy behind each exhibit is carefully worked out and the designers consult the keeping, horticultural, research, and educational staffs to ensure maximum impact of the display.

Some of the finest displays in Australian zoos are now being created for smaller animals, a group that has received so little attention in the past. The Misunderstood Creatures exhibit in Territory Wildlife Park is an arthropod and reptile exhibit with spiders, scorpions, millipedes, snakes, and lizards. It also includes a giant termite mound similar to those common in the northern part of the country. Termites are considered a major pest in the region, so the Wildlife Park uses this exhibit to inform visitors about the ecological importance of termites in recycling nutrients from plants.

In Melbourne, the Butterfly House was the culmination of 10 years planning by Alfred Butcher. He wanted to convey to the public the message that preservation of habitats and their invertebrate populations is vital to the preservation of all species. He was aware that it would be difficult to attract government funding and corporate sponsors for insect exhibits and decided to put his efforts into a display of the beautiful and delicate butterflies. The Butterfly House now embodies all four objectives with great success. It has a large research laboratory attached to it, signs on the visitors' walkway give information about the life cycle of butterflies and the role they play in the environment, and it is probably the most popular exhibit in the zoo.

With the emphasis on incorporating the objectives in exhibits, native fauna have also been given greater status in modern Australian zoos. In the 1960s, Ronald Strahan was keen to promote interest in Australian animals. He suggested that because so many animals, including kangaroos, were most active during the night, "Few people in the 1960s were aware of their rich variety and, due to this ignorance, there was little interest in their conservation." He experimented with artificial lighting in an old snake house and after about six months had arrived at a workable arrangement. He also placed koalas in an enclosure with a ramp for visitors that allowed them to look at the animals at treetop height. By 1972, Taronga had a platypus house, koala exhibition, rain forest aviary, waterfowl ponds, and a nocturnal house all displaying Australian animals.[49]

Today, native animals are also the subject of many of the conservation programs going on behind the scenes. It is estimated that, since European settlement in 1788, 18 of Australia's recorded 260 mammal species have been driven to extinction.[50] The zoo staff members realize they need to go beyond breeding animals to promoting concern for them within the community in order to maintain or even improve threatened habitats. This involves using all the educational means within their control. A greater interest in native wildlife is steadily becoming more apparent throughout the country. At Easter time, for example, the traditional chocolate rabbit is rapidly being replaced by the chocolate bilby, and a recent discovery of a colony of brush-tailed rock-wallabies was given wide coverage in the national media.

Today, each zoo is involved in a great number of conservation programs for large and small, foreign and native species. The Western Plains Zoo in Dubbo, for example, has a range of breeding programs; it has the largest breeding group of black rhinoceros outside Africa and one of the largest herds of Przewalski's horses in the world. It has also bred the Persian onager, has returned blackbuck to their native Pakistan, and, jointly with Monarto, has returned Przewalski's horses to Mongolia.

At the other end of the scale, in terms of size, the Dubbo zoo has a program for breeding the greater bilby, or rabbit-eared bandicoot, which was once widespread through most of inland

Australia and is now limited to a region in the Northern Territory, where they are under constant threat from feral cats and other predators. It is also involved in the interstate breeding program for the malleefowl. This bird, which incubates its eggs in large mounds of sand, was bred successfully in Melbourne Zoo at the turn of the century through the efforts of Dudley Le Souef. However, over the decades, its habitat in semiarid woodland has been cleared for agriculture. Wildlife authorities in four states and staff from various zoos met in Dubbo in 1988 to work out an approach to conserve the species. The Western Plains Zoo was given funds to breed up to 300 birds over four years.[51]

One of the major projects of the Adelaide Zoo involves the yellow-footed rock-wallaby. Native to the region, it was first exhibited there in 1883. The earliest birth on record for one of these wallabies in the zoo was in 1929. Since then, the yellow-footed rock-wallaby has been breeding steadily but, between 1929 and 1977, only two new animals were introduced to the group. In the meantime, threats to the animals' habitat grew because of agricultural, pastoral, and mining interests in the Flinders Ranges. Wild-caught animals were introduced to the captive population in the zoo to increase the genetic diversity before being tagged and released back to the wild, where they are monitored for up to a year. In the long term, the zoo hopes to supplement the wild population with zoo-bred animals.

The numbat, another threatened Australian species, was once common over the entire southern part of the continent and is now found in southwest Western Australia only. It is a termite-eating marsupial and very difficult to hold in captivity because of its diet of live termites. In 1969, Taronga received a pair; prior to that, none had lived in the zoo for longer than a month. The staff took the time and trouble to feed them properly and the numbats lived there for seven years. Although they bred, they failed to rear their young. Elaborate plans to develop a major exhibit failed to establish the ideal conditions because, on being transferred to it, the animals succumbed to humidity. In the later 1980s the staff member involved in caring for the numbats transferred to Perth Zoo, where, with the support of the World Wildlife Fund, a captive breeding program was created and an artificial diet that encourages breeding was developed.

Of course, this level of conservation work is dependent on the interest and ability of contemporary zoo staff. When Strahan joined Taronga Zoo, he restructured the staff to reflect levels of expertise from keeper-in-training to head keeper. A new wage scale for keepers was introduced and a training program prepared in conjunction with veterinary clinics, research laboratories, and other areas where live animals were handled. A certificate in animal care was created at Sydney Technical College in 1969 based on a three-year part-time syllabus. Even though the course was only open to people working in a relevant area, 72 students enrolled in the first year.

Strahan also appointed various professional staff, including a veterinarian who was to research causes of death of the animals, and two curators, one for the aquarium and one for the bird collection. Each of these individuals had relevant experience in zoos. The curators worked on rebuilding their respective exhibits to international standards. They also drew the zoo closer to professional staff in government customs and quarantine departments, and provided advice to the new state-based Parks and Wildlife Service.

The other zoos have since developed their own programs and the staff are now part of a professional organization, the Australian Society of Zoo Keepers (ASZK). In an important development for the zoo industry, ASZK and the more management-oriented association, ARAZPA, have recently merged. Zoo keeping is now a popular profession with keen competition for the few openings at a junior level. Technical certificates, such as the Certificate of Zoo Keeping from a Melbourne college, are usually required as a minimum qualification, although university degrees among staff are now more common. Both

Taronga Zoo and Perth Zoo have appointed research biologists with doctorates to work in collaboration with the Commonwealth Scientific and Industrial Research Organisation (CSIRO) and the universities in pursuit of zoo-related research programs. In Victoria, the Zoological Parks and Gardens Board has now set up a Conservation and Research Department, and its director is a professorial associate in the Department of Zoology, University of Melbourne.

As animal husbandry becomes more professional, demand for staff with specialized knowledge and expertise increases. In addition, staff members are required to work closely with all sections of the zoo community, from other keepers and vets to management and public relation staff. Secondments are also arranged from one zoo to another so that specialist skills and knowledge are shared.

Research into specific species has developed with this growing professionalism as individual staff members pursue projects relating to their charges. This can extend outside the zoos, as conservation programs require fieldwork as well as *ex situ* breeding. One of the Adelaide Zoo earlier research projects was in 1966 when the society received funding to study rare marsupials in outback South Australia. Two zoologists spent several months in the northern part of the state; they found some female rabbit bandicoots and gathered information about several marsupial mice, the yellow-footed and the black-flanked rock-wallabies. Further grants were received but, at every stage, it was necessary for the Zoological Society to add a considerable amount of its own money. After three years of research, the society decided it could not afford to maintain the field scientists and decided instead to offer its facilities to researchers from the University of Adelaide zoology department.

In the early 1970s, the staff in Taronga Zoo produced a research paper on the courtship behavior of the platypus based on observing the animal in the zoo. The life sciences division later organized a workshop to discuss the problems faced by platypus in the wild and the role a zoo might play in its long-term survival. Although they are not endangered, their delicacy and importance encouraged the zoo staff to preempt a crisis. Around that time, a veterinary-quarantine center was built with a post-mortem theater and records office; an operating theater and laboratories were added in 1978. This involvement continued with the opening of a Conservation and Research Centre as a repository of biological information made available to researchers in the zoo. Visiting researchers can also use it as a place to work.

A recent project concerning the application of human reproduction science to the rescue of endangered species has drawn zoos and universities into cooperation. Three zoos (Melbourne, Western Plains, and Taronga) and four universities (Monash, Sydney, Queensland, and Melbourne) established a new center called the Animal Gene Storage and Resource Centre. The center is part of a global network of institutions that have joined forces to collect genetic material from endangered species. The Australian Centre will focus on preserving sperm and eggs, and studying the reproductive systems of endangered marsupials, monotremes, birds, frogs, cats, and ungulates. Regional species already marked for attention include the northern hairy-nosed wombat, three rare types of wallaby, the New Guinea long-beaked echidna, and the greater bilby. The establishment of the center has recognized that programs designed to protect natural habitats and preserve species in the wild may not be sufficient to prevent the extinction of some species.[52]

These activities, among others, are contributing to a significant role for research in zoos, which still relies to some extent on the enthusiasm of individuals to spend extra time in collating their information. Certainly, the commitment to research is evident in the recent annual reports of each of the institutions and, gradually, the zoos' managers are finding ways to put resources into this aspect of their operations.

The excitement and pleasure of a day out is still the element that attracts the majority of Australians to zoos. Therefore, it is still necessary to provide additional amusements and facilities to maintain popular interest. Over the past 30 years, extra amusements have been altered to ensure they do not compromise the modern complex image of the institutions.

Melbourne Zoo stopped its elephant rides in 1962 as board members became aware of growing concerns about the safety of the practice. They conducted a rapid survey of zoos around the world and, upon discovering many had stopped providing rides, they too ceased them at the height of the summer season. The animal circus also came to a halt, although the fun-fair persisted until recently, largely because the family that organized it was of great help to the zoo during its darkest days in the mid-century. However, the site it occupies is now scheduled for development for coastal marine animals as part of the zoo's master plan.

In the mid-1970s, the Taronga Zoo staff faced up to the issue of extra amusements for children. In 1976, a Friendship Farm was created displaying domestic animals, tame parrots, and young marsupials to provide additional amusements for children. The farm displaced the miniature train and merry-go-round, a swap that concerned some members of the Zoological Parks Board, now the governing authority of Taronga. They felt mechanical rides and other diversions for younger children were necessary to maintain income. However, others felt this did not fit in with the new atmosphere in the zoo. In his report condemning the fairground in the zoo, the director, Peter Crowcroft, quoted Grzimek of Frankfurt Zoo as saying, "We are not Gypsies: we do not move on tomorrow."[53]

Such fun-fairs have since been proved unnecessary, as all of the zoos have developed highly satisfactory alternatives to mechanical amusements and animal circuses. Each zoo has beautiful picnic spots, while most have good refreshment facilities. Melbourne Zoo built a variety of restaurants to suit a range of needs. Besides its inexpensive restaurants and food stalls, it has a splendid reception area, with fine stained-glass features and high glass walls, that overlooks a lake and a Japanese garden. This development was the idea of Alfred Butcher who wanted to ensure that politicians and corporate sponsors could be entertained in suitable surroundings on a visit to the zoo. The area is also used for conferences, weddings, and other functions that produce a high return for the zoo.

Adelaide Zoo has converted its nineteenth-century monkey house into a restaurant. It is situated opposite a beautiful rotunda, which was built in 1884 and still has its original trimming of Victorian cast-iron lacework. The rotunda was fully restored in 1989 and, with the warm climate of South Australia, it has become a popular location for receptions, weddings, and other parties in conjunction with the restaurant. In 1991, Taronga Zoo also opened a dining and conference center. The Harbourview Terrace overlooking Sydney Harbor can accommodate 500 people for formal functions.

In creating such facilities, people who might not normally visit a zoo cannot help but notice their surroundings while attending a function. Many of these visitors may still harbor images of the old-fashioned zoos they were brought to when they were children. Now they are given an opportunity to glimpse the changes that have taken place and may, therefore, be encouraged to return for a recreational outing. Concerts of various sorts are also encouraging diverse groups of people to visit. Bush bands, Aboriginal music, chamber music, and jazz evenings have all been performed in the zoos in recent years. The Melbourne Zoo jazz concerts have been held on weekends during the summer months for a number of years and have been a huge success. They even attract visitors who begin their dinner parties on the grass in front of the bandstand before going on to nearby restaurants to continue their meal. In this way, the zoos are augmenting income and attracting new supporters without compromising the atmosphere and image of the institutions.

6.4.2 Visitors

Perhaps the most effective development with regard to visitors today is the creation of organizations to encourage supporters to take an active interest in their local zoos. These groups draw regular visitors into the daily life of a zoo. They provide members with a variety of privileges such as free entry to several properties, regular newsletters, special visits and tours, social functions, and lectures. In return, members provide a powerful body of informed and enthusiastic supporters who are aware of the activities of the zoos and, consequently, generate a considerable amount of goodwill for the institutions in the wider community. In material terms, they also raise funds for exhibits and act as tour guides, fulfilling a function the zoos could never afford on their staffing budgets.

Friends of the Zoos, or FOTZ as it is known, was established in 1980 to support the facilities of the Zoological Board of Victoria. Alfred Butcher, then chairman of the board, saw great value in creating an informed support group in the broader community. This group has grown from its initial membership of 160 to over 20,000. It produces a quarterly journal that contains a variety of academic and popular articles, as well as information about the activities of the zoos. It also operates a roster of highly trained voluntary guides to provide tours, to staff touch tables and information booths, to talk to groups around the city, and generally to be available to assist visitors. The guides are subject to a rigorous interview before they are selected for a lengthy and thorough training course. The commitment demanded of the trainees and guides is substantial, yet the standard of applicants and the success rate are both very high.

The other zoos also have their own visitors' associations that work with each of the facilities under their management umbrella. In South Australia, visitors are invited to become members of the Royal Zoological Society. In Western Australia, the Docent Association was established in 1982, and a separate fund-raising arm, the Perth Zoo Society, was also formed. The docents act as tour guides, undergoing a 10-week training course, entry for which is also heavily contested. In 1983, Taronga founded the Australian Association of Zoo Friends, a support group for Taronga and the Western Plains Zoo. In 1989, the group changed its name to the Association of Zoo Friends, NSW and, in that year, raised over 10% of the NSW Zoological Board's total income.[54]

The enthusiasm, knowledge, and interest of the members of these associations are now considered to be a vital part of the modern zoo. Through their fund-raising and journals, they are in a position to wield some influence over the direction of the zoos and, as the zoos are almost entirely reliant on their visitors for income, this is a healthy and positive relationship. Management has access to information about the interests and concerns of the regular visitors, and can use this to enhance its zoo for all visitors. It is now estimated that the zoos of Australia and New Zealand receive six million visits per year. Considering a total population of about 21 million, and even allowing for tourists, this demonstrates the popularity of zoos in this part of the world.

References

1. de Courcy, Catherine, *Evolution of a Zoo: A History of the Melbourne Zoological Gardens, 1857–1900*, M.A. thesis, University of Melbourne, 1991.
2. McCoy, Frederick, "Anniversary address … on acclimatisation, its nature and applicability to Victoria," in [First] *Annual Report of the Zoological and Acclimatisation Society of Victoria*, Zoological and Acclimatisation Society of Victoria, Melbourne, 1862, 34.
3. Rix, C. E., *Royal Zoological Society of South Australia, 1878–1978*, Royal Zoological Society of South Australia, Adelaide, 1978, 15–16, 49.

4. Prince, J. H., *The First One Hundred Years of the Royal Zoological Society of N.S.W. 1879 to 1979*, Royal Zoological Society of NSW, Sydney, 1979?, 9.
5. Zoological and Acclimatisation Society of Victoria, *Minutes*, October 6, 1876.
6. *Herald* (Melbourne), January 23, 1882.
7. *Argus* (Melbourne), December 22, 1876.
8. Zoological and Acclimatisation Society of Victoria, *Minutes*, March 1914.
9. Zoological and Acclimatisation Society of Victoria, *Minutes*, August 2, 1886 [Dudley Le Souef's report to the council on his return from Europe].
10. de Courcy, Catherine, *The Zoo Story: The Animals, the History, the People*, Penguin, Melbourne, 1995, 109.
11. *Age* (Melbourne), November 23, 1874.
12. *Argus* (Melbourne), December 2, 1876.
13. *Argus* (Melbourne), July 21, 1884.
14. [Second] *Annual Report of the Acclimatisation Society of Victoria*, Acclimatisation Society of Victoria, Melbourne, 1863, 32–33.
15. Zoological and Acclimatisation Society of Victoria, *Minutes*, October 6, 1857.
16. *Argus* (Melbourne), December 2, 1876.
17. *Herald* (Melbourne), October 28, 1885.
18. *Argus* (Melbourne), March 27, 1883.
19. *Argus* (Melbourne), October 14, 1893.
20. *Argus* (Melbourne), April 15, 1889.
21. *Herald* (Melbourne), November 28, 1888.
22. *Register* (Adelaide), July 27, 1900.
23. Moyal, Ann Mozley, Ed., *Scientists in Nineteenth Century Australia: A Documentary History*, Cassell Australia, Melbourne, 1976, 188–189.
24. *Argus* (Melbourne), July 21, 1884.
25. *Age* (Melbourne), October 13, 1893.
26. Rix, 1978, 181.
27. *Age* (Melbourne), January 23, 1875.
28. McCoy, 1862, 35.
29. Dugan, Kathleen G., "The zoological exploration of the Australian region and its impact on biological theory," in *Scientific Colonialism: A Cross-Cultural Comparison*, Reingold, Nathan and Rothenberg, Marc, Eds., Smithsonian Institution, Washington, D.C., 1987, 80.
30. Hoare, M. E., "Learned societies in Australia: the foundation years in Victoria, 1850–1860," in *Records of the Australian Academy of Science*, 1, 26, 1976.
31. Hoare, 1976, 11.
32. Strahan, Ronald, *Beauty and the Beasts: A History of Taronga Zoo, Western Plains Zoo and Their Antecedents*, Zoological Parks Board of New South Wales, Sydney, 1991, 23.
33. *Argus*, December 2, 1876.
34. Rix, 1978, 31–32, 34–35.
35. Jenkins, C. F. H., *The Noah's Ark Syndrome: One Hundred Years of Acclimatization and Zoo Development in Australia*, The Zoological Gardens Board of Western Australia, Perth, 1977, 29. See also Rix, 1978, 32.
36. Strahan, 1991, 42.
37. Zoological and Acclimatisation Society of Victoria, *Minutes*, June 1929.
38. Jenkins, 1977, 31, 34.
39. Rix, 1978, 205.
40. Strahan, 1991, 68.
41. Rix, 1978, 48.
42. Strahan, 1991, 80.

43. *Annual Report of the Zoological and Acclimatisation Society of Victoria*, Zoological and Acclimatisation Society of Victoria, Melbourne, 1928, 16.
44. Mason, Kevin, "Healesville Sanctuary: celebrating 60 years," in *Zoo News: The Magazine of Friends of the Zoos*, June, 10–15, 1994.
45. Strahan, 1991, 134.
46. Strahan, 1991, 83.
47. Strahan, 1991, 111.
48. Zoological Board of Victoria, *Proposal for an Education Centre at Melbourne Zoological Gardens*, Zoological Board of Victoria, Melbourne, 1974, 1–10.
49. Strahan, 1991, 79.
50. Giles, J. R. and Kelly, J. D., "Conservation and research program: proposals by the Zoological Parks Board of New South Wales," in *International Zoo Yearbook*, 31, 1, 1992.
51. Strahan, 1991, 144.
52. *The Weekend Australian* (Sydney), June 8–9, 1996, 44.
53. Strahan, 1991, 98.
54. Strahan, 1991, 104.

Su Lin, the first living giant panda brought out of China, resided at the Brookfield Zoo, Chicago, from 1937 to 1938. A second giant panda, Mei-Mei, was exported from China in 1937 and was also exhibited at the Brookfield Zoo. Su Lin's story is told in Chapter 5 on the zoological gardens of the United States. The giant panda has fascinated non-Asian naturalists since the species was first described for Western science in 1869. The animal was a popular diplomatic gift throughout the twentieth century. It has been a flagship species for China and for wildlife conservation, and one of the rarest exhibits at zoos. Photograph courtesy of Chicago Zoological Society/Brookfield Zoo.

7

Zoological Gardens of Asia

Sally Walker

7.1 Introduction

Asia is a huge, culturally diverse region. For many centuries localized royal classes, many of which had private menageries, controlled most of these regions. Their collections were occasionally mentioned in the reports of the earliest European sea captains and explorers, but they went unnoticed for the most part. During the nineteenth century European officials and naturalists supplanted many of these native collections with their own colonial collections, leaving the impression in the Western mind that these colonial collections were the first and only ones in Asia.

Views toward wildlife, as well as toward zoos and aquariums, vary from country to country, as one might expect in a region with such a diversity of cultural traditions. A number of zoo associations now exist in Asia, and in many cases the zoos and aquariums are making creditable efforts at improving their facilities, their education programs, and their wildlife conservation programs. Some cities find it difficult to improve their facilities or do not attempt to make the effort. In some cases countries are faced with a largely poor, uneducated, rural population, and their governments have more pressing needs. Often, wildlife is still considered more valuable as a commodity in the marketplace than as a conservation resource in the jungle. And then, there is the bureaucracy that stifles any effort at improvement and the wars that destroy those improvements that are made.

Each country and area has its own story to tell about its animal collections, although it is difficult to learn these stories because local information seldom reaches a broader audience and much of what is known has been brief glances which the occasional visitor to an individual zoo or aquarium has to offer. This information is a snapshot, with little information on what preceded and what followed. Moreover, communication with many of these countries — due to linguistic and governmental restrictions — is cumbersome, making the collection and verification of both historical and current information very difficult. It should be noted before proceeding on this sweep of Asia, that three of its major countries have been omitted because they are covered in other chapters: Russia, Japan, and India.

7.2 Southwest Asia

Southwest Asia, better known as the Middle East, is a biogeographic region having a harsh environment, dry climate, and similar cultures. Many of its wild animals are common throughout the countries making up the region. The history of zoos in this region is not an old one, since most of the zoos began in the 1960s. However, there were older, privately owned collections among the wealthy class, particularly collections of falcons and hoofed mammals. Ruling families still have private collections of native and exotic snakes, birds, and mammals, in addition to the public zoos.[1]

There are a number of zoos, breeding centers, rehabilitation centers, and wildlife reserves, which, if not as numerous as in other parts of the world, are equally effective. Overall, some countries in this region have developed conservation programs faster and better than other parts of the world and are leaders in successful reintroductions; however, other countries in the region still do not have a conservation culture, breeding centers, wildlife reserves, or even good zoos (such as Syria, Lebanon, and Yemen). Khalifs and sultans of Syria, Lebanon and Yemen are still known to keep animals for the hunt and a variety of personal entertainments.

One of the earliest successful wildlife reintroductions involved the return of captive Arabian oryx to the wild habitat of several Southwest Asian countries. The world herd, from which these animals were obtained for reintroduction, was gathered with the help of Western zoos and established first at the Phoenix Zoo in Arizona, United States. However, various Western countries, the local rulers, and desert tribesmen supported this effort. The reintroduction program became well established and successful because of this local support and because wealthy Arabs founded and maintained the breeding centers.

Arabian oryx, a native species, adapted easily to the region's desert habitat. Harsh environmental conditions that include heat, strong winds, and airborne sand can present difficulties

ARABIAN ORYX WORLD HERD

Arabian oryx (*Oryx leucoryx*) were exterminated from the wild on, or about, October 18, 1972. The eventual extinction of this oryx species was predicted in the early 1960s when the remaining population numbered some 100 to 400 animals. Nevertheless, hunting continued and on the date mentioned a hunting party from outside Oman found a herd in the Jiddat-al-Harasis region of Oman and killed them all. As it turned out, it was the last wild herd. Fortunately, in 1962, the Fauna Preservation Society had taken the warnings of a possible extinction seriously and funded an expedition that captured three males and one female from one of the last wild herds. One of the males died, and the London Zoo donated a wild-caught female. The two resulting pairs were shipped to the Phoenix Zoo in the United States to form the core of what was to become the world herd. Specimens soon joined these animals from the private collections of H.M. King Saud bin Abdul Aziz, King of Saudi Arabia, and H.H. Shaikh Jaber bin Abdullah al-Sabah, ruler of Kuwait. This brought the number of oryx in the herd to nine. By 1971, the herd had increased to 30 oryx and a group was sent to the San Diego Wild Animal Park. In 1975, the herd had increased to 45 and another group was sent to the Gladys Porter Zoo (Brownsville, Texas). Yet another group was established at the Los Angeles Zoo.

These zoos were chosen because they offered environmental conditions similar to the natural habitat of the oryx and they contributed to the cost of the program. Several other zoos in the United States joined these initial zoos during the late 1970s and early 1980s. By for managing captive exotic animals in this region. These extreme conditions limit work performance as well. Water is not plentiful, and native plants that can tolerate the climate, and which could be used for decorating exhibits, are often toxic to the animals. In addition, the procurement and transport of animal food is difficult. And construction is difficult as well, because of the many contract regulations and poor quality of the work.[2]

7.2.1 Afghanistan

Afghanistan is a dry and cold country with fauna and flora to match these conditions. The government recognizes its country's environmental stress — such as declining wildlife populations, loss of trees to people needing firewood, and soil erosion caused by the overgrazing of domestic stocks. However, the government has been preoccupied with military matters for many years, and this has been both a distraction and a further disturbance to the environmental quality of the country.

At some point in the 1960s, the Kabul University Faculty of Science began keeping an animal collection for research. When the public became interested in this collection, some thought was given to establishing a zoo. President HRH Prince Nader founded a Committee of Zoological Projects, which included members from the Royal Afghan government, the Municipality of Kabul, and zoologists from the Faculty of Science. The Municipality of Kabul provided a site on the bank of the Kabul River and the Kabul Zoo was inaugurated in 1967. A year later a zoological museum was also founded. The national and municipal governments supported the zoo financially, while the Faculty of Science and German zoologists provided the technical

---

1986, there were 172 oryx in 13 collections. Europe also had 30 oryx in six collections and the Middle East countries had 528 oryx in 10 collections. All of these, except for some in the Middle East collections, were from offspring of the world herd.

A reintroduction of oryx from the Los Angeles Zoo was made at the Negev desert, Israel in 1978. Another reintroduction of oryx from the Phoenix Zoo was made at the Shaumari Wildlife Reserve, Jordan in 1978–1979. The third reintroduction was made with oryx from the San Diego Zoo at Jiddat-al-Harasis, Oman in 1982. The latter reintroduction was made at the site where the last wild herd had been hunted out of existence 10 years previously. Seventeen oryx were shipped to Yalooni in the Jiddat-al-Harasis region in 1980–1981 and released into the wild in 1982. The Harasis tribe was employed to provide protection for the oryx, a function that suited its natural respect for the oryx and its habitat.

With the establishment of several successful herds in the wild and with a significant number of captive animals in many collections, the world herd cooperative agreement was dissolved and the remaining captive animals became the property of the zoos maintaining them. However, these zoos agreed to continue contributing animals to the reintroduction programs as needed and a studbook was established to ensure the captive herds remain genetically diverse and properly managed. And these animals may be needed. Between 1996 and 1999, poachers killed or captured 300 wild animals, reducing the Oman herds from 400 to 100. Of these remaining wild oryx, only 11 are females. So there is a continuing need for captive specimens and for improved conservation education and protection throughout the Arabian Peninsula.

Sources: Based on Anonymous, "Wild Arabian Oryx Faces Extinction," Environmental News Network, 22 April 1999 [reported on CNN Internet news]; Price, Mark R. S. Stanley, *Animal Re-Introductions: The Arabian Oryx in Oman*, Cambridge University Press, Cambridge, U.K., 1989.

and scientific expertise. This German connection began with Gunther Nogge, later director of Cologne Zoo in Germany, who was a professor in Afghanistan at the time he became involved with the zoo.[3]

Kabul Zoo was modern for its day, with moated outdoor exhibits. Most of the animals were native to Afghanistan and therefore habituated to the climate, so expensive buildings to house them were not necessary. The Faculty of Science, Kabul University, maintained a close connection with the zoo, resulting in a number of research publications and successful breeding programs. In 1972, there were 417 animals, including 32 mammal species, 85 bird species, and 4 reptile species. While most were native Afghanistan animals, there were also a lion, tiger, pheasants, and parrots from other countries.

Endangered species were also kept in the collection and successfully bred. Kabul Zoo maintained the Afghanistan leopard, which was only found in five other zoos. From his collection at Karez-i-Mir, near Kabul, the king of Afghanistan donated snow leopards. The zoo maintained a breeding group of rare Bactrian wapiti (*Cervus elaphus bactrianus*) in a sanctuary the king founded in Ajar Valley, central Afghanistan. There was also a small herd of goitered gazelle (*Gazella subgutturosa*), which in 1972 was reported to be nearly extinct as a result of uncontrolled hunting. Early attempts at ecological exhibits included an artificial mountain with Afghanistan red sheep (*Ovis ammon cycloceros*), marmots (*Marmota caudata*), and snowcocks (*Tetraogallus himalyensis*), as well as a pond enclosure resembling Afghanistan's well-known waterfowl lake, Ab-e-Istada, with its flamingos, spoonbills, and ducks. Native and exotic pheasants were kept and bred in a pheasantry.[3,4]

Currently, however, Kabul Zoo is not faring well. Its existence is at the whim of the municipal authorities, who are influenced by the social, political, and military unrest of the country. In 1998, Kabul Zoo had only two lions, six bears, one wild boar, two foxes, and six rabbits. Many of the animals were killed or maimed during the recent civil war, some being shot for food. Others succumbed to the cold, since there had not been power or fuel to heat the cages. The 60-year-old head keeper, who had worked at the zoo for decades, was taken from his hut and killed. Other keepers sneak out at night to find food for the animals. Taliban soldiers consider the remaining animals useful for sport, so the future of the zoo looks bleak.[5]

7.2.2 Cyprus

Limassol Municipal Zoo, located in Limassol, Cyprus, was founded in 1950. In 1974, it had 12 mammal species, 55 bird species, and 2 reptile species. There was also a small group of endangered Cyprus mouflon (*Ovis orientalis ophion*).[6-8] Cyprus also has a captive breeding center for the endangered Cyprus mouflon, the Stavros tis Psokas Forest Station, which began in 1950 with a group of three animals. Since then, 64 animals have been reintroduced.[8] Interestingly, Gerald Durrell almost chose Cyprus for his Wildlife Preservation Trust, when looking for a location in the 1950s. He decided instead to settle in Jersey (see Chapter 2 by Keeling on Great Britain).[9]

7.2.3 Iran

Iran (Persia) has a rich diversity of animal life. Persian royalty kept many of these animals in their *paradeisos* (the paradise gardens discussed in Chapter 1 by Kisling on ancient collections). Some animals, such as the Iranian lion, have become extinct and others, such as the Iranian cheetah, are endangered. Every once in a while, a cheetah is rumored to be held in captivity somewhere in the country, but the only ones known to be in captivity currently are at the Tehran and Mashhad Zoos. The Department of Environment has agreed to study and conserve this Asian subspecies.[10,11] Prince M. H. Dovlatshahi founded the Tehran Zoo on four acres of land at Shemiran in 1958. At the time of founding it had enough animals to require a staff of 50. By 1992, it had about 35 species of mammals, birds, and amphibians. This is the year it moved from the center of the city to a location west of the city and became more of a recreation center. Mashhad Zoo dates back to the 1960s and is privately owned. Pardisan Nature Park is under the administrative control of the Department of Environment and has a facility that serves as an orphanage for animals. There are several small zoos as well, such as the Babolsar Zoo in northern Iran, the Shiraz Zoo in Fars Province, and the Kish Island Zoo and a birdhouse in Isfahan.[11]

7.2.4 Iraq

Iraq has an ancient history of animal collections, which has already been told (refer to Mesopotamia in Chapter 1 by Kisling on ancient collections). A long interval of human intervention has caused many of its native species and subspecies to become extinct, including the rhinoceros, bison, Siva's giraffe (*Sivatherium*), water buffalo, elephant, wild cattle, stag, beaver, oryx, fallow deer, onager, cheetah, and lion. Several others are now endangered, such as the bear, leopard, wild sheep, wild goat, roe deer, gazelle, lynx, caracal, honey badger, marten, and squirrel. Iraq is not ready to support effective wildlife protection, and in areas lost to dense human population, there is no chance to establish wildlife reserves.[12] The Department of Range and Forestry operates the Saddam Hussain Park in Baghdad. It is not known when it opened and it may be the only zoo in Iraq. It was severely damaged in the Gulf War and many zoos and animal welfare organizations across the world offered it assistance.[13]

7.2.5 Israel

Israel has a large number of zoos for its size, with about 15 known zoos and aquariums. Four of these are found in the city of Eilat, including Eilat Aquarium (founded in 1957), Eilat Zoo (founded in 1962), Underwater Observatory Marine Park, and Hai-Bar (South) Nature Reserve. City authorities founded Eilat Zoo, while private donations support it and the Eilat Zoo Association manages it. Underwater Observatory Marine Park focuses on the marine life of the Red Sea. Hai Bar Nature Reserve has holdings of native animals and reintroduced species mentioned in the Bible. The Nature Reserves Authority in Tel Aviv manages this reserve, which is closed to the public.[6,14]

Hai Bar Carmel is another reserve and is located at Mount Carmel. Its specialty is endangered native wildlife, particularly those species mentioned in the Bible. Negev Desert Chai Bar Reserve, founded in 1971, is yet another wildlife reserve. These reserves maintain captive animals as part of their conservation programs and the latter reserve was the site of one of the early Arabian oryx reintroductions. Mount Carmel is also the location of the Zoological Garden of the Biblical Institute, which Pinchas Cohen founded in 1950.

Meir Segals Garden for Zoological Research is a project of the Department of Zoology at Tel Aviv University. Founded in 1961 (or perhaps earlier) as the Tel Aviv University Research Zoo, its primary purpose is breeding endangered species and behavioral studies, which the university students conduct (and is not open to the public). For about a decade, beginning in the mid-1980s, this research zoo was known as the Centre for Ecological Zoology and its specialty was the rehabilitation and release of birds of prey. Since the early 1990s, under its new name, it has focused on native Israel and regional species.

Tel Aviv had two other zoos, which have combined into one. Mordechai Schornstein, Chief Rabbi of Copenhagen, who emigrated to Israel in 1938, started one of these zoos. Unable to break into the closely knit Jewish religious network and fond of animals, he decided to open a pet shop to earn a living. Keeping pets was not common at the time in Israel, so to get people's interest in the animals, he exhibited the animals and charged an entrance fee. Although the birds and monkeys attracted attention, it was not a successful endeavor until Schornstein acquired a pair of lion cubs from the Cairo Zoo. Attendance increased as people came to see the first lions in Palestine since ancient times, when the last wild lions were exterminated. As the lions outgrew their cages and the townspeople became worried, the Tel Aviv municipality decided to establish a proper zoo and Schornstein's zoo became the Tel Aviv Zoo, with Schornstein as its director. It was eventually closed on December 30, 1980 and the remaining animals were moved to the Ramat-Gan Zoo.

The neighboring town of Ramat-Gan had a small zoo located in the Ramat Gan National Park beginning in 1964 (and perhaps as early as the late 1950s). Although only about 10 acres initially, it expanded in 1974 to 250 acres, covering about half the national park. In 1981, it received animals from the Tel Aviv Zoo, and evolved into a combined zoo–safari park which became the Zoological Centre Tel Aviv–Ramat Gan. Initially it specialized in African animals because the animal dealer Carr-Harley was a partner in its development and he thought the park could be a quarantine and breeding center for African game. This venture did not work and the collection expanded to include other exotic species.

There is now another zoo in Tel Aviv, in addition to the Meir Segals Garden for Zoological Research and the Zoological Centre Tel Aviv–Ramat Gan. It is the Botanic and Zoological Gardens of the Society for Protection of Nature in Israel, which specializes in Israeli species.[15–17]

Petah-Tiqya Zoo is a municipal zoo listed for the first time in 1966. In 1974, this zoo maintained 96 species and 820 specimens on only 2.5 acres of land, and specialized in gazelles and axis deer.[6] Haifa Educational Zoo, of the Biological Institute in Haifa, is a public nonprofit trust supported with aid from the city and the state government. Although a small zoo of just 10 acres, it has an education curator and a staff of 35 to care for 524 vertebrates of 169 species, as well as invertebrates. As its name implies, its emphasis is on educating the public.[6]

Tisch Family Zoological Gardens (New Biblical Zoo) in Jerusalem was begun in 1940 as the Jerusalem Biblical Zoological Garden. It began as an "animal corner" in the center of the city with a desert monitor (*Varanus griseus*) and a pair of monkeys. As the collection grew, it became necessary to find a new location, so the District Commissioner of Jerusalem assisted Professor Shulov, the founder, in finding an area, which became famous as the Mandelbaum Gate. With the War of Independence in 1947 it became necessary to care for the animals at night because daytime snipers made care of the animals during the day a life threatening task. Finally, the zoo moved to Mount Scopus, but wartime conditions continued to take their toll on the zoo, resulting in the loss of nearly all the collection. The director released harmless animals that could survive on their own.

With 18 of the original animals, the Biblical Zoo reopened in Shneller Woods, Romema, on October 5, 1950. This opening was made possible with financial support from the Ministry of Education, the Municipality of Jerusalem, and the Society of Friends of the Jerusalem Biblical Zoo. During the next two decades the collection grew to 700 animals. It featured species mentioned in the Bible, including species that have since become extinct in Israel. Again war brought havoc to the zoo with 110 animals killed during the Six Day War in 1967, but the animals were soon replaced and the zoo continued to develop. Following this war, Mount Scopus was cut off from the rest of Jerusalem, so the zoo moved to the Tel Arza neighborhood in West Jerusalem. In the early 1980s the Tisch family of New York made a sizable donation to the zoo under the auspices of the Jerusalem Foundation, which enabled the zoo to develop further. The zoo is now located in south West Jerusalem, renamed after its major benefactor (but also known as the New Biblical Zoo), and officially reopened on September 7, 1993. A nonprofit zoological society, with assistance from the city of Jerusalem, now supports the zoo. It maintains animals of the biblical period in natural environments, provides conservation programs for endangered species, and emphasizes education. It is also involved with reintroduction programs, especially with the griffon vulture and Persian fallow deer.[6,14,18]

7.2.6 Jordan

Jordan does not appear to have any zoos; however, one of its three wildlife reserves is, for all intents and purposes, a zoo. Azraq Wetland Reserve and Wadi Mujib Wildlife Reserve are managed as normal wildlife reserves, but Shaumari Wildlife Reserve is managed as intensively as a zoo. Shaumari's primary purpose is to serve as a captive breeding center for endangered or locally extinct species. The reserve has many contacts with zoos around the world and supplies Jordanian or Arabic animals for reintroduction. The reserve was founded in 1975 specifically for reintroduction of Jordanian indigenous species and was the site for the first reintroduction of Arabian oryx. The first of these animals (four males and two females) arrived in 1978 from the world herd and were held in captivity at Shaumari to facilitate acclimatization and to ensure breeding success. By 1983, there were 31 animals and these were released into the reserve. Other breeding programs have included endangered or locally extinct species, such as the onager. The Jordanian Royal Society for the Conservation of Nature manages these breeding programs. When new animals are obtained, they are maintained in semiwild captive environments with careful supervision. When these animals are sufficiently numerous, they are released to wander into the reserve. This kind of facility blurs the line between traditional zoos and (increasingly managed) national parks, not only in the Middle East, but throughout the world.

7.2.7 Kuwait

Kuwait has seen several of its desert species become extinct from hunting and other environmental problems; in particular, ostrich, hoofed stock like oryx and gazelles, and carnivorous mammals dependent on these species. Perhaps the first encounter of influential Kuwaiti citizens with a modern zoo occurred after World War I when the Sheikh's family visited London Zoo as state guests. Development of zoos in Kuwait, however, took a while. Sheikh Jaber Abdullah Al-Sabah founded Salwa Zoo in 1954. Then, in 1968 the Department of Agricultural Affairs and Fish Resources founded a government zoo, the Kuwait Zoological Garden. The Kuwait Zoo was completely destroyed during the Iraqi invasion in 1991 when invading Iraqi soldiers used it as a barracks. Animals were eaten and others were shot for sport. Out of 442 animals, only 30 survived. Kuwaiti volunteer caretakers collected food from stores around Kuwait and bribed Iraqi soldiers to let them feed the animals. With gifts of animals from other zoos and reconstruction, the Kuwait Zoo began anew in 1993.[6,7,19]

7.2.8 Oman

Along with Jordan and Israel, Oman provided one of the first sites for the reintroduction of Arabian oryx. This was the site where the very last wild herd of Arabian oryx existed. A hunting party from another country exterminated this last wild herd, an event that upset the Sultan and the country. When His Majesty Sultan Qaboos bin Said learned about the world herd and its associated reintroduction program, the Sultan, in cooperation with the region's Harasis tribesmen, welcomed an opportunity to participate in the program and to reintroduce these animals back into Oman. Protection was provided for these new animals so they would not suffer the same fate as the last herd. Yalooni Camp, established to make the reintroduction feasibility study and to conduct preliminary activities, eventually took on the appearance of a professional desert research station, with offices, laboratories, and holding pens. In 1976, it became the Oman Mammal Breeding Center and, in 1980, the first oryx arrived. These were released into the wild in 1982 and, by 1991, the original five animals had increased to 109 and the Oman Mammal Breeding Center was a firmly established zoological facility.[6,20]

7.2.9 Saudi Arabia

Although there are a number of small zoos and private collections in Saudi Arabia, there are only three "official" zoo-breeding centers. There is also a fenced nature reserve, the Mahazat Assayd Nature Reserve, which has all the elements of a very large zoo. It is an enclosed area that is closely managed and has breeding, research, and education programs. The reserve contains captive Arabian oryx from the world herd, as well as other species. The Arabian American Oil Company founded a short-lived private zoo at Dharan about 1970, but it closed a couple of years later for unknown reasons. Mention has been made of a zoo at King Faisal Military City, but nothing more is known about it.[6,21] Riyadh Zoological Gardens is a municipal zoo founded at Riyadh about 1965. This zoo donated two pairs of Arabian oryx to the original herd at Phoenix, which ultimately supplied animals to reestablish this species in its Saudi Arabian habitat.

The National Commission for Wildlife Conservation and Development manages two wildlife research centers in Saudi Arabia. One center is the National Wildlife Research Centre at Taif, founded in 1986 for the purpose of breeding endangered native species, to conduct research on native wildlife, to reintroduce native species into protected areas, and to promote wildlife conservation. Some of the species for which there are conservation programs include the houbara bustard (*Chlamydotis undulata*), Arabian bustard (*Ardeotis arabs*), ostrich (*Struthio camelus*, a relative of the Arabian ostrich, which is extinct), bald ibis (*Geronthicus ermitus*), Arabian oryx (*Oryx leucoryx*), dorcas gazelle (*Gazelle dorcas*), sand gazelle (*G. subgutturosa*), Nubian ibex (*Capra ibex*), and onager (*Equus hemionus onager*). Another center is the King Khalid Wildlife Research Centre, founded in 1987 at Riyadh. The original purpose of this center was to take over and manage the private animal collection of the late King Khalid. This collection included over 600 specimens of 20 species, including endangered Arabian species. Now the collection has more than 1,000 animals, mostly of Arabian origin such as oryx, dorcas gazelle, and sand gazelle. The center's research programs include native fauna and flora, including reptiles such as the desert monitor. Research areas include reproductive biology, ecophysiology, and animal health. It also conducts educational programs for the public.[22]

7.2.10 Turkey

Turkey is another Asian country with a long tradition of zoological collections, its sultans having had impressive collections since ancient times. Its modern zoos, however, are products of the post–World War II era. Hayvanat Bahcesi-Izmir Zoo is located in the Kulturpark of Izmir and was founded in 1939. A municipal zoo, it is now known as the Istanbul Belediyesi Gulhane Zoo. Major new buildings, such as a lion house, small mammal house, and aquarium were constructed in the 1960s, and it is known to have a good breeding record.[6,13,14]

Ankara Zoo in Ankara was also established around the same time, in 1940. It was part of an ambitious program to build a modern city alongside ancient Angora. The zoo is located a few miles outside the city on a large state farm known as Orman Chiflik (now Ataturk State Farm). Its objective has been to educate Turkish people about their animals, both wild and domestic. The zoo has also made an effort to provide farmers with information about improved breeds and therefore has had an extensive collection of foreign domestic breeds. Initially there were few foreign species — a few monkeys, a lion, and some birds. But now the zoo has 2,482 specimens of 142 species (mammals, birds, reptiles, and fish) and is considered Turkey's national zoo.[6,23]

Robert College Zoo and Natural History Museum in Istanbul began about 1966 as an educational facility. It was small and specialized in small mammals, birds, and reptiles. The zoo offered traveling educational lectures and displays, as well as educational films and guided tours for school groups. It is believed this facility only lasted a couple of years. A Turkish businessman founded Tuzcu Zoological Gardens in Istanbul about 1967. Mr. Tuzcu wanted to educate the Turkish people about their native fauna and make them aware of their declining natural resources. The zoo was to have a comprehensive collection of Turkish wildlife with an emphasis on breeding groups for endangered species; however, a lack of funds has slowed progress. There appears to be some connection between this facility and the Robert College Zoo, but more specific information is lacking.

7.2.11 United Arab Emirates

The United Arab Emirates have taken a number of conservation initiatives that are unique in Asia, insofar as these activities are coming from the private sector. There are two zoos and about a dozen private breeding centers. The breeding centers concentrate on Arabian oryx and other native species. The Zoological Gardens and Aquarium at Dubai opened about 1967 on a sterile piece of desert, which has since grown into a green animal oasis. O. J. Bulart, an Austrian engineer, established this private collection with the assistance of the Ruler of Dubai. This zoo was the first to breed Arabian oryx and Gordon's cat (*Felis silvestris gordoni*). It also has a good record of breeding Arabian wolf and other non-native species. It is a small zoo, but has plans for a new facility.[6,7,24,25] Al Ain Zoo and Aquarium (the National Zoological Garden) in Abu Dhabi was founded about 1970 with some of its own animals and some from the Dubai Zoo. This zoo has large populations of many species, including several oryx and gazelle species, the African spur-thighed tortoise, and several bustard species. The aquarium has a wide variety of marine and freshwater fishes, as well as sea lions and penguins.[6,25,26]

7.2.12 Yemen

Ancient Yemen was significant as a trade link between Egypt and the surrounding areas. Yemeni caravan men kept apes and monkeys as tame caravan pets. However, the general attitude of Yemeni people toward wildlife has been harsh rather than protective. Nevertheless, there is now some thought being given to establishing a game reserve to protect the dwindling population of leopards and other endemic Yemeni species. There are a number of small menageries, and the establishment of a national zoo is being considered.[27] Wild animals from Yemen, as well as exotics like lions, are kept in captivity in small cages for the entertainment of the people. A small display beside the Salah Palace in Taiz houses several lions in small, barred cages, along with a caracal, three hyenas, baboons, and other animals captured in the surrounding area. The lions are descendants of those that Emperor Haile Selassie of Ethiopia gave as a gift to the Imam of Yemen between 1948 and 1962. The palace is now a museum and visitors pay extra to see these caged animals. Even restaurants and movie theaters have displays of caged animals. Wealthy citizens, including the president of the country, keep leopards at their homes. Local environmental groups are buying these poorly managed animals and sending them to proper breeding centers, as well as working to establish wildlife reserves so that one day the animals can be released into the wild.[27]

7.3 South Asia

7.3.1 Bangladesh

The People's Republic of Bangladesh was formerly part of India, then became the East Pakistan Province (when Pakistan gained independence from India in 1947), and has been an independent country since 1971. Nature and wildlife conservation have not received sufficient attention, although ambitious projects have been attempted. The country has a rich diversity of species, although only one is known to be endemic. Two well-known wildlife areas include the mountainous Chittagong Hill Tracts with its wide variety of mammals and birds, and the Sundarbans with its tigers. A need for zoological and botanical gardens was felt at the very inception of Pakistan in 1947, and a government resolution to establish these was passed in 1950. Little action resulted from this effort; however, an advisory board was constituted in 1961 to plan and establish the Dacca Zoological Gardens, which finally got under way in 1964. Although progress has been slow, Dacca Zoo intends to provide its animals with natural habitat exhibits separated from visitors using wet or dry moats. Small, inexpensive structures were constructed first, however, so the zoo could be fully operational, and many of these early structures remain.

In addition to the exotic caged species there are many free-roaming native species that voluntarily visit the zoo. Visiting wildlife includes troops of monkeys, jackals, civets, wild cats, rabbits, many bird species, monitors, and butterflies. Of particular interest are the Gangetic dolphins which visit the bordering river Turag. These dolphin are seldom seen in captivity since they are very difficult to maintain, but can be seen at the Dacca Zoo in the wild. Since independence from Pakistan, the zoo has been poorly maintained, but considering the turbulent political environment after independence and the natural disasters that have racked the country, it is a wonder the zoo still exists. It even proved to be a major benefactor (with animal donations) to the Kuwait Zoo after the latter had been destroyed during the Iraqi invasion in 1991.[28,29]

7.3.2 Bhutan

Although a small country, Bhutan's diversity of habitats provides a rich fauna. Until about 40 years ago, the modern world had little influence on the country. In 1993, some 66% of its forest cover still existed and preservation of the environment is part of the country's tradition, as well as government policy. Motithang Zoo is in Thimphu, the capital of Bhutan. However, it is not known when the zoo began or much else about it. It has been reported to have a good collection of takin (*Budorcas taxicolor whitei*) and goral (*Naemorhedus goral*).[30]

7.3.3 Nepal

Nepal, best known for Mount Everest, has a diversity of wildlife that was particularly popular with British colonial bureaucrats. One of the best known of Nepal's colonial naturalists was Brian Houghton Hodgson, a British civil servant primarily interested in birds. However, he was also interested in mammals, giving detailed accounts of their behavior and morphology. He kept several of these animals in his private menagerie. Another British naturalist, William Henry Sykes, also kept a private menagerie, recorded observations, and made notes on the animals' care, feeding, and behavior. S. R. Tikell described many animals of Nepal, some of which he kept in captivity and as pets. Many more naturalists with private collections could be named. Of particular interest is a rhinoceros known to be kept on exhibit at Chittagong from 1868 to 1872 before it was sent to England, and another known to be exhibited at Katmandu for 35 years when Flower saw it in 1930.[31–34]

Prime Minister Judha Sumser Janga Bahadur Rana established the Central Zoo (originally known as the Katmandu Zoo) in 1932 using his personal collection of birds. In 1951, the zoo came under the administrative control of the Nepal government and opened to the public in 1956. Because the zoo has been constantly transferred to various government departments and has had many short-term directors, it has not fared well. Finally, on December 29, 1996, the zoo was transferred once again to the King Mahendra Trust for Nature Conservation, a private, nonprofit environmental organization. The trust has made a commitment to develop the zoo

7.3.4 Pakistan

Zoos in Pakistan are in transition, much like the Indian zoos were in the 1980s when there was not even a complete list of zoos. There is no sense of a zoo community and no zoo association. There has not even been a meeting of its zoo directors or other personnel.[37,38] Ahmad, Director of the Zoological Survey of Pakistan in the early 1990s, describes Pakistan zoos as operating under a "primitive pattern," using "old styled" cages and management. The zoos are primarily for recreation, with public conservation awareness and education as occasional by-products.[39] The average urban citizen in Pakistan regards wild animals with fear, animosity, distaste, and at best curiosity, while rural citizens who have more contact with wild animals regard them as pests or objects for sport. Shariff, Director of Jungle Kingdom Leisure and Animal Parks (of Asia), feels that zoos in Pakistan are little more than menageries of days gone by, with enclosures that are still barred cages and concrete floors, with poor signage, and with no education programs. These zoos do nothing to change the attitude of citizens toward wildlife.[38] Interaction between Pakistan zoos and the world zoo community has been minimal. Some of the zoos have developed only recently, from the 1980s. Many are small and several are private facilities. Several of the newer facilities are aviaries only, while others are more like wildlife parks. In all, there are now about 30 zoos in Pakistan.[37]

Pakistan's oldest zoo is the Lahore Zoological Gardens, founded in 1872 under the administrative control of the Punjab government and with a subsidy from the Lahore municipality. This zoo was transferred to the Deputy Commissioner of Lahore in 1923, to the Livestock and Dairy Development Department in 1962, and to the Wildlife Department in 1982.[40,41] Lahore Zoo is an excellent example of colonial architecture in a garden setting. When Flower visited the zoo in 1913, it was free to the public and was located in the 112-acre Lawrence Gardens. There was an inner section for the office, primate and carnivore cages, aviaries, reptile house, and goldfish tanks. A waterfowl pond occupied the second section. An outer section consisted of large paddocks for ungulates and poultry. The zoo also had an aquarium, veterinary hospital, taxidermy office, workshop, and restaurant.[40] The zoo has undergone a modernization, completed in 1988, to provide natural habitat enclosures and to upgrade the remaining cages. Landscaping was improved and public amenities were added.[42] Yet more improvements are needed, both in animal exhibits and in education, and were recommended in internal reports issued during 1997–1998.[41,43]

Karachi Zoological Gardens is the second oldest zoo in Pakistan, beginning as a government garden in 1881 (or 1885).[39] The grounds of the Karachi Zoo, as part of the gardens, go back to 1799 when they surrounded a factory of the East India Company. The gardens went through many changes and served a number of purposes. In 1881, the municipality took control of the gardens and, as was common with colonial gardens, began an animal collection. Little mention of the zoo can be found prior to 1913 when Furrell and Ludow[44] wrote that while Karachi was a young city with no attractions of interest to tourists (who, unless they stayed with friends, would not desire to stay any longer than obliged to do so), the zoological garden was well worth a visit. In the very early part of the century, the zoological garden and surrounding botanical garden were a very popular meeting place on Sundays for all classes of the community and for all ranks of society.

Karachi's own residents donated much of the early animal collection. By 1890, there were 93 mammals and 465 birds. However, a "mixed and somewhat irresponsible body of gentlemen who worked with sporadic zeal but without much knowledge" managed the collection until a paid superintendent was appointed in 1889. W. Strachan served as the superintendent until 1899 when another European superintendent was appointed. The new superintendent did not do a good job and was soon let go. Then the first local individuals served in this position, the brothers Ali Mahomed (who served as superintendent until he died in 1911) and Ali Murad (who served as overseer of the animals until he became superintendent when his brother died).[44] Widely spaced reports have commented on the Karachi Zoo. Flower described his visit in 1913, Qurashi (Director of the Zoo) submitted information to *International Zoo News* in 1955, a plan to create an entirely new zoo was published in 1963, construction of new facilities occurred in 1968, and Van den Brink (an animal dealer) visited in 1979. Development of a new zoo, as planned in the 1960s, was not carried out; however, many new facilities have been built since then. The zoo now covers 35 acres and has native and exotic animals.[39,40,45–48]

In addition to an aquarium at the Karachi Zoo, there was a Karachi Municipal Aquarium founded on the beach at Clifton in 1962. It may have been established as a result of the Pakistan Aquarium Exhibition of 1953. The Pakistan Aquarium Society organized this exhibition and the president of this society used the exhibition's popularity to make an appeal for a large public aquarium to exhibit fish from all over the world. In 1974, the aquarium exhibited 3 marine reptile species, 125 fish species, and 43 marine invertebrate species. It had a central hall with four wings containing both marine and freshwater tanks. Labels not only described the organisms, but had information on water, oceans, and related subjects for education of the visitors.[6,14,48,49]

There was also a crocodile pit in Karachi, which visitors commented upon as early as 1838 and which was still there when Flower visited in 1913. Known as Mugger Pir, it was originally a swamp with hundreds of mugger crocodiles, but was later drained and turned into a tank with a high wall. Flower saw two dozen crocodiles measuring 6 to 12 feet, along with 70 to 80 young animals born there; however, because of neglect, the number of crocodiles has dwindled to a few dozen.[40]

Peshawar Zoological Garden particularly impressed Flower in 1913. Founded in 1909, it had a full staff, including director, veterinarian, keepers, assistant keepers, head gardener, water carrier, and grass cutter. Its exhibits were some of the cleanest Flower had seen during his 1913 trip. Animal houses were well designed and the aviary was well arranged and landscaped. And there were animals seldom seen in zoos then, such as the black Celebes apes and a striped hyena that Flower thought was an undescribed subspecies. Unfortunately, this well-cared-for zoo is no longer in existence and a description of its demise has yet to be found.[40]

Of the newer Pakistan zoos, the Jungle Kingdom Theme Park at Rawalpindi is both a traditional zoological garden and a wildlife-theme family amusement park. A private venture begun in 1992, many of its animals had been confiscated or rescued, and the zoo uses the conservation effort that saved these animals for educational purposes. Private individuals, rescue centers, and governments around the world have donated the animals, but most have come from Asia where the trade in illegal species is most severe. The theme park attracts visitors and complements the zoo. In addition, the company managing Jungle Kingdom has plans for nine large projects throughout Pakistan. These projects include a Scientific Captive Breeding Centre, a Rescue Centre, a Research Centre (for wildlife studies), a Training Centre (for training zoo and wildlife personnel), a Conservation Education Centre, a Pets Advisory Centre, an EcoTourism Centre, a Natural History Museum, and a Public Library.[38]

7.3.5 Sri Lanka

Sri Lanka (formerly Ceylon) is another Asian country with a long history of royal collections and the keeping of wild animals as pets. Robert Knox, writing in 1681, says King Rajasinghe of Kandy had a menagerie of fish, hawks, other strange birds, dogs, deer, tigers, elephants, and other beasts.[50] Travelers throughout the eighteenth and nineteenth centuries related stories of snake charmers, various native animals as house pets, and the domestication of the elephant. Colonial administrators also kept various native animals as pets.[51–54] An early zoo was associated with the Colombo Museum. This zoo was already operational in 1909 and was closed at the beginning of World War II in 1939 (an event that was felt even in faraway Ceylon).[55,56] The early 1900s also saw the start of the Zoological Gardens Company's private menagerie, which John Hagenbeck owned. John was the brother of the well-known animal dealer, animal trainer, and zoo owner Carl Hagenbeck of Germany. John exported animals for Carl, and exhibited them as well. The Zoological Gardens Company was liquidated in 1936, the Company's property confiscated as "enemy property," and Hagenbeck interned as an enemy alien for the duration of the war.[50,57]

W. C. Osman Hill resided in Ceylon for some 15 years during this pre–World War II period. Osman Hill maintained his own private menagerie, which included native and exotic species, primarily primates and parrots. It was an impressive collection, including star tortoises, leopard tortoises, elephant tortoises (these being from eggs laid by a group of adults belonging to the former Governor of Ceylon, Sir Reginald E. Stubbs, K.C.M.G., who was a benefactor of the London Zoological Society), various cockatoos, hawk-headed parrots, eclectus parrots, ruddy mongoose, and a wide variety of primates. When Osman Hill left Ceylon in 1944 he sent some of the animals to the London Zoo and others to the Dehiwala Zoo in Colombo.[56]

When Hagenbeck's Zoological Gardens Company liquidated in 1936, the government acquired much of the collection and added it to the Dehiwala Zoo collection (also known as the Colombo Zoo or Zoological Gardens of Ceylon). Although the Dehiwala Zoo officially began in 1936, there already existed an animal collection at its location, part of the Hagenbeck company's holding area that the public could visit. Major Aubrey Weinman, Director of the Dehiwala Zoo for many years, was an important zoo professional and naturalist. Of Dutch descent, Weinman was a citizen of Ceylon by domicile. He had a number of pets as a child and retained a life-long interest in animals. He was an officer in the Indian Army during World War I and the Afghan War, after which he worked as a librarian at the Colombo Museum. While at the museum, he developed its animal collection, mentioned earlier. He then became Director of the Dehiwala Zoo and was, apparently, the one who persuaded the government to acquire Hagenbeck's collection. World War II broke out just as Weinman had given the zoo some shape and order. He served in Malaysia and was a prisoner of war there for more than three years (he would return later to help design Zoo Negara in Kuala Lumpur). Weinman returned to Dehiwala Zoo after the war and was responsible for modernizing the exhibits and the animal husbandry. He was honored with an OBE (Order of the British Empire) in the early 1960s and in the mid-1960s wrote a series of articles entitled *My Personal Ark*.[50,58,59]

Dehiwala Zoo, under the guidance of Weinman, recovered after World War II, building new exhibits and having very successful breeding results. Weinman promoted occupational therapy (behavioral enrichment as it is now known) and cruelty-free training for the zoo's elephants. Native species were of primary importance to the collection, as was education and conservation. The zoo developed a conservation exhibit that resembled a miniature national park and educated the visitor about the animals and their habitat.[60–62] Weinman retired in 1962 and Lyn de Alwis became director of Dehiwala Zoo. de Alwis was a particularly successful director, under whose guidance the zoo reached its zenith. He designed the Singapore Zoo and created the night safari concept using ecological exhibits of mixed species from the same habitat. de Alwis did not like the taxonomic menagerie style of exhibit, preferring instead the full spectrum of species found in the same habitat. He also continued the tradition of exhibiting native species. In 1969, half of the collection consisted of native species, with virtually all of the mammals represented. In 1973, the zoo maintained 158 mammal species, 259 bird species, 56 reptile species, and 7 fish species (3,190 specimens in all). In 1979, the zoo became involved with the development of a new zoo at Aranayake, some 40 miles from Colombo. This new zoo was designed to contain large numbers of a relatively few local species, with an emphasis on breeding and conservation. The zoo also worked with the local university on a reptile breeding research project. However, not much has been reported on the zoo since the early 1980s. Since de Alwis's retirement, the zoo's directors have changed frequently. A renovation committee (that includes de Alwis) is currently struggling to bring the zoo back to its former glory.[48,63,64]

Sri Lanka also has specialized facilities for elephants, turtles, and reptiles. Pinnawala Elephant Orphanage (opened in the 1970s) and Udawalawe Elephant Orphanage (opened in the 1990s) house both young and mature elephants caught in pits, injured, or otherwise in trouble. Many are in poor physical condition and need to be nursed back to health. Mahouts care for these elephants and take them for daily baths in a nearby river. They are kept in simple, but enormous open-sided bamboo and wood structures. At Pinnawala there are about 60 elephants, ranging from a few weeks to a few years old.[50,65] There were also about 25 turtle hatcheries across the country which usually exhibited three or four species of marine turtles in various stages of development.[50] Ranil Senanayake, with the help of some German collaborators, started Gampaha Snake Farm in the early 1980s; however, it was not in operation for long. Some hotels also keep animals, such as the star tortoise, peacocks, and pythons, as tourist attractions. Boy's Town Hendela maintains a private zoo with many mammals (including lions and bears), birds, reptiles, and fish. Even some Buddhist priests keep animals. One priest, Rev. Aluthnuwara Anomadassi at the Keraminiya Temple, had a rather large collection of mammals, birds, reptiles, and fish in 1992. Ahungalla Zoo is a more recent private zoo, begun in the early 1990s. Although the zoo was popular, conservationists and academics did not think well of it. Eventually one of its lions attacked and killed a visitor and the government closed the zoo. Its mammals were sent to other zoos and the crocodiles were released into sanctuaries.[50]

7.4 Southeast Asia

Southeast Asia comprises countries from the Indochinese and Malay peninsulas, and the island of Borneo. The South East Asian Zoological Parks Association cooperatively organizes its zoos; however, effective coordination is difficult in a region with so many political, cultural, linguistic, and economic differences. The first meeting of the association was held in Malaysia in 1988 on the occasion of Zoo Negara's 25th anniversary. The Sultan of Selangor hosted the inaugural meeting of the association after Y. B. Tan Sri Dato V. M. Hutson, Chairman of the Malaysian Zoological Society, developed the idea for an association following a visit to the Philippines and Singapore to see their zoos and discuss common problems. Membership in the association includes the countries of Southeast Asia, as well as the countries of Indonesia.[66,67]

General D. Ashari, who was the association's founding president and remained president until recently, has been a very influential individual in all matters pertaining to wildlife conservation in Southeast Asia and Indonesia. Ashari was a freedom fighter during the struggle for Indonesian independence during 1948–1949, ending his military career as a general. He served as a cabinet minister, as well as ambassador to Japan and to the United States. He then became the Chairman of the Indonesian Parks Association and the South East Asian Zoological Parks Association. He has been active in promoting the zoo profession and the zoo associations in both Indonesia and Southeast Asia.

7.4.1 Brunei

Considering the wealth of Brunei and the threatened status of its wildlife, the country's zoos are not what one would expect, and it is particularly difficult to obtain either historical or current information about these zoos. These zoos include the royal family's private collections, three facilities open to the public, one that is not open to the public, and one public "national" aquarium. The sultan and his family have had a collection of wild animals, perhaps for a long time if it is similar to other royal collections of Asia, but its current status cannot be confirmed.

Batang Duri Mini Zoo (Tamburong Zoo), founded in 1986, is an interpretation center for the Batu Apoi National Park. The Tamburong District administrative office manages the zoo and the Agriculture Department provides veterinary services. What visitors probably will not see in the dense park forest, they can see in the zoo. The zoo exhibits native species such as thick-spined porcupine, small-clawed otter, binturong, Malay civet, red muntjac, sambar deer, monkeys, and various birds. In addition to serving park visitors, the zoo provides an educational experience for local school groups. Wildlife conservation is being encouraged in a country where the people view wildlife with fascination, but without empathy. Local people normally do not keep wild animals as pets or for pleasure, and they feel Allah determines the animals' fates, not themselves. This makes the implementation of conservation measures difficult because it is not easy to gain public support.

Mohd Louis Bin Abdullah started the Louis Mini Zoo in the early 1980s as a hobby. His first collection comprised both rare and common birds from Borneo and Indonesia. Now the zoo contains more than 1,000 animals, exotic and native. It opened to the public in 1988. Since the zoo has a good breeding record and accepts all donations, its one-acre site is overcrowded. The zoo is trying to acquire a suitable site and has hopes the government will help it do so. Hassanal Bolkiah Aquarium opened in 1972 with 122 tanks. It is under the administrative control of the Fisheries Department, Ministry of Industry, State of Brunei. The director of the Fisheries Department is also the director of the aquarium. Because of this governmental control, it is considered the national aquarium.

7.4.2 Cambodia

Both the French and the Cambodian Wildlife Department established modern zoos in the larger cities of colonial Cambodia, but these zoos were destroyed during recent wars. One zoo emerged after these wars, the Phnom Tamao Zoo, founded in 1995 under the management of the Department of Forestry and Wildlife. This zoo is located outside the capital, Phnom Penh, on a large site divided into a carnivore park, herbivore park, and bird park. These exhibits are integrated into the site's natural environments and contain 56 species of native animals that have been donated or confiscated locally.[68,69]

7.4.3 Laos

Laos (Lao People's Democratic Republic), like many of its neighbors, is a rural country that recent wars have devastated. Two Thai individuals, who were in the animal trade, founded the country's primary zoo, the Vientiane Zoo, at the capital in 1992. There are also other zoos associated with hotels or that exist as private menageries.[70]

7.4.4 Malaysia

Malaysia has a 12-member Malaysian Association of Zoological Parks and Aquaria, founded in 1996. The Malaysian government also encourages good management and maintenance of zoos through its Wildlife Protection Act of 1973, and is considering a National Zoo Policy that will provide guidelines with minimum standards for safety, animal welfare, veterinary medicine, and enclosure design.[71]

Zoo Negara Malaysia, Malaysia's "National Zoo," can trace its roots to 1957 when the Malayan Agri-Horticultural Association organized a small collection for an annual exhibition. The popularity of this animal exhibit was such that the association continued to exhibit animals at its annual exhibitions. Between 1957 and 1963, the exhibition animals were kept in the five-acre garden of V. M. Hutson who organized the shows on behalf of the association. Hutson's garden was located at Bangsar Estate (now Bukit Damansara), some distance from Kuala Lumpur. Nevertheless, during weekends people from the city would drive out to see the animals. In those early years the collection included a Malayan tiger, three orangutans, six estuarine crocodiles, and other animals.[72] In 1961, it was decided to establish a permanent zoo site for these animals and a working committee was appointed to form a society and establish the zoo. This decision resulted in the formation of the Malayan Zoological Society in 1962 with plans for a small zoo in or near Kuala Lumpur. The society would fund and manage the zoo, but the Malaysian government would support the effort. After much searching for a location, the Society selected a 42-acre site at Ulu Klang, with another 100 acres earmarked for future expansion. An initial 36 acres of secondary jungle was converted into a zoological garden, which opened November 14, 1963. Surrounded by jungle and rubber estates, it was called the "zoo in the jungle." This location, eight miles from the center of Kuala Lumpur, was considered too far out of the city; nevertheless, the zoo enjoyed good attendance, and by 1997 the city had grown around and past the zoo site. With successful fund-raising and government support, the zoo developed an additional 22 acres, built an aquarium, and created a five-acre drive-through Safari Lion Park. However, the latter was closed in 1983 and converted into an African savannah exhibit. A veterinary hospital and modern reptile building have been added and the zoo has the beginnings of an excellent educational program.[72–74]

Two older zoos existed at Johor Baru: the Johor Baru Zoo founded in 1928 and the private collection of the Sultan. The sultan's collection was discontinued and transferred to the Johor Baru Zoo (a public zoo that the Sultan also supported, even though the Johor State Government operated it) in 1962. Renovations and expansion occurred at the zoo and the new zoo was completed in 1965.[74,75] Taiping Municipal Council founded the Taiping Zoo (Taman Mergasma Idris Shah) in 1962 to attract tourists. It is one of the more naturalistic zoos in the country and its 1995 master plan put an emphasis on conservation, research, and education programs.[75] Melaka Zoological Garden was founded in 1964 to attract tourists as well. The Department of Wildlife and National Parks Malaysia administers this zoo. With its large open enclosures and systematic breeding programs, it is one of the best zoos in the country. Zoo Melaka also takes in problem tigers (because they prey on livestock) the Wildlife Department captures.[75,76]

Other zoos in Malaysia include bird parks, reptile parks, butterfly parks, and a number of roadside zoos. Most of these are private and small, but like their counterparts elsewhere, these poorly maintained collections detract from the better facilities that are members of the Malaysian Association of Zoological Parks and Aquaria.[77] One other unusual collection existed that demonstrates how wild animal collections take many forms. From 1937 to 1941, the Botanical Garden in southern Malaysia near Singapore maintained a small group of Southeast Asian pig-tailed macaques for the purpose of collecting botanical specimens for the gardens. The Malays trained the macaques by having them retrieve coconuts. However, the program was abandoned and the gardens' macaques "liberated" when Singapore fell at the beginning of World War II.[78]

Sarawak is a Malaysia state on the island of Borneo. While smaller than the country of Kalimantan (Indonesian Borneo), it offers one of the few efforts on the island to establish zoos and aquariums, although orangutan rehabilitation centers exist in Kalimantan. The second white Rajah, Sir Charles Brooke, began one of the few early natural history museums on the island, the Sarawak Museum, in 1886 or 1888. In the letter appointing the first museum curator in 1888, Sir Charles stated that the curator was in charge of both the museum and the aquarium. However, the official record does not acknowledge the existence of an aquarium and one is not known to have been established until 1961. In this year the museum curator, in cooperation with the country's Director of Agriculture, installed three rows of aquariums in the old building of the museum. These aquariums proved to be a great attraction, exhibiting not only fish, but an ever-changing collection of living butterflies, scorpions, turtles, lizards, and snakes. The curator, Tom Harrisson, and his wife also raised orphan orangutans at the museum, some of which were sent to European zoos. By 1972, only one orangutan was still being held at the museum, along with a sun bear and a few macaques.[79,80]

7.4.5 Myanmar

Myanmar (Burma) has a rich variety of wildlife that British colonial civil servants and others have intensively studied. In 1906, the Rangoon Zoo was founded on the site of the Victoria Memorial Park. The municipal government provided the site and the citizens subscribed funds to create a memorial to Queen Victoria. The Prince of Wales opened the zoo when he visited Rangoon on his last India tour. The park, with its hilly terrain and ponds, attracted local wildlife that supplemented the captive wildlife, which itself was mostly native species. The zoo also had a "white" elephant, which was symbolically significant to Asians (because it was thought to be the incarnation of the Hindu God Vishnu, one of the three primary Hindu gods). R. M. Sen, superintendent of the zoo, wrote a pamphlet in 1910 on the management and improvements needed at the zoo, but this work and much of the zoo's historical documents were lost. In the 1970s the zoo built a natural history museum and aquarium. There are plans to build a new, more naturalistic zoo outside of Rangoon.[33,40,55,68,81,82] Another extensive collection was maintained at the Buddhist Theyboo Monastery near Mandalay in the early 1900s. It included large aviaries that were well landscaped and contained many species of small native birds, as well as some exotic species. There were large enclosures with pools for waterbirds, an area for parrots and pheasants, and a marsh for ducks, insects, and reptiles. Freshwater ponds abounded with fish and there was a large body of water containing a dozen gharials. A section for mammals contained about 60 enclosures, which housed a large variety of species, mostly from India and Southeast Asia.[82]

7.4.6 Philippines

The Philippine archipelago abounds in wildlife species, many of which are endemic or restricted in distribution to undisturbed forests. In a country where there is rapid deforestation, where cockfighting is a national sport, and where there is immense hunting pressure on the wildlife, it is difficult to promote the conservation needs of animals. There are three types of captive wildlife facilities in the Philippines: zoos, breeding centers, and rescue centers. The efforts of these facilities to conserve Philippine wildlife is important, but most of the facilities are of recent origin.[83–87]

Before the current Manila Zoological and Botanical Garden was founded, there was a public garden in Manila that exhibited some mammals, birds, and reptiles. And even before this, the Spanish colonial government maintained a collection of animals. By the early 1900s, however, these gardens and the few animals they exhibited were destroyed during the Philippine–American War. The Manila Zoological and Botanical Garden was founded on July 25, 1959 as a municipal facility. Since its opening, this new zoo has remained virtually unchanged. A redevelopment and master plan were developed in 1995 because the zoo had not had any major renovations, had utilized only 50% of its available land for animal exhibits, and did not provide proper visitor services and amenities. Unfortunately, these plans, which should have been completed in 1997, have not been implemented.[14,55,88,89]

Other zoos include the Clark Zoo (St. Angeles City), Ceby City Zoo (Ceby City), and Malabon Zoo (Manila). The Philippine Eagle Foundation (Davao City) is a private foundation founded to conserve the Philippine Eagle, but is open to the public. Birds International is a commercial parrot-breeding farm, but also maintains other bird species (such as pigeons, the Palawan peacock, and pheasants) and mammal species (such as orangutans, binturongs, and mouse deer). Wildlife Rescue Centre DENR is an official government center managed by the Department of Environment and Natural Resources. Founded in the early 1970s, it was initially intended as a minizoo for the recreation and education of the public, but was later converted into a rescue center. In 1997, it housed 775 specimens, nearly half of which were endemic to the Philippines.

Conservation breeding centers include the Center for Tropical Conservation Studies (Silliman University, Dumaguete City), Ecosystems Research and Development Bureau Centre (Los Banos), NFEFI Conservation Center (Negros Forests and Ecological Foundation, Bacolod City), Mari-it Conservation Park (West Visayan State University, Panay), and the Captive Breeding Centre (Mindoro). These centers have programs to conserve the Philippine spotted deer, Calamian deer, Visayan warty pigs, Philippine fruit bats, and other endemic species. These centers also participate internationally, providing breeding groups of endemic Philippine species to zoos outside the Philippines. Wildlife protection legislation dates back to 1916, but the laws have had weak penalties. These penalties have been strengthened with recent legislation, but enforcement is difficult and largely ineffective. In 1995, the Department of Environment and Natural Resources began an accreditation process for zoos and private wildlife facilities, more as an effort to control the illegal wildlife trade than to improve standards. As of September 1997, there were 44 accredited facilities.[87]

7.4.7 Singapore

The Republic of Singapore is an independent city-state consisting of a main island and 50 adjacent islands off the southern tip of the Malay Peninsula. Along with Hong Kong, it is one of Asia's most important seaports, financial centers, and manufacturing districts. Much of its land is developed, with only a small area of the central hills remaining under natural jungle cover. Singapore has a large population and tourist industry to support its six zoos. Singapore's original colonial menagerie began as part of a botanical garden that the Agri-Horticultural Society maintained. This society was formed in 1859, but soon lacked sufficient funds and public support to maintain its botanical garden. In 1874, the society turned its property over to the government, which appointed a superintendent and issued the first official report on the menagerie in 1876. At that time, the collection included a number of large mammals, such as a rhinoceros, a sloth bear, and kangaroos, as well as many birds. In 1878, the government decided to send these larger animals to the Calcutta Zoo and maintain only the smaller mammals, along with the birds. From 1888 to 1902 the menagerie collection grew, with the Malay peninsula fauna becoming an important part of the collection. The public donated many of these native specimens. In 1902, an admirer of the menagerie recommended improvements to the enclosures, which prompted the staff to prepare an estimate for improving the housing. But in 1903 the government turned it down and closed the menagerie instead.[90]

After the closing of the zoological section of the botanical garden, a Singapore Naturalists Society formed in 1921, in part to pursue the founding of a zoological garden and aquarium. After much discussion, the society formed a committee to investigate the issue further, but no zoo or aquarium was forthcoming.[91] It was 1955 before the Ministry of National Development opened the Van Kleef Aquarium in Central Park. This aquarium remained open until about 1993.[92,93] Currently, the only aquarium is Underwater World, opened in 1991 on Sentosa Island, which houses a collection of primarily local tropical marine animals.[93] Another major undertaking of the Singapore Naturalists Society was the zoo display at the Malayan–Borneo Exhibition in 1922. The exhibition showcased the region's natural resources and the products made from these resources for the benefit of the visiting Prince of Wales. One of the most popular exhibits was the zoo, which was in a small area, but had an enormous variety of animals. Interested individuals from the outlying areas of the region contributed these animals. Central to the zoo was a large aviary fitted with small pools, rocks, and tree branches. Adjoining the aviary was a paddock and an open enclosure for deer. A shallow tank held reptiles, while 50 smaller cages contained a variety of small mammals and larger cages exhibited the stronger animals. These animals included tiger cubs, clouded leopards, Malay bear, binturongs, civet cats, proboscis monkeys, gibbons, birds, snakes, turtles, crocodiles, and Oriental insects. After the exhibition closed, most of the collection (some animals died and some were returned to their contributors) was presented to the Prince of Wales and shipped to the London Zoo.[91]

Major zoological facilities did not appear until the 1970s. Jurong Bird Park, which the Singapore Minister for Defense conceived in 1968, and the Ministry for Finance founded, opened in 1971 on Jurong Hill, Jurong Town. This park consists of several exhibits, as well as a five-acre walk-through aviary containing 1,500 free-flying birds. There is also a World of Darkness exhibit and a Falconry Arena. The collection includes over 7,000 birds of 600 species, with a specialization in Southeast Asian birds.

Singapore Zoological Gardens incorporated in March 1971 and opened in June 1973. During the entire time since the earlier Singapore menagerie closed in 1903, the idea of founding a proper zoo never really died. However, it was 1968 before a successful effort was made to establish a new zoo. Ong Swee Law, Chairman of the Public Utilities Board, was the one who decided it was time to have a zoo, so he visited zoos in Europe, the United States, Australia, and Asia in order to consult with their administrators. Law's report was submitted to the Singapore government. As a result of this report, the government formed a nonprofit company (one with charitable status, but which the government owned) and the zoo was incorporated in 1971. Law was appointed to the zoo's board of directors and Lyn de Alwis, Director of the Dehiwala Zoo (Colombo, Sri Lanka), was hired to design the zoo. The specialization of the zoo has been in mammals and reptiles, since the nearby Jurong Bird Park contains an extensive collection of birds and the Van Kleef Aquarium (still open at that time) had a large collection of fish. By 1990, there were 1,600 animals of 170 species at the Singapore Zoo.[92,94–100]

Singapore Night Safari is a facility separate from the Singapore Zoo, but they both have the same parent corporation. The safari specializes in tropical, nocturnal species and possesses several tropical geographic habitat exhibits. These exhibits are furnished with the appropriate flora as well as fauna. Lighting, which simulates moonlight, took on a special importance requiring years of experimentation (and a specialist in stage lighting from London to get it just right) so the visitor can see the animal, but the animal still thinks it is nighttime. Bernard Harrison, Director of the Singapore Zoological Gardens, considers the night safari concept a quantum leap in zoo design and as important as the Indian Forest Service's biological park concept. Not only does it provide natural exhibits, but does so using nocturnal species, which make up a significant segment of the region's native species and which are rarely seen in captivity or in the wild.[92,101–103]

7.4.8 Thailand

Sultans, or kings, of Thailand (Siam), have had animal collections since ancient times and this situation remained true right up to the turn of the twentieth century. During the nineteenth century princes of Southeast Asia rivaled Indian rajahs in the splendor of their animal collections. Among these splendid collections was the king of Siam's collection in Bangkok during the mid-1800s, which had 800 elephants that were used to transport material and perform combat. A few of these elephants were the greatly respected "white" elephants, thought to be the incarnation of the Hindu God Vishnu. These sacred elephants were adorned with gold bracelets, gold necklaces, amulets covered with precious stones, and wore covers made of velvet. Each of these sacred elephants had its own stall, 10 keepers to care for it, and its meals served on finely engraved platters made of precious metals. The king's collection, which also included mammals, birds, crocodiles, fish, and rarely seen local animals, existed up through the early 1900s and was open free to the public on a restricted basis. Coexisting with these royal collections were several hundred Buddhist pagoda (forest monastery) collections. These pagodas were refuges for animals, which the Buddhist monks protected. Surplus pets and pious donations embellished these collections with dogs, cats, pigs, monkeys, birds, tortoises, fish, and other animals.[82,104,105]

Currently, the establishment of zoos in Thailand has taken two tracks. One track is the official zoo effort, which came into being through royal decree on February 15, 1954, incorporating the officially recognized institutions into the Zoological Parks Organisation of Thailand. Another track is the array of unorganized facilities that have been established independently of any controlling influences. Despite the organization's good intentions, the Thai government decided in the 1990s that the Zoological Parks Organisation of Thailand was not performing its intended function and that its member institutions were not of particularly good quality. A study of these zoos was done in 1991 by a team consisting of Ron Tilson (Minnesota Zoo, United States and member of the Conservation Breeding Specialist Group), Pisit na Patalung (World Wildlife Fund-Thailand), and Philphat Thaiarry (Chulalongkorn University). Their report, released in 1993, contained recommendations for improvements in zoo master planning, animal management, animal health, education, and conservation. The Thai government and the Zoological Parks Organisation approved these recommendations, and they are currently being carried out for the five zoos that form the Zoological Parks Organisation of Thailand.[106]

Although the king of Thailand apparently discontinued his animal collection sometime in the early 1900s, he continued to maintain a botanical garden which was attached to the royal residence. This garden was turned over to the Thai government and, in 1938, the Dusit Zoo was established on this property, which is now inside the capital city of Bangkok. The zoo is located across from the Royal Palace and the historic Parliament Building; it now has 1,500 animals of 200 species, a natural history museum, and programs in education and research.[106]

Other early Thai collections include the Pasteur Institute Snake Farm, founded in 1923 at the Queen Saovabbha Memorial Institute (an operation of the Science Division of the Red Cross of Thailand). In addition to exhibiting snakes, the snake farm collects venom for use in snakebite vaccines and serums.[107] Bangkok Reptile Grove was founded in 1930 as a privately operated collection of reptiles.[108] Chiang Mai Zoo, a municipal zoo which the Chiang Mai City government operates, was founded in 1954. It is a natural habitat zoo located adjacent to contiguous forests within the Doi Suthep National Park.[106] A crocodile farm was founded in 1960 in Samut Prakarn Province (outside of Bangkok). A private individual interested in breeding and conserving several species of crocodiles started this facility.[109] More recent collections include the Khao Kheow Open Zoo at Chonburi, founded in 1974. It is adjacent to the Khao Kheow Wildlife Sanctuary and has a collection of 5,240 animals of 212 species.[106] Nakorn Ratchasima Zoo was founded in 1989 at Korat in northeast Thailand.[106] Songkla Zoo, founded in 1989 (but which, due to development delays, did not open to the public until 1998), is located between the cities Songkla and Hat Yai near the Gulf of Thailand.[106,110] Safari World, founded in the late 1980s, is a large safari park with cage enclosures located in a leisure complex. It also has a marine park and three large aviaries.

There are also several Gibbon Rehabilitation Projects, all of which the Wild Animal Rescue Foundation of Thailand manages: Bang Pae Forest Reserve, Phuket Gibbon Rehabilitation Project, and Ko Tai Island Gibbon Rehabilitation Project. The gibbon is as important to Thailand as the panda is to China and the orangutan is to Borneo. Several gibbon species are found in, or adjacent to, Thailand. These species are endangered and are often illegally captured for trade or food. Many of these illegally obtained gibbons are confiscated by authorities, who have nowhere to keep them. Therefore, T. D. Morrin founded the gibbon rehabilitation projects in 1992 to care for these animals and to prepare them for release back into the wild. The first of these projects was in the Bang Pae Forest Reserve. Visitors are allowed to see the gibbons while they are in their enclosures, but not after the gibbons begin the reintroduction process since this period is when they must learn to be independent of people.[111]

7.4.9 Vietnam

Royalty in Vietnam (Cochin China), like their Asian counterparts, have kept animal collections for centuries, if not since ancient times. The king still maintained his collection in the mid-1800s and used some of its animals for combat exhibitions. There were also 60 elephants in the main royal collection at that time, with smaller provincial collections of 20 to 30 elephants each.[112] Currently, there are two main zoos, along with several small ones that businesses and private individuals own.[113] There are also several breeding and rescue centers for conserving endangered species and reintroducing animals back into the wild.

French colonialists founded Saigon Zoo at Ho Chi Minh City in southern Vietnam in 1864. As it was intended to be both a zoological and botanical garden, it was modeled after the Jardin des Plantes menagerie in Paris. French botanist J. L. Pierre managed the botanical garden, while the zoo was under the management of French army veterinarian Louis A. Germain. Germain also directed the zoo's growth, planned its roads, built its cages, and collected its animals. It was occupied by Japanese troops in 1944 to 1945, French troops in 1945 to 1955, and then U.S. and South Vietnam troops during the 1960s to 1970s. In 1963, the site increased from its original 30 acres to 83 acres, with plans to relocate outside the city on a 750-acre site. Reconstruction of the zoo (replacement of most cages with large outdoor enclosures) on its city site occurred in 1965 and its name was changed to Thao Cam Vien Saigon (Saigon Zoological–Botanical Garden). During 1968, in the midst of the Vietnam War, the zoo was clean and well run, attracting large crowds. Animal cages had labels in Vietnamese, French, and English, along with the Latin scientific name. Another effort to create a new zoo outside the city occurred in 1990, but a plan to renovate the existing zoo superseded it. Once again, in 1996 a master plan was developed to improve the existing zoo and to improve its conservation programs.[14,114–119]

In the early 1900s there was a zoo in the Hanoi botanical gardens. Thu Le Park Zoo, also in Hanoi, was either a different facility or was the new location for the botanical garden zoo. The present Hanoi Zoo was founded in 1976 with an emphasis on exhibiting and conserving the indigenous fauna, and its collections are part of the captive breeding programs at the Institute of Economical and Biological Resources, National Center for Scientific Research.[55,113,114]

As with other Asian countries, conservation is beginning to receive serious attention in Vietnam. One outcome of these efforts has been the establishment of breeding and rescue centers. Eakao Breeding Station, founded in 1986 (Dac Lac Province, South Vietnam), has programs for Vietnam sika deer, Asian elephant, brow antlered deer, kouprey, binturong, and slow loris. Bavi Breeding Station, founded in 1990 (on an island in Suoi Hai Lake near

Hanoi, North Vietnam), has programs for Vietnam sika deer and Owston's palm civet. The Endangered Primate Rescue Centre, which the Forestry Ministry founded in 1993 at Cuc Phuong National Park, is one of several the Vietnam government intends to establish to reintroduce confiscated primates. Another project is the Deer Farm at Cuc Phuong National Park, which is working to reintroduce endangered deer, as well as to distribute specimens to zoos around the world. Hanoi Rescue Centre, founded in 1997, is the second government center to rescue endangered species. An effort is also being made to conserve a group of Javan rhino, only discovered in 1989 and previously thought to exist only in one national park in Java. These five to seven animals are being intensively managed in semiwild conditions at the Cat Loc Nature Reserve, which is near their future home, the Cat Tien National Park.[120–124]

7.5 Indonesia

Indonesia is the world's largest archipelago (the Malay Archipelago), extending between the continents of Asia and Australia. It consists of 17,068 islands, only 6,000 of which are inhabited by people. It has tremendous biodiversity with influences from both neighboring continents. It also has great cultural diversity and has been a major cross-cultural trading region since ancient times, especially because of its tropical spices. Arab, Chinese, Asian, and African trading ventures existed for centuries prior to the arrival of Europeans. These activities produced wealthy regional rulers, some of whom had private menageries. Captains of the earliest English East India Company ships to arrive in the region were witnesses to these collections. At the time of these seventeenth- and eighteenth-century European excursions, local rulers had collections of buffalo, tigers, rhinoceros, and elephants that they would use to stage fights.[33,82,125] Unfortunately, very few visitors commented on these early collections, and when they did they described the spectacles and not the collections that housed the animals used in the spectacles. These royal, and later colonial, collections existed throughout the Indonesian region, even in areas where there are no known modern zoos. Timor (the largest and easternmost of the Lesser Sunda Islands), for example, had a collection in the public gardens of its capital as late as 1901. This collection contained monkeys, deer, and birds, including cassowaries.[55]

Javan rhinos are the rarest of the rhinoceros species and have been exhibited less than any of the other species. However, Javanese princes and nobles kept this species in their private menageries from the seventeenth to the nineteenth centuries, and probably since well before this period. The private collection of the Javanese nobleman Soesoehoenan von Mataram contained many Javan rhinos in the years 1648 to 1654. Batavia (Jakarta) was one of the earliest Dutch colonial administrative centers and had one of the earliest botanical collection stations (established sometime in the 1600s). Like many of these colonial stations, it acquired a menagerie and in 1751 this collection had a Javan rhino. Another Javan rhino was kept at Surakarta around 1815 to 1821 and another in a collection near Batavia in 1875. Early twentieth-century zoos and private collections contained Javan rhinos as well.[33,126]

Indonesia is making great efforts to organize and improve its modern zoos. These efforts resulted in the establishment of an Indonesian Zoological Parks Association in 1969. This association is a nonprofit, nongovernmental organization which the Ministry of Forestry recognizes as a management partner and the Indonesian Scientific Institution recognizes as a scientific partner. Indonesian Zoological Parks Association aims include the improvement of Indonesian zoos through relocation when necessary and through improved standards. In addition, the association has played a larger role in the Asian region by encouraging the establishment of the South East Asian Zoological Parks Association, organizing regional conferences, and providing training workshops for personnel dealing with endangered species management.

Indonesian zoos represent a great diversity in status, size, and management authority. These 28 zoos comprise 17 zoological gardens, 3 public aquariums, 2 bird parks, 4 reptile parks, and 2 butterfly parks. The distribution of these zoos is concentrated on three islands, with 4 zoos in Bali, 15 zoos in Java, and 9 zoos in Sumatra.[127,128] Extant Bali zoos are of recent origin and are specialized. Citra Bali Bird Park, Bali Reptile Park, and Bali Butterfly Park were founded in 1995, while the Indonesia Jaya Reptile and Crocodile Park was founded in 1997. The first three are private facilities, while the last is a branch facility of the one in Jakarta (Java).[129]

7.5.1 Java

Ragunan Zoo in Jakarta can trace its history back to a menagerie opened in 1864 as part of an amusement complex on land the Javanese artist Raden Saleh donated to the city. Saleh was originally from Arabia and was educated in the Netherlands. He then traveled throughout Europe, Persia, Australia, and Africa, settling at Batavia (Jakarta) in 1852 to paint portraits, landscapes, and animals. The Dutch colonial administration began a menagerie on Saleh's donated property, which received private support from the Society for Plants and Animals. The amusement complex, including the menagerie, was popular as it was open to all residents of the city, not just to Europeans, which was unusual for that time.[130] A few years after Indonesian independence, in 1949, the name of the Society for Plants and Animals was changed to Kebun Binatang Cikini, and the name of the zoo to Cikini Zoo. During the zoo's centennial year, the government of Jakarta agreed to move the zoo to a more spacious site. This was done in 1966 when it moved to land that had been used for fruit tree research and development, and the zoo became the Taman Marga Satwa Jakarta (Animal Park of Jakarta).

A few individuals, including Governor Ali Sadikin, Mr. and Mrs. Benjamin Galstaun, and Mrs. Ulrike von Mengden, spearheaded the effort to move and improve the zoo. The Galstauns had traveled to zoos throughout Australia and Europe and were familiar with zoo design and management. Mr. Galstaun was honored with the Magsaysay Award in 1977, an award the Philippine government presented to him for his outstanding work on behalf of Asian communities. After his retirement, Mr. Galstaun helped with the planning of other zoos in Asia. Mrs. Von Mengden maintained a compound for ill and orphaned animals, primarily orangutans. Together with others, they helped build the zoo and encouraged its development.[131,132]

Again, in 1974, the name of the zoo was changed, this time to Kebun Binatang Ragunan (Ragunan Zoo). The zoo was enlarged again as well. By 1977, the zoo had over 1,600 mammals, birds, and reptiles. Natural assets of the zoo site, such as its rain forest and wetlands, provide thematic areas that are used as ecosystem exhibits. Ragunan Zoo has had outstanding breeding successes, including several endangered Indonesian species, such as orangutans, Komodo dragons, and mouse deer. The zoo has hosted training workshops and has developed an excellent education program. Sahabat Satwa (Friends of the Zoo) is probably the most active and successful friends of the zoo group in Asia. Currently, the zoo has 104 mammal species, 185 bird species, 46 reptile species, 5 fish species, and 5 invertebrate species, for a total of 4,348 specimens.[127,131,133]

Another early zoo, Taman Wisata Satwa Taru Jurug (Satwataru Zoological Park in Surakarta, Central Java), was begun in 1878, but little is known about it. A Dutch journalist, H. F. Kommer, founded Kebun Binatung Surabaya (Surabaya Zoological Gardens in Surabaya, East Java) in 1916. He obtained funds for the zoo from a variety of donors, but it was really a way for him to pursue his own hobby of collecting animals. The entire site was converted to animal exhibits by 1940. The zoo now has a collection of 84 mammal species, 133 bird species, 21 reptile species, and 175 fish species, for a total of 3,546 specimens. The zoo's Captive Propagation Centre was used as a model in the design of a pre-release training center at Bali Barat National Park. Several birds from the zoo's breeding programs were released from this center into the park. Breeding programs exist for several endangered Indonesian species, such as Komodo monitors, Bali mynahs, Javanese green peafowl, Jalak Bali pelican, and babirusa (a wild pig endemic to Celebes Island).[127,134–138]

Kebun Binatang Tamansari Bandung (Bandung Zoological Garden in Bandung, West Java) was founded in 1933 with support from the Yayasan Margasatwa Tamansari foundation. Margaraya Tinjomoyo Semarang (Semarang Zoological Park in Semarang, Central Java) was founded as a government zoo in 1951. Gembira Loka Zoological Garden (Yogyakarta Zoo in Yogyakarta, Central Java) was founded in 1953. Initially, its purpose was to provide facilities to supplement the education of zoology students from local schools. In 1966, the aims of the zoo were broadened to include the education of other school groups and adults. Yogyakarta Zoo has also been involved in conservation programs, breeding several species such as Komodo monitors, Javan deer, muntjac, and banteng. Animal behavior and reproductive biology research is conducted, and the zoo maintains apes for the Indonesian Air Force space flight research program.[127,139]

Oceanarium Gelanggang Samudra Jaya Ancol (Ancol Oceanarium) opened in 1974. A privately operated aquarium, it displays a variety of mammals besides marine mammals, such as hippopotamus and sun bears. It also provides daily shows using dolphins, sea lions, and multiple species (for example, sun bears, birds, and river otters). Taman Mini Indonesia Indah (East Jakarta) was founded in 1976. Its name means "beautiful Indonesia in miniature park" and it is intended for enjoyment as a bird park, as well as an educational exhibit of 249 bird species (3,925 specimens) from throughout Indonesia. The bird collection is divided into two parts representing eastern and western species, in keeping with the division of the region according to the Wallace Line (the zoogeographical boundary between Australian species and Asian species which Alfred R. Wallace, co-originator of the origin of species theory, established during his nineteenth-century studies in Indonesia).[127]

Several other zoos, aquariums, reptile parks, and insectariums were established in the 1980s and 1990s. A family of circus owners, headed by Mr. Manasang, began Taman Safari Indonesia (Indonesia Safari Park in Jakarta) in 1986. After working for a European circus in the Philippines, Manasang started his own circus with family members as the star attractions. Wanting to switch from circus acts to animal exhibits, the family bought 150 acres of mountain land near the west Java tea plantations and began his Safari Park. The Safari Park consists of several drive-through safaris, one with ungulates, one with tigers and lions, and one with bears. Other species, such as primates and birds, are kept in these safari areas on islands and in enclosures. The Safari Park has also developed several endangered species programs, supports the Indonesian Center of Reproduction for Endangered Wildlife, and has opened a second facility in Pasuruan (East Java).[127,140]

7.5.2 Sumatra

In Sumatra, local governments have started a number of small zoos: Taman Marga Satwa dan Budaya Kinantan (Zoological Park Kinantan at Bukittinggi, West Sumatra) in 1929, Taman Hewan Pematangsiantar (Pematangsiantar Zoological Park at Pematangsiantar, North Sumatra) in 1936, Kebun Binatang Medan (Medan Zoological Park at Medan, North Sumatra) in 1973, and Kebun Binatang Taman Ria Aneka Rimba Jambi (Zoological Park Anera Rimba Jambi at Jambi, Sumatra) in 1973. Several other zoos were established in the 1980s and 1990s, either through local governments or private foundations.[127]

A number of orangutan rehabilitation centers have been established in the last few decades in Sumatra and Borneo. Barbara Harrisson organized the first center at Bako National Park (Sarawak) in 1960. Since then, the Indonesian Nature Conservation Service established one at Ketambe (North Sumatra), and Biruté Galdikas established Kalimantan Rehabilitation Station at the Tanjung Puting Wildlife Reserve (Kalimantan). Four others have been established; one at Bohorok (North Sumatra), two near the Tanjung Putting Reserve (Kalimantan), and one in Wanariset (Sumatra).[127,141]

7.6 East Asia

7.6.1 Hong Kong

With a history all its own, Hong Kong has played an important role in the fate of Asian wildlife. As a preeminent commercial center and seaport, millions of animals have passed through Hong Kong, most on their way to other countries; however, hard data are so scanty that quantitative assessment is out of the question. Nevertheless, the data that do exist on reported transactions are shocking, although only a small portion of the total animal trade is accounted for. While zoos in Southeast Asia could be making an effort to help minimize this wildlife trafficking through education, they seem reluctant to do so. Rescue centers, which are being established in Southeast Asian countries, are now setting an example which, it is hoped, the zoos will follow.

The first section of what is now the Hong Kong Zoological and Botanical Gardens opened to the public in 1860, although the official literature cites 1871 as the founding date. Originally, it was a colonial botanical station that eventually included a menagerie. Exactly when this occurred and the history of its existence was not recorded until after World War II. By 1975, the animal collection had expanded so much, the name of the gardens was changed to reflect its zoological content. The Urban Council of Hong Kong manages the gardens, which are free to the public. A subway connects the two parts of the garden, which are divided by the Albany Road. The "old garden" contains a large bird collection with 280 species, of which 100 species have reared young. The "new garden" contains the mammal collection. Although a small zoo, it is active in education and conservation programs. It also has breeding programs for a number of endangered species, including Bornean orangutan, red-cheeked gibbon, lion tamarin, lion-tailed macaque, white-naped crane, white-winged wood duck, and Rothschild's myna.[142]

Edward Youde Aviary is located in Hong Kong Park, along with landscaped gardens, a conservatory, and other public facilities. The park was developed in 1979 and its large walk-through aviary was constructed about the same time. Using a tropical rain forest theme, the aviary exhibits species from Southeast Asia, Indonesia, and New Guinea. There is also a lake for waterfowl, three large cages for the larger birds, and an education center.[143] Lai Chi Kok Zoo (Kowloon, Hong Kong) was a small private zoo open between 1951 and 1993. It was poorly maintained, so the government closed it.[144] Ocean Park is an entertainment complex with a full range of mammals, birds, reptiles, fish, and invertebrates. Opened in 1977, it continues to expand its facilities with the addition of the Shark Aquarium in 1989, the Bird Paradise in 1986, the Flamingo Pond in 1991, and the Peacock Garden in 1993. It is actively involved with education and conservation programs.[145–148] Kadoorie Farm and Botanic Garden is a private facility that changed its focus in 1994 from agriculture to conservation. It is located in a nature reserve and maintains a rescue and rehabilitation program for native birds and other animals. This rescue center returns native species to the wild and exotic species to their country of origin, and uses disabled animals in its education program.[149]

7.6.2 Korea

Korea comprises the Democratic People's Republic of Korea (North Korea) and the Republic of Korea (South Korea). Korean zoos are not part of the interactive international zoo community and little is known about them. Six zoos are known, but there is reason to believe there may be many more small provincial zoos. A report on the critically endangered Amur leopard (*Panthera pardus orientalis*) illustrates this situation. It states there is an "unknown number of animals in the Pyongyang Zoo," which are not registered in the studbook and "may not be available" for a coordinated breeding program.[150] Pyongyang Zoological Gardens, the primary zoo in North Korea, opened in 1960. Chang-gyeong Weon Zoological and Botanical Gardens (Seoul Grand Park Zoo), the national zoological gardens of South Korea, opened in 1909. This zoo was completely destroyed during the Korean War in 1950, but has now been reconstructed. Its master plan of 1973 included areas for Korean, African, Asiatic, American, and Australian faunas. Also included were a botanical garden with tropical bird aviaries, a reptile building, and a nocturnal building.[14,151]

7.6.3 Macau

Yi Lung Hau Gung Yuen (Two Dragon Throat Public Garden, Macau) is a little-known facility, even to those concerned with the region's animal welfare issues. In 1997, the animals were in temporary cages while new exhibits were being built. It is a very small zoo with a few mammals and some doves, but appears to be the only zoo in this small country. Other zoos have existed, but have closed, according to residents.[144]

7.6.4 Taiwan

Taiwan (Republic of China) is another Asian country that has historically been involved with the wildlife trade, both locally and internationally. Here, as in Hong Kong, wildlife rescue centers are taking the lead in overcoming illegal trade and the rehabilitation of confiscated animals. Although these rescue centers only retain the animals temporarily until they can be returned to the wild or to an appropriate zoo, they are a significant aspect of the modern zoo community. In particular, those responsible for the wildlife rescue effort in Taiwan have taken a lead in trying to develop standards, to establish communication among centers in South and Southeast Asia, and to develop a network of centers and zoos to further the conservation of Asian species. Taipei Zoo, founded in 1915, is the oldest extant zoo in the country. In 1986, it was moved from its location in northern Taipei to a larger site in a southeastern suburb of the city. It now has 2,500 mammals and birds of 250 species, as well as invertebrates and a butterfly aviary. Program emphases are on wildlife conservation and environmental education. It also functions as a rescue center for confiscated reptiles.[152]

Several other zoos, aquariums, bird parks, and rescue centers have been established in Taiwan, but most are recent. Pintung Rescue Centre was founded in 1993 with funding from the Council of Agriculture to provide temporary accommodation for confiscated and rescued wild animals. In 1997, the center housed 186 carnivores, primates, birds, and reptiles, and is the largest of four animal rescue centers which the government supports. Other zoos include Kaohsiung Zoo (Shousan Park, southern Taiwan), Hsinchu City Zoo (a small zoo with a surprisingly wide variety of species), Leefoo Zoo (Leefoo Park, south of Taipei, designed in the safari park style), Far East Animal Farm (Ba-Jya Village, Gwei-Zen Hsian, Tainan, a private farm with caimans and cobras), I-Lan Wild Animal Rescue Centre (I-Lan Institute of Agriculture, I-Lan City, a government-supported center for primates and birds), Feng-Huang-Ku Bird Park (Luka, Nan Tou, a privately owned zoo for birds), Ocean World (Yehliu Wanli, a privately owned aquarium that also has tigers and bears), and the National Museum of Marine Biology/Aquarium (currently under construction at Kaohsiung).[153–155]

7.6.5 China

China's cultural heritage, including its zoological collections, has a rich and ancient history (see Chapter 1 by Kisling on ancient collections). However, China's public zoological gardens are a modern phenomenon, beginning after the turn of the twentieth century, with many established after World War II. In 1974, the *International Zoo Yearbook* listed only 20 zoos in China. By 1995, the number listed had jumped to 130. In 1997, the Chinese Zoo Association had 181 member institutions.[6,156]

Prior to 1950, there were only a handful of zoos in China, and these were in miserable condition as a result of the political instability of the previous three decades. Major zoos in some 40 cities were founded after the establishment of the People's Republic of China. Even at that time, in 1949, some five to eight million people visited Peking Zoo annually.[157] In 1965, Al Oeming (owner of a game farm in Alberta, Canada) visited China and remarked that he would like to have Chinese zookeepers at his game farm because they did their work so carefully. He was told these keepers could not be spared because so many new zoos were being planned for the immediate future in China.[14] A regular visitor to China has remarked:

PÈRE DAVID'S DEER AND GIANT PANDAS

Several Asian regions have key species that are symbolically and environmentally important: the orangutan in Borneo, the rhinoceros in Java, the gibbon in Thailand, the tiger in India, and the Komodo monitor in Indonesia. For China, the key species have been Père David's deer and the giant panda. Their stories are a contrast: Père David's deer was quietly brought back from the brink of extinction, while the giant panda is being rescued from a threatened status with great fanfare. But the stories also have a common denominator, the intervention of zoos to save these species from extinction.

Père David's deer (*Elaphurus davidianus*) was already extinct in the wild when Armand David (1826–1900) accidentally discovered it. During the mid-1800s, when the Lazerist Order of the Catholic Church in France was establishing missions in China, the Director of the Muséum d'Histoire Naturelle in Paris, Henri Milne-Edwards, was working with the Church to have its missions help the museum with its scientific endeavors. This combination of duties was ideally suited to David's abilities as a missionary and a naturalist. David met with Milne-Edwards and David promised his cooperation in helping the museum. In 1862, David went to China and immediately began zoological, botanical, and geological surveys. He mastered the language and became the natural history teacher at the mission school. In 1863, he sent his first shipment of specimens to the museum. These specimens were of such high quality that he received additional support and encouragement from the museum to make further studies throughout China. In addition,

continued

Ever since the 1950s the Chinese authorities have realized that zoos are an excellent way of keeping the people healthfully occupied on their days off. It is a matter of prestige for each municipality to have one. I have rarely gone to a town and not found a zoo .... Even if there is not a proper zoo, many public parks have a few cages in a corner with a few tigers, lions, monkeys, etc. and big towns often have several animal collections.[144]

The Chinese Association of Zoological Gardens, founded about 1985, is an attempt to organize the growing number of zoos. Its six regional branches are attempting to coordinate the zoos' conservation efforts, publish scientific and educational materials, make and implement animal protection regulations, establish technical standards for animal husbandry, establish breeding facilities for endangered species, coordinate research on captive breeding

---

PÈRE DAVID'S DEER

David founded a natural history museum in Peking, which was popular with foreign visitors as well as Chinese aristocrats.

David's famous discovery of the deer now bearing his name happened in 1865. David came upon the deer in the Hae-Dze game park, about two miles south of Peking. Foreigners were not welcome in the park, but David climbed a wall and saw what he thought was some sort of elk or moose. The Chinese name for this animal was milu (or sibuxiang) which means, according to one interpretation, "neither deer, nor cow, nor goat, nor horse, but an animal with features of all four." The herd of some 100 animals that David saw was the remnant of a species that had been extinct in the wild for more than 300 years. David managed to obtain the hide and bones of two animals, and the carcass of another specimen. The latter animal had been presented alive as a diplomatic gift, but had died. These materials were sent to the Paris Muséum where Milne-Edwards described and scientifically named the species.

Before leaving China, David obtained some live deer and exported them to Europe, the first ones going to the London Zoo in 1869. Three other deer arrived at the Berlin Zoo in 1876. Not many deer survived the journey, but they were ultimately better off in European collections because none of the Chinese specimens survived. A flooding of the Yellow River in 1894 brought down the walls of the park and destroyed large portions of it. This disaster was followed by the Boxer Rebellion of 1900, when peasants slaughtered the animals in the park for food. The surviving deer were taken to a park in Peking, but by 1920 the species had died out in China.

Captive deer in the European collections were gathered together by the 11th Duke of Bedford at Woburn Abbey during 1893 to 1895 to further the propagation of Père David's deer. Beginning with an initial herd of 18 animals, the success of the propagation program brought the herd up to 88 animals in 1914 and 255 by 1948. In 1986, the Zoological Society of London estimated there were about 1,500 deer in more than 100 collections around the world. In 1956, the Peking Zoo acquired four specimens from England, and, in 1985, two groups of deer were released into Chinese reserves: one group from various zoos into Da Feng, Jiangsu Province, and another group from Woburn Abbey into the Imperial Deer Park (Nanhaize Park), the original site where David had discovered the deer. As of 1998, there were about 400 deer at Da Feng, 200 at Nanhaize Park, and 130 at another site (Shishou in Hubei Province). Without much fanfare, the dedicated efforts of the captive herds' keepers have allowed these animals to once again live in their native habitat.

Giant pandas (*Ailuropoda melanoleuca*), although known locally since ancient times, were not introduced to Western science until 1869. Once again, it was the French missionary Père David who discovered this animal. On three separate occasions during March 1869, David

programs, provide training for zoo personnel, and cooperate with international organizations on conservation matters.[158–160]

Beijing Zoo was founded in 1906 or 1908. It is the first modern zoo in China. It is said to have been built for the amusement of the dowager Empress Tse Hul of the Ch'ing dynasty, serving as an experimental farm cultivating silkworms and cotton. When the Chinese communists gained power, the zoo began a new phase of development. In 1952, the zoo built a new elephant building in place of the cotton fields. A few years later, in 1955, it was named the Peking Zoological Gardens. During the 1960s, the zoo controlled all affairs related to the country's zoological gardens. The zoo had knowledgeable officials, an impressive entrance, attractive informational notices, and skillful landscaping. There

---

obtained a panda skin, the first from local people in Baoxing Xian (Sichuan Province), the second a young panda that had died, and the third the corpse of an adult panda. David named the species, described it, and sent these materials to Milne-Edwards at the Muséum d'Histoire Naturelle in Paris. David had classified the panda as a bear, but Milne-Edwards decided otherwise and named it with its current scientific name. Various descriptions of the panda were published in the late 1800s, and in the early 1900s several expeditions managed to locate and obtain specimens. Few specimens were caught alive and none survived the journey away from its natural habitat.

It was not until Ruth Harkness obtained a young panda in 1936 and returned with it to the United States that a living panda was brought to the non-Chinese public's attention. It was the first panda to be exported from China, and with its transfer to the Brookfield Zoo (Chicago) in 1937, it was the first to be exhibited in a zoo. Harkness returned to China in 1937 and obtained another young panda, which she also presented to the Brookfield Zoo in 1938. Roy Spooner obtained a young panda for the New York Zoological Park in 1938. And, in 1938, Floyd Tangier Smith (who it is believed owned the original panda that Ruth Harkness obtained; see Chapter 5 on U.S. zoological gardens), brought five pandas to London. During the 1930s to 1940s several more pandas were exported to foreign zoos, but it was not until 1953 that a panda was finally exhibited in a Chinese zoo, first at the Chengdu Zoo, and then in 1955 at the Peking Zoo.

After World War II, Chinese scientists began warning those who would listen that the giant panda was on the verge of extinction. Nevertheless, the Chinese government continued to export pandas as goodwill ambassadors to various zoos throughout the world from this time up through the 1980s. Finally, in the 1990s, zoo officials, government officials, and conservation organizations began cooperating on serious conservation efforts.

Sources: Deer: Based on Loisel, G., *Histoire des ménageries de l'antiquité à nos jours*, Octave Doin et Fils and Henri Laurens, Paris, 1912 [translation in preparation provided by J. C. Edwards]; Brambell, M., "The Peking Zoo," in *Great Zoos of the World: Their Origins and Significance*, Lord S. Zuckerman, Ed., Westview Press, Boulder, CO, 1980; Beck, Benjamin B. and Wemmer, Christen M., Eds., *The Biology and Management of an Extinct Species: Père David Deer*, Noyes Publications, Park Ridge, NJ, 1983; Bedford, (12th) Duke of, "Père David's deer: the history of the Woburn herd," *Proceedings of the Zoological Society of London*, 121, 327, 1951; Reichenbach, H., "Der David vom Davidshirsch: Armand David (1826–1900) und die europäische Entdeckung chinesischer Tiere," *Bongo*, 28, 23, 1998 ["The David of Père David's deer: Armand David (1826–1900) and the European discovery of Chinese animals"]; and Thouless, C. R., Chonggui, L., and Loudon, A. I., "The Milu or Père David's deer (*Elaphurus davidianus*) reintroduction project at Da Feng," *International Zoo Yearbook*, 27, 223, 1988. Pandas: Based on Jing, Z. and Yangwen, L., *The Giant Panda*, Science Press and Van Nostrand Reinhold, New York, 1980; Schaller, G. B., *The Last Panda*, University of Chicago Press, Chicago, 1993.

was also a large statue of the giant panda that gave birth to the first giant panda cub in captivity in 1963.[14,157]

Many native species were maintained at the Beijing Zoo in the 1950s and 1960s, including many of the endangered species China's first modern wildlife preservation laws of 1958 protected. These included giant panda, golden monkey, takin, wild yak, Bactrian camel, Manchurian tiger, Thorolds deer, kiang, Japanese black-necked cranes, Asiatic white cranes, Chinese alligators, and others. Exotic species from Asia and around the world were not neglected. Neither was the oldest collection of the zoo neglected, the goldfish collection. Rows and rows of half barrels are still maintained, each containing a different variety of goldfish (black, red, silver, blue forms; long-tail forms; long-fin forms; ones with great pouches under their chins; ones with huge protuberances on the head; etc.).[14,157] Indications are that the Beijing Zoo continues to expand its exhibits at the expense of its natural areas.[144]

Shanghai Zoo was founded in 1953. It shares the city with six zoos, a safari park, two park menageries, a dolphinarium, and a circus. The zoo is still popular despite its competition, having over 2.3 million visitors in 1994. Over the years, from the 1970s to 1990s, the zoo has been impressive, with high standards of care. Although it has had a circus display and a dog and cat section, it nevertheless has exhibited a number of rare animals, such as several species of crane, giant panda, Tibetan lynx, and golden monkeys.[14,144]

Guangzhou Zoo (formerly known as the Canton Zoo) was founded in 1956 and opened in 1958. In the mid-1960s, it was already undergoing a complete rebuilding and had a rather indifferent collection of common, nonendemic Chinese animals. Substantial improvements were made by the 1970s; however, in 1994, the ubiquitous Chinese zoo circus had been added.[14,144]

Chengdu Zoological Gardens was founded in 1954 and rebuilt in 1973. While the zoo contains a dog section and a theme park, it also has a successful giant panda–breeding center. Chengdu Giant Panda Breeding Research Station is located outside Chengdu and holds both giant and red pandas. The station has a museum with one floor devoted to pandas, one for a butterfly exhibit, and one with meeting facilities, including accommodations for visiting scientists. Chinese scientists put a great deal of importance on high-tech reproductive techniques, particularly for giant panda.

In the last decade there has been a proliferation of breeding centers in China, in addition to the Chengdu Giant Panda Breeding Research Station. These centers include Chongging Endemic Animal Breeding Centre, Shenyang Crane Breeding Centre, Wuzhou Tonkin's (Snub-nosed) Langur Breeding Centre, Xian Golden Monkey and Takin Breeding Centre, Suzhou South China Tiger Breeding Centre, and Wuhan Golden Monkey Breeding Centre.[158] A number of musk deer farms have also been established since 1958 in Szchuan, Anhui, and Shaanxi Provinces.[161] China, always a land of contrasts with a culture to match, has fine conservation centers and zoos juxtaposed with many small, poorly equipped zoos. One hopes its many basic zoos will not be lost in the glamour of the flagship institutions.

References

1. Khan, R., "Gulf states, Arabian peninsula," *CBSG News* [Newsletter of the Conservation Breeding Specialist Group], 4, 4, 1992.
2. Jones, D. M. and Kock, R. A., "Management problems for zoos in the Middle East," in *Proceedings of the Fourth Symposium on Zoo Design and Construction*, Stevens P. C. M., Ed., PZG, Paignton Zoological and Botanical Gardens, Paignton, 1994.
3. Nogge, G., "Kabul Zoo the show-window of Afghan fauna," *Outdoorsman Monthly*, 3, 1972.
4. Nogge, G., "Kabul Zoo," *International Zoo News*, 20, 114, 1973.
5. Grey, S., "Wounded animals at bay in Kabul's zoo of horror," *London Times*, January 25, 1998, 14.

6. *International Zoo Year Book*, Zoological Society of London, London [Annual serial, 1960+].
7. Khan, R., "Arabian peninsula report," *CBSG News* [Newsletter of the Conservation Breeding Specialist Group], 4, 4, 1993.
8. Shackleton, D. M., Ed., *Wild Sheep and Goats and Their Relatives. Status Survey and Conservation Action Plan for Caprinae*, IUCN/ SSC Caprinae Specialist Group, IUCN, Gland, Switzerland and Cambridge, U.K., 1997.
9. Durrell, Lee, personal communication, 1998.
10. Ganji, M. H., "Flora and fauna," in *Iran – Elements of Destiny*, Benny, R., Ed., Collins, London, 1978.
11. Jackson, Peter, personal communication, 1998.
12. Hatt, R. T., *The Mammals of Iraq*, University of Michigan Press, Ann Arbor, 1959.
13. Swengel, F. B., Hill, T. J., and Sullivan, E. J., Eds., *Global Zoo Directory*, Conservation Breeding Specialist Group, Minneapolis, 1996.
14. Kirschofer, R., *The World of Zoos: A Survey and Gazetteer*, Batsford, London, 1968.
15. Mendelssohn, H., personal communication, 1998.
16. Sivan, Pnina, personal communication, 1998.
17. Terkel, Amelia, personal communication, 1998.
18. Anonymous, *Guide to Jerusalem Biblical Zoo*, Jerusalem, n. d.
19. Anonymous, "News in brief," *Newsletter of the International Primate Protection League*, 18, 26, 1991.
20. Spalton, A., "Recent developments in the reintroduction of the Arabian Oryx (Oryx leucoryx) to Oman," *CBSG News* [Newsletter of the Conservation Breeding Specialist Group], 2, 8, 1991.
21. Nader, I. A., "Rare and endangered mammals of Saudi Arabia," in *Wildlife Conservation and Development in Saudi Arabia, Proceedings of the First Symposium, 1987*, Abuzinada, A.H. and Goriup, P. D., Eds., 1989, n. p., n. l.
22. Anonymous, "Gazelles (*Gazella saudiya*) given to Wildlife Research Centre," *CBSG News* [Newsletter of the Conservation Breeding Specialist Group], 1, 18, 1990.
23. Burr, M., The Ankara Zoo, *Zoo Life* [Bulletin of the Zoological Society of London], 3, 10, 1948.
24. Anonymous, "UAE makes extra effort to protect wildlife," *Arabian Wildlife*, 1, 22, 1994.
25. Ommer, Naseer, personal communication, 1998.
26. Anonymous, *CBSG News* [Newsletter of the Conservation Breeding Specialist Group], 3, 5, 1992.
27. Vigne, L. and Martin, E., "Captive leopards and lions in Yemen need help," *International Zoo News*, 45, 22, 1998.
28. Dacca Zoo, *Dacca Zoo Guide Book*, Dacca Zoo, Dacca, 1979.
29. Lockwood, I., "Bangladesh's declining forest habitat," *Sanctuary Asia*, 18, 1998.
30. Wollenhaupt, H., "The mishmi takin (*Budorcas taxicolor whitei*) and the goral (*Naemorhedus goral*) in Bhutan," *Caprinae Specialist Group Newsletter*, 7, 3, 1993.
31. Cocker, M. and Inskipp, C., *A Himalayan Ornithologist: Life and Work of Brian Houghton Hodgson*, Oxford University Press, New York, 1988.
32. Flower, S. S., "Contributions to our knowledge of the duration of life in vertebrate animals. 5. Mammals," in *Proceedings of the General Meetings for Scientific Business of the Zoological Society of London*, 145, 1931.
33. Reynolds, R. J., "Asian rhinos in captivity," *International Zoo Yearbook*, 2, 17, 1961.
34. Tikell, S. R., *Mammals of Nepal*, The Zoological Works of Samuel Richard Tickell Collection, London Zoo Library, London [unpublished manuscript].
35. Lewis, D., Dezonne, F., Pringle, C., Fairfield, S. and Sanday, J., *KMTNC Central Zoo Progress Report — Master Plan Team*, Central Zoo, Katmandu [unpublished typescript, ca. 1996].
36. Nogge, G., personal communication, 1998.
37. Rao, A. L., *Proposed Zoo cum Botanical Garden in Islamabad*, n. p., n. d.
38. Shariff, Z. A., "Establishment of a new zoo using rescued and confiscated wild animals: the Jungle Kingdom, a multi-dimensional conservation project. Proceedings of the first international workshop on the management of wildlife rescue centers in south and southeast Asia," *Zoos' Print*, 13, 29, 1998.

39. Ahmad, M. F., *Zoos in Pakistan* [unpublished typescript, ca. 1991].
40. Flower, S. S., *Report on a Zoological Mission to India in 1913*, Cairo Government Press, Cairo, 1914.
41. Toosy, A., *Lahore Zoo (Past, Present, Future): Recommendations* [unpublished manuscript, ca. 1997].
42. Qureshi, Z. A., "A report of Lahore Zoo," *Zoos' Print*, 7, 16, 1992.
43. Bailey, T., *Education at Lahore Zoo*, Lahore Zoo, Lahore, 1998.
44. Farrell, H. P. and Ludlow, F., *The Municipal Gardens and Zoological Collection, Karachi*, Times Press, Bombay, 1913.
45. Anonymous, *International Zoo News*, 11, 105, 1964.
46. Anonymous, *International Zoo News*, 15, 4, 1968.
47. Qureshi, Z. A., "Note on Karachi, Pakistan," *International Zoo News*, 2, 6, 1955.
48. Van den Brink, J., "Three Asian zoos," *International Zoo News*, 10, 8, 1980.
49. Anonymous, *Souvenir Pakistan Aquarium Exhibition*, Pakistan Aquarium Society, Karachi, 1953.
50. de Silva, Anslem, personal communication, 1998.
51. Anon [An Officer, Late of the Ceylon Rifles], *Ceylon: A General Description of the Island*, W.H. Allen, London, 1876.
52. Cumming, C. F. G., *Two Happy Years in Ceylon*, Chatto & Windus, London, 1893.
53. Tennent, J. E., *Ceylon: An Account of the Island, Historical and Topographical with Notices of its Natural History, antiquities and Production*, Longman, Green, Longman, and Roberts, London, 1860.
54. Raven-Hart, R., Ed., *Travels in Ceylon, 1700–1800*, Associated Newspapers of Ceylon, Colombo, 1932.
55. Flower, S. S., "A list of zoological gardens of the world," *The Zoologist*, May, 1909.
56. Osman Hill, W. C., "My zoo in Colombo," *Zoo Life*, 1, 1946.
57. Hagenbeck, C., *Beasts and Men: Being Carl Hagenbeck's Experiences for Half a Century among Wild Animals*, Longmans, Green and Co., New York, 1909.
58. Anonymous, "Profile of Major Aubrey Weinman," *International Zoo News*, 4, 1957.
59. Weinman, A., "My personal ark," *International Zoo News*, 12, 1965 [appeared in serial form in *Asia Magazine* under the title, "My Animal Friends"].
60. Weinman, A., "Breeding results at the Dehiwala Zoo," *International Zoo News*, 4, 1957.
61. Weinman, A., "Report of the Zoological Gardens, Dehiwala-Ceylon," *International Zoo News*, 9, 1960.
62. Weinman, A., "Elephant techniques at Colombo Zoo," *International Zoo Yearbook*, 2, 8, 1961.
63. Hahn, E., *Zoos*, Secker and Warburg, London, 1968.
64. Schmidt, C., "The Zoological Gardens of Ceylon," *International Zoo News*, 17, 77, 1970.
65. Rich, M., "Elephant orphanage," *International Zoo News*, 151, 1978.
66. Anonymous, "Formation of an association of ASEAN zoos," in *Upacara Perasmian Perayan Ulangitahon*, Zoo Negara, Kuala Lumpur, 1988.
67. Ashari, D., "South East Asian Zoological Parks Association (SEAZA)," *CBSG News* [Newsletter of the Conservation Breeding Specialist Group], 6, 12, 1995.
68. de Alwis, W. L. E., "The next zoo revolution: keynote address for the seventh South East Asian Zoological Parks Association Meeting," *Zoos' Print*, 13, 27, 1998.
69. Sokhonn, U., "Establishment of Phnom Tamo Zoo in Kingdom of Cambodia," in *Proceedings of the Sixth Conference of Southeast Asian Zoological Parks Association*, Southeast Asian Zoological Parks Association, Melaka, 1996.
70. Ware, D., personal communication, 1998.
71. Anonymous, *Proceedings of the Sixth Conference of Southeast Asian Zoological Parks Association*, Southeast Asian Zoological Parks Association, Melaka, 1996.
72. Anonymous, "How it all started: a short history of the zoo," *National Zoological Park Zoo Guide*, Zoo Negara Malaysia, Kuala Lumpur, 1964.
73. Anonymous, "Note on Weinman retiring in 1962 and going to Malaysia for national zoo," *International Zoo News*, 11, 1964.

74. Sims, K., personal communication, 1998.
75. Lazarus, K., "The history of Malaysian zoos with particular reference to Taiping Zoo," in *Proceedings of the Ivy Zoo Symposium*, Budapest, 1996.
76. Abidin, Z. Z., Abdullah, T., Rahman, M. A., and Yusif, E., "Large mammals in peninsular Malaysia," in *The State of Nature Conservation in Malaysia*, Kiew, R., Ed., Malayan Nature Society, Kuala Lumpur, 1991.
77. Malik, Idris S., personal communication, 1998.
78. Corner, E. J. H., "The pig-tailed monkey as a plant collector," *Zoo Life*, 1, 89, 1946.
79. Harrisson, T., "Second to none: our first curator (and others)," *Sarawak Museum Journal*, 10, 17, 1961.
80. Kniveton, T., "Concerning the Sarawak Museum in Borneo," *The Bartlett Society Newsletter*, September, 1998.
81. Aung, S. H., "A brief note on dugongs at Rangoon Zoo," *International Zoo Yearbook*, 7, 221, 1967.
82. Loisel, G., *Histoire des ménageries de l'antiquité à nos jours*, Octave Doin et Fils and Henri Laurens, Paris, 1912 [translation in preparation provided by J. C. Edwards].
83. Anonymous, "Vietnam-Philippine working group minutes," *CBSG News* [Newsletter of the Conservation Breeding Specialist Group], 3, 18, 1992.
84. Lepiten, M. V., "The status of and conservation programs for five Philippine endangered wildlife species," in *Proceedings of the Fourth Conference of Southeast Asian Zoos*, Hong Kong Park, 1995.
85. Oliver, W. L. R., "Threatened endemic mammals of the Philippines: an integrated approach to the management of wild and captive populations," in *Creative Conservation: Interactive Management of Wild and Captive Animals*, Olney, P. J. S., Mace, G. M., and Feistner, A. T. C., Eds., Chapman and Hall, London, 1994.
86. Oliver, W. L. R. and Heaney, L. R., "Biodiversity and conservation in the Philippines," *International Zoo News*, 43, 329, 1996.
87. Toledo, S. U., de Leon, J. L., and Lim, M. S., "Management of Wildlife Rescue Centre and problems associated with maintaining confiscated wild animals in the Philippines: Proceedings of the First International Workshop on the Management of Wildlife Rescue Centers in South and Southeast Asia," *Zoos' Print*, 13, 19, 1998.
88. Friends of the Manila Zoo Foundation, *WHO Cares about the Manila Zoo?* Friends of the Manila Zoo Foundation, Manila, 1996.
89. Manila Zoo, *Manila Zoo Redevelopment Concept*, Manila, 1995.
90. Ridley, H. N., "The menagerie at the botanic garden," *Journal of the Straights Branch of the Royal Asiatic Society*, 46, 1906.
91. Anonymous, "Public aquarium," *The Singapore Naturalist*, 3, 1924.
92. Harrison, B., *A Study of the Planning and Design Principles Involved in Development of Mammal Exhibits in a Tropical Zoo*, masters thesis, National University of Singapore, Singapore, 1985.
93. McKay, Bruce, personal communication, 1998.
94. de Alwis, W. L. E., "From rain forest to zoological garden," in *Singapore Zoological Gardens Twentieth Anniversary Commemorative Programme*, Singapore Zoological Gardens, Singapore, 1993.
95. Anonymous, "Bird park Singapore," *International Zoo News*, 17, 1, 1970.
96. Harrison, B., "The past twenty years," in *Singapore Zoological Gardens Twentieth Anniversary Commemorative Programme*, Singapore Zoological Gardens, Singapore, 1993.
97. Strahan, R., "New zoo at Singapore," *International Zoo News*, 18, 103, 1971.
98. Walker, S., "How they started the Singapore Zoo," *Gnu's Letter*, 2, 12, 1984.
99. Walker, S., "Zoo interview with Bernard Harrison," *Zoos' Print*, 4, 4, 1989.
100. Walker, S., "'Breakfast with an orangutan' makes behavioural research possible," *Zoos' Print*, 4, 7, 1989.
101. Harrison, B., personal communication, 1997.
102. Greatz, M. and Corder, S., "The Night Safari four years after — a post occupancy review" [unpublished paper presented at the Paignton Zoological Garden's Symposium on Zoo Design, 1998].

103. Kumar, personal communication, 1997.
104. Flower, S. S., "Notes on a second collection of reptiles made in the Malay peninsula and Siam," *Proceedings of the General Meetings for Scientific Business of the Zoological Society of London*, 600, 1899.
105. Flower, S. S., "On the mammalia of Siam and the Malay peninsula," in *Proceedings of the General Meetings for Scientific Business of the Zoological Society of London*, 306, 1900.
106. Tilson, R. L., Garland, P. G., and Phipps, G., *Thai Zoo Master Plan for Conservation*, IUCN SSC Captive Breeding Specialist Group and Zoological Parks Organisation of Thailand, Apple Valley (Minnesota) and Bangkok, 1993.
107. Goldsberry, T., "Of Bangkok, bazaars and bantengs," *International Zoo News*, 23, 26, 1976.
108. Siah, Y., "Bangkok Reptile Grove," *International Zoo News*, 10, 53, 1963.
109. Chuenkamrai, S., *The Crocodile Farm, Thailand*, Allied Promotions Co., Ltd., Bangkok, 1991.
110. Satrulee, C., personal communication, 1998.
111. Tiyacharoensri, C., "Gibbon rehabilitation project of the Wild Animal Rescue Foundation of Thailand: Proceedings of the First International Workshop on the Management of Wildlife Rescue Centers in South and Southeast Asia," *Zoos' Print*, 13, 22, 1998.
112. Le Fevre, D., "Details respecting Cochin-China," *Journal of the Indian Archipelago and Eastern Asia*, 1, 56, 1847.
113. Richardson, D., personal communication, 1998.
114. Banks, C. B., "Hanoi and Saigon Zoos, Vietnam," *International Zoo News*, 41, 51, 1994.
115. Jones, M. L., "Tham Cao Vien, the Saigon Zoological Gardens," *International Zoo News*, 16, 13, 1969.
116. Son, V. D., "Saigon Zoo and Botanical Gardens, Ho Chi Minh City, Vietnam," *International Zoo News*, 40, 56, 1993.
117. Than, N. Q. and Lam, P. V., "The zoo-botanical park of Ho Chi Minh City and the duty of protecting and conserving natural resources of south Vietnam," in *Proceedings of the Third Conference of SEAZA and Meeting on the Establishment of South East Asian Zoological Parks Association*, 1990, 93.
118. Tilson, R., Banks, C., and Richardson, D., *Saigon Zoo Master Plan for Conservation*, IUCN SSC Conservation Breeding Specialist Group, Apple Valley, MN, 1997.
119. Van Dam, G. T., "The zoo at Saigon," *International Zoo News*, 10, 152, 1963.
120. Adler, H. J. and Peter, W. P., "An international effort to save Vietnam's primates," *International Zoo News*, 41, 40, 1994.
121. Adler, H. J. and Wirth, R., "Species conservation priorities in Vietnam and the potential role of zoos," in *Creative Conservation: Interactive Management of Wild and Captive Animals*, Olney, P. J. S., Mace, G. M., and Feistner, A. T. C., Eds., Chapman and Hall, London, 1994.
122. Foose, T. J., personal communication, 1998.
123. Huynh, D. H. and Dang, N. X., "Captive breeding of endangered species in Vietnam," *CBSG News* [Newsletter of the Conservation Breeding Specialist Group], 3, 27, 1992.
124. Nadler, T. and Rosenthal, S., "Wildlife rescue centers and problems of confiscated wild animals in Vietnam: Proceedings of the First International Workshop on the Management of Wildlife Rescue Centers in South and Southeast Asia," *Zoos' Print*, 13, 5, 1998.
125. Keay, J., *The Honourable Company: A History of the English East India Company*, Macmillan, New York, 1991.
126. Knight, C., *The English Cyclopaedia: Natural History*, Bradbury and Evans, London, 1854.
127. Ashari, D., personal communication, 1998.
128. Anonymous, *Indonesian Zoological Parks Association (IZPA): Organisation, Status, Activities*, Jakarta, 1990.
129. Bintituah, J., personal communication, 1998.
130. Abby, personal communication, 1998.
131. Hoffman, L., Ed., *Kebun Binatang Ragunan Zoo*, Jl. Harsono R. M. Pasar Minggu Jakarta, 1993.
132. Lorentz, P., personal communication, 1989 [interview with Mrs. Von Mengden].

133. Kreeger, T. J., Ed., "Zoo biology course given in Indonesia," *CBSG News* [Newsletter of the Conservation Breeding Specialist Group], 1, 18, 1990.
134. Anonymous, "Surabaya Zoo, Indonesia," *International Zoo News*, 37, 41, 1990.
135. Prawiradilaga, D. M., "Conservation of Javanese green peafowl (*Pavo muticus*) by the Indonesian zoos," in *Proceedings of the Third Conference of the South East Asian Zoological Parks Association*, 1990, 47.
136. Rachim, A., "The Sourabaja Zoo," *International Zoo News*, 16, 129, 1969.
137. Reksowardojo, D. H. and Wahyuni, E., "The influence of feed quality and serum gonadotrophin injection on the reproductive performance of Babyrusa babyrusa celebensis Deniger," in *Proceedings of the Sixth Conference of Southeast Asian Zoos, Southeast Asian Zoological Parks Association*, 1996.
138. Van Balen, B. and Gepak, V. H., "The captive breeding and conservation programme of the Bali starling (*Leucopsar rothschildi*)," in *Creative Conservation: Interactive Management of Wild and Captive Animals*, Olney, P. J. S., Mace, G. M., and Feistner, A. T. C., Eds., Chapman and Hall, London, 1994.
139. Osman, H. H., "The function and aims of Jogjakarta Zoo," *International Zoo News*, 13, 120, 1966.
140. Walker, S., "Commercial zoo with conservation potential," *Newstime*, August 19, 1990.
141. Anonymous, "Endangered orangutans cannot exist without their rainforest home," *AWI Quarterly*, 41, 10, 1992.
142. Anonymous, *Hong Kong Zoological and Botanical Gardens*, Urban Council of Hong Kong, Hong Kong, n. d.
143. Anonymous, *The Edward Youde Aviary, Hong Kong Park*, Urban Council of Hong Kong, Hong Kong, n. d.
144. Wedderburn, John, personal communication, 1998.
145. Leatherwood, S., "A conservation challenge for Asian zoos and aquariums," in *Proceedings of the Fourth Conference of Southeast Asian Zoos*, Hong Kong Park, 1995.
146. Ng, T. S. K., "Development of education programmes in Ocean Park, Hong Kong," in *Proceedings of the Fourth Conference of Southeast Asian Zoos*, Hong Kong Park, 1995.
147. Steward, M. D., Chan, C. K., Cheng, C. L., Keung, S. F., Mak, K. M., and Lai, T. M., "Captive reproduction of elasmobranches at Ocean Park," in *Proceedings of the Fourth Conference of Southeast Asian Zoos*, Hong Kong Park, 1995.
148. Wong, S., "Avian breeding programme at Ocean Park 1993-4," in *Proceedings of the Fourth Conference of Southeast Asian Zoos*, Hong Kong Park, 1995.
149. Ades, G. W. J., "Wildlife rescue and management of the Kadoorie Farm and Botanic Garden Rescue and Rehabilitation Centre in Hong Kong: First international workshop on the management of wildlife rescue centers in South and Southeast Asia," *Zoos' Print*, 13, 7, 1998.
150. Kreeger, T. J., Ed., "Status of the Amur leopard," *CBSG News* [Newsletter of the Conservation Breeding Specialist Group], 4, 15, 1993.
151. Chang-young, O. and Moon Hong-sik, "Seoul, South Korea," *International Zoo News*, 159, 40, 1979.
152. Chu, See-wu, "1995 SEAZA at Taipei Zoo," in *Proceedings of the Fourth Conference of Southeast Asian Zoos*, Hong Kong Park, 1995.
153. Agoramoorthy, G., "Rehabilitation and captive management of endangered species in a wildlife rescue centre in Taiwan," *International Zoo News*, 45, 71, 1995.
154. Agoramoorthy, G., Pei, K., and Lin, V., "Wildlife rescue and the management of Pingtung Rescue Centre in Taiwan ROC: Proceedings of the first international workshop on the management of wildlife rescue centres in South and Southeast Asia," *Zoos' Print*, 13, 1998.
155. Hsu, M. J. and Agoramoorthy, G., "Wildlife conservation in Taiwan," *Conservation Biology*, 11, 834, 1997.
156. Qinguo, Z., "CAZG regional report," *CBSG News* [Newsletter of the Conservation Breeding Specialist Group], 8, 10, 1997.
157. Brambell, M., "The Peking Zoo," in *Great Zoos of the World: Their Origins and Significance*, Lord S. Zuckerman, Ed., Westview Press, Boulder, CO, 1980.

158. Menghu, W., "Relationship between conservation of endangered wildlife in China and species conservation in Chinese zoos," in *Proceedings of the Fourth Conference of the Southeast Asian Zoos*, Hong Kong, 1995.
159. Qinguo, Z., "CAZG regional report," *CBSG News* [Newsletter of the Conservation Breeding Specialist Group], 6, 5, 1995.
160. Shuling, Z. and Qingguo, Z., "Chinese Association of Zoological Gardens," *CBSG News* [Newsletter of the Conservation Breeding Specialist Group], 7, 9, 1996.
161. Baoliang, Z., "The domestication of musk deer, Shan VI Province, PRC," *Zoo Zen*, 6, 68, 1991.

Additional Sources

1. Anonymous, "Wild Arabian Oryx Faces Extinction," Environmental News Network, April 22, 1999 [reported on CNN Internet news].
2. Beck, Benjamin B. and Wemmer, Christen M., Eds., *The Biology and Management of an Extinct Species: Père David Deer*, Noyes Publications, Park Ridge, NJ, 1983.
3. Bedford, (12th) Duke of, "Père David's deer: the history of the Woburn herd," *Proceedings of the Zoological Society of London*, 121, 327, 1951.
4. Jing, Z. and Yangwen, L., *The Giant Panda*, Science Press and Van Nostrand Reinhold, New York, 1980.
5. Price, Mark R. S. Stanley, *Animal Re-Introductions: The Arabian Oryx in Oman*, Cambridge University Press, Cambridge, U.K., 1989.
6. Reichenbach, H., "Der David vom Davidshirsch: Armand David (1826–1900) und die europäische Entdeckung chinesischer Tiere," *Bongo*, 28, 23, 1998 ["The David of Père David's deer: Armand David (1826–1900) and the European discovery of Chinese animals"].
7. Schaller, G. B., *The Last Panda*, University of Chicago Press, Chicago, 1993.
8. Thouless, C. R., Chonggui, L., and Loudon, A. I., "The Milu or Père David's deer (*Elaphurus davidianus*) reintroduction project at Da Feng," *International Zoo Yearbook*, 27, 223, 1988.

8 Zoological Gardens of India

Sally Walker

8.1 Introduction

Indian civilization spans thousands of years with a wealth of historic documentation on the close relationship enjoyed by people with animals, dating to prehistoric times when the rishis kept tame animals such as deer and doves in their *ashramas*.[1] There is such a luxury of fascinating material throughout Indian culture and history that writing about it is like being a kid with a penny in a candy store. What, of so many delicious and important items, to choose?

Mogul emperors and other ancients kept wild animals for work, war, and their own pleasure, both culinary and comic.[2] *Pinjarapoles* (safe houses) for animals date to the fifth century B.C., when religious communities set up shelters for all animals, including wild animals, and even insects.[3,4]

From very early times into the twentieth century, wild animals were kept for combat. Princely menageries, maintained at great cost for the pleasure — sometimes even for intellectual pleasure or scientific inquiry — of maharajas, nawabs, rajas, and zamindars predated public menageries and zoos. Kings and princes often allowed the public in their menageries, gardens, and parks and also staged animal fights for their guests. Animal trade is another activity that has occurred from ancient times and goes on even now, albeit in a much diminished and clandestine form.[1]

In India, a country with a history of extreme kindness to animals on the one hand and inordinate cruelty on the other, there were until recently well in excess of 300 collections of wild animals in captivity registered with the Central Zoo Authority — zoos, deer parks, traveling menageries, captive breeding centers, specialist collections, nature centers, and biological parks.[5] Their story is complex.

Public gardens and natural history cabinets (early forms of museums) with live animals preceded zoos and served as sites of zoological investigation. Afterward, at some of them, the animals were retained and a zoo established; others were sent elsewhere for another animal collection and perhaps a zoo. In addition to the nawabs and maharajas, ordinary people, such as expatriates in the military, visiting and resident civil servants, wives, amateur naturalists, local people, etc. — whoever wanted to do so — collected and kept wild animals (both live and preserved) as curiosities, pets, mascots, or objects of study. Sometimes these collections were truly impressive. Wild animals in captivity were hardly an oddity and listing all the collections would be impossible. As a result, one has to conclude that the line between what is a public and what a private zoo, or between a zoo and an animal collection in India was, until recently, very fine.

In Indian history many of the elements of a modern zoo, such as scientific investigation, open-moated enclosures, attention to the welfare of animals, and so on, were in place long before what is considered the "modern age" of zoos emerged.[6-8] A close look at the Indian zoo scenario may yield the view that India had more influence on Western zoos than otherwise, at least until recently. From the history of captive wild animals in India, and perhaps other colonies as well, one could conclude that much of the impetus and inspiration for systematic wild animal collections, which had their beginnings in Europe, came as a result of animal collections in the colonies. According to John C. Edwards, a historian of European collections and the London Zoo, "Raffles and other zoologists of the early nineteenth century drew much inspiration from collections in India and elsewhere in Asia, rather than the other way around."[9] This concept contradicts the theory that Mullan and Marvin propose, claiming the zoological garden was imposed on India.[10] It is true that the colonizers influenced India. However, it would take far more than a few hundred years of British colonialism to override (more than temporarily) the thousands of years of culture that is India.

8.2 Ancients and Invaders — Vedics, Guptas, Moguls, Europeans

Indian history, in terms of wildlife, zoos, and natural history, can be divided into five distinct eras, each with its own character; first was the Vedic period, when rishis kept animals such as birds and deer in their ashramas (beginning about 3000 B.C.); second was the Gupta period (467–320 B.C.), when Hindu emperors governed India, maintaining game parks and promulgating laws relative to animals, both domesticated and wild; third was the Mogul Empire, when Muslim rulers dominated India, maintaining fabulous *Shikar khanas* (hunting departments) and modernizing principles of animal husbandry and veterinary medicine; fourth was the European colonial period (beginning in the early 1500s), when those who were interested in the utilitarian and scientific aspect of animals colonized India; and finally, over the last 50 years, there has been a period of independence, when India has tried in every aspect of life, including zoo management, to forge its own path and be its own country.

8.2.1 The Vedic Period — Spiritual and Mystical Values

Not a great deal is known about the Vedic Indians, except through their hymns, but there are theories they were aborigines from Central Asia. Some of these were the Shariats who believed the souls of animals, in partnership with them, would help them to achieve realization. The art of these aborigines was based on animal forms that were transformed in their drawings to represent the divine.

Ruins left from the Indus Valley civilization indicate people of that time were familiar with, and had affection for, some wild animals. Elephant, tiger, rhinoceros, buffalo, wild boar, and deer figures were carved expressively on archaeological relicts of this period, demonstrating the positive relationship Indians of this period enjoyed with animals. Basham, in *The Wonder That Was India*, relates how the seals or amulets "usually depicted animals — bulls, buffalo, goat, tiger, elephant ... which were very detailed and delineated with powerful realism and evident affection."[2] This was around 2300 B.C. when the Harappa people made naturalistic models of animals, including terra-cotta cattle with movable heads, monkeys that would slide down a string, and whistles shaped like birds.

The rishis kept a variety of gentle animals in their ashramas. Gautama Buddha, it is written, gave his first discourse in a deer park. This is described in the *Dhammacakkappavatana-sutta*: "Thus I have heard at the time the Lord dwelt at Benaras at Isipatana in the Deer Park." Deer parks were commonly maintained in ashramas.[11]

The relationship of people with animals during the Vedic period, as indicated in their art, was an affectionate and respectful partnership. Aviaries and barnyards of the Brahmanas contained a variety of wild animals, which had Sanskrit names: *Syena* (hawk, falcon), *Hamsa* (swan), *Kalavinka* (sparrow), *Tittiri* (partridge), *Vaja* (falcon), *Kapinjala* (hazel cockatoo).[2] In fact, the world got its chickens from the red jungle fowl, which was domesticated in India during ancient times.

Ayurvedic texts list medicinal animals and plants that can be identified today. Later Sanskrit literature involving animals such as the *Panchatantra* are fables that describe human behavior through the personification of different animals.[12] Early zoological references to animals of the Vedic Hindus describe animals of other-worldly proportions and abilities, fit companions and protectors for gods.[2] Modern Indians worship the same gods today in animal forms, or with animal consorts; there is even a zoo in the temple city of Tirupati with animal gods as a theme. That is how it was for 3,000 years in India, and that only brings Indian history to the beginning of the Christian era.

8.2.2 The Gupta and Mogul Periods

Greeks writing about India lent great insight into the attitudes and practices toward animals. Megasthenes describes the pageantry of court festivals during the Gupta period when kings kept large collections of animals for parades. "Collections of animals of all kinds were great features ... in addition to panthers, lions and other animals, there were great wagons carrying whole trees to which a variety of birds bright in plumage and lovely in song were attached."[1]

Animals were a prestigious form of gift or offering to royalty. "The Indians do not think lightly of any animal, tame or wild." They apparently accepted all kinds, not rare ones only, but even cranes, geese, ducks, and pigeons. "Or one might bring wild ones, deer and antelopes or rhinoceroses." The same people, of course, who did "not think lightly of any animal," arranged animal fights, in which butting matches between rams, or between wild bulls, or between rhinoceroses, and fights between elephants, were common.[1]

It was during this period that Buddha and Mahavir attained enlightenment and became highly influential spiritual leaders. Before the time of these holy men, many people of India — even during Vedic times — were committed carnivores. Mahavir and Buddha achieved recognition as spiritual leaders about the same time. Although their doctrines were entirely different, they had in common the philosophy of *ahimsa*, nonviolence, which precluded the taking of animal life for food. Previous to this, even the sacred cow of India was not completely banned from the menu. Strictures against killing it only existed because the cow was valuable for its continuous gifts of milk, butter, curds, urine, and dung, all of which then had their practical uses, as they do today. Mahavir's followers are called Jainas or Jains, the most zealous of which — even today — wear a mask to avoid inadvertently consuming an insect.

The enlightened Emperor Chandragupta (reigning ca. 321–297 B.C.) was a warrior and hunter when he was a young man, but later embraced nonviolence and passed laws to that effect. In about 300 B.C. Chandragupta embraced Jainism, a religion based on *ahimsa*. He and his grandson Asoka (reigning ca. 273–232 B.C.), having given up war and sports hunting, together gave us the world's first edicts against hunting for sport and for protecting wildlife. Ashoka set up hospitals for animals called *pinjarapoles*, and passed conservation laws. They also made regulations restricting the slaughter of animals for food, especially on occasions of festivals and public shows.[1,3]

Pinjarapoles span a great swatch of history and are included because these institutions provide insight into the profound cultural bias toward protecting life in all its forms. They are another type of captive wild animal facility, welfare home, or hospital for animals. Pinjarapoles still exist in the same name and form today, although they rarely keep wild animals. Individuals and institutions aspiring to spiritual "credit" for doing good works or simply wanting to help animals have kept them. The word *pinjarapole* is derived from the Sanskrit *pinjara* and *pala*, which mean "cage" and "protector," respectively, and reflects its role as a refuge for any animal species where medical treatment is provided.

One of the first printed references to pinjarapoles after their creation during Ashoka's reign occurs in 1583 when a European traveler named Burnes described the *jivat khan* or insect room in a pinjarapole: "By far the most remarkable object in this singular establishment is a house on the left hand side of the entrance about 25 feet long with a boarded floor elevated about 8 feet. Between this and the ground is a depository where the deluded *Banias* (shopkeepers) throw in quantities of grain which give life to and feed a host of vermin, as dense as the sands of the seashore and consist of all the various insects found in the abodes of squalid misery."[4] A number of early travelers in India, from the fifteenth to twentieth centuries, give accounts of having visited pinjarapoles, including S. S. Flower during his inspection tour of Indian zoos in 1913.

In 1913, Colonel S. S. Flower, Superintendent of the Giza Zoo in Egypt, toured wild animal facilities in India to learn techniques he could use in his zoo at Giza. Flower visited what was considered the primary pinjarapole in India at the time and described it as consisting of several courtyards and sheds in which an "enormous number of animals are herded together in no particular order" and without much sanitation. He comments that it was a "very, very sad sight to see the large number of crippled or aged animals which were kept there alive instead of being put out of their misery [as they would be] in England." At first he thought it a cruelty, but later decided it was a case of "misdirected kindness." Flower was an astute observer and recorder, and his comments about different animal collections in India can be trusted.

According to Flower, the majority of mammals were domestic oxen and buffalo, but there were also sheep, goats, horses, dogs, and a tame white rat. The wild animals included macaques, a hedgehog (which Flower noted, he "did not see in any Indian zoo"), gazelles, nilgai, and domestic or semidomestic birds — pigeons, peafowl, poultry, and ducks. Flower also described an "insect room" which employed a man to go in and allow the parasites to feed on his blood. He was tied to the bed and not allowed out until the time of his contract was over.[6]

Lodrick writes as recently as the 1950s: "In Ahmedabad the jivat khana still survives. In a building especially constructed for the purpose three rooms are set aside for insects." Lodrick also distinguishes between the *goshala*, which means a "place for cows" and indicates its traditional function in providing shelter for old and disabled cattle, and the pinjarapole, which is for all animals. In India today, the words *goshala* and *pinjarapole* are often used interchangeably, or even together, to describe the same institution, causing confusion regarding their nature and function. However, the traditional distinctions are still valid and it is in the pinjarapole that one finds both wild and domestic creatures.[13] The very existence of pinjarapoles from ancient times to the present underscores the very long tradition of protecting animals for their own sake — as well as the tenacity of concepts — in India.

The coming of the Moguls, or Muslims, provided serious competition to the philosophy of *ahimsa* and the practice of animal protection for its own sake for several millennia. Moguls observed their own protection system in their forests, but it may have been as much to ensure sufficient game for their court than for any moral or aesthetic principle. Some of their practices mimicked sustainable utilization, to which we aspire so impossibly today.

Akbar (1542–1605) and Jahangir (1569–1627) maintained thousands of animals and were the first animal managers and naturalists. The *Akbar Nama* and *Ain-e-Akbari* — highly significant scientific works coming out of Muslim courts in the Persian or Urdu languages — deal with court and household care, including proper medical treatment, training, and nurture of the elephant, cheetah, and horse. *Ain-e-Akbari*, a three-volume record of practices of the times (much of it about animals), may be considered among the first modern wild animal husbandry and veterinary literature in the world.

Much is known about these times because the literary and artistic custom of the day was to chronicle the lives of rulers and kings for posterity, and because of the tradition of court painting. Abul Faz'l was a famous court chronicler of the time and much of what is known about animal management of those times comes from his books. The beautiful Mogul paintings that have survived also provide insight into the importance and integration of animals in daily life.[14]

During Akbar's period, which was the Golden Age of the Mogul Empire (roughly from 1555 to 1600), hunting was a privilege of kings for whom whole forests or vast parts of forests were kept for their use. These royal game preserves were managed so there would always be plenty of sport for the rulers, their families, and their entire court, all of whom participated. Mogul rulers relaxed from the tension of war and recovered from the tedium of court life in this manner.[1] Emperor Barbar, Akbar's grandfather, is said to have hunted rhinoceros, which still frequented the banks of the Indus River during the middle of the sixteenth century. Barbar records the occurrence of lions in Banaras District during his day, along with wild elephant, rhinoceros, and wild buffalo.[15]

Emperors also used hunting to assess the condition of their lands, the populace, and the army. Emperor Jahangir is said to have spent more than three months at a time hunting, and Akbar went on long hunting expeditions.[16] Hunting was a major pastime and experimentation seems to have been common. The animals kept for hunts formed a diverse menagerie. Not only did Akbar hunt with the usual elephants, cheetahs, and falcons, but also, as Abul Faz'l describes, he hunted with deer, dogs, caracals, and even with frogs and spiders.[17]

One of the most famous and fascinating methods of hunting involved the use of cheetahs for hunting ungulates. Using cheetahs to hunt did not originate in India, but it reached its zenith there with the emperor Akbar who kept as many as 1,000 of these beautiful animals at one time at his Shikar khana. Mutamid Khan, a Mogul chronicler, claims that Akbar collected at least 9,000 cheetahs during his half-century reign. There were 200 keepers and trainers in charge of the *khasa* (imperial cheetahs). Two chapters in *Ain-e-Akbari* are devoted to cheetahs, describing their trapping, hunting, classification, food, keepers' wages, and other information. The imperial cheetahs were named according to their talents and given rank, with costumes to match their status. Because of Akbar's fascination and passion for these animals and the hunt, they became faddish, with drawings of them appearing everywhere and on every kind of article: bookcovers, textiles, carpets, metalware, and other objects.[18]

Emperor Jahangir (reigning 1605–1627), son of Akbar, wrote so intelligently about different species of animals in his possession or sight that we can consider him an early naturalist. His memoirs include detailed descriptions of 33 species of animals, 11 species of plants, and incidental references to 36 animal species and 57 plants. This fascinating work, *Jahangir the Naturalist*, which Alvi and Rahman compiled and annotated, pools in one resource Jahangir's descriptions and paintings of animals, plants, and other natural phenomena of his period. The authors comment on Jahangir's information, saying it "includes distinctive characters, ecology, anatomical notes, habits, local names, weights and measurements." Jahangir maintained a "big aviary and a menagerie, carried observations, tests and experiments. Often he would order a specimen dissected in his presence, keep records for ascertaining long range phenomena and take down measurements and weights."[19]

Jahangir's interest in animals was well known, and as a result visitors to his court brought gifts of animals from several continents: a zebra from Africa, a loriquet from Malaysia, a mottled polecat from Afghanistan, Himalayan pheasants, Tibetan yak, and even a European domestic turkey. One of the great contributions of Jahangir and his artist is a painting of a dodo, which may have been one of two birds in the East India Company's Surat factory, or even a third one in Jahangir's possession. As far as his science goes, Jahangir did not fare badly for his time. "His descriptions are precise enough to make possible the correct scientific identification of the objects described. While the idea of scientific classification was unknown to Jahangir, an understanding of groups is evident in his use of a particular word for indicating comparable affinities of animals we classify under the same family and sometimes in the same genus."[19]

During the same period of 1656 to 1668, a Frenchman named François Bernier traveled in India and described the military pomp with which the emperor traveled. Two private camps (so one could go ahead and be set up to await the king) were transported using more than 60 elephants, 200 camels, 100 mules, and 100 human porters. Animals for hunting, show, and sport were taken along: dogs, leopards for catching antelopes, lions, and rhinoceroses brought merely for parade, large Bengal buffaloes to attack lion, and tamed antelope to fight in the presence of the king.[20]

8.2.3 The European Period — Utilitarian or Mechanistic Values

Christian tradition espouses a utilitarian or mechanistic philosophy that developed in European countries and is reflected in some of the first European zoos. The Zoological Society of London, for example, is considered by many to have established the first truly "scientific" zoo (1828). It was not so much scientific as utilitarian, however. The early founders were interested in domestication and farming. There was not, in those days, so much of a distinction between wild and domestic animals. Every animal was a potential food source, as food supplies were seen to be insufficient in a burgeoning human population, and therefore a problem.[9]

The prospectus of the Zoological Society of London reads: "For the general advancement of Zoological Science ... the object of which will be the collection of such living subjects of the Animal Kingdom as may be introduced and domesticated with advantage ... as it is impossible to attain all the objects of the Society on its first establishment, those of utility will engage its earliest attention, and the more scientific views will be attended to as the means of the Society admit."[21] Other zoos started in Europe during these early years bear this out. Motives of the early zoos in temperate countries included acclimatization of foreign species to restock hunting preserves, crossbreeding and hybridization of foreign animals to improve local breeds, zoological knowledge for purposes of comparison with already domesticated ones, and general instruction and entertainment of the public.

By contrast, menageries and animal collections of wealthy rulers in India, even the Moguls, were for their private aesthetic pleasure — a more mystical model, which was later extended, again solely for pleasure, to their public. According to Sri Fatesingh Rao Gaekwad, himself the Maharaja of Baroda and descendant of an owner of a large royal menagerie: "The Maharajas indulged themselves in possessing exotic creatures not just for power and prestige, but because they liked them." They liked looking at them and interacting with them in the hunt; they liked breeding them and beautifying their royal parks; they liked having them to give to important friends, visitors, and political colleagues; and the attitude of admiration for wild animals prevailed over utility.[22]

The genuine interest of the Indian princes in wild animals has not been highlighted historically. According to Rosita Forbes, the Raja of Dantheal loved looking at animals quite as much as shooting them, but tigers had to be destroyed because of the harm they did. He also wanted his guests to have a good hunt.[23] Although hunting was an established practice of Indian rulers, it was sometimes as much an excuse to inspect the kingdom as it was a hunt. Hunting does not seem to have been performed on the scale practiced after the coming of Europeans; it is not in the Indian tradition to kill more than what can be used immediately.

Early captive collections and wildlife (or game) parks in India were for pleasure, aesthetic and humanitarian, as well as for power and the hunt, but certainly not for profit. Europeans introduced utilitarian or mechanistic values, which led to the exploitation of natural resources for export and commerce, as well as for sport. Some Indian rulers assimilated and imitated these values. It was in this way, not in the "imposition of the zoological garden," that Anglo-Saxon values impacted India.

8.3 Nineteenth-Century Indian Zoos

Harrison distinguished between a menagerie and a zoological garden as "the original intention of the display. A menagerie displays its animals simply for public viewing, the enclosures being more for the convenience of the spectators than for the welfare of the animals. A zoological garden is more scientific in its approach to the display of animals, considering their physical and psychological comfort amongst other aspects." Harrison confirms that the first "real zoo" was London Zoo, but he quotes Fisher as stating the first Indian zoo was the Bombay Zoo. Information not available to the Western zoo community about India and other parts of the East can show that the definition of "zoo" is not so simple and that zoos have been around in India a great deal longer than most people think.[24,25]

The first phase of zoos in India, as they are known today, consisted of those the Maharajas started for the sake of their subjects and those the British societies, colonial states, or municipal governments started for instruction and entertainment. The second phase, the "post-independence stage," of Indian zoos can be divided into different types: those of the Indian central or state governments, those of the forest departments (with their plentiful space and money), and those without the benefit of either. The third phase, which is just beginning in India, is the "post–Zoo Act Amendments stage" in which zoos will not be allowed to start without permission of the national government, a process that will involve a careful review of the prospectus, prospects, and master plan of the zoo.[26]

8.3.1 Calcutta's Wild Animal Collections — Four Early Ones

Calcutta was the seat of government for most of the three centuries the British ruled India. Because Calcutta is a port, and also because of its proximity to the northeast of India, which was a door over land and sea to the rest of southern and southeastern Asia, there was a steady flow of animals into Calcutta. Perhaps these factors account for the phenomenon that four of India's most interesting menageries originated there. The four collections were the Barrackpore Menagerie, founded about 1801 and closed about 1876;[27] the Marble Palace Zoo, founded 1854 and continuing, but in decline;[28] the private park of Wazir Ali Khan, begun in 1856 and ending upon his death in 1889;[29] and Alipore Zoo, founded in 1875 and continuing under renovation.[30]

The very interesting fact about all these early zoos, which sprang up in the same city in the same century, is that compared with most other zoos and menageries of the nineteenth century, and even much later, they were all started with a serious purpose. That may indicate that most of the intellectual life of British India centered around Calcutta, just as did its political life, its commercial life and its social life. However, these zoos represent initiatives of both Indians and colonials, and even collaborative efforts. Barrackpore was purely British, Marble Palace and Wazir Ali Khan's Park were purely Indian, and Calcutta Zoo was a mixture.

Barrackpore Menagerie was the laboratory for a scientific endeavor of Richard Wellesley, then Governor General of India (1797–1805). Barrackpore township, on the banks of the Hooghly River, was a military cantonment which later became a rural holiday estate for British governors and their guests. It formed the "weekend seat" of government during that time.

Barrackpore menagerie, Calcutta, 1920. © Oriental and India Office Collections, The British Library.

Barrackpore residence, as a government institution, lost its significance only when the government moved to Delhi in the early part of the twentieth century.[31]

French historian Loisel remarked (incorrectly) of Barrackpore, that it was one among many menageries the British governors set up to imitate the pomp and power of Indian royals in their Calcutta residences.[9] This remark is interesting in the context of the theory that zoological gardens were imposed on India. However, the motivating factor governing the founding of the Barrackpore collection was purely, and overwhelmingly, scientific investigation.[32] Even Lord Curzon, in his *History of the British Empire in India*, mourned the ignorance of the real story behind the founding of Barrackpore menagerie, the memory of which had completely died out even while he was still in India. He writes: "It seems to have been generally believed that the collection of animals was due to the whim of some early Governor General assisted by the presents which Indian Chiefs are or were in the habit of making to the head of the Government. Such was not the way in which the Menagerie was founded, though it afterwards formed a principal source of its replenishment."[31]

The significance of the Barrackpore facility, both in zoo history and the history of natural history, has been completely overlooked. Aside from being a fascinating story involving some of the more eccentric and sensitive individuals of the British colonial period, it is significant as a transitional institution. As one learns more about Barrackpore, the more it seems likely to have influenced more famous, and longer-lived, institutions. Barrackpore as a zoological institution came about as a result of Wellesley's proposal for an advanced college at Fort William, with a natural history institute attached to it. The college was to have been located in Garden Reach and it was envisioned that the most learned scholars and pundits from all religions and languages of India would be invited to teach young *babus* from England who came to administer India. Wellesley thought, and rightly so, that a background in Indian culture would improve the standard of local government, but his superiors in the British East India Company did not agree and refused to approve the college as Wellesley proposed.[33]

Wellesley, characteristically assuming any idea of his would be supported, had already begun the natural history project. A number of wild animals had been collected for the institute and records exist of funds being spent on their upkeep between 1800 and 1804.[34] According to some science historians, Wellesley was sympathetic to scientific research because he felt financial benefits would accrue to the company from knowledge of Indian natural history. However, Wellesley, although practical and even mercenary, was genuinely interested in natural history. During his tenure in India he collected more than 2,660 paintings of animal and plant life. He also kept some wild animals himself, including five cheetahs, which various supplicants and friends presented to him. Wellesley was well known for his patronage of the arts and sciences in India.[17,33]

Wellesley provided his personal physician, Francis Buchanan (a doctor and natural scientist of repute, albeit in botany), with a mandate to begin an "official study" of the natural history of India and to create an establishment at Barrackpore "where the quadrupeds and birds which may be collected for Dr. Buchanan will be kept until they have been described and drawn."[35] This was a highly ambitious plan to illustrate, describe, and classify all the birds and quadrupeds of southern Asia. It was the first systematic study of Indian zoology, and well in advance of its time. Zoology was not yet the equal of botany.[33] This Natural History Project preceded the formation of the Zoological Club of the Linnean Society by 18 years, the Zoological Society of London by 22 years, and the first Indian journal of natural history by 36 years.[36,37]

In July 1805, Wellesley having been transferred, a new governor general took over and he wanted to close the menagerie as it was a drain on company finances. Buchanan and his assistant William Gibbons prevailed, however, and the research continued for a time.[34] Somehow, the menagerie at Barrackpore was maintained for 75 years, although on a "one-day-at-a-time" basis, even after the collapse of the Natural History Project. Its quality was entirely dependent upon the interest of the reigning governor general or his family. It persisted, in highly fluctuating stages of disrepair and refurbishment, until about 1878 when, after the start of the Alipore Zoological Gardens (the Calcutta Zoo), the entire collection was handed over to this new zoo.[27]

During the short period of Wellesley's advocacy and Buchanan's zoological and taxonomic skill, the Natural History Project made an extraordinary beginning toward indexing the natural history of India. Buchanan and his successor Gibbons described, and commissioned detailed paintings of, dozens of Indian mammals and birds. Unfortunately, the material was never published (but a project to annotate and publish these works is now in progress).[38,39]

Barrackpore Park was beautifully landscaped with lovely trees, shrubs, and flowers, which were a marvelous mixture of Indian tropical and English garden varieties. It was considered one of the "finest specimens of dressed and ornamented nature which taste has ever produced ... combining the grandeur of Asiatic proportions with the picturesqueness of European design." It is obvious from the glowing descriptions of visitors, such as the above, that Barrackpore was a refreshing treat after the hot and dusty city of Calcutta. It is possible to track the status of Barrackpore Menagerie's collection through the letters and diaries of visitors over the decades.[31]

By 1810, there was a pelican, sarus crane, flamingo, ostrich, cassowary, Java pigeon, two tigers, and two bears in the menagerie. The Indian species had all been described by Buchanan and Gibbons, so they were no longer needed for the Indian Natural History Project. The presence of the ostrich, cassowary, tigers, and bears indicates the menagerie was being kept for people as well as the project. A black leopard had been added by 1814. Between 1817 and 1819 a new aviary was constructed and by 1822 a new menagerie was built with the animals confined in structures which were "partly of Gothic and partly of classical style" — very typical of the times.[31] A watercolor of the menagerie dated 1820 shows a bear, lion, and pelican.[40]

In 1824, the adventurous Fanny Parks visited Barrackpore and commented on the "remarkably fine tigers and cheetahs."[41] In 1828, Reginald Heber, Bishop of Calcutta, paid a visit to Lord and Lady Amherst and enjoyed his first elephant ride. He described "several uncommon animals": the gayal, a very noble creature of the ox or buffalo kind; an African wild ass; and two lynxes "led each in a chain by his keeper, one of them in body clothes, like an English greyhound, both perfectly tame and extremely beautiful." Other animals were two or three tigers and leopards, two kinds of bears (a sun bear and probably three sloth bears), a porcupine, a kangaroo, monkeys, mouse-deer, birds, and a tame hyena.[42]

In 1829, the French naturalist Jaquemont visited Barrackpore and wrote about the "magnificent park." He also commented that Lord Bentinck, who was governor general at the time, was neglecting this "princely menagerie," as he had just given away to some rajas a royal tiger, some African lions, and several cheetahs trained for hunting. Remaining were "bears of two species, a caracal, a wild ass, a gibbon, some musk deer, an ostrich, a cassowary and some fine large waterbirds. About ten elephants wandered freely in the park and a tame one-horned rhinoceros was attached with a long chain to a tree beside a pond where it stayed the whole time." Incidentally, Jaquemont was assured during his visit that mountain peasants on the other side of the Ganges used rhinoceroses for agricultural purposes.[9] The new governor, Lord Bentinck, however, "gave away birds and beasts without remorse and permitted the buildings to decay" until the collection in his tenure sunk to "two miserable bears." Apparently the menagerie had a prestigious, as well as recreational status, as Lord Lawrence wrote that "Anglo-Indian society thought this the end of Empire."[43]

As aggressively as Bentinck let the menagerie go to seed, Lord Auckland and his two intrepid sisters, Fanny and Emily, swiftly rectified the situation. They took "the liveliest interest in this miniature zoo which henceforth flourished greatly." Emily, in her numerous letters home recounted having "two rhinos, a tiger, two black bears, two cheetahs, a white monkey, three sloths, a baboon, a giraffe, and some birds." Lord Auckland himself started a new aviary for some golden pheasants from China.[44] By 1841, naturalists of the area were again using Barrackpore to some extent for the purpose it was intended. The minutes of the Calcutta Natural History Society for July of that year recorded the receipt of animals from Captain Bogle and Lieutenant Phayre, the latter a familiar name to contemporary naturalists. These animals were shipped in two consignments, the first consisting of a rhesus monkey and a hoolock, which died on the way; several specimens of *Lemur tardigradus*; several *Rhizomys* or bamboo rats, and monitors. The second dispatch consisted of a young hoolock, three lemurs (part of a consignment imported from Africa), a martin, two young swine, a *Rhizomys*, a porcupine, a small cat, and two spotted deer.[45] It was also recorded that the latter animals were in a "sickly state" on their arrival in Calcutta and could not bear the additional travel to Barrackpore, so they were retained. The hoolock gibbon, wild cat, and lemurs died, while those that recovered were sent to Barrackpore. Members of the society reached agreement with Captain Bogle that cold weather was best for transporting animals. Bogle requested that the cages be returned to Arrakan so he could, during the next cold weather, "renew the collection."[46] Such comments lead to the conclusion that the menagerie at Barrackpore was very much part of the scientific endeavor of the Calcutta Society, which had developed considerably in the past decades.

In 1847, an anonymous writer listed tigers, leopards, bears, ostriches, monkeys, rhino, giraffe, waterbirds, white egrets, flamingos, China pigeons, and Monal pheasants in the zoo. In 1848, the private letters of Lord Dalhousie reported that Barrackpore had an aviary and a menagerie. A marvelous little story of a child and an ostrich in 1864 includes information that Barrackpore had an official "gamekeeper" and refers to a "Government Aviary."[43] However, by 1866, an article in the *Calcutta Review* entitled, "The Indian Museum and the Asiatic Society of Bengal," advocates starting a zoological garden in Calcutta as the collection in Barrackpore "had deteriorated." The article further comments that the collection at Madras was "well known to be comparatively worthless, while that at Barrackpore is not deserving of the name."[47] There were no more saviors for Barrackpore. During 1874 to 1878 Lord Lytton handed over the remaining animals to the Calcutta Zoo, which had been opened at Alipore in 1875. Some years later Lytton's wife procured a few spotted deer for the park, but, as cages had been removed, the zoo did not accumulate animals again.[27]

An Indian children's book has designated the menagerie at Barrackpore the "first Zoological Garden in Calcutta." However, the author says (incorrectly) the zoo was set up only for British people and "the general public were not allowed to visit this Garden."[48] There is, to the contrary, an interesting watercolor by Charles D'Oyly dated 1820 clearly showing Indians, as well as Europeans, in the garden looking at the animals.[40] In 1824, when Fanny Parks visited Barrackpore, her *ayah* asked to accompany her to the zoo, which indicates that Indians were comfortable going to the menagerie and that it was not off-limits to them.[41] The reference from the *Calcutta Journal of Natural History* in 1841 gives every indication that Barrackpore was viewed seriously as a collection and that it was taken for granted the collection would continue to be renewed. Finally, a small book by Pal entitled, *The Park at Barrackpore*, confirms that Barrackpore evolved into a public zoo. Pal was superintendent of the park in the 1930s and comments, "The zoological collection was a great attraction to all classes of visitors to the park and specially to the juvenile section whom it supplied with a fund of merriment. Even now we meet old people who sigh for the days of childhood they enjoyed in the company of the denizens of the forest in the Park."[27] Barrackpore seems to have been, without a doubt, a "real zoo" as much as London Zoo was a "real zoo," and deserves its rightful place in a list of zoos of the world.

The Marble Palace Zoo, which Rajah Rajendro Mullick Bahadur founded at Chorebagan, Calcutta in 1854 still stands today in the depths of Calcutta on the bustling Muktaram Babu Street (although it is in the process of closing, having been refused Central Zoo Authority recognition). Rajah Mullick opened it as a public service for the education and entertainment of people, much in the same spirit that he fed beggars and indigents, but also because of his keen interest in animals. Mullick was a student of natural history. The menagerie in his house contained birds and mammals collected from different parts of the world, as well as India. His biographer, Dinabandhu Chatterjee, opines that his must have been an important animal collection, even before opening to the public.[28]

Marble Palace Zoo, Calcutta. The zoo's (and India's) oldest existing cage, dating from the 1850s. © Photograph by Sally Walker.

Marble Palace Zoo, Calcutta. The zoo's traveling cages (although very old, their age is not known). © Photograph by Sally Walker.

Mullick gave hundreds of animals to other institutions. He donated numerous animals to the Zoological Society of London, and ultimately the society made him a corresponding member in 1863, as well as according him other honors.[49–51] He sent many valuable animals to several European zoos and received honors, as well as animals, in return. Despite the fact that these activities were truly appreciated, it was unusual for an Indian to be so honored in those days.[28] Mullick was one of several persons who pushed through, and funded, the long-delayed Calcutta Zoo, which was discussed for years before it could finally come about. Mullick gave the zoo valuable animals and the zoo named its first house after him, "Mullick's House."[43,52] Marble Palace Zoo was not merely another plaything, but was nurtured by persons who admired and cared deeply about the dramatic, beautiful, and interesting creatures from around the globe. Purnendro's (Mullick's great-grandson) description of the enclosure for hispid hare is evidence that much thought went into the keeping and display of the collection's animals. It was known hares required privacy, so they were kept in a very long cage where the animals could retreat to the back, away from the public. In addition, the exhibit included a bamboo partition and provided sand for burrowing. These enclosures can be seen today, although most of them are empty and the zoo itself has sustained a miserable state of decline, prompting its closure.[53]

A significant, although short-lived, animal collection of Calcutta which deserves a place in zoo history is the enormous menagerie of Wajid Ali Shah. It has been described in a book entitled *Lucknow: The Last Phase of an Oriental Culture* by Abdul Halim Sharar, relative of the exiled King of Lucknow.[30] Wajid Ali Shah kept an enormous establishment of fighting animals in Lucknow, but after losing a war, he was exiled to Calcutta. His nephew has written about his many interests, and King Wajid Ali Shah had myriad hobbies, one of which was animals. Even taking into consideration the extreme exaggeration that Mogul biographers employed in flattering their heroes, the story of this animal collection is significant.

According to Abdul Sharar, Wajid Ali Shah maintained a large estate called Nur Mahal, which was enclosed by an iron fence. Hundreds of spotted deer, buck, and other animals ran free in this compound, where passers-by could observe them. An area in the middle of the estate was full of partridges, ostriches, turkeys, sarus cranes, geese, herons, demoiselle cranes, ducks, peacocks, flamingos and hundreds of other birds, as well as tortoises. Sharar claims so much attention was given to sanitation that droppings and shed feathers could not be seen at any time. Surrounding the pool in this area were caged tigers and a meadow with cages full of a great variety of monkeys from every part of the world. Of the monkeys, Sharar comments they "performed comic antics which people could not help lingering to watch," indicating that the menagerie was not solely for the pleasure of the king.

In addition to several fish pools, there was a large tank with four steep and slippery sides containing thousands of large snakes. There was an artificial hill in the center of this tank where snakes crawled to the top and down to the bottom to catch frogs supplied for them. There was also a moat around the hillock where snakes would swim and catch the frogs. Sharar comments that "it was quite safe for people to stand by the tank and watch what was happening." So again, there is an indication the menagerie had visitors of some sort. Shah's biographer was rather enamored with the snakes and opined that it was "unlikely that arrangements for keeping snakes in captivity had ever been made anywhere before and Wajid Ali Shah was the first person to think of it."[30] When Nawab Wajid Ali Shah died, he left behind 51 mammals, 205 birds, and 100 reptiles, most of which were given shelter in the Calcutta Zoo under the care of Sanyal. Later, the government sanctioned funds for purchasing only selected animals from this collection.[54]

Although Barrackpore Menagerie continued, there were people who, as early as 1841, felt Calcutta should have a "proper" zoo. The July 1841 issue of the *Calcutta Journal of Natural History* relates Mr. Raleigh's proposal for the formation of a Zoological Garden. Although the Jardin des Plantes had been formed less than five decades before in France and the Zoological Society of London about 15 years before in London, these institutions had such an impact that in a relatively short time it was deemed "necessary" that a city have a zoo. In 1842, again in the *Calcutta Journal of Natural History*, Dr. McClellan advocated the establishment of a zoo in the city, which still did not elicit any response.[55]

In 1866, the establishment of a good zoo in Calcutta was still merely a subject of discussion, but the following year Sir Joseph Fayrer, President of the Asiatic Society of Bengal, proposed a scheme to start the zoo, and even went so far as to get funds subscribed by Indian royals. The Maharaja of Burdwan, the Maharaja of Vizianagaram, and the Rajah Rajendro Mullick (who had already opened his zoo to the public 14 years previously) pledged a total of 39,000 rupees — quite a bit of money in those days. An appropriate site could not be located, so the zoo did not materialize during that decade, but the idea, as John Anderson expressed in his Calcutta zoo guide, was "fully recognized and never lost sight of, but only held in abeyance."[43]

In 1873, a subcommittee formed for the purpose of seriously discussing the zoo. Carl Louis Schwendler headed the subcommittee, which was composed of members of the Asiatic Society and the Bengal Agri-Horticultural Society. But again, the idea had to be deferred for want of a site. Finally, in 1875 after more than three decades of discussion, debate, and defeat, the government of Bengal in conjunction with the public established a zoological garden for Calcutta. The government supplied land and secured its maintenance and management, while the public contributed funds for the grounds and buildings. Schwendler turned over his private collection of animals to the new institution. Anderson gives credit primarily to Indians, the "native section of the community," for their contributions plus their subscriptions of 1867 which were "at once paid and even increased."[43]

It is interesting to note that although the primary source of income for starting the zoo was from Indians, there were no Indians on the first managing committee, which the lieutenant governor appointed. This did not deter Indians from taking part in the growth and development of the zoo, however, as the lists of animal and money donations indicate. When the Reptile House was constructed, money was collected from 22 persons, all of whom were Indian, most of them royals.[52,56–58]

RAM BRAHMA SANYAL — THE FIRST ZOO BIOLOGIST

Ram Brahma Sanyal. © Photograph by Sally Walker.

Calcutta Zoo cannot be discussed without paying tribute to Ram Brahma Sanyal, a man who distinguished the Calcutta Zoo, as well as the Indian nation, by writing the first zoo management book in the world. Sanyal was the first Superintendent of Calcutta Zoo and author of *A Handbook of the Management of Wild Animal in Captivity in Lower Bengal*.* The British journal *Nature*** acknowledged the book as the "first of its genre" at the time of its publication in 1892, and it is so listed in the article, "Libraries and Archives in American Zoological Parks and Aquariums."****

According to Sanyal's biographer, Dilip Kumar Mitra, the history of Calcutta Zoo centers around Sanyal, whose tenure at the zoo extended from 1876 to his death in 1908. A scientific director was needed to guide the activities of this new zoo. After some years of searching they could not find a suitable person, either from India or from England. One of the members of the zoo's managing committee, George King, Professor of Botany in the Medical College (who was responsible for landscaping the zoo), engaged as an ordinary worker a young Bengali medical student, unable to continue his studies due to problems with his eyesight — Sanyal.

Although it made every effort, the committee could not find a European director with a scientific background who would take the position, and ultimately the committee had to

John Anderson, who was a member of the committee and a founder of the zoo, merits recognition for the detailed *Guide to the Calcutta Zoo* (1883), which is one of the earliest zoo guides in the world. In its extensive coverage (181 pages), Anderson takes visitors on a step-by-step tour and describes practically every animal minutely, including, for example, the nerves in the bill of ducks and geese, and the toes of true ducks as opposed to ducks that frequent the sea. Four species of rhino are described. Rusty-spotted and marbled cats, as well as caracals, are among the 17 species of felids enumerated. No zoo today offers its visitors such detailed and scientific material.[43]

Calcutta Zoo is still very much a going and growing concern today, although it is not without its problems. What was a quiet tropical suburb in Alipore during the days of the inception of the zoo is today an intensely crowded part of the metropolis of Calcutta. Visitation is reflected in the growth of the city's population: 20,000 visitors a day is not unusual for the 45-acre zoo, and on special days the count reaches 100,000. This takes an enormous toll on the animals, as well as the staff and facilities.[59] For a long time the Calcutta Zoological Society, which now functions under the Department of Environment of the State of West Bengal, has planned to depend solely on an Indian, R. B. Sanyal, "whom they trained up according to their need." Sanyal worked hard and rose swiftly. In a few months he was promoted and given more supervisory responsibilities, including the nutrition and medical care of the growing number of animals. He exhibited such a facility for management and documentation that he was named acting superintendent, and then superintendent by default, as no more suitable person could be found.[†]

If we consider Heini Hediger the father of zoo biology, then Sanyal might well be considered the great-grandfather of that science. He contributed significantly to the development of a professional identity and was probably the first to practice zoo management as a holistic, interdisciplinary scientific endeavor. He was certainly the first to write about it as such.[††]

Sanyal began his scientific observations into the biology and behavior of captive wild animals in 1877, recording his experiences in the *Daily Register* of the Zoological Gardens. It was from this detailed register that he took material for the *Handbook*. He was the first person in the world to write specifically about zoo enclosure designs, even including dimensions and furniture (or props) required for the animal's general welfare and biological requirements. According to Mitra, "Sanyal never believed that the objective of the zoo could ever be simply to put animals on exhibit; showing naturalistic behavior was the real goal."

Under the heading of "Treatment in Health," suggestions based upon experiences gained in the Gardens are given on the best way of housing, breeding, feeding, and transporting animals in captivity. Regarding nutrition, Sanyal gives details of food, including quantity and occasional changes in the animal's food habits. He has also pointed out the food habits of some exotic animals under Indian conditions as well, distinguishing between their natural diet and what might be fed in captivity that is readily available in India.

Breeding animals was dealt with in detail, including the reasons for failures in breeding. In considering this, it is important to remember that these were the days of menageries, when most people managing captive animal collections were not particularly concerned about breeding. Sanyal was the first to deal with breeding as a serious subject, and seemingly as an objective of the institution. Sanyal was the first to initiate a collection policy with the help of Calcutta Zoo committee members to control the "haphazard manner in which the animals had been acquired." In this way he ensured that the zoo's emphasis switched from

continued

shift the zoo to a larger area so open, spacious enclosures could be provided for the animals, but, as before, difficulties finding an available and appropriate site have delayed the project.[59] Now there is a Forest Department of West Bengal plan to establish a grand modern zoo. One can only hope the old zoo will be preserved and will never lose its historic importance.[60]

8.3.2 Old Madras State Zoos

For many years, People's Park in Madras was thought to be the first zoo in India. Historical evidence indicates that this is not so, but the origin of the People's Park has many unique features and is a story worth telling. A talented and sensitive English surgeon, Edward Balfour, founded the Madras Government Central Museum in 1851, which included a natural history section. In the natural history section, Balfour attempted an "experiment," keeping a live tiger cub and a live leopard for visitors to see. While visitors observed the animals, Balfour observed the visitors, maintaining records of their numbers and even recording comments about their behavior. The addition of live animals coincided so precisely with a dramatic rise in visitation to the museum, Balfour concluded that a "living collection" was popular with the public. Before

RAM BRAHMA SANYAL

showing off the "maximum number of first importations to sustaining breeding groups." Sanyal organized several successful breeding programs for mongoose lemur, short-spined porcupine, agouti, tiger, and leopard.††

Mitra writes eloquently of the tremendous versatility of Sanyal in meeting challenges the managing committee gave him. "Due to his medical background he was a good pathologist and sanitarian — and gardener as well. He played a very important part in establishing the Garden and in maintaining its reputation, both with the local public and the scientific institutions. His personal influence among Indians gained the zoo numerous donations and benefactions."††

Sanyal's efforts did not go unrecognized, in his own lifetime at least. Soon after the publication of the *Handbook*, immense appreciation came from Sir Charles Alfred Elliot, Lieutenant-Governor of Bengal (1890–1895). The Zoological Society of London made Sanyal a Corresponding Member (CMZS) in 1893 as a result of the *Handbook*, and Sanyal later visited London and other European countries to see their zoological gardens. The Committee of Management of the Calcutta Zoo bestowed a special honor when they requested the government of India to include Sanyal as a member of the committee in 1902. He was invited to Bombay and Rangoon where his help in planning their zoos was much appreciated.††

The *Handbook* continued to be held in much regard. In 1907, Gustave Loisel, the French historian, was impressed by the superintendent, "a Hindu, Ram Bramha Sanyal who

Calcutta Zoo's bear cage, dating from the late nineteenth century to the early twentieth century. © Photograph by Sally Walker.

making this deduction, however, he removed live exhibits and observed that attendance fell. He replaced them and watched attendance increase again. Only then did he record his conclusion in the annual report of the museum, which was probably the first recorded instance of a zoo visitors study. On the basis of these observations, Balfour started a small zoo in a corner published a Handbook which is still admired today by Zoo Directors."††† Very little appeared about Sanyal from then until Lee Crandall, Curator at the Bronx Zoo, cited the *Handbook* 17 times in his book of a similar title published in 1964.⁋ Copies of the *Handbook* were almost unobtainable for many years, although a few of the older U.S. and European Zoos had copies in their libraries. Calcutta Zoo itself had only one copy. In 1986, an Indian zoo magazine, *Zoos' Print*, began featuring excerpts from Sanyal's book, and in 1995 the Central Zoo Authority of India reprinted it.⁋⁋ As a result, modern zoo personnel are now beginning to appreciate the book. Sanyal was unquestionably the first truly modern zoo man — and the last for many decades.

* Bengal Secretariat Press, Calcutta, 1892.
** Anonymous, "Ram Bramha Sanyal on the management of animals in captivity," *Nature*, 46, 314, 1892.
*** Kisling, V. N., Jr., in *Encyclopedia of Library and Information Science*, Volume 57 (Supplement 20), Marcel Dekker, New York, 1996.
† Anonymous, *Zoological Gardens, Calcutta*, Calcutta Zoological Gardens, Calcutta, 1900.
†† Mitra, D. K., "Ram Bramha Sanyal and the establishment of the Calcutta Zoological Gardens," in *New Worlds, New Animals*, Hoage, R. J. and Deiss, W. A., Eds., Johns Hopkins University Press, Baltimore, 1996.
††† Loisel, G., *Histoire des ménageries de l'antiquité à nos jours*, O. Doins et Fils, Paris, 1912 [*History of Menageries from Antiquity to the Present*, translation by Edwards, J. C.].
⁋ Crandall, Lee, *The Management of Wild Mammals in Captivity*, University of Chicago Press, Chicago, 1964.
⁋⁋ Sanyal, R. B., *A Handbook of the Management of Wild Animals in Captivity in Lower Bengal*, Natraj Publishers, Dehra Dun, 1995 [reprint of 1892 edition].

of the museum compound, prevailing upon the Nawab of the Carnatic (for whom he was political agent) to "donate" his entire animal collection to the museum. This small animal collection was the nucleus of the People's Park, which was founded in 1855.[61]

Colonel S. S. Flower visited the Madras Zoo, as well as the Government Central Museum. He remarked on the useful *Guide to People's Park, Madras*, published in 1876. The Municipal Zoological Garden occupied one end of the park, which was 116 acres and was open free to the public. Animals of note included a large, dark, male orangutan; a female Asiatic two-horned rhinoceros (*Dicerorhinus sumatrensis*), which had lived in the garden about 14 years at the time of Flower's visit; a male Malayan tapir; and two great black-headed gulls, which Flower said were rarely seen in menageries. Flower was far more effusive about the Government Central Museum, which he described as a "great institution." It is interesting to note that the Madras museum still kept a collection of live animals, doubtless hailing back to Balfour's success, and at the time of Flower's visit had in their hall gerbils, owls, pigeons, jungle fowl, tortoises, lizards, 10 species of snakes, fish, and scorpions.

Flower also visited the Madras Aquarium, which exhibited 54 species of fish. Flower included a table of aquariums he had visited during the previous six years and the number of forms of fish displayed to indicate how rare it was for an aquarium to display so many forms. Only the aquariums at Frankfurt and Amsterdam had more species (out of the 13 institutions he visited). Flower describes in detail the layout and filtration system of the Madras Aquarium, with which he was quite taken. The aquarium was the first in India, opened in October 1909 with the objectives of providing an attractive display of living fish and furnishing material for the scientific study of fish and other marine organisms of the Tamil Nadu coast. The Madras Government published *A Guide to the Marine Aquarium* in 1912, which explains in detail how the aquarium works, and from which Flower quotes extensively. The Madras Aquarium seems

The older Madras Zoo lion building. © Photograph by Sally Walker.

not to have been renovated or improved since 1909 and the same guide, dated 1912, is still being given to visitors (or at least this was so as of the early 1980s).[6]

Madras Zoo (the People's Park) was surrounded by a growing city and well before a century was over the zoo was crowded, outdated, and polluted. This zoo was closed in the late 1970s or early 1980s and the animals shifted to a new site outside Madras (now Chennai) where the Forest Department of Tamil Nadu developed one of the biggest and best zoos, or "biological parks," in the country. Arignar Anna Zoological Park in Chennai covers more than 1,100 acres and is designed on the principles of the biological park, an Indian innovation in zoo design.[62] Degraded land, which contained only eucalyptus plantations, was allowed to regenerate and then planted profusely. In just over 15 years the zoo has become lush and green. Giant walk-

The new Madras Zoo (Arignar Anna Zoological Park) lion enclosure. © Photograph by Sally Walker.

through thematic aviaries with bird species from different bird sanctuaries in the state of Tamil Nadu are a unique feature of this zoo, but certainly not the only one. The zoo has outstanding educational facilities, a good research reputation, and is planning the country's first invertebrate house.

8.3.3 Kerala Trivandrum (Old Travancore State) and Trichur Zoos

The Maharaja of Travancore, Marthanda Varma, had a great interest in animals and kept a collection. In 1855, a painting of the Durbar shows a live giraffe on parade with the state elephants. Colonel Flower relates records of two giraffes in the possession of the Maharaja of Tranvancore at that time. Trivandrum Museum and Public Gardens, part of which was the menagerie, was founded in 1857 or 1859 (according to different records). The institution was given great importance and a curator from Kew Gardens was invited to plan the layout.[63] Flower was very complimentary of the layout and general appearance, which has the advantage of an undulating landscape with verdant tropical vegetation, even today.

Flower commented on the longevity of the animals at that time and two species deserve special mention. One is a Malabar civet (*Viverra civettina*), received in 1887 and alive at the time of Flower's visit more than 25 years later. This species is particularly interesting today because it is possibly extinct in the wild, with no individuals in captivity in any zoo in the world. Another interesting observation, simply by virtue of the fact that no zoos in India keep, breed, or exhibit amphibians, is that in 1912 Trivandrum Zoo kept an Indian tigrine frog (*Rana tigrina*). It was caught in 1904 and did not eat for nine months. When it showed signs of emaciation it was force-fed on fish and continued to survive on this diet until 1912 when Flower saw and marveled at it. Flower also mentions a male orangutan, a Nilgiri langur, and some tigers with good longevity records. He comments favorably on the reptile house, which had cages constructed so the sun never fell on the reptiles.[6]

P. R. Chandran, a long-time director of Trivandrum Zoo (since retired), credits Allen Brown, Fellow of the Royal Society in Astronomy, with insisting that Trivandrum have a zoo, and it was he who put the idea to the Maharaja of Travancore. In early colonial days, he says, the British were chosen as directors and, when they left, very senior civil servants were selected to take their places. Chandran himself took over from P. Kesavan Nair (director for 25 years) who held a Ph.D. in environmental studies from the University of London and was a student of the famous J. B. Haldane. "Imagine, a Ph.D. in environment in those days ... and he was my zoo guru."[64]

In 1937, the zoo and garden were separated from the museum, and in the annual report that year there is a curious statement about the zoo and garden originally coming into being just for the purpose of serving as an inducement for people to visit the museum and benefit from a study of its valuable exhibits. Apparently, in those days dead specimens were considered more valuable than the real thing. Another interesting comment in the same report is one about visitors from far and near admiring the Mendelian breeding experiments in the carnivore section between *Panthera leo* and *P. tigris*. Although these experiments were tried, there was no result, as there is no record of a tigon, or any such animal, born at Trivandrum Zoo.[65]

Trichur Zoo was another early zoo, founded in 1885, and under the same administration as Trivandrum Zoo. Today, both zoos are congested, in the middle of the city, and need to move to outlying areas with more space. Citizens of Trivandrum objected when a serious effort emerged to move that zoo, but the Trichur community is ready. The present director of the Trichur Zoo is leading a movement to shift the institution, which has fallen into a sad state of disrepair. Both zoos are fraught with problems, many of which stem from the fact that they are under the Department of Zoos and Culture. The director of the zoo is, as often as not, a historian rather than a zoologist, and, in addition, is transferred quite often.

8.3.4 Sakkarbaug Zoo and the Gir Lions

The Nawab of Junagadh, Mahebatkhanji Babi II, founded the Sakkarbaug Zoo in Junagadh as a way station to house injured and infirm lions from nearby Gir Forest. It may be the only zoo founded to service a protected area, which it still does. The original construction of the zoo, computed in 1863, was only three to four cages made of rough stones for the Gir lions exclusively. Later on, other animals such as tigers, panther, deer, antelope, and birds were added. The Garden Department of Bombay managed the zoo during the nawab's time. After independence, the Education Department managed it, after which it was handed over to the Agriculture Department (in 1956), and then to the Forest Department. From the beginning the most important feature of this zoo has been the breeding and care of Asiatic lions. Badly wounded and injured lions from Gir have been, and still are, brought to the zoo, treated successfully, and released back to the wild. Lions that had to remain captive too long, and therefore could not be returned to the wild, served as founders of the Asiatic lion-breeding program.[66]

Sakkarbaug Zoo entrance. This zoo was founded by the Nawab of Junagadh to hold lions from the Gir Forest needing veterinary or special care. © Photograph by Sally Walker.

In 1956, the governments of Saurashtra (now Gujarat) and of Vindhya (now Uttar Pradesh) made an early effort to introduce Gir lions to an alternative habitat. The Executive Committee of the Indian Board for Wildlife accepted an offer from the Uttar Pradesh government to introduce lions in Chakia forest, and so an area of about 37 square miles was selected and prepared. As preparations took some time, the lions captured from Gir Forest were held for nine months at Sakkarbaug Zoo and finally transported by rail, 1,400 hot and arduous miles through Rajasthan. They were kept in a holding area at the new Chandraprabha Sanctuary and were officially released on December 2, 1957. For some time the lions bred and their number increased to 11 by 1965. After this, they disappeared from Chandraprabha and were never sighted again, neither in the sanctuary nor elsewhere.[15,67] Forty years later there is again a plan for a "second home for lions," this time in an area of Palpur Kuno in the state of Madhya Pradesh. Much more is known about how to carry out such delicate wildlife operations and a very conservative and careful approach has been recommended.[68]

8.3.5 Maharastra State and the Bombay Zoo

Bombay Zoo (now Jijabhai Bhosle Udyan Zoo) was founded as a botanic garden with animals in 1873. Today, the popularity of the zoo has overtaken the botanic garden, although the garden is well maintained. It is one of India's most problematic zoos because of high visitation, city pollution, and problematic administrative organization. The zoo is in the process of renovation and is replacing dreary cages with open, moated enclosures.

Colonel Flower visited the Bombay Zoo in 1912 and gathered the following information. In 1862, a botanical garden opened in Bombay and during the next decade a small menagerie was constructed in the corner. In 1889, it was decided to make the institution a combined zoological and botanical garden known as Victoria Gardens and to have it open free to the public. Flower reports that the zoo received several donations from wealthy individuals and industries for constructing cages, in addition to occasional grants from the municipality. The Bombay Tramway Company gave 10,000 rupees for new bear and parrot cages in 1890, Sir D. M. Petit donated 1,500 rupees for an aviary in 1891, the Nawab of Junagadh donated 2,500 rupees toward a cage for large carnivores in 1894, and the Maharaja of Bhavanagar gave 4,000 rupees for a small carnivore house in 1899. These were very substantial amounts in those days. Flower particularly noted trees with name labels and gave substantial detail about Bombay Zoo expenditures, enclosure designs, and staffing patterns. At that time Bombay Zoo held 64 mammals, 83 birds, and 3 reptiles (which were giant land tortoises, two of which were extremely large specimens).

In reporting on his trip, Flower also describes the tremendous variety of animals for sale at the famous Crawford Market at Bombay. For the most part, wildlife authorities have now tamed the notorious Crawford Market, but at one time the variety of animals held there rivaled almost any zoo. Among the animals for sale during Flower's visit were guenons, macaques, baboons, marmosets, black- and white-ruffed lemurs, red- and white-fronted lemurs, mynah birds, goldfinches, crested larks, short-toed larks, roller, large numbers of Indian parrots, Moluccan and Australian bird species, Chukor partridges, jungle fowl, pheasants, herring gull, and many more.

Flower also visited the Bombay Natural History Society, which maintained an interesting collection of live animals at that time, a "small but select menagerie ... of such mammals, birds, reptiles and fishes" which are not too large for indoor life. At that time there was a tame female Malabar squirrel (a male had lived 17 years there but had died just prior to Flower's visit), and numerous birds, such as white-eye, red-billed hill tit, three species of bulbul, black-headed myna, Baya weaver-bird, rose finch, black-headed bunting, Indian palm dove, rain quail, and rock bush-quail. A great hornbill lived in the secretary's room, where it had done so for nearly 19 years. Reptiles were kept as well, such as *Geomyda indopeninsularis*, Indian chameleon, Indian python, long-nosed tree snake, Indian cobra, and a tree viper.[6]

8.3.6 Princely Zoos

Jaipur Maharaja Savai Ram Singh founded Jaipur Zoo in 1877 (according to modern records; 1868 according to Loisel) "during the great famine which desolated India" as a distraction and an entertainment for the poor workers of the city. Loisel describes the administration having an "eccentricity" in recruiting its staff from the child apprentice gardeners who were brought up and trained in the same garden. At the time the raja did not maintain any captive animals in his palace except crocodiles, which guarded the approaches to the palace.[9]

Jaipur Zoo, with the help of the Public Works Department, constructed a tiger enclosure with an enormous moat soon after the founding of the zoo.[8] The number of tigers increased quickly, however, and in 1901 Cameron visited the Jaipur Museum and commented that "tigers and other animals occupy cages in an adjacent plot grounds."[69] Apparently, the zoo surplus was accommodated at the museum. In 1912, Flower paid a visit to Jaipur Zoo and learned it had been laid out by Dr. de Fabeck. Among the outstanding features, according to Flower, was the Great Paddock, "an enormous undulating piece of ground with a nulla running through it." Flower commented, "There is nothing to compare with it in size in any zoological garden I know except some of the enclosures for the Duke of Bedford's animals at Woburn Abbey in

England." The paddock was then inhabited by blackbuck, sambar, and wild peacocks. Flower also remarks on the "flying cage," which was large with much rockwork and water containing the following species: 4 purple coots, 1 white pelican, 1 Dalmatian pelican, 17 cormorants, 3 Indian darters, 5 black-headed ibises, 9 white storks, 2 white-necked storks, 2 Indian yabiru storks, 18 painted storks, 1 grey heron, and 2 cattle egrets. There was also a waterbirds aviary containing a large pond with 31 flamingos, 24 bar-headed geese, 12 species of ducks (each species represented by dozens of individuals), grey cranes, and demoiselle cranes.

Another outstanding feature of Jaipur Zoo that Flower described was the "Monkey Tree." This consisted of a banyan tree made into a home for captive monkeys. Individual kennels were fitted onto the branches of the tree, one for each animal. Each animal was chained to a ring that slid on perpendicular rods allowing the monkeys to "come and go as they please from the kennels and branches to the ground." Primates kept in the Jaipur monkey tree were Bengal monkeys, rhesus macaques, and young *Papio hamadryas*.

In contrast to Loisel 50 years later, Flower reported there was indeed an "official menagerie" in Jaipur, which kept five tigers and two leopards. A row of stone and iron-barred cages with small holding areas in the back made up the enclosures. Flooring was stone, but a resting shelf of wooden boards was provided for the animals to lie upon. Flower makes particular mention of the protection the maharaja gave to birds and beasts, which allowed large numbers of monkeys, squirrels, blackbuck, and a great variety of birds to roam freely, with great confidence that they would not be harmed.[6]

Maharaja Sajjan Singh of Mewar State founded the Udaipur Zoo at Rajasthan in 1878. The maharaja was interested in medicinal plants and attempted to stock the garden with flora that had some medicinal value. A specialist in horticulture helped and worked at the garden from 1882 to 1920. To stock the menagerie, many animal dealers were contacted and a great variety of species from around the world were kept for display. In addition, local trappers were sent into the nearby jungle to capture animals for the Udaipur Zoo. Any person who presented some curious or rare animal to the maharaja was suitably rewarded. Animal combat between tiger and lion or tiger and wild boar was arranged for entertainment. After some time, the zoo held valuable animals like black leopard, rhino, ostrich, hoolock gibbon, zebra, tiger, lion, leopard, mouse-deer, other deer and antelopes, and birds such as rare pheasants and parrots. After independence, however, the Udaipur Zoo animals were gradually shifted to other zoos and the Udaipur Zoo has never regained its importance.[70] Cameron visited Udaipur during his north India tour and reported that in the foothills of the Khas Odi Lake hundreds of wild boar were fed daily, undoubtedly by the royal family. "It is," said Cameron "a curious sight to see these wild denizens of the forest trooping at the appointed time to receive the Indian corn which is thrown down to them from the roof. A herd contained numerous tuskers of great size, with bristling mane from crown to rump, but they were of all sizes and apparently all degrees of temper." Another feature of the place was a small arena of tigers and other animals which were "occasionally let loose among the pigs."[69]

The Gaekwar of Baroda (a maharaja) maintained a large garden containing a collection of wild animals which was open to the public for years. It was an inspiration and source of animals for the Sayyaji Baug Zoo, officially opened in 1879. In 1923, Rev. Reginald Heber, Lord Bishop of Calcutta, visited the Gaekwar and saw two "fine hunting tygers [sic] in silver chain" as well as a mahout riding a tame rhinoceros "quite as patiently as an elephant."[42] The gaekwar also maintained an elaborate stable of animals for fighting purposes, about which many travelers of long ago have written. Gustave Loisel in 1870, as well as Louis Rousselet in 1878, relate that the gaekwar kept numerous male elephants trained for fighting. They were prepared three months in advance on a mixture of sugar and butter, which excited them for battle. In addition, other animals were kept for combat, such as hyenas and leopards to fight with buffalo, donkeys, and wild pigs, as well as a rhinoceros (which was painted red before the fights).[9]

Lut'f'ullah, according to Lockwood Kipling, described the populace of Baroda regularly spending its leisure time in the well-known "animal yard" there and its wondering interest in the rhinoceros. Kipling comments in 1892 that "these arenas are still haunted by the people and will probably change gradually into zoological gardens but there can be no doubt that the beast fight is popular today."[71] Another French traveler, Dr. Arbel, visited the court in 1901 and saw a great number of animals kept for fighting, combats between rams and cocks, vultures and sparrow hawks used in falconry, and cheetahs and caracals used to hunt large birds such as bustards and cranes. In addition, Arbel reports that "besides these animals, the Rajah of Baroda had and still maintains today in a park open to the public, a few ornamental animals; these last are tamed and permit themselves to be harnessed to little carriages for children as we do with goats."[9]

Sayyaji Baug Zoo was founded on the banks of the river Vishvamitri, which the gaekwar selected "to make a vast garden which would include a zoo." On January 8, 1879 the park was declared open, displaying animals from the maharaja's private collection of Indian and exotic animals.[72] Colonel Flower visited Sayyaji Baug Zoo in 1912, which he described as a "large and beautiful garden containing a collection of wild animals which is open free to the public. The main feature of this garden is a deep nullah [a large ditch], with steep picturesque banks, which winds through the grounds and is crossed by several bridges." He goes on to relate his visit to the zoo with its "monkey tree" (modeled on the one in Jaipur) and open-air enclosure for lions. The iron fence surrounding it was patterned after that of the Bombay Zoo. The most valuable animal was a Malayan tapir which shared a paddock with Nilgai and caused an anxious comment from Flower. There was also a large "flying aviary" planted with vegetation and containing 12 species of waterbirds.[6] Today, the zoo in Baroda sprawls over a rather large area which also contains a nullah and bridge.

For some time in the last few decades, Baroda has had two zoos. In 1956, Lt. Colonel Dr. Fatesingh Rao P. Gaekwad set up the Maharaja Fatesingh Zoo. Gaekwad commented on the motivation of the maharajas starting zoos. He says they did so for various reasons, "some laudable and others for reasons bordering on lunacy." Some started zoos because they loved wildlife, some because they wanted to concentrate on breeding of a particular species. Educating and entertaining their people were certainly motives for some royals. And, to be sure, some of them started zoos because the neighboring prince had already started one.[73]

It is likely the origin of Shivaganga Gardens Zoo (1882–1994) in Tamil Nadu owes something to Raja Serfagee and his animal collection. Serfagee was, like Jahangir, an early Indian amateur naturalist who kept animals in a palace menagerie, had paintings made of them, and described them himself in a most charming manner. The British recognized Raja Serfagee of Tanjore as the ruler in 1798. Serfagee had been educated at the Lutheran Mission and had an interest in natural history. Although his English was not perfect and his comments naive, Serfagee learned some of the elements, and certainly much of the enthusiasm, of scientific description. He described a shell as a "kind of sea-worm cage" and wrote of "tygers" that they have "more strength than all the animals and its nature without provocation is fierceness and cruel … tygers in Mysore … is 9 feets long from his snout to the tail and 5 feet in height … but Tyger drawn in this picture was got in the eastern part of the Country and its Eight feets long from snout to the tail and 4 feet in height." In truth, others have commented that tigers from Sunderbans in eastern India are smaller than other tigers.[74] However, Serfagee speculates there was a different reason: "It has been taken when it was young and continually confined in a strongest wooden cage." Serfagee had dozens of natural history paintings done of mammals, birds, reptiles, amphibians, and even insects and marine invertebrates. His private collection is significant because of these drawings and his attempts to be scientific in 1800.[17]

By 1990, Sivaganga Gardens was a poor example of a zoo, with a few tiny, cramped cages containing wretched animals trying to find a corner away from the sun, and reports of deer dying in great numbers due to panic and dogs.[75] The municipality was in charge of the zoo, but letters to them querying different aspects of the zoo even drew denials that a zoo existed. India's Central Zoo Authority denied recognition and Sivaganga Gardens closed about 1994.

8.3.7 Old Mysore's Zoos

While there are several zoological–botanical garden combinations in India today, few people in India are aware of the zoological history of Lalbaug, one of their most famous botanic gardens. Lalbaug Botanical Gardens held an impressive variety of wild animals, including tigers, lions, monkeys, kangaroos, an orangutan, and a rhinoceros, all of which were great attractions for visitors to the garden.[6,76] According to official sources in the centenary volume of the *Glass House*, the Lalbaug Botanical Gardens were established in the very early years of the nineteenth century. Colonel Flower, writing in 1913, however, was informed that Hyder Ali "caused the garden to be made … sometime before his death in 1782 … and planted with mango trees, less than a mile east of Bangalore Fort." The earliest mention of an animal at the Lalbaug Menagerie was in 1862, when a black panther was purchased.[6,76] A special feature of Lalbaug, according to Flower, was the Great Paddock, which was "so large that field glasses were necessary to see the animals in it." At the time of his visit there were blackbuck, gazelles, chital, sambar, barking deer, kakar, and emu.[6]

Cameron took charge in 1874 and ran the "menagerie and aviary" for more than 30 years. He brought a great interest in animals to the garden and enhanced the collection considerably. Despite this, he agonized over the fact that the expense of maintaining fauna might cause loss to the flora. His main orientation was horticulture and his contributions in this field were great (so much so that a major road in Bangalore is named after him). Cameron kept copious notes on his management of the gardens. These notes form a tattered document in the M. H. Marigowda Library entitled the "Cameron Report" covering the period 1884 to 1904. Cameron mourns the difficulty in procuring a lion or tiger for the menagerie, stating that despite having applied in the most likely places in the country, no cubs were forthcoming. The existing lioness died of dropsy, which must have left the menagerie without a lion or sufficient carnivorous animals for the interest of the public. The problem was solved in part in 1890, however, when the Maharaja of Mysore donated a tiger. Both Europeans and Indians donated other animals, including loris, peacock, spotted deer, sambar deer, fox, hyena cub, sloth, antelope, and waterbirds.

In 1892, visitation was particularly high, especially the "native public," probably due to the acquisition of an orangutan. In their enthusiasm to see the animal the crowd became "a little unmanageable at times" until a small posse of policemen was employed to regulate the crowds and maintain order. The orangutan was a male, nearly full grown, which Sanderson had imported from Sumatra.[67]

Until 1919, Lalbaug Botanical Gardens maintained a good animal collection, but in 1920 many of the large animals were transferred to Mysore Zoo, which had been founded in 1892. The policy of the Mysore government changed to concentrate more on the Mysore Zoo, and so only a few deer, birds, rabbits, and pigeons were left in Lalbaug. A small menagerie continued, at least until 1937, as the annual report for that year states the birds and deer in Lalbaug were well taken care of, and a pair of lemon-crested cockatoos, a peacock, a chital, a blackbuck, and a sambar were received as gifts. Animal areas were gradually eliminated, the old monkey house converted into a picnic place, and the rodent caves turned into rest areas.[77] A deer paddock is still maintained at Lalbaug, although poorly, and is one of the main attractions.

Chamarajendra Zoological Gardens at Mysore opened officially as a public zoo in December 1892, but many of the buildings and structures had been there as part of H. H. Chamarajendra Wodeyar's private collection. It is said that the present meeting hall at the zoo used to be a sumptuous bar.[78] The zoo began with a scant 10 acres, but added bits of land over the years and now the zoo covers about 350 acres. The zoo has additional land nearby, which is being made into a safari park for large ungulates and other animals.[79] Mysore Zoo is the only old zoo in India to keep pace with the times and to continue renovating its outdated enclosures. The city has grown up around the zoo, but Mysore is such that this is an asset, not a liability.

Until Indian independence, Mysore Zoo was the private property of the maharaja and kept at his expense. When Flower visited the zoo in 1912 he was highly impressed with the lovely ponds and their pink lotus, as well as the collection of animals, which included an albino crab-eating macaque and albino jackal. Interesting normal-colored animals included native ones such as tigers and leopards, as well as exotic ones such as mandrills, drills, and two polar bears. The polar bears had been in the zoo more than seven years and caused Flower to comment that it is not the size of the cage that determines the health and well-being of the animals, but the care and personal attention the keeper gives to them. Flower also commented on the red pandas, which had been in the zoo for a year.

Mysore Zoo Museum, which was originally part of the palace zoo. © Photograph by Sally Walker.

Mysore Zoo was later administered by the Department of Horticulture, which became a proper government department after independence. For some years M. H. Marigowda ran the zoo. Marigowda was a brilliant administrator who was Director of Horticulture in the state for many years. Afterward, C. D. Krishne Gowda, a young zoology graduate, ran the zoo. Gowda had been handpicked by Marigowda and followed in his footsteps. Mysore Zoo was then transferred from the Horticulture Department to the Forest Department. In about 1980, the Forest Department of Karnataka created a separate administrative unit for the zoo, so decisions could be made quickly and development of the zoo could proceed without the burden of bureaucracy. Originally, all the zoos and protected areas of the state were to come under this autonomous body, which would have facilitated the management to a greater extent; however, this did not take place and the body was renamed the Zoo Authority of Mysore.[80] Under the Zoo Authority of Mysore's efficient system, which built upon the work of Gowda, the zoo has flourished and is unrecognizable as a century-old zoo today.

8.3.8 Early Collections Not Normally Mentioned

Maharaj Baug Zoo in Nagpur is perhaps the only zoo in the world set up and run under the authority of a veterinary department at an agricultural university. In 1894, some enthusiastic wildlife lovers must have thought this would be a wonderful way to further their knowledge of wildlife diseases. This zoo is still operated by the veterinary department, but the hopes with which it was started were not realized, as it is by far one of the worst-managed zoos in all of India. The zoos of Maharastra include other oddities. Sholapur has two zoos on either side of its small township. One is a vegetarian zoo, which keeps only herbivorous animals, and the other is a nonvegetarian zoo, which keeps carnivorous animals. That such a small town would go to so much trouble to provide appropriate zoos for their religious residents is remarkable. Today, however, these zoos do not figure in the official list of zoos.

There were many more fascinating animal collections in these early days, which are not normally recognized, but which deserve mention. These were not formal zoos, but animal collections that were, according to Flower, "open to the public" and deserving of mention. Flower lists these as Burdwan, Nagpur, Secunderabad, and Jabalpur. He also mentions that several maharajas had private menageries, or a few animals, in their public gardens, as at Alwar, Gwalior, Indore, and Kolhapur.[6] Facilities at Nagpur, Gwalior, Indore, and Kolhapur have continued, perhaps in a different form, and are registered as zoos with the Central Zoo Authority today.

In Bhurtpoor (probably Bharatpur), Char Baug, which means "the Four Gardens," is indeed surrounded by four small flower gardens, which the raja laid out solely for the use of English visitors. Nearby this tribute to sycophancy was an orchard in which the rajah kept rams, cocks, antelopes, and quails for fighting, as well as numerous favorite birds for the amusement of his wives. Near the gardens some tigers and six leopards were kept, trained to hunt antelopes.[81]

Several travelers have mentioned the Botanical Gardens at Sarahanpur, but Fanny Parks comments on the animals: the garden "is an excellent one and in high order; some tigers were there growling over their food, several bears and a porcupine."[41] Saharanpur Botanical Garden employed an artist as well, who painted some of the animals.

There is no easily accessible record of Alfred Park in Allahabad, but it is known as a significant collection because of the large number of animals given to Calcutta Zoo over the years from Alfred Park. Allahabad is located in the state of Uttar Pradesh and is today an important political center, but has no zoo.

Although an animal collection in Alwar did not continue as a zoo, its history is particularly interesting. Captain Leopold von Orligh visited in 1840 and met the Rajah of Alwar. The rajah invited Orligh to come and amuse himself and promised him the "best opportunity for shooting tigers, wild boars and antelopes." The raja had dogs that would attack and kill a tiger and he brought a tiger and a dog with him to exhibit this talent to his honored guest.[81]

Colonel Flower, visiting some 70 years later, described the "Company Baug" as a "beautiful public garden" in Alwar where there were three live tigers, Himalayan bears, and tanks in the greenhouse with "countless" goldfish. In the maharaja's private garden, there was a tiger house so remarkable that Flower described it in detail. The interior of the house was carefully constructed so that keepers had convenient access and the "retiring rooms" had light, air, and a bench for the animals. The truly remarkable aspect of this structure was its outside arrangement, done so that persons driving to the palace or strolling in the park saw the tigers "apparently at liberty." The enclosure was surrounded with a ditch or moat around which was planted a hedge so it could not be seen. An elaborate security mechanism was also provided in case the animals fell into the moat. Another interesting feature of Alwar City was, like Jaipur, that all birds were protected. Flower reports seeing large numbers of different species all over the town, so confident of their safety they were almost tame. Flower also remarks on the large number of blackbuck outside the city, and even a female wandering inside the bazaars.[6] As far as is known, there is no zoo in Alwar today.

8.4 Twentieth-Century Indian Zoos

8.4.1 Modern Zoos and the National Zoological Park

Lucknow has an interesting animal history, which includes animal collections and combats that defy belief. It was also the center of a thriving animal trade for many years. In 1824, Bishop Heber visited Lucknow and described a menagerie with a large number of rare and curious animals owned by the maharaja of Oudh. He relates that it was in "far worse order" than the collection at Barrackpore, which he apparently saw during one of its many travels. The king's menagerie was kept in a well-wooded park on the "other side of the river" and contained a large collection of cows (many different varieties), camels, deer, and rhinoceros. The rhinos, according to Heber, were gentle and quiet (but for one with an intense dislike of horses). He reported that they bred in captivity "without reluctance." Heber also describes a "deer park" kept at the site of the king's summer palace, with nilgai, "noble red deer," and tame monkeys. The monkeys were fed in a cage in the middle of the park every two or three days. Heber speculated that pious Hindus fed them as there was a statue of Hanuman, the monkey god, in front of the cage.[42]

Mrs. Fanny Parks described the same Dil Bahar ("Heart's Delight") in 1827. It was "filled with game — deer, nilgai, antelope, bears, tigers, peacocks, etc.," which "his Majesty visits often for shooting." The king of Oudh, Nusuruddin Hyder, entertained Parks with some animal combat of which she remarks, "I saw some good elephant fights, a rhinoceros against wild buffaloes, some indifferent tiger fights, battles of every sort; some were very cruel and the poor animals had not fair play." She describes quail fights as "the best."[41]

Two decades later, Captain Leopold von Orligh related how the King of Lucknow had organized an elephant fight on the bank of a lake opposite to that of the observation area. He describes this as a "singular but very dangerous amusement, for only male animals with tusks are selected which are driven by the mahouts till they rush against each other." Orligh also describes the king's menagerie which was "not far from Goomty [River]." This consisted of a large menagerie in a large square court surrounded by piazzas. His party saw 13 tigers, many monkeys, fighting antelopes and rams, and even fighting quails, like those to which Fanny Parks referred. The quails, it seems, were put on the table to fight for the amusement of dinner guests. Orligh was fascinated by a Beejee or Indian ichneumon (mongoose) which, he related, can be easily domesticated when young and can kill most poisonous snakes. Orligh comments further on the menagerie: "I am sorry to say the value of this menagerie has been very much decreased by the removal of several animals — six rhinoceroses and several other remarkable animals having taken them to adorn various sepulchral monuments."[81]

A collection of the Mogul king of Lucknow deserves further comment. Sharar described the park in Calcutta where Nawab Wajid Ali Shah maintained thousands of animals. In the same work Sharar devotes two chapters to the animal combats of the deposed king's earlier Lucknow establishment, in which he analyzes the practice of animal combat and the personality of the Moguls who promoted and supported it. He attributes the obsession with animal combat in Lucknow to two factors. First, the Moguls were fighting men who had given up territorial conquests and the ambition of battlefields, so they sublimated their warrior inclinations through watching, and providing the opportunity for others to watch, animal combats. His other theory is that, particularly in Lucknow, the British friends of Ghazi ud Din Haidar influenced his interest and more or less corrupted him. There are, however, records of Mogul animal fights much earlier.[30]

In any case, according to Sharar the elephants trained for fighting numbered 150 and these fights were the most popular. Other animals made to fight were tigers, panthers, cheetahs (which were often cowardly), camels, rhinoceroses, stags, rams, horses, and a great variety of bird combinations. The animal combats in themselves are not so important in a history of zoos, although the collections they came from were clearly precursors of zoos, but what is very interesting is the skill and courage of the keepers of these poor animals that were made to fight. Sharar describes them in some detail.

As the king and his visitors watched the fights in safety and comfort from across the water, animal keepers had to manage the fights at close quarters. Making the animals fight was not difficult, according to Sharar, but controlling the fights, pulling the animals apart, and herding them back to their cages after the fight was both difficult and dangerous. At the end of the fight these keepers took care of both "victor and vanquished," which undoubtedly meant treating wounds and serious injuries. Sharar says European travelers have commented in their diaries that "there were none in the world better at looking after, training and controlling the wild animals than these keepers."[30] If any of these keepers, or a court recorder, has written about these experiences and the treatment of the animals, it would be a valuable addition to wild animal veterinary and training lore, but thus far such records have not been found, at least in English.

In 1921, the Prince of Wales visited India and in the same year a zoo was founded with his name at Lucknow. The prince inaugurated the Prince of Wales Zoo, which is located in the center of the city.[82] On one side such a narrow street divided the zoo from residences that it was possible for people living in these houses to throw things from their first floor verandah into the enclosures. During the 1990s, Lucknow zoo officials undertook a renovation program, moving or modifying most of those enclosures so they are away from the dwellings.

At one time there was a proposal to shift the Prince of Wales Zoo to a location just outside of Lucknow, which the Forest Department of Uttar Pradesh owned, called Kukrail Park. Kukrail Park extends over 1,000 acres and houses a few animals such as deer, birds, and small mammals, in addition to the crocodile- and turtle-breeding centers that once supplied animals for reintroduction. An Endangered Species Project for jungle fowl and pheasants was maintained at Kukrail for some time, and now there is a plan for an otter-breeding project. The space and lush vegetation, as well as access to water and a quiet atmosphere, would provide an excellent environment for a zoo, but the townspeople of Lucknow will not hear of it. They like having the zoo convenient to them for casual visits and morning walks, so Lucknow zoo may stay within the city for years to come. Besides, the state of Uttar Pradesh has a "forest zoo" a couple of hours away, the Kanpur Zoo.

Hill Garden Zoo in Kankaria is something of an anomaly in Indian zoos. Reuben David started the zoo in 1951. The zoo and its founder became very well known in India and, to some extent, in the world. David was a great personality; he was a wrestler and then a self-made veterinarian, as in those days one did not necessarily need a degree to practice. When a traveling animal show wanted to leave behind some birds and small mammals in Ahmedabad, the Ahmedabad municipality asked David to look after them. Thus began a life-long love affair between David and his zoo. David wanted lions for his zoo but the conservative Hindu Gujarati *babus* would never hear of purchasing animal flesh to feed the lions. As conservative about money as about their religious beliefs, the zoo board members were told by David that the lions were a gift when actually he had purchased them with his own money. The parsimonious municipal governors could not say no to something free, so the Ahmedabad Zoo had its lions. David was concerned about conservation and breeding before any other Indian zoo director. He succeeded in breeding many animals past the second generation and had the best breeding record in India. He took personal interest in each and every animal that came into the zoo and ensured that it was looked after as well as could be done. David was not a modern zoo man and he had peculiarities that prevented him from being a true zoo biologist in the contemporary sense. He was fascinated with freaks and white animals — two-headed calves, zebra–mule hybrids, and albinos. He once refused a white tiger because it was not an albino, as he was interested only in true albinos. He kept albino crows, porcupines, spotted deer, monkeys, and others. In addition, he achieved remarkable breeding success with species that other zoos found difficult, such as flamingos and mouse-deer.

Reuben David was a showman and a teacher, as well as an animal man. He founded the Balvatika, a children's museum that also had live domestic animals doing simple tricks. He wrote hundreds of articles on conservation, but his contribution to zoo conservation was flawed because of the tremendous affection he had for his animals. He could never part with an animal. Before his death he was awarded the Padma Shree, an Indian national award for outstanding service. It was the first time a zoo man had received such an honor and reflected well on the whole zoo community.

Assam State Zoo is another biological park that the Indian Forest Service and the Forest Department of Assam developed in 1958. The zoo covers 320 acres of very lush tropical forest area for which northeastern India is well known. It also includes in its center a large botanical garden. The land is dramatically undulating and includes a lake. Assam State is where the maximum number of Indian rhinoceros is found, in the large Kaziranga National Park. Every year during the monsoon, several young rhino and elephant calves are stranded and separated from their mothers by the rain. The forest department has no recourse but to pick them up and turn them over to the zoo. Assam State Zoo, therefore, has more Indian rhinos and Indian elephants than any other zoo. Many of these have been given to zoos outside of India. Assam Zoo has also bred several Indian rhino calves over the years. Its other important species are the hoolock gibbon, golden langur, and capped langur, for which there is a special off-exhibit breeding facility. There have also been programs for pigmy hog and white-winged wood duck, both critically endangered species in the state, but they have not been successful to date.

A recent effort to breed pigmy hog involves a collaboration between the Forest Department of Assam, the Ministry of Environment and Forests, the Pigs and Peccaries Specialist Group of IUCN, the Jersey Wildlife Preservation Trust, and the Darwin Initiative. At this stage the project does not involve a zoo but maintains a captive breeding facility in a separate area close to the natural habitat of the animals. The project has gone well so far for this delicate, difficult, and critically endangered species, and may represent its last hope.[83]

Nandankanan Biological Park in the state of Orissa, declared open in 1960, is one of India's most interesting and attractive zoos, as well as one of the very early biological parks. It is named after *nandavanam*, the pleasure garden of the gods. Situated in a natural forest setting covering more than 1,200 acres, Nandankanan Zoo combines very large, open, moated enclosures with beautiful forest cover and undulating landscape. The zoo has a large lake, with a botanical garden on the other side. The area was selected with expansion for the biological park in mind. Nandanakanan Zoo has been a leader in training, education, breeding endangered species, maintaining studbooks, and developing innovative enclosure designs. Although this park does not restrict itself to indigenous animals, it exhibits and breeds many important Indian species. The park is adjacent to Chandaka Wildlife Sanctuary and animals have wandered into the zoo from the sanctuary.[84]

The State of Bihar is known more for its controversial animal events and facilities than for its good zoo, which is located in the state capital, Patna. Sanjay Gandhi Biological Park is one of the zoos founded (1973) after the opening of the National Zoo, and its design owes much to the National Zoo. Located adjacent to Raj Bhavan, the Patna Zoo enjoys both the protection of the local government and the threat of being overtaken should the government require more space. The zoo is beautifully forested, but unfortunately level, the designers having flattened the once attractively undulating landscape. It attracts many rural visitors and has attempted to cater to the portion of the public that is illiterate by erecting a zoo map with animal illustrations instead of words. Patna Zoo has had very good breeding success with Indian rhinoceros and with both large and small cats. Zoos in India rarely breed small cats easily, but Patna has bred clouded leopard and innumerable leopard cats. Patna is also the first zoo to have a really good aquarium.

Just outside of Patna is a whole community of people whose main occupation for centuries has been the capture and trade of wild animals. Mir Shikar Tola — literally "place of chief hunter" — used to be notorious for the illegal trading of highly endangered species, as well as the more common birds small traders carried in flimsy wire cages on the back of their bicycles. Before the passage of the Wildlife Protection Act in 1972, and for some years afterward as well, the animal dealers of Mir Shikar Tola received an untold number of animals from different parts of India and sold them to zoos, traveling menageries, and circuses. While there were many other dealers located throughout India, particularly Bombay, Calcutta, Lucknow, and Meerut, the names of dealers from Mir Shikar Tola were known to all.

Perhaps because of the village of Mir Shikar Tola, Bihar was also the state with the greatest number of traveling menageries, although Uttar Pradesh also had a good number. These tiny zoos travel from village to village with every sort of animal, from a loris to a lion. Small traveling cages make up the entire environment for the unfortunate animals in these mobile menageries. They move from village to village where visitors pay a small amount for the dubious privilege of seeing and shouting at these miserable animals. The Central Zoo Authority deliberately dignified these facilities with the name "zoo" in the Zoo Act (of 1991) so they would come under the legal purview of the government and be required to be registered and inspected. On the eve of closing these zoos, the Minister for Environment and Forests, who coincidentally was from the same state as most of the owners of these zoos, must have come under pressure and rushed to their rescue, stating the traveling zoos provided the only opportunity for villagers to see wild animals. However, in 1998 they were closed altogether and the animals given to other zoos.

Chhatbir Zoo is officially known as the Mahendra Chaudhary Zoological Park. It is named after the governor of Punjab, who inaugurated it in 1977. It is considered to be one of India's best zoos in one of India's best-planned cities. The zoo is situated in a natural forest, so despite the hot, dry climate, it is green and pleasant, a welcome relief from the parched landscape of Punjab. Chhatbir Zoo has safari areas where visitors drive through in a bus and see lions in one forest area and then deer in another. It also has a walk-through aviary holding a variety of attractive birds. The zoo is well maintained, provides training for zoo personnel, maintains links with veterinary colleges, conducts research, and serves as a proper example for other Indian zoos.

In the State of Uttar Pradesh there is another zoo, in addition to those in Lucknow, a number of deer parks, and some minizoos. Kanpur Zoo, also known as the Allen Forest Zoo, is one of the most beautiful in the country. One story goes that Robert Allen donated the land as a park, but it was not developed, deteriorating into a haven for bad individuals and activities. There was a plan for the city to develop it as a commercial area, but before it could do so, the forest department rescued it and planned a zoo. Kanpur Zoo covers about 250 acres of beautifully rejuvenated forest with a lake right at the entrance and many beautiful, naturalistic enclosures. The zoo has had good success in breeding rhinoceros, as well as other animals.[85]

One of the Kanpur Zoo rhinos was used for the first rhino reintroduction project. This was done at Dudhwa National Park where all the rhinos were either translocated from other rhino areas or born in the park from translocated animals. This was a successful conservation project the government of India undertook to repopulate Dudhwa Park with rhinoceros, which had become extinct there some years before. Rhinos were brought from Nepal and Assam, kept in a stockade for some months, and then released into the park where they bred and prospered.[85] Unfortunately, the introduction of the male from Kanpur Zoo some years later was not successful. The resident male caught his scent and broke through the stockade in which he was being kept and attacked him. The young zoo rhino was saved but only just, and had to be returned to the zoo.

The history of zoos in India cannot be told without a special mention of the Forest Department of Andhra Pradesh and its outstanding biological parks. Bernard Harrison, Director of Singapore Zoo and a student of exhibit design, opines that the Indian Forest Service contributed a unique "zoo form" to the world's zoo culture, which he calls the "biological park." According to Harrison, a biological park is distinguished from a zoological garden (or park) and a biopark by the fact that the biological park is designed to replicate forests that once flourished on the site where the institution now sits. Very large, moated enclosures with acres of rejuvenated forest are designed on the area, taking into consideration the best natural features and contours of the land. Generally, indigenous animals are the showpieces of these zoos, and they are displayed for the most part in forest areas where their ancestors roamed a few decades previously.[60] In India, poor, and even middle-class, persons may not be able to visit a sanctuary or national park in their lives, but the biological park provides a substitute experience.

Harrison credits the Indian Forest Service with the creation of the biological park, and the Andhra Pradesh Forest Department was the leading proponent. In India some of the state forest departments are more "zoo and wildlife wise" than others. The Forest Department of Andhra Pradesh is one that attempts to retain officers who are trained or experienced in wildlife and zoos, rather than transferring them often (as is the common practice). As a result of this policy, Andhra Pradesh has three outstanding biological parks, each unique and exceptional in its own way. Sri Pushp Kumar of the Indian Forest Service served the zoos for many years and can be credited with much of the success of the state's zoos.

Nehru Zoological Park, opened in 1959 at Hyderabad, was the first and the most traditional biological park, and contains another innovation that has become a standard in large Indian zoos, the safari park. Safari parks are areas of perhaps 5 to 50 acres where animals roam free and visitors see them from a vehicle. Usually the park is just forest with a path for vehicles curving around different areas of vegetation and points of topographical interest. Today, Nehru Zoo has several dramatic safari parks for lions, tigers, bears, and even gaur. Other zoos have emulated this feature with varying degrees of success.

Indira Gandhi Zoological Park, opened in 1970 at Vishakapatnam, is one of the most dramatically placed zoos, being right on the coast of the Indian Ocean. Dominated by attractively undulating landscape, this zoo boasts an aviary constructed in a huge gully, such that it is below ground level. The landscape of the zoo, therefore, is not spoiled with an awkward chain-mesh aviary construction, and the design is such that, once inside, neither the visitors nor the birds are aware of being below ground level.

Sri Venkateshwara Zoo in Tirupati is a sprawling 3,000-acre area with a religious theme to match the famous Tirupati temple, which thousands of persons visit daily. The animal consorts of the Hindu gods are displayed beside enormous enclosures surrounded by the lovely hills of Tirupati.

In addition to these zoos, Andhra Pradesh also boasts an outstanding deer park, which the Central Zoo Authority has classified as a zoo. Most of the 100-plus deer parks in India are sorry affairs where some land was available from the forest department, municipality, or industry, and a few common deer were procured to fill it, or a few unwanted deer were relocated there. Sri Mahavir Harina Vanasthali Deer Park, opened in 1975, was founded to claim several hundred acres of degraded land a few kilometers from Hyderabad City, which might otherwise have been developed into a shopping complex or factory site. Pathways for a viewing bus were carefully marked out and water holes, salt licks, and feeding areas judiciously placed. A beautiful interpretation center made completely with waste timber stands just inside the entrance and a special "hide," or viewing tower, made of rough wood, overlooks a salt lick where animals gather. This deer park is a model facility that has improved with time, instead of falling into a state of shabbiness, as have most of the other 100-odd deer parks in India.

As idealistic and beautiful as they are, there are problems with biological parks, which are now sprinkled throughout India. For example, the enclosures are so large the public cannot always see the animals, and special arrangements have to be made. Also, the concept of these biological parks presumes a far more active interest and motivation than most visitors actually have. People are expected to look at these beautiful forested areas and learn to appreciate nature. While this is a laudable objective and noble in theory, it has been demonstrated that most visitors need a great deal of interpretation provided in order to be affected.[86] This notwithstanding, the biological park is still evolving and is a magnificent experience for those who can, and will, appreciate it.

Van Vihar National Park, opened in 1981, is actually a zoo legally designated a national park to provide stronger legal oversight of visitor behavior. Van Vihar in Bhopal, Madhya Pradesh, covers 1,000 acres on the bank of a five-mile-long lake. The main viewing road runs along this bank so that visitors always have a view of the lake. This dramatic zoo displays only animals indigenous to Madhya Pradesh in enclosures ranging from 10 to 50 acres. A variety of deer and antelope have the run of the entire area, while carnivores — bears, lions, tigers, otters, and gharials — occupy large enclosures. Gharials have an attractive, large water body in which to swim and banks on which to bask. When the zoo wanted to build a large aviary, then Director S. M. Hasan could not bear to clutter the landscape with a structure, so he created a marsh in the lake, which became a heronry. Van Vihar is one of the most dramatic zoos in India, and perhaps the world, today.[87]

Madhya Pradesh has other zoos which, although not particularly interesting or attractive in terms of their layout, have other qualities that deserve mention. Indore Zoo (1974) is located in a city with the dubious distinction of having the heaviest vehicular traffic in India, and is shockingly shabby to look at, but ironically had (until recently) perhaps the most creative and active education program of all the zoos in India. There is always something going on for the public at Indore Zoo. In addition, Indore Zoo has purebred tigers and, of all the zoos, has taken the Indian Tiger Breeding Programme the most seriously, sending animals to other zoos, keeping very accurate records of breeding, and promoting the tiger breeding program both in the press and on enclosure labels.[88,89] Gwalior Zoo is also not much to look at and cannot boast of anything in the area of animal care. However, Gwalior Zoo was the main site and inspiration for the first doctoral dissertation on the management of zoos written in India. While the conclusions of the dissertation could be challenged, the precedent of the School of Studies in Zoology, University of Gwalior in allowing a student to undertake a Ph.D. on zoo management is unique and admirable.

National Zoological Park

The Indian Board for Wildlife (IBWL) felt that a good zoo should be founded on modern principles with open, moated enclosures and naturalistic displays to serve as an example to other zoos. Although there were many such enclosures in Indian zoos already, no zoo had been planned from the beginning with only these enclosures in mind. Carl-Heinrich Hagenbeck was asked to design what was then called the Delhi Zoological Park. The name was later changed to the National Zoological Park to reflect more correctly the purpose of this institution — one the central government administers and finances to provide a model for other zoos in the country. The scheme for establishing a zoo in Delhi was formulated as early as 1952 and the zoo was inaugurated in 1958.[90]

The National Zoo was designed on the then-popular concept of continental areas and in its first few years came close to fulfilling its promise as a "model zoo" for the country. The zoo was well known for its attractive design, dramatic site (next to the historic Red Fort), white tigers, lion-tailed macaque, and Manipur brow-antlered deer breeding. Over the years, however, government agencies responsible for it have not been able to address its problems in a timely manner. The Delhi Public Works Department is the only agency that is permitted to do needed repairs or construction, and the zoo is a very low priority for the department. Decisions about almost any small matter must be approved at the incredibly busy and complex Ministry of Environment and Forests, which has administrative responsibility for the zoo. Too many duties devolved on the zoo's director, who cannot fill supportive posts due to bureaucratic delays and obstructive procedures. Labor problems in the nation's capital have also proved difficult. Nonetheless, the National Zoo has played a major role in modern Indian zoo history.

The National Zoo today is considered a problem for which privatization may be the only solution. The Central Zoo Authority has made an attempt to privatize the National Zoological Park since, over the years, the National Zoo has deteriorated rapidly and is felt to have become literally unmanageable. It is not just the National Zoo that suffers from this, but many of the municipal and state-run zoos as well. If the National Zoo could be transformed by privatization, then some of the state and local government zoos might follow the example.[91]

8.4.2 Indian Crocodile Project and Specialist Zoos

One of the more innovative, ambitious, successful, and unacknowledged conservation projects in the world is the Indian Crocodile Project. Conceived more than 20 years ago as a collaborative project with assistance from the government of India, FAO, and UNDP, the Indian Crocodile Project included all three species of Indian crocodiles: mugger, saltwater crocodile, and gharial. The project contained all the elements of modern conservation: survey, conservation breeding and rearing, release back to the wild, monitoring, education, sustainable use, and cooperation between states, zoos (both nationally and internationally), breeding centers, universities, and wildlife agencies.

The project base was situated in the Nehru Zoological Park and included a Crocodile Research Centre and Training Centre, which was the initial stage of the Wildlife Institute of India, the premier training institute for wildlife management in India. Rehabilitation of gharial started in 1975 and continued under a "grow-and-release-programme" of the Indian Crocodile Project. Crocodile and gharial eggs were collected from the wild, artificially hatched, reared until they were old enough or large enough to fend for themselves, and then released into suitable areas. The young crocodilians were marked with a notching system so they could be monitored.

Eight protected areas at Chambal, Ken, and Son in Madhya Pradesh; Kateriniyaghat and Corbett in Uttar Pradesh; Satkosia in Orissa; Jawaharsagar in Rajasthan; and Papikonda in Andhra Pradesh were selected as sites for release of gharials, although some of those did not become active sites. Five Captive Rearing Stations were established, as well as Captive Breeding Centres. The Nandankanan Zoo breeding program was particularly interesting because it was a gharial from the Frankfurt Zoo that provided the fertile male the Indian zoo required.[92-94]

Although the gharial project has been successful, the results of a Population and Habitat Viability Assessment Workshop held at Gwalior University in 1993 indicated that the status of gharial was "conservation dependent." If conservation measures, such as continued monitoring, protection, and periodic restocking should come to an end, the gharial would be very much at risk again.[95]

Several "specialist zoos" have evolved in India, two kinds of which are snake parks and high-altitude zoos. Probably all Indian snake parks are indebted to some extent to the first, the Madras Snake Park, which set standards for breeding reptiles in captivity, for research both *in situ* and *ex situ*, and for the development of sustainable-use activities, particularly those assisting Irula tribals whose specialty is snake catching. Madras Snake Park began a venom extraction unit which became a model facility and later, having shifted to a facility farther from town called the Crocodile Bank, the zoo started a creative rat-catching project as well. Another project to breed Indian crocodiles in captivity, both for conservation (for restocking depleted areas) and for use (including export of their leather), floundered due to an about-face in the Indian government's policy regarding export of animals. Madras Crocodile Bank now supports many field research projects in India, its islands, and even other countries. There are several snake parks in India today, some well run and others very badly run. Some of these provide services for the public, such as catching snakes people find in their houses and releasing them into a wild area.[96]

High-altitude zoos specializing in indigenous high-altitude animals have been constructed at three hill stations in India: at Kufri near Shimla, at Nainital in Uttar Pradesh, and at Darjeeling in West Bengal. These zoos are in cold areas and concentrate on keeping and breeding animals which are, for the most part, from these areas. Padmaja Naidu Himalayan Zoological Park was founded in 1958 and today can boast of breeding Siberian tiger and snow leopard, as well as having a well-defined conservation project for red panda. The Red Panda Project includes a field component that should result in a reintroduction exercise involving several zoos in the northeastern part of India, as well as the United States and Europe. Foreign zoos have donated several red pandas to the Indian program to enhance the genetic diversity of the animals and to otherwise strengthen and facilitate the breeding program.[97] Zoos in Shillong, Meghalaya, and elsewhere in the northeast are still organizing thematically and may become high-altitude zoos.

8.5 Indian Zoos and Wildlife

Twelve Indian zoos still open at the end of the twentieth century were founded in the nineteenth century. Between 1900 and 1950 five zoos were founded that remain in some form today. In the latter half of the twentieth century, more than 300 zoos were founded. Those figures are dramatic. In the 50 years after Indian independence in 1947, there was a quantum leap in the founding of zoos, which proliferated without control until 1991 when the Zoo Act was passed.[98–101]

In the 15 years between 1936 and 1950, however, there is no record of any new zoos being founded in India. Perhaps this was due to the impending turmoil during which wild animals in captivity were not a priority. However, even during those years, in fact particularly during those years, people were increasingly concerned about the decline in numbers of wild animals. In 1935, Lord Willingdon organized the Delhi Conference to discuss their concerns. Subsequently, the All Indian Conference for the Preservation of Wildlife was founded to effectuate the resolutions of the Delhi Conference. An All India Convention was drawn up during the conference and referred to as the "Magna Carta" of wildlife in India. It was conceived along the lines of the "famous London Convention" held two years earlier, and a journal was started as the official organ of the conference. Major Jim Corbett, the famous hunter of man-eating tigers, headed the editorial board. An Indian, Hasan Abid Jafry, was managing editor, and another Indian provided the major financial support for the initiative. The president, Khan Bahadur Raja of Mahmudabad, and his brother contributed heavily to the conference and also to the Association for Preservation of Game in Uttar Pradesh. He gave an annual grant, provided a set of rooms for the library and offices in Lucknow, and allowed the use of his printing press for printing thousands of leaflets.[102]

From the language used in the journal, it sounds as though the individuals responsible for the convention and association were more concerned about the loss of game (for sport) than the loss of wildlife. Colonials, and Indians with the habits and outlook of colonials, dominated the convention. Both the convention and association are important, however, because their objectives included provisions for encouraging individual states to preserve fauna and flora in India, establishing a library, popularizing natural history in schools, establishing a national park, assisting in establishment of sanctuaries, and publishing books on Indian natural history. These activities undoubtedly provided a structure for the Indian Board for Wildlife, founded when the new government was formed after independence. With respect to zoos, there is only one mention in the journal. Under a photograph of a tiger in a cage, there is a heartrending poem by Major W. R. Lawrenson titled "A Prisoner for Life," which does not support zoos.[102] This attitude had changed considerably, however, by the time the Indian Board for Wildlife first met.

8.5.1 Indian Board for Wildlife and Wildlife Protection Act

Indian independence came in 1947, and in a record time of five years after the new government began the Indian Board for Wildlife (IBWL) was founded in 1952. The IBWL was supportive of zoos in theory, but also very particular that they maintain good standards. Interest in wildlife soared and the opening of animal facilities mushroomed; however, not all facilities maintained good standards. Minutes of the IBWL reflect a very different atmosphere than that which existed at the Delhi Conference. While mention of game and hunting still cropped up in the meetings, for the most part the deliberations of the board were focused on saving wildlife as part of India's cultural heritage. The board was made up almost completely of Indians. Most of the British had, by that time, returned to England, and few of those who remained enjoyed the trust of the Indians. It is clear that the national will, reflecting concern for the mystical rather than the mechanistic, had reasserted itself in very short order.

In 1951, the Bombay Wildlife Act included a provision for supervision over zoos and their licenses.[103] In 1955, the first All Indian Zoo Superintendents' Conference was held at Madras. The objectives were to discuss problems of zoo administration; find ways to improve existing zoos; and to assess the scientific, educational, recreational, and aesthetic value of zoos in the community life of the nation. Important resolutions and recommendations included exchange of animals between zoos, formation of an All India Zoo Association, expansion of the scope of zoos to include rehabilitation of "denuded areas" (by rearing important species in zoos and introducing them into these areas), providing a training center for zoo personnel, and more. The need for a list of zoos in India was quickly filled and appeared with the proceedings. At that time 20 zoos were listed, although there were a few more small ones. At the third meeting of the Executive Committee of the IBWL, held about three weeks later, the recommendations of the All Indian Zoo Superintendents' Conference had a place on the agenda and the committee approved them.[104,105]

A year later, the IBWL formed a special Zoo Wing, which met for the first time on May 16, 1956 and then reported to the regular meetings of the IBWL. Deliberations of this Zoo Wing, like those of the Zoo Superintendents, were astute and current with the rest of the world. They were concerned with welfare, the size and amenities of accommodation, the provision of company — if not mates — for social animals, prevention of inbreeding, education promoted with labels and publications, adequate representation of indigenous — as opposed to exotic — animals, research, nutrition, inventory and records, transport, and even breeding rare species for reintroduction. In particular, cheetahs were discussed, with a view to importing African cheetahs as companions for the single male at Mysore Zoo. At that time there seemed to be an awareness of subspecies in cheetahs, and there is no question that the animal at Mysore during the 1950s was an Indian cheetah.[106,107]

Passage of the Indian Wildlife Protection Act in 1972 made provisions for zoos and museums, allowing capture of animals for zoos and keeping of skins and trophies for museums. Between the passage of the Wildlife Protection Act and 1999, more than 250 zoos were founded under 14 different administrative heads. On average, this is about nine zoos per year. Even for a country the size of India, this is a formidable number.

There are several reasons for this growth of zoos. Many were founded, no doubt, simply to keep up with the next city or state; zoos had become a token of prestige. Some of them were outgrowths of animal-holding centers established to keep animals that had been rescued or trapped for different reasons, or confiscated from persons holding them illegally. These holding areas naturally attracted the public, and the step to making them a zoo was a short one. Other facilities were started simply because a politician or public servant such as a minister, a forest officer, or a district collector wanted to do something for the public, and for the individual's own record of achievements as well. A few zoos were started as privately operated traveling zoos that go from village to village and earn a meager living for their owners. Some zoos were founded to replace or utilize old zoos or royal collections. Some specialist collections, such as snake parks, were founded as a result of individuals particularly interested in certain kinds of animals. Finally, a few good zoos were initiated with foresight and out of a genuine conviction that zoos were part of the conservation process. In general, however, it is safe to say many of the zoos were set up for the wrong reasons. Time proved this to be the case when they deteriorated, lacking the continued personal interest of the individual who set them up, and without a mechanism to ensure their proper upkeep.

8.5.2 Management of Zoos in India

In 1975, there was great concern about the quality of zoos in India, which, instead of improving, seemed to be deteriorating. An expert committee formed and toured the major zoos in India. This committee issued a comprehensive report with recommendations. There is no question that the committee pinpointed all the pertinent problems facing Indian zoos and suggested reasonable and effective solutions for these problems.

One of the recommendations was for a central coordinating body to monitor and unify the zoos. It emerged that the zoos were isolated from one another, and even zoos in the same state and under the same administrative unit were unable to coordinate an animal exchange or emulate a good design used by the other. At the time, the National Zoological Park was asked to serve as the coordinator of the zoos.[108] The director of the National Zoo was expected to carry out this formidable task, as well as run the zoo. Coordinating the zoos meant coordinating the exchange of animals and species breeding programs, as well as tasks such as procuring drugs and equipment for all the zoos, organizing training, initiating the association of zoos that the first All Indian Zoo Superintendents' meeting had recommended, and even publication of a national zoo bulletin.

In 1982, the Department of Environment (later to become the Ministry of Environment and Forests) issued a list of zoos, botanical gardens, and sanctuaries in India.[109] The list consisted of only 44 zoos. In 1987, two lists of zoos were produced almost simultaneously. There was a simple list of about 115 (increased later to 160) that the Zoo Outreach Organization had compiled and published in its journal *Zoos' Print*. There was also a detailed directory the Nandankanan Biological Park had published.[110,111] These documents clearly showed that there were many more zoos than those listed in the official Department of Environment list. It was simply impossible to ignore the fact that there was a population explosion of zoos. When the Indian Zoo Act was passed in 1991 and it became legally incumbent on zoos to register, more than 350 institutions were listed.[5]

J. H. Desai, then Director of the National Zoo and who was valiantly trying to implement the zoo's mandate to coordinate the zoos of India, called a meeting of directors on May 24–25, 1983 at the National Zoo. One of the significant outcomes concerned captive breeding of rare and threatened species. Zoos were given responsibility for certain rare species, which included approved breeding programs implemented with 100% funding from the central government. Other important outcomes included formation of the Association of Indian Zoo Directors, creation of a zoo service, and workshops and seminars on important themes.

At this meeting it was also decided that a recommendation from the 1972 report, *Management of Zoos in India*, concerning the establishment of a central coordinating body for zoos, should be pursued. The body established for this purpose was called the Central Zoo Authority of India and was given authority to decide norms for important issues affecting zoos in India.[112]

It had been about 30 years since the last official meeting of zoo directors (the All Indian Zoo Superintendents' Conference) was held in Madras in 1955. In 1987, a meeting was finally held in connection with the 35th anniversary celebrations of National Zoo. In 1988, another meeting of Indian zoo directors was held in Trivandrum. Recommendations from this meeting included many of the same suggestions as in previous meetings; however, one new suggestion called for a joint director for zoos in the Department of Environment to facilitate better coordination among the zoos and to attend to zoo matters. This suggestion was implemented forthwith and, largely because of this, subjects of concern to zoos began to receive far better and more efficient attention than previously.[113] S. C. Sharma, former Director of Kanpur Zoo, was appointed to this post and ultimately became the first head of the Central Zoo Authority.

In 1990, the Zoo Consultancy Project was initiated in the Ministry of Environment. Desai headed this project, under the auspices of the Wildlife Institute of India. The Zoo Consultancy Project was multifaceted and was to carry out all the tasks that had been set at the Zoo Superintendents' Meeting in 1955, in the report of the Expert Committee of 1975, and in the resolutions of the several meetings of zoo directors held since 1982. The activity and output of the Zoo Consultancy Project had a profound influence on the pace and direction of the zoo movement in India even before its comprehensive report on the status of zoos in India was submitted. Standards and guidelines were drawn up on management, health care and disease control, preparation of master plans and management plans, education and interpretation programs, and conducting research. The project also made recommendations for achieving the objectives of conservation breeding programs and for the structure, role, and function of the proposed central coordinating body for zoos, the Central Zoo Authority of India. Finally, the project developed and initiated training programs for professionals and technicians.[114]

Under the project, nearly all the zoos were visited. At that time there were 205 listed facilities keeping wild animals in captivity: 120 zoos, 49 deer parks, 6 snake parks, 24 nature education centers, and 6 aquariums. The Training Course for Zoo Personnel was developed and the first course was held at Nandankanan Biological Park in 1990. Every year since, a course for directors or technical-level zoo personnel has been conducted, since 1993 in collaboration with the Central Zoo Authority. This is probably the only systematic, regular, fully indigenous zoo management training course of any tropical country in the world.[115]

An Indian Zoo Association was recommended as early as 1955 at the First All India Zoo Superintendent's Meeting. Finally, at a meeting of zoo directors held in association with the Symposium on the Role of Zoos in Wildlife Management at Sakkarbaug Zoo, a set of bylaws was approved and a president elected who would also be responsible for registering the association. S. K. Patnaik, then Director of the Nandankanan Zoo and who was subsequently elected President of the Indian Zoo Directors Association, registered the association with the Registrar of Societies in the State of Orissa on July 13, 1991. Some of the association objectives are to promote cooperation between the zoos; to exchange information, experience, and knowledge relevant to techniques and management of zoos; to aid in acquisition, exchange, or loan of animals between Indian zoos and foreign zoos; to stimulate zoological research and studies; to educate the public; and to publish periodicals devoted to management of all aspects of zoos.[116]

In 1991, the Indian Parliament enacted the Amendments Act, including the Zoo Act.[117,118] The Zoo Act, probably the strongest piece of zoo legislation in the world to date, provided for the autonomous Central Zoo Authority with a membership of 12 persons, chaired by the Minister for Environment. The membership consists of about 50% official and 50% nonofficial members. The main functions of the authority are, briefly, to specify minimum standards for housing, upkeep, and veterinary care of zoo animals; to evaluate zoos according to these standards and recognize or "de-recognize" them; to identify endangered species of Indian wildlife for breeding and to coordinate acquisition, exchange, and maintenance of studbooks; to identify themes and priorities with regard to display; to coordinate training, research, and education programs; and to provide technical and other assistance.[98]

In August 1992, the Recognition of Zoos Rules, which laid down standards for assessing zoos, became law. An inspection committee consisting of a minimum group of a manager, veterinarian, and educator was established and directed to inspect the zoos of India systematically. No zoo would be allowed to operate without recognition of the Central Zoo Authority. Zoos would be thoroughly inspected and assessed according to the legal norms and standards, and recommendations given for bringing the zoo into accord with the norms. Zoos that are clearly incapable of meeting a decent standard could be requested to close. In this case, the zoo would be given six months to "show cause" why it should not be closed. The zoo would be given time to sell off the animals, and some funds to compensate their workers. The Central Zoo Authority would also help find another zoo to take the animals.[5]

Much of India's zoo culture is governed by the country's religious, social, and even political mores. Hindu, Buddhist, and Jain religions honor life and preclude the taking of any life for almost any reason, even — except in very extreme cases — to end suffering. The management practice of euthanizing or culling surplus or ill animals is fraught with danger in India, not just for zoo directors, but for their superiors in the forest department or municipality, for their superiors in the ministry, and even for senior politicians of the party in power. Regional and religious practices are part and parcel of politics in India and no one wants to take an action that could provide an opposition party with a volatile issue. Consequently, closing zoos cannot be undertaken lightly for a variety of reasons, not the least of which is disposal of the animals.

In the Zoo Act the definition of "zoo" was deliberately contrived so it would include practically any animal facility for the public, even traveling menageries. In several other countries, a zoo is legally defined as a stationary institution. In the case of the Indian Zoo Act, however, it was desirable to ensure that the Act could be applied to all substandard facilities which, if the definition excluded movable zoos, it could not. Therefore, at least in theory, these traveling menageries, which number in the dozens, can be inspected, found wanting, and requested to close. In practice, this proved to be more complicated than anticipated, again because of politics.[119]

In the few years since its inception, the Central Zoo Authority has inspected all 300-plus zoos and has made good progress in the reinspection and according of recognition. The Central Zoo Authority recommended that a number of zoos be closed and some of them complied. Others have used a variety of ruses to continue operating, but few recommended for closure have improved. By the first part of 1999 the Central Zoo Authority had granted recognition to 56 zoos classified as "large," "medium," and "small" according to criteria concerning area, number of visitors, number of animals, etc. In addition, 106 "mini-zoos" (which include deer parks and other animal facilities) had been granted recognition. The Central Zoo Authority has closed 46 of the mini-zoos and 18 more were refused recognition, which effectively means closure after a grace period in which the zoos have the opportunity to improve themselves. A remaining 18 mini-zoos are yet to be evaluated. Perhaps the most dramatic achievement of the authority has been the closure of more than two dozen traveling menageries after a long and bitter court battle. Now the challenge is placing the animals from these facilities, as well as some 300 carnivorous animals from circuses as a result of legislation prohibiting the use of wild animals for performances.

The Central Zoo Authority has given large grants to zoos whose administrative authority would match the amount, and with these funds outdated enclosures have been renovated and new ones constructed, labels have been improved and education programs initiated, veterinary facilities have been upgraded, and equipment has been purchased. The Central Zoo Authority has conducted workshops, training, seminars, and conferences, both on its own and in collaboration with different agencies. One of the objectives of the Central Zoo Authority has been to bring together experts from the different fields and disciplines which are required to make zoos more meaningful for conservation and more humane for the animals. One of the more ambitious collaborative projects the Central Zoo Authority has funded is in the area of biotechnology — DNA fingerprinting, artificial insemination, etc. — with help from the Centre for Cellular and Molecular Biology (CCMB), Hyderabad. Now there is a plan to set up a Zoo Biology Institute in collaboration with CCMB where research specifically related to zoo and wildlife conservation problems can be conducted and training given.[5]

8.5.3 The Future of Indian Zoos

The authors of *Zoo Culture* claim "Indian wildlife was raided to supply animals for European menageries and later for zoological gardens. As a result of European intrusion into India as part of the colonial process, which brought Anglo-Saxon cultural forms, the zoological garden was imposed on India. The result of this was that an alien cultural artifact was used to frame animals within their country of origin in a way which was artificial for the indigenous observers. Thus the colonizers had attempted to reshape Indian views of their own wildlife."[10] Aside from the fact that India and other countries were indeed raided for their wildlife, the authors' view of what occurred with regard to zoos as a result of colonialism is not correct. Indian facilities were as much a factor in the shaping of European zoos as the opposite, if not more so.

What is truly ironic, however, is what is occurring now, for it seems there is an attempt to justify acquisitions of the past with current conservation efforts. Now, Western zoos are trying to reshape the wildlife management views of tropical zoos in such a way that the tropical zoos feel they need Western expertise to carry out high-tech programs, when what they really need is to improve their basic animal care. Sarah Christie of the London Zoo, who has visited Southeast Asia many times on training and technology exercises, is adamant on this point. She claims that too many zoo personnel are much too quick to see biotechnology as a cure-all for their problems, when they have not established rudimentary animal care facilities. Western zoos, understandably, want tropical zoos to use Western-bred tropical animals (or their genetic material) for reintroduction programs when this may not be the most effective way to conserve species. On the other hand, Indian zoos, as a tropical example, have been unable to sustain a single scientific breeding program thus far. To a great extent they require some of the animals bred in the West to avoid taking animals from the wild for these programs. It is a conundrum for which no one has a good answer.

Whereas India has pioneered a comprehensive zoo management book (1872), innovative enclosure style (the biological park — 1970s, but based on nineteenth-century colonial zoo exhibits), and powerful zoo legislation (1991), Indian zoos have not lived up to their enormous conservation potential. Despite a large number of zoos (many of them with vast areas compared with their Western counterparts) with many species of their own rich biodiversity, there are no viable, systematic, scientific breeding programs. A few threatened species have been bred in respectable numbers in the zoos, but in a haphazard and ad hoc manner without management to ensure genetic and demographic viability. Moreover, Indian zoos are overcrowded with common species. A 1997 analysis of Central Zoo Authority animal records for mammals in all the zoos revealed that less than 3% of them were threatened species. The reasons for this situation are complex and are related to politics, religion, the transfer system, a lack of cooperation, and a lack of interest or understanding of conservation priorities — just to name a few. India's policy makers are aware of these problems and have tried to solve them, but the problems are deeply buried in the country's cultural and administrative system.

India may consider seriously whether it is worthwhile or workable to try to force its zoos to fit the global zoo conservation blueprint. Perhaps Indian zoos should just concentrate on educating their public to protect its remaining habitat rather than expend valuable resources on breeding and reintroduction programs that are uncertain at best, and highly risky at worst. Changing the entire administrative system of the country will be required to make the kind of changes needed to ensure good zoo conservation practices. Indians have to decide for themselves what is good and right for their country's conservation needs, considering all its cultural, social, economic, and political forces.

References

1. Rapson, E. J., Ed., *The Cambridge History of India* (Volume 1, Ancient India), Cambridge University Press, Cambridge, 1935.
2. Basham, A. L., *The Wonder That Was India: A Survey of the Culture of the Indian Sub-Continent before the Coming of the Muslims*, Grove Press, New York, 1954.
3. Nikam, N. A. and Mekeon, R., Eds., *The Edicts of Asoka*, University of Chicago Press, Chicago, 1959.
4. Purchas, S., *Kakduytus Posthumum: or Purchas His Pilgrimes, Contayning a History of the World in Sea Voyager and Land*, Volume 1, Macmillan, New York, 1905–1907, 170, 182 [Reprint of 1625 edition].
5. Sharma, S. C., *Central Zoo Authority Third Annual Report*, 1993–94, New Delhi, 1995.
6. Flower, S. S., *Report on a Zoological Mission to India in 1913*, Cairo Government Press, Cairo, 1914.
7. Mitra, D. K., "Ram Bramha Sanyal and the establishment of the Calcutta Zoological Gardens," in *New Worlds, New Animals*, Hoage, R. J. and Deiss, W. A., Eds., Johns Hopkins University Press, Baltimore, 1996.
8. Walker, S., "Interview: R. D. Yadav, Director, Jaipur Zoo," *Zoos' Print*, 4 (2), 2, 1989.
9. Loisel, G., *Histoire des ménageries de l'antiquité à nos jours*, O. Doins et Fils, Paris, 1912 [Unpublished typescript, translation in preparation provided by J. C. Edwards].
10. Mullan, B. and Marvin, G., *Zoo Culture*, Weidenfeld and Nicolson, London, 1987.
11. Thomas, G. I., "Sainyutta-Nikaaya, verse 420," in *Early Buddhist Scriptures*, Regan Paul, London, 1935.
12. Banerje, S. C., *Flora and Fauna in Sanskrit Literature*, Naya Prokash, Calcutta, 1980.
13. Lodrick, D. O., *Sacred Cows, Sacred Places: Origins and Survival of Animal Homes in India*, London, 1954 [Also, University of California Press, Berkeley, 1981].
14. Das, A. K., "Imperial cheetahs in Akbar's Shikarkhana as seen through paintings of his time," *India Today*, Thomson Living Media-India, New Delhi, 1985.
15. Negi, S. S., "Transplanting of Indian lion in Uttar Pradesh State," *Cheetal*, November, 98, 1969.

16. Blochmann, H., *The A In-I Akbari by Abul-Faz'l*, New Taj Office, Delhi, 1989 [First reprint of 1873 edition].
17. Faz'l, Abdul Allami, *Ain-i-Akbari*, New Taj Office, New Delhi, 1989 [Reprint edition].
18. Archer, M., *Natural History Drawings in the India Office Library*, Her Majesty's Stationary Office, London, 1962.
19. Alvi, M. A. and Rahman, A., *Jahangir the Naturalist*, Indian National Science Academy, New Delhi, 1969.
20. Bernier, F., *Travels in the Mogul Empire, AD 1656–1668*, Oxford University Press, London, 1914 [translated and edited by T. A. Constable based on 1890 edition].
21. Raffles, S., *Prospectus for Forming the Zoological Society of London*, London, 1925 [privately published leaflet].
22. Gaekwad, F. R., "Why did the Maharajas start the zoos," *Zoos' Print*, 10, 1, 1988.
23. Forbes, Rosita, *India of the Princes*, J. Gifford, London, 1939.
24. Harrison, B., *A Study of the Planning and Design Principles involved in Development of Mammal Exhibits in a Tropical Zoo*, National University of Singapore, Singapore, 1985.
25. Fisher, J., *Zoos of the World: The Story of Animals in Captivity*, Natural History Press, Garden City, NY, 1967.
26. Sharma, S. C., personal communication, 1997.
27. Pal, A. C., *The Park at Barrackpore, A. D. 1785–1931*, Private Secretary's Press, Calcutta, 1931.
28. Chatterjee, D., *A Short Sketch of Rajah Rajendro Mullick Bahadur and His Family*, Kumar Nagendro Mullick, Calcutta, 1917.
29. Sharar, A. H., *Lucknow, The Last Phase of an Oriental Culture*, Oxford University Press, Delhi, 1994 [translated and edited by E. S. Harcourt and F. Hussain].
30. Schwendler, C. L., "The establishment of a zoological garden for the town of Calcutta," *Proceedings of the Natural History Committee of the Asiatic Society of Bengal*, Calcutta, 1873.
31. Curzon, M., *British Government in India*, Volume I and II, Cassell and Company, London, 1925.
32. Martin, M., Ed., *Minute on Foundation of College of Ft. William*, W. H. Allen, London, 1937.
33. Vicziany, M., "Imperialism, botany and statistics in early 19th century India, the surveys of Francis Buchanan (1762–1829)," *Modern Asian Studies*, 20, 625, 1989.
34. Anonymous, Boards Collection 3360–3410, Extract of Public Letter, Minute from Governor General, June 18, No. 303, Calcutta, 1805.
35. Wellesley, A., Extract from public letter minute 58, Minute from the governor-general appointing Buchanan as Director of the Natural History Institute, Fort William, July 26, 1804.
36. Scherren, H., *The Zoological Society of London, A Sketch of its Foundation and Development*, Cassell, London, 1905.
37. Mitchell, P. C., *Centenary History of the Zoological Society of London*, Zoological Society of London, London, 1929.
38. Buchanan, F. and Gibbons, W. L., *Descriptions of Birds and Quadrupeds in the Barrackpur Menagerie, Calcutta*, OIOL, London [unpublished manuscript, ca. 1906].
39. Walker, S., Ed., *Newsletter of the Society for Promotion of History of Zoos and Natural History in India*, 1 (1), 1996+.
40. D'Oyly, C., *Barrackpore Menagerie* [Painting], Paintings and Drawings Section, India Office Library, London, 1820.
41. Parks, F., *Wanderings of a Pilgrim in Search of the Picturesque, during Four and Twenty Years in the East with Revelations of Life in the Zenana*, Pelham Richardson, London, 1850.
42. Heber, R., *Narrative of a Journey through the Upper Provinces of India from Calcutta to Bombay*, B. R. Publishing, Delhi, ca. 1825.
43. Anderson, John, *Guide to the Calcutta Zoological Gardens*, Honorary Committee of Management, Calcutta, 1883.
44. Dunbar, J., *Golden Interlude: The Edens in India 1836–1842*, John Murray, London, 1955.
45. Anonymous, *Minutes of the Calcutta Natural History Society*, Calcutta, 1841.

46. McClellan, E., Ed., "Collections," *Calcutta Journal of Natural History*, 2, 297, 1841.
47. Anonymous, "The Indian Museum and the Asiatic Society of Bengal," *Calcutta Review*, 1866.
48. Chakraborty, S., *Chhotoder Kaliketa Tinsha*, Mousami Prakashani, Calcutta, 1990 [Children's Calcutta Zoo]
49. Anonymous, "On additions to the menagerie," in *Proceedings of the Zoological Society of London*, March 22, 138, 1864.
50. Dunn, W, "Extract of a letter from Mr. W. Dunn," in *Proceedings of the Zoological Society of London*, November 10, 370, 1863.
51. Anonymous, *Proceedings of the Zoological Society of London*, November 14, 820, 1867.
52. Anonymous, *List of Animals in the Zoological Gardens, Calcutta*, Bengal Secretariat Press, Calcutta, 1878.
53. Walker, S., "Interview: Purhendro Mullick, Director, Marble Palace Zoo, Calcutta," *Zoos' Print*, 4 (5), 15, 1989.
54. Mitra, D. K., "First practical handbook: more history of R. B. Sanyal," *Zoos' Print*, 2, 7, 1988.
55. McClellan, E., Ed., "Proposal to form a zoological garden in Calcutta," *Calcutta Journal of Natural History*, 2, 295, 1841.
56. Anonymous, *Report of the Honorary Committee for the Management of the Zoological Gardens*, Calcutta Zoological Gardens, Calcutta, 1886–1933 [an annual report covering 1885/86–1932/33].
57. Anonymous, *List of Animals Which Have Been Included in the Collection of the Zoological Garden, Calcutta*, Bengal Secretariat Press, Calcutta, 1890.
58. Anonymous, *Zoological Gardens, Calcutta*, Calcutta Zoological Gardens, Calcutta, 1900.
59. Das, A. K., personal communication, 1995.
60. Bahuguna, N. C., personal communication, Darjeeling, 1997.
61. Anonymous, *Madras Zoo Centenary Souvenir*, Madras Zoo, Madras, 1955.
62. Harrison, B., "The evolution of zoological gardens," in *Proceedings of the Meeting of the International Union of Directors of Zoological Gardens at Singapore*, IUDZG, Singapore, 1991.
63. Chandran, P. R., personal communication, 1986.
64. Walker, S., "Interview: P. R. Chandran, Director, Kerala Zoos," *Zoos' Print*, 2 (1), 3, 1987.
65. Pillai, N. K., *Report on the Administration of the Public Gardens and Zoo for the Year 1111 M. E.*, Trivandrum, 1937.
66. Chavan, S. A. and Raval, P. P., "The role of Sakkarbaug Zoo in the management of Gir Sanctuary and National Park, Session I.4: in situ and ex situ cooperative management," *Zoos' Print*, 5, 15, 1990.
67. Anonymous, *Proceedings of the IV Meeting of the Executive Committee held at Sasan Gir, Saurashtra*, Indian Board for Wildlife, Ministry of Food and Agriculture, Government of India, New Delhi, 1956.
68. Ashraf, N. V. K., Chellam, R., Molur, S., Sharma, D., and Walker, S., Eds., *Report of the Population and Habitat Viability Assessment for Asiatic Lion*, Zoo Outreach Organisation and Captive Breeding Specialists Group, Coimbatore, 1994.
69. Cameron, C., *The Cameron Report*, M. H. Marigowda Library, Lalbaug Botanical Gardens, Bangalore [unpublished manuscript].
70. Tehsin, Raza, "Udaipur Zoo, a brief history," *Zoos' Print*, 2, 6, 1987.
71. Kipling, J. L., *Beast and Man in India: A Popular Sketch of Indian Animals and Their Relations with the People*, Macmillan, London, 1892.
72. Jadeja, V., Ed., *Souvenir of International P.H.V.A. Workshop on Asiatic Lion*, Zoo Outreach Organisation, Coimbatore, Baroda, 1993.
73. Walker, S., "Interview: Fatesingh Rao Gaekwad, President, WWF – I," *Zoos' Print*, 3 (6), 3, 1988.
74. Das, A. K., personal communication, 1988.
75. Namboodiri, N., personal communication, 1995.
76. Anonymous, *Glass House — The Jewel of Lalbaug*, Mysore Horticultural Society, Bangalore, 1991.

77. Anonymous, *Correspondence from the Superintendent, Government Gardens, Forwarding His Report on the Administration of Government Gardens and Hill Stations for the Year 1931–32*, Government of H. H. Maharaja of Mysore, General and Revenue Departments, Bangalore, 1933.
78. Hughes, A. C., *Sri Chamarajendra Zoological Gardens, Mysore, Report for the Year 1911–12 with a Guide for Visitors to the Zoo*, Mysore, 1912.
79. Raju, R., personal communication, 1996.
80. Gowda, C. D. K., personal communication, 1982.
81. Orligh, L., *Travels in India including Sinde and the Punjab*, Longman, Brown, Green & Longmans, London 1845 [translated by H. Evans Lloyd].
82. Russell, W. H., *The Prince of Wales in India*, Discovery Publishing House, London, 1877.
83. Oliver, W. and Narayan, *Pigmy Hog Conservation Project Report*, Guwahati [unpublished manuscript, ca. 1996].
84. Mishra, R., "Nandankanan Biological Park, Orissa," *Indian Zoo Bulletin*, 1, 2, 1972.
85. Bhadauria, R. S., personal communication, 1988.
86. Walker, S., "An informal historical perspective of zoo exhibitry in India," *Zoos' Print*, 9 (8), 11, 1994.
87. Hasan, S. M., personal communication, 1988.
88. Mahodaya, A. K., "Birth of wild origin tiger cubs at Indore Zoo," *Zoos' Print*, 8, 16, 1992.
89. Walker, S., "Profile — the Indore Zoo conservation and education program," *Zoos' Print*, 12 (5), 16, 1997.
90. Anonymous, *Proceedings of the Third Session Held at New Delhi from 14th to 16th February 1958*, Indian Board for Wildlife, Department of Agriculture, Ministry of Food and Agriculture, Government of India, New Delhi, 1958.
91. Sharma, S. C., personal communication, 1996.
92. Choudhury, R., "Mugger breeding successful in crocodile breeding farm at Tadoba," *Zoos' Print*, 1, 17, 1986.
93. Anonymous, "The crocodile conservation plan," *Zoos' Print*, 1, 28, 1986.
94. Patnaik, S. K., personal communication, 1988.
95. Rao, R. J., Sharma, B. B., Molur, S., and Walker, S., Eds., *Report of the Population and Habitat Viability Assessment for Gharial*, Zoo Outreach Organisation and Captive Breeding Specialist Group, Coimbatore, 1995.
96. Walker, S., "Interview: Romulus and Zai Whitaker," *Zoos' Print*, 2 (4/5), 3, 1987.
97. Bahuguna, N. C., "Darjeeling Zoo receives red pandas for captive breeding," *Zoos' Print*, 8, 7, 1993.
98. Anonymous, "Indian wildlife amendments: the zoo act," *Gazette of India*, October, 1991.
99. Sharma, S. C., *Central Zoo Authority Second Annual Report, 1992–93*, New Delhi, 1994.
100. Walker, S., "The first real zoo in India … in the world?, a vastly abridged version of a work in progress on Barrackpore, precursor of the modern zoo," *Zoos' Print*, 9 (7), 1, 1994.
101. Walker, S., "Historical listing of Indian zoos according to Central Zoo Authority registration forms," *Zoos' Print*, 9 (11), 29, 1994.
102. Corbett, J., Morris, R. C., and Jafry, H.A., Eds., *Indian Wildlife, Official Organ of All-India Conference for the Preservation of Wildlife*, 1 (1), July 1936+.
103. Walker, S., "Interview: birdman of Bhavnagar, K. S. Dharmakumarsinji," *Zoos' Print*, 1, 14, 1986.
104. Mathur, I. P., Ed., *Proceedings of All India Zoo Superintendents' Conference and Minutes of Third Meeting of the Executive Committee of the Indian Board for Wildlife*, Government of India, Delhi, 1955.
105. Mathur, I. P., Ed., *Proceedings of the Third Meeting of the Executive Committee of the Indian Board for Wildlife Held in Tern Hill Palace, Ottacamund, from the 6th to 9th May, 1955*, New Delhi, 1955.
106. Anantharamaiah, M. V., personal communication, 1985.
107. Gowda, C. D. K., "Endangered species of human beings," *Gnu's Letter*, 1, 2, 1983.
108. Anonymous, *Management of Zoos in India: Report of the Expert Committee on Zoos*, Indian Board for Wildlife, Department of Agriculture, Ministry of Food and Agriculture, New Delhi, 1975.
109. Anonymous, *National Wildlife Action Plan*, Department of Environment, New Delhi, 1982.

110. Walker, S., "How many zoos?," *Zoos' Print*, 2 (4/5), 7, 1987.
111. Patnaik, S. K. and Acharjyo, L. N., *Directory of Indian Zoos*, Nandankanan Biological Park, Orissa, 1989.
112. Desai, J. H., *Proceedings of the Meeting of Directors of Indian Zoos held on 6th November 1985*, Government of India, National Zoological Park, New Delhi [unpublished manuscript].
113. Walker, S., "Report of the All India Zoo Directors' meeting held at Trivandrum Zoo," *Tiger Paper*, 15 (3), 15, 1988.
114. Desai, J. H., *Proposal for Establishment of Central Coordinating Authority for Zoos in India*, 1990 [document approved at All-India Zoo Directors meeting January 15–16, 1990, Junagadh]
115. Walker, S., Poster presented at VIth Conference of Breeding Endangered Species in Captivity, Jersey, C.I., 1992.
116. Patnaik, S. K., personal communication, 1991.
117. Anonymous, *First Draft Zoo Policy for India*, Department of Environment, Forests and Wildlife, Government of India, New Delhi, 1987 [unpublished document].
118. Sharma, S. C., Ed., Minutes of meeting held on 10 September 1989 at FRI, Dehra Dun to finalise the Zoo Policy of India, New Delhi [unpublished manuscript, ca. 1989].
119. Sharma, S. C., personal communication, 1991.

Additional Sources

1. Sanyal, R. B., *A Handbook of the Management of Wild Animals in Captivity in Lower Bengal*, Bengal Secretariat Press, Calcutta, 1892.
2. Anonymous, "Ram Bramha Sanyal on the management of animals in captivity," *Nature*, 46, 314, 1892.
3. Kisling, V. N., Jr., "Libraries and archives in American zoological parks and aquariums," in *Encyclopedia of Library and Information Science*, Volume 57 (Supplement 20), Marcel Dekker, New York, 1996, 292.W
4. Crandall, Lee, *The Management of Wild Mammals in Captivity*, University of Chicago Press, Chicago, 1964.
5. Sanyal, R. B., *A Handbook of the Management of Wild Animals in Captivity in Lower Bengal*, Natraj Publishers, Dehra Dun, 1995 [reprint of 1892 edition].

9
Zoological Gardens of Japan

Ken Kawata

9.1 Introduction

More than 160 live animal exhibit facilities exist today in Japan. These animal collections often differ in many respects from those in other nations; differences between Japanese zoos and their foreign counterparts stem from the cultural uniqueness of the Japanese. The focus here will be on selected topics concerning zoos in Japan, drawing comparisons with Japan's closest Western ally, the United States.

9.2 Historical Overview

9.2.1 Pre-Restoration Era

Modernization of Japan, which included the development of zoos, began during the era of the Meiji Emperor and is known as the Meiji Restoration. This transformation began after power was seized from the feudal lords in 1868. Exotic animal exhibits, however, date to the periods prior to this modernization. Roadside animal exhibits were seen as early as the seventeenth century in Kyoto. A book from this era vividly illustrates how the public paid admission fees and marveled at the tiger, porcupine, and peafowl that were kept in crude cages. Also, a 1716 publication described sideshows in Kyoto displaying a dog act, a monkey act, birds, and fish. During the Kansei Era (1789–1801), tea houses with live animal exhibits emerged in large cities such as Edo (later named Tokyo), Osaka, and Nagoya. Under various titles, such as peacock tea house, deer tea house, and the like, they featured these native animals and other animals from exotic lands. Customers would pay a fee, enter a yard, sit on chairs, and while sipping green tea, they would watch exotic animals in wire-mesh enclosures.

In June 1796, for example, a peacock tea house in Osaka opened, exhibiting not only peafowl, but also golden pheasant, ring-necked pheasant, and white-naped crane. These tea houses were equipped to accommodate patrons in rainy weather, admission fees were low, and they became extremely popular with elderly people, women, and children. According to some, these tea houses were the forerunners of amusement parks and zoos that blossomed after the Meiji Restoration.[1–5]

Seventeenth-century roadside exhibit in Kyoto. These exhibits displayed live animals such as tigers, porcupines and peafowl for a paid admission fee. This period painting depicts a tiger in a crude cage. Photograph courtesy of Mr. Akio Takizawa.

9.2.2 The Emergence of Modern Zoos

As early as 1865, an idea for constructing an animal exhibit building at Osaka was considered. Four years later, Yoshio Tanaka, known as the father of Japanese museums, developed another plan for a zoo at Osaka. However, neither of these plans materialized. Tanaka was, nevertheless, instrumental in the birth of the first zoo in his country. On March 20, 1882, Ueno Zoo (pronounced WAY-no), the first modern zoo in Japan, opened its gate in Tokyo. The national government established it as a division of the museum and patterned it after the Menagerie du Jardin des Plantes in Paris. The second modern zoo in the country opened in Kyoto on

Ueno Zoo gate, ca. 1907. The zoo opened in 1882 in heavily wooded Ueno Park, Tokyo, as part of the national museum, and was the first zoo in Japan. Photograph from *Ueno Zoo: the 100 Year History*. © Ueno Zoo.

Zoological Gardens of Japan

Renovated Kyoto Zoo gate, ca. 1907. Kyoto Zoo was the second zoo in Japan, opening in 1903 to commemorate the wedding of the crown prince, who later became emperor. Photograph courtesy of Mr. Akio Takizawa.

April 1, 1903. The mayor conceived the idea of building this zoo as a commemorative project to celebrate the wedding of a crown prince. The birth of the third zoo followed on January 1, 1915 in Osaka, a municipal facility commonly known as Tennoji Zoo. Thus, the major population centers that arose during the centralized feudal system prior to the Meiji Restoration — Edo (Tokyo), Kyoto (the imperial capital), and Osaka (the commercial center for western Japan) — housed the first zoological gardens in the country after the restoration.

Chimpanzee tea ceremony at Tennoji Zoo in Osaka. Rita (left), who arrived in 1932, became a national celebrity with charismatic popularity. She was later joined by a young male, Lloyd (right). Photograph from *Osaka Tennoji Zoo: The 70 Year History*. © Osaka Tennoji Zoo.

In much of the rest of the world, a zoo's name often includes the name of the city where it is located, such as Zoo Berlin or the San Diego Zoo. So, foreigners in Japan are likely to call the zoos in Tokyo and Osaka, "Tokyo Zoo" and "Osaka Zoo," but no such zoos exist. These zoos should be properly called Ueno Zoo and Tennoji Zoo, respectively. The official name for the Ueno Zoo is Onshi Ueno Zoological Gardens. Onshi means the "imperial gift," since the imperial family gave the zoo to the City of Tokyo in 1924 to commemorate the wedding of another crown prince. Other zoos in Japan do include the name of the municipality in their names. The zoo in Kyoto was officially known for decades as the Kyoto Commemorative Zoo ("commemorative" to mark, again, the wedding of a crown prince, but the term was dropped in 1964). However, many citizens call the Kyoto Zoo the Okazaki Zoo because it is in Okazaki Park.

Each of these first three zoos had its origin, to varying degrees, in an exposition. As Japan was awakening as a modern nation, the need to absorb Western civilization, industry, and technology was quickly apparent. Expositions in large cities were an educational vehicle to inform the populace of the world outside of Japan. In Kyoto alone, at least 34 expositions were held between 1871 and 1903. Many of these public exhibitions contained menageries, large and small. In some cases, as expositions ended, their grounds were turned into future zoo sites and the animals in makeshift cages were transferred to permanent facilities at newly built zoos. Expositions continued to provide a cradle for these zoos well into the 1970s. For example, part of the Okinawa International Oceanic Expo became a permanent facility, opening its gates as an aquarium on August 1, 1976.

As more cities built zoos, there was a move toward a national organization. The Japanese Association of Zoological Gardens and Aquariums (JAZGA) was established in 1939. The list of member institutions contained 16 zoos and 3 aquariums. However, those 16 zoos included one in Seoul, Korea and another in Taipei, Taiwan, both countries under Japanese control at the time. Regarding the establishment of the first aquarium in Japan, there have been conflicting opinions. A small freshwater aquarium was opened on September 20, 1882 at the Ueno Zoo. Kobe had temporary saltwater aquariums on two occasions as a part of expositions; the first opened in 1895 and the second was open from September 1 to November 30, 1897.[1,2,4,5]

9.2.3 World War II and Beyond

In the late 1930s, when Japan was at war with China, zoos were faced with difficulties that included shortage of animal feed, problems in acquiring animals, lack of staff, and the possibility of air raids. Carnivorous animals suffered from dwindling supplies of meat and fish, so attempts were made in zoos across the country to formulate meat substitutes. Also, decreasing supplies of metal made it difficult to build guardrails and animal cages.

With its sudden attack on Pearl Harbor, Hawaii in December 1941, Japan plunged into World War II. In the early days of the war, zoos still enjoyed large public attendance. Soon, however, air raids, evacuation from cities, and accelerated shortages of essential goods cast the nation deeper into wartime difficulties. As employees were drafted by the armed forces, zoos faced the challenge of maintaining the status quo with fewer people. Lack of coal, the main fuel for the heating system in the animal houses, also became a serious problem.

The tide of war rapidly turned in favor of the Allies. Thanks to military propaganda, the Japanese populace was euphoric over the "victory over victory" reports; however, those on the war front in the vast areas of Asia were aware of the grave reality. Maintaining zoo animals was not a priority at this time of grim reality for the nation. On the home front, decreasing food supplies for animals began to affect not only carnivores, but the herbivores as well. In Osaka, two elephants, the consumers of the largest amount of foodstuff, died of gastroenteritis and

As Japan plunged into World War II, zoos were involved in the war efforts. Air raid drill participants at Osaka's Tennoji Zoo included an elephant. Photograph from *Osaka Tennoji Zoo: The 70 Year History.* © Osaka Tennoji Zoo.

malnutrition in January and March of 1942 as a result of inadequate feeding. Zoo staff planted vegetable gardens on the premises for the animals, although such measures did not yield enough produce. As the war situation went from bad to worse, the governing authorities developed a plan to destroy so-called dangerous zoo animals, such as the large carnivores and elephants. The stated rationale behind such an extreme step, as part of a wartime contingency plan, was to prevent the escape of these animals in case bombing should damage the cages and release the animals. However, there were covert reasons as well. One was to forestall any false rumors of zoo animals running loose in the cities. A second was the hope that, by laying responsibility for this atrocity at the feet of the Allies, citizens' grief over the loss of their favorite zoo animals would be redirected to mounting hostility toward the enemy.

Some zoo officials resisted the orders to destroy animals; others pleaded to delay the killings as long as possible. But they were fighting a losing battle. Desperate plans were made in vain to move the animals targeted for death away from the large city zoos to zoos in the smaller cities, which would presumably not be bomb targets. In the capital city of Tokyo, large carnivores and elephants were destroyed in 1943 using various methods, including starvation. At that time the Ueno Zoo had three Asian elephants. It took John, a male, 17 days to die of starvation; the females expired after 18 and 30 days. They were later immortalized by the public and news media as victims of war.

Zoos in other cities reluctantly followed suit. Between September 1943 and March 1944, a total of 26 carnivorous mammals of 10 species were destroyed at the Tennoji Zoo. The methods varied from poisoning with strychnine to strangulation by rope. In Kyoto, army authorities issued an order to destroy "dangerous zoo animals" on March 12, 1944. Between the 13th and 25th of that month, bears and large felids were destroyed by various methods, including being shot with rifles and strangulation by wire. In Nagoya, the zoo continued to keep "dangerous animals" despite pressure from a "vigilance" committee, which consisted of hunters. When 70 B-29 bombers raided Nagoya on December 13, 1944, angry hunters forced the zoo director to action. Before the day was over, two lions, two leopards, a tiger, and two bears were shot to death.

Unlike their European counterparts, few Japanese zoos became victims of air raids. Not all zoos were lucky, however. In Osaka, the massive air raid of March 13, 1945 resulted in 2,000 incendiaries being dropped on the zoo grounds, killing 33 birds of 10 species. For ordinary citizens, this was no time for visiting the zoos. Zoo attendance fell sharply nationwide. In 1942, paid attendance at the Ueno Zoo had surpassed three million. In the following year, it dropped to fewer than two million, further decreased to 360,893 in 1944, and bottomed at 288,832 in 1945. Some zoos were closed, either temporarily or permanently. When the war ended in August 1945, there were hardly any popular animals left in the zoos. Farm animals filled the space once occupied by lions and tigers. At the time Japan plunged into World War II, there were about 20 elephants in zoos. Only three were left in the entire country when the war ended. These three elephants headed the short list of surviving popular mammals, followed by four giraffes and one chimpanzee. Because the military had requisitioned guardrails and metal bars, many zoos lay in ruin. In Kyoto, the occupation forces occupied one third of the zoo grounds from 1946 to 1952, necessitating temporary closure of the zoo.

Severe shortages of feed took its toll as well, killing valuable animals that survived wartime contingency actions and bombing. In Kyoto, a giraffe, an elephant, and a camel, among other animals, died of malnutrition shortly after the war ended. In Osaka, the zoo animal population declined from 447 specimens of 127 species at the war's end to 223 specimens of 99 species a year later. The winter of 1946 was exceptionally harsh in Osaka, causing additional loss of undernourished animals. It was a time when there was not enough food to go around for people, let alone zoo animals. When the war ended, there were 11 red-crowned cranes at the zoo in Osaka. Literally skin and bones and covered with lice, they perished one by one. After a year and a half, only one was left. The harsh reality was that, between 1946 and 1949, nearly half the animal feed at Ueno Zoo consisted of table scraps from soldiers of the occupational forces.

World War II divided the history of Japanese zoos into two distinct and unmistakable segments: the pre-war and postwar periods. The bombing of cities, the atomic devastation, and the first-ever occupation of Japan by a foreign army took an immense emotional toll on the Japanese. More than half a century later, the death of elephants at the Ueno Zoo is still a painful memory of the war. However, a new period of unprecedented prosperity for the zoos and aquariums began shortly after the end of the war. In 1949, the Hogle Zoo in Salt Lake City (Utah, United States) contributed animals to the Japanese zoos, which the citizens enthusiastically welcomed. On April 17, the Ueno Zoo received the first postwar shipment of these animals, consisting of four mud turtles and four box turtles. Shipments later in the year included lions, pumas, coyotes, striped skunks, and macaws. These animals were distributed to various zoos. A group of animals, including a lion and a puma that arrived on June 23, was the first major postwar shipment of animals to the Kyoto Zoo, another gift from the Hogle Zoo. As of July 1949, there were 12 zoos operating in Japan. According to Yamamoto, the fact that two elephants survived through the hard times at the Higashiyama Zoo in Nagoya "was a landmark in Japanese zoo history."[6] Children, eager to see those elephants, visited Nagoya on specially arranged trains from other cities. Children in Tokyo wanted an elephant so much they wrote to Premier Nehru of India for one. Their wishes were granted, and in 1949 Nehru sent a female elephant to the children of Japan. Named Indira after his daughter, the 15-year-old elephant arrived at the Ueno Zoo, her permanent residence, on September 25.

In September 1949, Premier Nehru of India sent Indira, a female elephant, to the children of Japan. At a Tokyo pier she was greeted by Dr. Tadamichi Koga, Ueno Zoo Director (in dark suit). Photograph from *Ueno Zoo: the 100 Year History*. © Ueno Zoo.

Imports of exotic animals continued in 1950. The arrival of elephants symbolized the revitalization of a zoo in a community, reassuring citizens that peace had finally become a reality. Eleven elephants, all Asian females, that landed on Japanese soil between 1950 and 1958 were still living in nine zoos as of March 1997. Some zoos that had been established in the pre-war years rebuilt, expanding their premises by annexing surrounding areas. By about 1952, most zoos had revitalized themselves by bringing in major crowd-pleasers, such as large felids and elephants. In 1959, the Tennoji Zoo was close to its pre-war census of 1,231 specimens of 222 species.

Ueno Zoo celebrated its 70th anniversary in 1952. Among the commemorative events was a circus featuring a lion show that included a trainer, his assistant, and eight lions. Juro Hayashi, who later became the zoo's director, traveled to Kenya and brought back 48 specimens, including a black rhinoceros, hippopotamuses, giraffes, and cheetahs. This anniversary marked the end of the postwar reconstruction period, and the beginning of the expansion period.

The 1950s saw an unprecedented zoo construction boom across Japan. As the number of zoos mushroomed, they spread from large cities to medium-sized and smaller cities. In 1951 alone, four sizable cities built new zoos. Yamamoto commented that "the years 1950 to 1960 may be regarded as the most important period of zoo building, because more than 30 zoos were opened during them."[6]

Several factors contributed to this zoo construction boom. Thanks to postwar economic prosperity, municipalities and businesses established a financial basis for such costly endeavors as building and operating a zoo. The need for recreational facilities for the citizenry, who was freed from the hardships of wartime, as well as the need to serve an increased population caused by the postwar baby boom, contributed to the birth of many a zoo. Additionally, a political and social metamorphosis had taken place in Japan in reaction to the American-styled democracy transplanted by General Douglas MacArthur. This dramatic change brought about a system for the public election of local administrators, and gave women the right to vote. In some cities, candidates for political offices appealed to women voters by promising to build zoos. The zoo construction boom continued into the 1960s and 1970s, albeit at a slower pace, with new zoos and aquariums opening across the country.[1,2,4,5]

9.3 Institutional Overview

9.3.1 The Setting

JAZGA listed a total of 92 zoos and 59 aquariums as member institutions in 1990. The majority of these institutions are distributed in cities located on the Pacific Coast, as well as in the central and western parts of the country. There exists a heavy concentration of zoos and aquariums in the Kansai district, which includes such large cities as Osaka, Kyoto, and Kobe. According to the 1989 statistics, from the 88 zoos whose figures were available, the total attendance was 58,644,131. Attendance figures for aquariums totaled 44,587,526. The combined total attendance, both paid and free, was therefore 103,231,657, which means that a number equivalent to 84% of the country's population visited zoos and aquariums during the year.[7,8]

Still, major Japanese zoos may not be immune to the worldwide trend of decreasing urban zoo attendance. Paid attendance at the Tennoji Zoo fell below the one million figure after 1976, dropping to about 750,000 in 1996. In nearby Kobe, Oji Zoo's attendance fell from 1,270,000 in 1982 to 1,030,000 in 1996. This national trend has not affected aquariums. The Osaka Aquarium Ring of Fire, which opened in July 1990, brought in 5,130,000 visitors during the first year despite a rather steep admission fee. In Kobe, even prior to the reconstruction, the Suma Aqualife Park enjoyed an annual attendance of 640,000. Its reconstruction boosted attendance to more than two million visitors in 1989.

9.3.2 Zoos

Many zoos and aquariums are accredited as museums under the provision of the Museum Act, a national law. That this accreditation provides a level of legitimacy for their operation is evidenced by some institutions publicly announcing their accredited status. For the sake of statistical interest, JAZGA member zoos are classified on the basis of certain criteria. The method used has been patterned, with modification, after an early American Association of Zoological Parks and Aquariums (AAZPA, now AZA) method.[9] The 92 JAZGA member zoos (as of 1990) are first divided into generalized and specialized zoos. Generalized zoos include any zoo that exhibits representatives of two or more classes of vertebrates without special emphasis on a particular class, geographic region, or any other theme. Specialized zoos include institutions that emphasize one type of animal (such as primates, birds, or reptiles) and other animal collections that are part of a much larger entity (such as an amusement complex). These are zoos that may not fit into a traditional zoo definition.

Generalized zoos are subdivided into four categories by animal collection size. Classifying zoos based on a specimen–species count, the S-S count, may appear simplistic and anachronistic. This measurement, however arbitrary, is needed to make the compiled information useful; it is certainly not intended as a "yardstick" for the quality of the institution.

Category A consists of zoos that have more than 1,800 specimens, of which 1,300 are tetrapods (mammals, birds, reptiles, and amphibians) of at least 300 species. Two zoos fall into this category. Tokyo's Ueno Zoo has the largest number of tetrapod species with 417 specimens of 71 mammal species, 506 specimens of 177 bird species, 257 specimens of 70 reptile species, and 157 specimens of 28 amphibian species, for a total of 1,337 specimens of 346 species. However, Higashiyama Zoo in Nagoya holds the Japanese record with a total of 1,740 specimens of 338 species.

The 13 zoos in category B are also relatively large. Listed geographically from the north, they are Maruyama Zoo in Sapporo, Tobu Zoo in Saitama Prefecture (a prefecture is an administrative district similar to a province), Chiba Zoo in Chiba, Tama Zoo in Tokyo, Nogeyama Zoo in Yokohama, Nihondaira Zoo in Shizuoka, Kyoto Zoo in Kyoto, Tennoji Zoo in Osaka, Oji Zoo in Kobe, Tobe Zoo in Ehime Prefecture, Asa Zoo in Hiroshima, Fukuoka Zoo in Fukuoka, and Hirakawa Zoo in Kagoshima. Municipal or prefecture governments own and operate all but one of the zoos in categories A and B, combined. The exception is the Tobu Zoo, which a private railway company owns. Only one, Nogeyama Zoo, has no admission charge. Although the other zoos charge admission, it is a common practice in many zoos not to charge admission fees for children.

Category C consists of zoos that have about 500 specimens, or at least 400 tetrapods of at least 100 species. The 18 zoos in this category might be called Japan's medium-sized zoos. Four of them are privately owned, although a nonprofit organization manages one of them. Municipalities own and operate the others. All 18 zoos charge admission fees. The 29 zoos of category D are the small zoos. These smaller collection sizes range from fewer than 100 to several hundred specimens. Eighteen of them charge admission fees. With the exception of two privately owned zoos, they are all tax supported.

Among the 30 specialized zoos, municipalities own 4 (a private organization operates 1 on contract), nonprofit corporations run 6, and 20 are private. Of the latter, the railway industry owns 5. Collections specializing in bears, goat antelopes, primates, birds, reptiles, and alpine animals characterize some of the holdings in this category. Also included are safari parks and collections that are part of amusement parks. One of the 30 zoos did not provide information on attendance or admission fees. All of the 29 others charge admission.

In terms of acreage, Japanese zoos are typically small, and many are located near the center of a city. Older zoos, such as the Ueno Zoo, with 35 acres in its premises and the Kyoto Zoo with a mere 10 acres, tend to be quite small. By comparison, newer institutions were allocated relatively larger areas. For example, Nagoya's Higashiyama Zoo (opened in 1937) has 74 acres and Tama Zoo in Tokyo (opened in 1958) has 130 acres.

9.3.3 Aquariums

During the pre-war years, aquariums were uncommon. They were more academically oriented: perceived as a laboratory, part of a university, and geared toward studies of marine biology and fishery science. Postwar aquariums became commercialized as showplaces for marine life, with private firms controlling many of them. The first large-scale private aquarium was probably Enoshima Aquarium, which opened in 1954. Located in a well-known resort area south of Tokyo, this aquarium became a popular tourist attraction. During the 1960s, following the zoo construction boom of the 1950s, aquariums proliferated. In 1967 alone, eight new facilities were opened. In the early 1990s, there were approximately 80 aquariums, including non-JAZGA-member institutions.

Aquarium popularity is partially a cultural phenomenon. Japan is an island nation, surrounded by oceans. Marine life and the aquatic environment have had a profound influence on every aspect of Japanese life, including language, culture, and food habits. The Japanese are typically fascinated by marine life such as porgies, sea breams, and octopi at aquariums.

In contrast to zoos, more aquariums are privately owned than are tax supported. Private firms own and manage 30 of the 59 aquariums, including 5 the railway companies own. Of the 23 tax-supported aquariums, prefectures own 4, the national government owns 1, and local governments administer the rest. Private organizations operate 4 of the tax-supported aquariums on contract. Nonprofit corporations own and manage 4 aquariums, and universities govern 2 aquariums. Thus, ownership and modes of operation of aquariums are more diverse than those of zoos.

The following data use an S-S count based on fish as an arbitrary criterion. Among the 59 aquariums, 5 have more than 10,000 specimens of 300 or more species: these are, beginning from the north, Otaru Aquarium in Otaru; Joetsu Aquarium and Teradomari Aquarium, both in Niigata Prefecture; Sunshine International Aquarium in Tokyo; and Toba Aquarium in Mie Prefecture. Four others have at least 10,000 specimens of fewer than 290 species: Niigata Aquarium in Niigata, Notojima Beach Park Aquarium in Ishikawa Prefecture, Hiyoriyama Aquarium in Hyogo Prefecture, and Suma Aqualife Park in Kobe. Six others have sizable collections but with relatively fewer species or specimen counts: Aburatsubo Marine Park Aquarium in Kanagawa Prefecture, Marine Science Museum of Tokai University in Shizuoka Prefecture, Takeshima Aquarium in Aichi Prefecture, Shima Marineland in Mie Prefecture, Katsurahama Aquarium in Kochi Prefecture, and Nagasaki Aquarium in Nagasaki. Data have been compiled for these 15 institutions only.

9.3.4 Traveling Menageries and Safari Parks

The history of traveling menageries dates back centuries. Although they are not categorized as zoos in a traditional sense, they exhibit exotic animals as zoos do and so are discussed here. Traveling menageries imported a large number of various exotic species from pre-Restoration days through the 1950s, leaving colorful episodes behind. In 1907 an orangutan escaped from a traveling menagerie in a small town in central Japan. This animal enjoyed a brief sojourn in a dense grove of trees adjacent to a Shinto shrine, building a nest in a tall zelkova tree, from October 6 until it was brought back to its cage October 9. More orangutans were brought to Japan for sideshows and traveling menageries during the period from 1912 to 1920, reaching a peak in 1919 and 1920. Ten adults and fourteen or fifteen young were imported a year. In one instance two mothers with infants were brought in.

The wholesale importation of exotic animals in the 1950s was partly the result of demands from large traveling menageries, which characterized the decade. Although menageries were a passing phenomenon in the history of animal exhibition, they introduced species to the public that had never before been kept in zoos. During the 1950s, there were three large traveling menageries. One of them, the Japan Traveling Zoo, imported some of Japan's "firsts," including an African elephant, a juvenile that arrived in March 1953, and gorillas, a juvenile pair that arrived at Haneda airport in Tokyo on December 23, 1954.

The surge of drive-through animal parks, or safari parks, that swept through Europe and United States came to Japan in 1975 when a park opened its gates at Miyazaki in southern Kyushu on November 1. During the first year, it brought in 1.8 million visitors. By mid-1978, five safari parks were in operation across Japan, while six more were said to be either under construction or on the drawing board. Behind the rush to build safari parks lay the real estate industry. During the days of rampant land speculation, firms purchased land for later use as golf courses and for housing development. Recession and tighter land-use ordinances necessitated some realtors to look for other ways to market their land. The lands were then diverted for use as safari parks, which, with a few exceptions, was how safari parks were born. In several years the safari park construction boom began to wane. As of the summer of 1992, eight safari parks were in operation, including one non-JAZGA-member park.

Some mammalian species that had not bred in traditional urban zoos began to reproduce in safari parks for the first time in Japan. In all the following cases, each park initially brought in a sizable multimember animal group and maintained the animals in a large enclosure. Undoubtedly, these were the contributing factors for successful reproduction. Kyushu African

Safari in Oita Prefecture received 40 adult cheetahs (21 males and 19 females) from southwestern Africa in May 1976. Two cubs were born on January 5, 1977. Located not too far from this park on the same island was Miyazaki Safari Park. Twelve white rhinoceroses (four males and eight females) arrived there in 1975. These animals had been captured in South Africa in 1972 and kept in Europe for three years prior to departure for Miyazaki. Five calves (two males and three females) were born between April 12 and July 16, 1978.

The first African elephant birth in Japan took place at the Gunma Safari Park in Gunma Prefecture. The park acquired two male and six female African elephants in 1979, the year the park was opened. A female died on September 2, 1981, and one male and one female were sent to South Korea on October 29, 1982. (This female later gave birth to a calf in Korea, establishing the record for the first African elephant birth in the Far East.) One male and four females remained at Gunma, and one of the females, Sakuve, gave birth to a male on January 31, 1984, but the calf died after 30 minutes. On May 5, 1986, she gave birth to a healthy male calf.

9.3.5 Japanese Association of Zoological Gardens and Aquariums

Established in 1939, JAZGA listed membership of 97 zoos and 65 aquariums by 1997. H.I.H. Prince Akishino, the second son of the Emperor and a trained biologist, is the President of JAZGA. Headquartered in Tokyo, JAZGA is engaged in three types of activities: conferences, awards, and publications.

An annual conference is held for four days, mainly focusing on business and administrative issues. JAZGA also conducts a two-day seminar on business administration every year. With regard to biology, the association conducts three symposia per year, for two to three days each, on general animal management, aquarium management, and marine mammals. Also, each region holds a one-day symposium on general animal management and aquarium management.

In 1957, JAZGA established the First Breeding Award for the first successful reproduction of a species in the country. This award has two categories: parent-raised and hand-raised, so the award may be given twice for the same taxon. The Special Achievement Award began in 1960. One to three recipients are selected each year from authors who contribute articles to the *Journal of Japanese Association of Zoological Gardens and Aquariums*. The Koga Award, established in 1986 and named after Dr. Tadamichi Koga, a former director of the Ueno Zoo, is a third award. This award may be given to up to three recipients a year for outstanding achievements in animal reproduction.

Various publications, which are almost entirely in Japanese, reveal the most tangible results of the association's efforts. Periodicals include the *Journal of Japanese Association of Zoological Gardens and Aquariums*, a quarterly technical publication started in 1959, and a monthly *Newsletter*. Veterinarians' case reports and marine biologists' observations filled pages of the *Journal* in the early years. More recently, topics have become diverse, reflecting the broadening scope of authors. Some articles also have English summaries. Other publications include a zoo and aquarium directory, which is published every decade, and an animal keepers' manual for zoo and aquarium personnel.

In terms of vital information on Japanese zoos and aquariums, the most noteworthy publication of the association is the *Japanese Association of Zoological Gardens and Aquariums Annual Report*, which was initially published in 1951. Summarizing the business of member institutions for the year, this soft-bound two-volume work consists entirely of statistics with no editorials or narrative descriptions of events. The content, entirely in Japanese, is divided into nine categories, including business-related subjects, such as finances, attendance, and new buildings and exhibits, and biological statistics, such as births and hatchings, animal longevity, mortality, and animal inventory. The inventory, which is the second volume, covers everything from mammals to insects and gives the number for each taxon its member institutions maintain.

Users of documents from Japan should be aware of some peculiarities. For example, the fiscal year in Japan begins on April 1 and ends on March 31 of the following year. This means the 1996 *Annual Report* presents data as of the end of March 1997, not December 1996. The animal inventory, however, is an exception to the rule, listing animals as of December 31.

9.4 Administrative Overview

9.4.1 Governing Authorities

Japanese zoos, be they large or small, are usually operated by local goverments. Whereas the most advanced and scientifically operated zoological collections in the United States are under the jurisdiction of a zoological society, this mode of operation was nonexistent in Japan before the 1980s. Local governments own and manage the mainstream Japanese zoos. By contrast, as mentioned above, aquariums are in many cases privately owned and operated.

In the mid-1980s, discussion about the privatization of municipal zoos dominated the zoo scene, whereby zoos would be subsidized by nonprofit, private organizations, in most cases zoological societies. The intention was that such groups would take over zoo operations partially or entirely. On the surface this proposal may appear to be a move in the right direction, potentially unchaining a zoo from a bloated bureaucratic system and allowing the zoo to pursue its essential missions unfettered. The truth of the matter was that this plan was simply another attempt at local government administrative reform, downsizing the workforce. In other words, it had nothing to do with the modernization of zoos, and some believed it was a way for the municipality to unburden itself of the need to operate a zoo. In Hiroshima, the privatization

---

RAILWAYS BUILDING ZOOS

Japan's rapid growth in finances and industries in the early twentieth century stimulated the evolution of population centers, which in turn allowed the emergence of a new breed of middle class citizens, the urban commuters. This brought about the development of railway and housing industries in the outskirts of large cities to accommodate the commuters' needs. In the new suburbia, private electric railway companies promoted economic development along the railways, building large department stores and amusement parks. Exotic animals were kept in those amusement parks as a part of the marketing strategy. Thus, the marriage of zoo and amusement park was established. An early example may be traced to a park opened at the turn of the last century, but it was short-lived. Hanshin Park, which opened in 1929 as part of a huge entertainment complex between Osaka and Kobe, was the most prominent of all the amusement park–style zoos that the railway companies managed. By and large these amusement parks have been successful, catering to the need of the populace for entertainment. The success of these kinds of zoos has profoundly affected the historic course of zoos in Japan. It is probably no exaggeration to state that amusement park zoos have helped formulate the concept of a zoo as a showplace of exotic animals, rather than a kind of natural history museum or educational facility.

of Asa Zoo became a local controversy in 1986, and citizens petitioning against such a measure collected more than 400,000 signatures. Despite this effort, the zoo was transferred to the zoological society as planned.

An interesting characteristic of the governance of Japanese zoos is the role that privately owned electric railway companies have played. Local railway companies have established at least nine zoos and five aquariums. These facilities were mostly designed as part of amusement complexes near train stations to attract passengers and to promote economic development of the area along the railroad. Moreover, the influence of these railway companies often escapes published zoo statistics. Owning only nine zoos and five aquariums may not seem significant; however, there is more there than meets the eye. Even in cases where other governing bodies nominally own zoos, whether municipal or private, railway companies may have an indisputable influence and control over the management.

For instance, the Japan Monkey Centre (JMC), which operates a primate zoo in west-central Japan, is by definition a nonprofit, private organization. Visitors approaching JMC will notice that JMC is within a minute's walk from a train station that the Meitetsu railway company owns. Once inside the administrative office of the primate zoo, the visitor will find that the business staff of this zoo is actually a part of the railway company. As a matter of fact, the railway built JMC and exerts unquestionable control over its operation.

Another example of a railway company's influence is the Tama Zoo, a rapidly growing, large suburban zoo in Tokyo. Tama Zoo is a municipal institution; however, from the very start, the Keio railway company was heavily involved in the construction of the zoo. In the spring of 1955 an area of approximately 74 acres of forested, rolling hills was donated to the Tokyo metropolitan government as an initial zoo site with the cooperation of Keio. Moreover, Keio donated a sum of 140,000,000 yen for its construction. When the zoo was opened in 1958, it was virtually impossible to visit the zoo without taking a train that Keio operated. From a business point of view it was a shrewd move, indeed. The railway enjoys a lucrative monopoly of transporting passengers to and from the zoo, while leaving the responsibility for managing the expensive zoo in the hands of the Tokyo metropolitan government.

Staff

According to 1990 statistics, the 92 zoos then listed by JAZGA employed a total of 3,950 persons, ranging from 5 to more than 100 per institution, while the 59 aquariums then listed employed a total of 2,321 persons, from 5 to nearly 200 per institution. Thus, the JAZGA member institutions had a combined total of 6,271 employees, including administrative, clerical, animal care, and maintenance. However, personnel listed in the *Annual Report* may not reflect the actual numbers working in certain areas. For example, the lists may or may not include part-time or temporary employees. Also, depending on individual institutions, such work divisions as food service or ground maintenance may be contracted-out to concessionaires.

Included in the 6,271 employees were animal collection personnel: 1,527 for zoos and 694 for aquariums. The number per institution ranged from 2 to 64 for zoos, and from 2 to 38 for aquariums. In traditional tax-supported urban zoos, animal collection personnel make up a substantial part of the total staff — between 43 and 67% in zoos in such cities as Tokyo, Osaka, Nagoya, Kyoto, and Kobe. In the private sector, by contrast, particularly in commercialized operations and amusement parks, personnel involved with the animal collection constitute a small segment of those employed. For instance, out of 116 employees in one amusement park, only 5 were charged with caring for the animal collection. As for aquariums, in one private institution 51 out of 134 employees were in the catering division, while only 15 were animal care personnel.[7,8]

Statistics aside, attention should be given to the organizational structure and daily work of the staff for more insight. Superficially, zoo workers may appear to be functioning in the same way anywhere in the world. Closer examination into the inner fabric of the institutions may reveal unexpected findings.

There are three basic positions in a zoo that affect management of the animal collection: zookeepers, curators, and directors. The quality of zookeepers is generally high in Japan. Many gather data on the animals under their care and publish papers with valuable, original information. Keepers are a bread-and-butter component of any zoo, but in Japan they play a larger role. Dedicated and capable zookeepers take initiative and commence animal management programs, at times without clear directions from their superiors. Emily Hahn once observed such an example, a remarkable occurrence at the Higashiyama Zoo. Rikizo Asai, a zookeeper, trained three western lowland gorillas as part of a management program, which later became extremely popular:

> People lifted children to their shoulders, or just pummeled their neighbors in the scramble, as a tall, lanky man in overalls and a floppy hat came through the door, followed by a sort of chain gang of three gorillas, single file, the second and third in line with their hands on the shoulders of the one ahead. What was striking was that they weren't baby gorillas or even toddlers; they were nearly full-grown. Once in a great while some animal trainer will put on an act where he is at close quarters with a gorilla of reasonable size, but it happens very seldom, and I'd never before seen one man on such familiar terms with a trio. Chimpanzees are generally said to be difficult actors after the age of eight, or even six, but gorillas are believed to be difficult at any age. Yet these three went through their paces without any friction at all .... After my visit to Nagoya I bored people to death, talking about the gorilla act. Some of my hearers didn't know anything about the subject and so couldn't appreciate the wonder of it; others just didn't believe me and said that what I'd seen were not gorillas, but chimpanzees — a remark calculated to send me into frenzies of indignation.[10]

As for middle management, it must first be noted that the curatorial system in the European and American sense does not exist in Japanese zoos. When European and American zoo workers visit Japanese zoos and meet with managers, an automatic assumption is made that Japan, too, has the curatorial system. Likewise, when Western zoo personnel are introduced to veterinarians, it may be assumed that the Japanese veterinarians share the same educational social status as their overseas counterparts.

According to Hediger, "Today the wild animal is considered to have cultural value; it is regarded as part of our heritage, to which the whole mankind and particularly future generations, has a legitimate claim. Zoological gardens, to which these living items of culture are entrusted, therefore represent cultural institutions."[11] This view elevates a zoo from an animal showplace to a more sophisticated institution. Although animals represent the most conspicuous component, in essence they are just one of the materials that constitute a zoo; it is people, not animals, that make or break the zoo. It is therefore vital to take a moment for a close look at the human dimension of Japanese zoos, with special reference to the management staff.

Instead of the curatorial system, Japanese zoos have supervisors of varying ranks who are usually in a linear chain of command. These supervisors oversee keepers and the day-to-day care of the animals. Their scope of supervisory duties hardly extends beyond the realm of maintaining animals for public display. Animal dealers, who stay in the shadow of the zoo directors, may often make decisions on the zoo's animal collection structure, albeit unofficially. These animal dealers may handle the selection of species, since the zoo's supervisors often lack basic knowledge about wildlife and its management in captivity. More about the role of animal dealers later. The supervisors' concerns seldom extend beyond animal feeding, animal housing, and treatment of diseases. The burden of other critical decisions concerning animal husbandry, such as establishing a breeding program, often falls on the keepers' shoulders. In addition, scientific work and the technical aspects of animal health care are in the hands of veterinarians. Discussion of the role of veterinarians is crucial to understanding management in Japanese zoos.

"MacArthur should have upgraded our education system, but he did not," remarked a Japanese veterinary professor back in the late 1950s. After Japan lost World War II, General Douglas MacArthur of the U.S. Army marched into the arc-shaped archipelago and began waves of reform that profoundly affected all aspects of people's lives. Many sweeping changes, including educational reform, were patterned after the U.S. system. However, some areas remained relatively untouched. Veterinary education was left in a less advanced stage, especially compared with its American counterpart; correspondingly, the social status of the veterinarian stayed at a considerably lower level than that of the American veterinarian. To become a veterinarian in Japan then required only four years of post-secondary education, in contrast to eight or more in the United States. It was common for a Japanese zoo to hire a nationally certified veterinarian as a zookeeper, a practice unthinkable in the United States. In the late 1990s, there were still veterinarians who worked as keepers. Veterinary education reform did come in the 1980s, when the veterinary degree was upgraded to an equivalent of a U.S. master's degree, or six years of post–high school education. But even after the change, the training and education to become a veterinarian lag behind the American standard. For this reason alone, zoo veterinarians in Japan may not be viewed on the same level as their American counterparts, although some exert immense authority and power over Japanese zoos.

In the United States, veterinarians and veterinary students who are interested in zoos choose this field in the course of their studies and commit to a career in this specialty. They may relocate from one zoo-related employment to the other, but stay in the field as an integral part of the zoo. Such is not the general rule in Japan, particularly in the public sector, although some individuals stay in zoos indefinitely. In municipalities, veterinarians are frequently transferred between a zoo and other local government departments totally unrelated to zoos at the whim of the administration. There are veterinary students who wish to specialize in zoo animal medicine, and with a little luck they may find employment in zoos and take a deep interest in wild animals. However, after a couple of years they may be sent to a totally different field — perhaps public health, dog pound, or meat inspection — and replaced by transferees from other municipal departments, who may have little interest in zoos. Or they may be selected for higher positions in a zoo, with responsibilities far removed from animal care. Under such a system, fundamental errors can easily be made in the treatment of wild animals. Also, under the circumstances, it is difficult to maintain professional continuity and to develop a core group of persons dedicated to upgrading their expertise.

Tremendous challenges await those veterinarians who are capable and fortunate enough to stay in zoos. Veterinarians form practically the only technical staff in their workplace, yet they are often overwhelmed with responsibilities other than animal health. For example, one veterinarian even had the management of amusement rides as one of his responsibilities. This may be an extreme example, but, in general, the more the duties of veterinarians are spread among diverse areas, the more difficult it is for them to grow into specialists or to set up long-term programs in animal management or research. This is in stark contrast to American zoo veterinarians, who often specialize in certain areas and who play increasingly prominent roles in the cutting edge of international species survival programs and related activities.

Subordinate to veterinarians in the power structure are managers who have majored in livestock science, and these two groups with agricultural backgrounds are the chief operators of Japanese zoos. Historically, and to this day, graduates with degrees in biology or zoology

Tadamichi Koga — Japan's Mr. Zoo

Over the past century, a few interesting characters, scholars, and visionaries may be found in the ranks of directors at Japanese zoos and aquariums. Thus far, however, Japan has produced but one towering giant, and to this day he stands out as the nation's only zoo man of international stature. In terms of personality, leadership, administrative skills, intuition in biology, and, above all, vision, he has no peer in his own land. Dr. Tadamichi Koga (1903–1986) was born into the family of a former *samurai* on the southern island of Kyushu. After graduation from the prestigious University of Tokyo College of Veterinary Medicine in 1928, he took a position at the Ueno Zoo and became director in 1932, serving in that capacity until 1962. Examples of his work include the propagation of cranes and polar penguin care. Earlier in his career he became the spokesman for zoos, and thus established himself as Japan's "Mr. Zoo." Although the war interrupted his productive years, he became a leading conservationist in postwar Japan. Within the restrictions of a municipal government he managed to modernize Ueno Zoo. His legacy continues at the Ueno and Tama Zoos, both in Tokyo. Despite ups and downs over the decades, these zoos maintain a level of professionalism that is seen nowhere else in Japan.

have been virtually absent from the zoo scene. In contrast, scientific aspects of aquarium management are in the hands of marine biologists, who maintain a higher level of technical expertise. Regardless of their background, young and capable persons who enter the zoo world will soon find that there exist few mentors or role models who can nurture and develop their potentials. Even if they stay in the zoo field with determination and tenacity, opportunities for promotion and advancement are limited. Upper managerial positions, such as assistant director and director, are usually reserved for political appointees with no training or experience.

In the private sector, zoo directors are often businessmen from other divisions of the parent company. In the public sector, the typical zoo director's position is usually a convenient and brief stopover for a senior official. Most zoo directors occupy the position for just a few years — younger ones eagerly waiting for higher positions and older ones comfortably waiting for retirement. In Kyoto, during one period from the 1930s to 1960s, the zoo had eight directors at the helm, with an average term of office a little over three years. These directors often represent an alien element to the zoo profession and are often not familiar with the many facets of the zoos' programs. It is small wonder that Japanese zoo directors, by and large, rarely show a grasp of issues regarding wildlife and zoos on a national level, let alone on a global level. After all, this is a nation where a zoo director's achievements are frequently measured by increased public attendance. Often, a zoo's primary mission is entertainment.

Sheltered Complacency

Federal laws and regulations abound in the operation of American zoos, particularly with regard to animal health, quarantine, and importation. In addition, American animal protectionist groups closely scrutinize zoos. And in more recent years, precarious economic situations in urban communities have put pressure on municipal zoos to be financially independent. Such is not the case in Japanese zoos, where lax quarantine restrictions allow exotic animals to be imported at will. On a social level, animal protectionist groups have yet to gain influence, resulting in fewer laws and regulations pertaining to zoo operations. Economically, the continuing growth in Japan allows generous expenditures, in most cases, for zoos. As a result of these various conditions, Japanese zoos operate with fewer legal, social, and economic pressures, compared with their American counterparts. Unfortunately, such conditions may lead to a sense of secluded complacency and stagnation.

Publications

In addition to the various JAZGA publications already mentioned, individual zoos, especially the large and medium-sized institutions, often have their own publications. Most major zoos and aquariums produce guidebooks, maps, and postcards. These zoos often publish magazines containing articles of interest to both the general public and zoo professionals. The Japan Monkey Centre, Tennoji Zoo, Oji Zoo, and Asa Zoo are among the zoos that have such periodicals. Tokyo Zoological Park Society publishes *Insectarium*, a popular monthly journal on entomology. These journals are entirely in Japanese, with the exception of *Animals and Zoos*, a monthly of the Tokyo Zoological Park Society, which occasionally includes English summaries. Like American and European zoos, some Japanese zoos publish annual reports.

9.4.2 Marketing and Events

Japanese affection for, and attachment to, cherry blossoms is legendary, taking on an almost religious fervor. The family outing in spring during cherry blossom season is an important annual event. Some zoos have groves of cherry trees on their premises, providing a popular attraction.

Once the major marketing tool for Japanese zoos was bringing in rare and unusual animals. These zoos, however, can no longer rely on an indiscriminate importation of exotic animals from overseas now that the government has ratified the Convention on International Trade in Endangered Species of Wild Fauna and Flora (CITES). As public zoo attendance in large urban areas begins to decrease, diversification of zoo marketing schemes for both children and adults is mandatory. Private zoos have been pioneers in this area, especially those the railway companies manage. For example, private zoos routinely hold floral exhibits, such as chrysanthemum shows in the fall. Other examples of special events include world expedition exhibits, animal postage-stamp shows, transportation fairs, Wild West shows, fairy tale exhibits, and animal circuses.

New Year's Day is the most important holiday in Japan, with families paying the first visit of the year to the Shinto shrine. Some zoos participate in this tradition by taking animals, such as chimpanzees, to the shrine. Other unique annual events in Japanese zoos are ceremonies to honor the souls of dead animals. Foreign visitors may notice what appears to be a tombstone in Japanese zoos. As a rule an epitaph is engraved on the stone, and one may see a Japanese zoo visitor offer a bouquet of flowers. Religious ceremonies are conducted, usually during the autumnal equinox, in front of this small monument. Not only people attend such an event, but also an animal representative such as an elephant. What appears to be a tombstone is actually a cenotaph, and the annual religious event is a mass for the repose of the animals' souls. This ceremony is a product of Japanese spiritualism that originated in animism. It is a common practice for the Japanese to pray for the peace of animals that died for the well-being of humans. Memorial services for dead animals are held both on a regular basis and on special occasions. After the so-called dangerous animals were destroyed during World War II, zoos held services for the animal war victims.

Some zoos have had a long history of children's summer classes as annual events. In August 1937, the Itozu Zoo in Kyushu commenced "a school in the woods," a summer event for children on a variety of subjects including arts, folklore, and physical education. The subject of animals was added the following year. More recently, education programs have been on the increase in Japanese zoos.

9.4.3 Philanthropy

In Japan, tax levies funds new construction projects in municipal zoos; private owners pay for these projects in private zoos. In contrast, in the United States, philanthropists, foundations, and business corporations often provide donations for such undertakings. Philanthropy is not the rule in Japan. As a matter of fact, there is a profound difference in the concept of "donation" in the two nations.

Various organizations host fund-raising events in United States. Wealthy citizens commonly make contributions to universities, art museums, hospitals, and zoos. Exhibits, buildings, libraries, scholarships, and faculty positions may be named for the donors. In addition, wealthy individuals may bequeath a large share of their estates to a public institution. Similarly, when a child of a wealthy family dies, parents may donate a large sum of money in memory of the child. The U.S. income and inheritance tax laws powerfully encourage such practices.

The custom in Japan is quite different when it comes to philanthropy, but the lack of such acts of giving may not necessarily indicate Japanese insensitivity to a social cause. Rather, it is largely due to the difference in tax laws. When an American donates to a public institution, the amount given is deducted from taxed income or inheritance. In sharp contrast, even if a wealthy Japanese citizen wishes to donate to a zoo, there is no tangible tax benefit and hence no financial incentive.

9.4.4 Animals as Commodities

Municipalities own most mainstream zoos in Japan. In municipal operations, animals are first and foremost a municipal property, rather than a biological resource that professionals collectively manage. It is a common practice in municipal zoos to inventory animals in terms of monetary value. Transferring animals outside of the municipal premises, even on a breeding loan basis, may require an incredible amount of bureaucratic red tape. Animal ownership, firmly rooted in a rigid bureaucratic system, could hinder the advancement of national and international species survival programs, an issue that will be discussed later.

Another unique aspect concerning Japanese zoos is the role of animal dealers. The zoo director may have experience with livestock management; however, after six months at the helm, he is hardly familiar with zoo animals. He may have little interest in wildlife, and is in no position to make a reasonable decision or take initiative in selecting species for his zoo. In the absence of a curatorial staff, he may not have input from middle management either. Enter an animal dealer with years of experience in and knowledge of exotic animals who can offer a selection of attractive species that will draw crowds and consequently increase attendance. It is no exaggeration to say that these animal dealers profoundly influence the structure of animal collections in Japanese zoos. This further strengthens the position of animals as a commodity, which is measured in terms of their monetary value.

Thumbing through a 1989 annual report of a large zoo in JAZGA category B, one may discover just how animal dealers have had their fingers "in the pie." During this fiscal year under "notable animal acquisitions," 100 mammals and birds were listed, ranging from gorillas, marmosets, cranes, flamingos, and macaws. Animal dealers brought in 64 of these animals under such arrangements as exchange, purchase, loan, and compensation. Of the other 36 specimens, other zoos donated 7 and 4 arrived as exchanges with other zoos. In all, 131 mammals and birds, including marmosets, colobus monkeys, a Malayan tapir, cranes, and macaws were in the "notable animals shipped" column. Of these, 103 animals (or nearly 80%) were given to animal dealers as exchanges or returns. Only 14 were donated to other zoos. Typically, very few animals are traded directly between zoos; animals usually move from zoo to zoo by way of animal dealers.

9.5 Animal Collections

9.5.1 Marveling at Giraffe

Many Japanese people have an intense sense of curiosity. Confined in an island nation, surrounded by water, many have cast a keen eye on things exotic for centuries. Carl Hagenbeck, an internationally known German animal dealer, circus man, and zoo man, played a vital role in the infancy of Japanese zoos by supplying rare and unusual animals. At the turn of the twentieth century, Hagenbeck supplied a great many exotic animals, including species that were "firsts," such as giraffe, llama, rhea, and Andean condor. Another shipment of animals from Hagenbeck came in 1933, bringing in other "firsts," such as South American tapir and black rhinoceros. The rhinoceros, however, was too expensive and left Japan without finding a permanent home.[12]

Lorenz Hagenbeck, who toured Japan in 1933, was impressed with the interest of the Japanese people. His show was scheduled to be in Nagoya during the rainy season. "The most you can put your show on at Nagoya is four days," he was told. He then recalled, "We had a five weeks' run at Nagoya, with two performances daily, at two and at five, both always sold out to the last seat. In addition, on many days we counted no less than twenty-five thousand visitors to the menagerie. The giraffes were invariably surrounded by classes of school children busily drawing, by photographers and by astonished sight-seers." He went on to say, "Behind us, as living mementos of our six months' stay on tour of Japan, we were also leaving polar bears and those fairy-tale animals the giraffes, for Tokyo's Ueno Park Zoo had made sure of those powerful attractions for itself."[13]

Gradually, Japanese animal dealers began to import zoo animals actively. A newspaper clipping from East Africa in the early 1930s described this episode: "There is desolation among the juvenile residents of Upper Parklands, but relief among proud owners of gardens — Sally and Bronco are now on their way to Japan, in company with cheetahs, hyenas and chameleons." Sally and Bronco were two giraffes, which the white hunter, H. R. Stanton, sold to Isamu Kagawa, a Japanese animal dealer (who, incidentally, became director of a zoo in Fukuoka after World War II). The article stated that Stanton made a specialty of supplying zoos with giraffes. Kagawa obtained the giraffes, 3 cheetahs, 6 hyenas, and 100 chameleons. "On the way to Nairobi railway station from Stanton's grounds in the Lower Kabete road, the number of the latter was increased — a female chameleon gave birth to a small family, probably half a dozen or so, a long way short of the record of 29."[14]

9.5.2 Animal News Makers

During its heyday, Osaka's Tennoji Zoo enjoyed great popularity. A decade beginning in the early 1920s was marked by the introduction of rare species. Representative of this era were a Sumatran rhinoceros, the only specimen of its species recorded in Japan, which arrived in June 1921, and a female chimpanzee named Rita. Rita, imported by the above-mentioned Mr. Kagawa, was extraordinarily well trained for public entertainment. She made her debut in 1932, and soon became a national celebrity.[2] Her enormous fame boosted zoo attendance to an all-time high of more than 2.5 million in 1934. In comparison, the visitor attendance at Tokyo's

Ueno Zoo for the same year was 1.68 million. Rita, because she was highly individualized, if not humanized, was probably recognized as Rita first, and as a chimpanzee second. In that respect, her legendary popularity far exceeded that of the postwar animal craze of such charismatic megamammals as the giant panda and the koala. Stimulated by Rita's success as an attraction, chimpanzees were soon purchased for zoos in Nagoya, Kyoto, and Kobe.

Some institutions had an early start in the exhibition of live cetaceans. Around 1928, Nakanoshima Aquarium, the forerunner of Izu Mito Sea Paradise, kept the minke whale and the Gill's bottle-nosed dolphin. In 1935, a pilot whale was placed on temporary display at an exposition in Yokohama and drew great popular interest. Another noteworthy episode took place at Hanshin Park, a zoo in Nishinomiya, where 10 false killer whales were maintained for eight months in 1937. Small toothed whales became more common in captivity across Japan in the late 1950s.

The Japanese military often brought animals from the Asian front into zoos. In January 1939, ponies, mules, donkeys, Bactrian camels, and dogs that the military used in China arrived at the Ueno Zoo. Ceremonies were held to honor these animals. During the Russo-Japanese and the Sino-Japanese Wars, the military had also brought in animals. During World War II, unusual species began to arrive from South Asia. Two Komodo monitors, which the Imperial Navy captured on Komodo Island, arrived at the Ueno Zoo on October 30, 1942. Unfortunately, because of the fuel shortage, the zoo was unable to keep the building warm enough for them. The first two Komodo monitors in Japan died on February 16 and June 12, 1943.

Recovery from the war years necessitated that zoos discover new attractions, especially in the Kansai district where competition was greater among neighboring zoos. The populace was ready for something more exciting. One marketing scheme concocted was to hybridize large felids, thereby "creating" a new breed of ferocious beast beyond the imagination of the ever-curious public. Some orthodox zoo officials shunned such showmanship, but the thrill of the plan was too tempting. In the early 1950s the race for new attractions was already on for zoos.

The first postwar wave of large felids to Japan had already produced offspring, enabling zoos to put together juvenile lions and tigers as future mates for the project. While other zoos attempted to create ligers and tigons, Hanshin Park decided in 1953 to aim for a leopard–lion cross. Two cubs born in the park, a lioness and a male leopard, were introduced to each other in December 1955. The first litter of leopons, consisting of a male and a female, was born on November 2, 1959. This was believed to be the first such hybridization in the world. The second and the last litter of one male and two females was born on June 29, 1961. According to Emily Hahn, "All are spotted: the older male has longer hair around his neck, like a rudimentary mane, and all five take after their mother rather than their father in being large."[10] At the time, Hanshin Park beat out all other competitors, but the race to create more "artificial" beasts accelerated after the birth of the leopons.

Zoo efforts to cross a leopon and a tigress were in vain. The first leopon, a male, died on April 10, 1977; the last of the five died in 1985. There had been another episode in the saga of the hybrid felids. On September 8, 1975, three female cubs were born to a tigress mated with a lion at Tennoji Zoo. All those ligers died within a week. By this time the public opinion concerning zoo hybrids had shifted; criticism of a "side show mentality" was mounting, pressuring zoo officials to separate the lion–tigress pair.

The 1970s will be remembered for the panda craze that swept Japan. On September 29, 1972, Japan's Chief Secretary of the Cabinet, while visiting China, made an announcement of the gift of a pair of giant pandas from the people of China to Japan. Ueno Zoo's switchboard was deluged with calls from the news media and citizens, asking about those animals. Everyone rightly assumed the pandas would be sent to Ueno, but totally unaware of any details, the zoo staff was unprepared for this flood of inquiries. And this was just the beginning of the craze. The first giant pandas to arrive in Japan, four-year-old female Lan Lan and two-year-old male Kang Kang, arrived in Tokyo on October 28 of that year and were housed at the Ueno Zoo. On November 5, the day of their public debut, zoo enthusiasts waited in a mile-long line to see them. The annual attendance of the zoo for 1973 reached 9.19 million, more than a 50% increase over the previous year.

Goodwill animal exchanges had frequently taken place between Japanese zoos and foreign countries. The postwar animal exchange with China began in 1955, even before official diplomatic relations were established. In the early 1970s, the two giant pandas opened the door to even more traffic of animals between Japan and the People's Republic of China, as Tokyo and Beijing tried to broaden their diplomatic relationship. For example, in 1974, the Tennoji Zoo sent a pair of sea lions and two pairs of Humboldt penguins to the Shanghai Zoo, and received a pair of wolves (melanistic form) and a pair of white-naped cranes in return. Also, two pairs of blue-and-yellow macaws and a pair of red kangaroos were sent to Beijing Zoo, which in turn shipped a pair of Mongolian gazelles and a female red-crowned crane to Osaka. Ever since, zoos across Japan have received many animals from China including red pandas, Francois leaf monkeys, cranes, and pheasants.

Lan Lan died on September 4, 1979 during pregnancy and Kang Kang died on June 30, 1980. They were replaced by the female Huan Huan (who arrived January 29, 1980) and the male Fei Fei (who arrived November 9, 1982 and died on December 14, 1994). Attempts at artificial insemination resulted in three births by Huan Huan: the first calf, born on June 27, 1985, died after 43 hours; the second, Tong Tong, a female born on June 1, 1986; and the third, a male You You, born on June 23, 1988.

Arrival of the koala in the 1980s followed the panda craze. Zoo investments escalated, in terms of exhibit construction costs and staff time, for this small marsupial. Six koalas, the first of their species in Japan, arrived in Tokyo on October 25, 1985 and were divided among the Tama Zoo, Higashiyama Zoo in Nagoya, and Hirakawa Zoo in Kagoshima. By September 1997, nine Japanese zoos had a total of 88 koalas.

The faddish acquisition of a particular species is not limited to zoos. From time to time, certain types of marine mammals have generated a high level of popularity, setting new trends in the aquarium field. Toothed whales have great potential as charismatic crowd-pleasers. Two killer whales arrived at Kamogawa Sea World in Chiba Prefecture on September 4, 1970, in time for its opening on October 1. This was the first exhibition of the species in a public display facility. Other first imports to Japan included a female dugong at Oita Ecological Aquarium in Oita on December 17, 1969; three belugas at Kamogawa Sea World on September 19, 1976; and three walruses at Izu Mito Sea Paradise in Shizuoka Prefecture on April 29, 1977.

In recent memory, however, no other marine mammal has caused as great a sensation as the sea otter. On October 3, 1983, the Toba Aquarium acquired one male and three female sea otters from Alaska. During the same period Izu Mito Sea Paradise also acquired sea otters. It was no accident that attendance figures at these facilities jumped dramatically in one fiscal year, nearly 100% and over 50%, respectively. Other aquariums scrambled to catch up. By 1987, 15 institutions had sea otters and by 1996 the number had climbed to 117 sea otters in 28 institutions.

9.5.3 Exhibits

From time to time, zoos endure periods of reconstruction. Under the buzzword "modernization," exhibit styles from past decades are replaced with new architectural styles. Beginning in the late 1920s, Ueno Zoo underwent a massive remodeling project, including Japan's first barless, moated polar bear exhibit, which opened to the public in 1928. By the late 1930s, most of the older animal facilities in Ueno had been replaced. Younger zoos, such as the zoo in Nagoya that opened in 1937, brought in more modern design concepts. Its barless lion enclosure was the first of its type in the country. Legend has it that Lorenz Hagenbeck, while on a Nagoya tour in 1933, made recommendations on the design of this lion enclosure. It was after World War II, however, that the application of moated exhibits with a zoogeographic arrangement was adopted on a large scale. Misaki Park Zoo (opened in an area south of Osaka in 1957) and Tama Zoo (opened in a Tokyo suburb in 1958) are prime examples.

The global trend to maintain a group of animals in more naturalistic exhibits and for improved breeding has begun to take root in Japan. However, many institutions still subscribe to the "postage-stamp collection" exhibit scheme, illustrated by social or gregarious species being represented by one or two specimens. Examples abound. According to the 1996 statistics, 46 zoos are listed as holders of chimpanzees. Of these, 3 have singles, 12 have two each, and 19 have five or more each. As for ungulates, among 24 zoos that keep the Grant's zebra, 2 have singles, 7 have two each, and 10 have more than five each. There are 186 giraffes in 56 zoos. Tama Zoo has four males and 13 females (but this is an exception), 9 zoos have singles, 18 have two each, and 10 have more than five each.[15]

In 1964 a prototype of the safari park opened at the Tama Zoo. A popular feature of this zoo is the moated lion yard in which lions roam at liberty and visitors are confined to tour buses. However, the zoo's most noteworthy exhibit is its insectarium.

Insectariums

Unlike most Americans, many Japanese people traditionally have had a special fondness for insects. For centuries they have been fascinated with the colors, unique body shapes, and songs of insects. In 1954, a live insect collection was added to the Takarazuka Zoo in the Kansai district. This is believed to be the first live insect exhibit in the country. Previous insect exhibits had consisted of preserved specimens rather than live animals.[16] The first bona fide insect exhibit facility was developed at the Tama Zoo. It began on April 1, 1961 as a modest entomology laboratory with three staff members on the third floor of the elephant building. Although the main function of the laboratory was to propagate grasshoppers as food for the zoo animals, it represented the first attempt in Japan to include live insects in a tax-supported zoo collection. The small exhibits in this embryonic insectarium consisted of a series of terraria, similar to its counterparts in European zoos.

Over the years the insectarium kept growing. Relocated to a separate area, it expanded with new exhibit buildings, adding more staff and developing insect culture techniques along the way. Particularly popular were the butterflies and fireflies. Gradually, the insectarium gained public support for further expansion. On April 26, 1988, a huge exhibit complex, with a construction cost of 770,000,000 yen, opened its doors as a commemorative project for the zoo's 30th anniversary. Called the Insect Ecological Land, it has a total floor space of 2,480 square meters and maintains more than 54,000 insect specimens. During the design process, efforts were made to free insects into a larger, naturalistic setting, instead of confining them in small, glass-fronted containers as if they were jewels in a showcase. Within a year of the opening of this facility, there was a 20% increase in the zoo's visitor attendance compared with the previous year. The popularity altered the visitor traffic flow, requiring a review of the design of the immediate area, such as the placement of gift shops.[16] Insect exhibits are now on the increase in Japan. As of 1995, there were about 30 live insect exhibit facilities in the country.[17]

Graphics

Graphics in Japanese zoos are largely limited to labels, or to be more specific, to animal identification and theme signage. In contrast, improved interpretation and visitor communication have become vitally important in the American zoo and aquarium exhibit development process. More interpretive and interactive graphic devices are used in American zoos and aquariums to put the message across to the public on a variety of subjects, including natural history, animal behavior, wildlife conservation, and behind-the-scenes information. Japanese zoos and aquariums must plan a more sophisticated approach to their graphics, not only to improve exhibit aesthetics and general appearances, but also to improve the connection between the public and the educational value of the exhibits.

9.5.4 Breeding Programs

Sad stories abound regarding short-lived animals and their failure to breed in captivity in the pre-war years. The first great ape birth took place on July 23, 1940, when the Tennoji Zoo's chimpanzee Rita had a stillbirth. Rita's death on the same day marked the end of Tennoji's prewar prosperity. As a result of improved husbandry techniques and housing in the postwar years, species that were previously difficult to maintain, such as the great apes, not only began to live longer, but also to breed. The first orangutan birth was recorded at the Ueno Zoo on May 29, 1961, when a female, later named Hatsuko, was born. On August 5, 1971, she made history again by giving birth to a female at Ueno, the first multigeneration breeding for the species in Japan. The Kyoto Zoo had the honor of having the first gorilla birth in Japan when a male was born on October 29, 1970.

Similarly, in recent decades, native species that were notoriously difficult to keep alive began to reproduce routinely. The Japanese serow, a native ungulate that is under the protection of the national government, illustrates this improved situation. Between February 1946 and April 1962, a total of 25 serows were brought to the Ueno and Tama Zoos. With the exception of one female that lived for 9 years and 11 months, all of them died within a very short period of time; 16 of them lived for less than two months. The records in other zoos were also unimpressive. In eight institutions, a total of 42 serows were kept between 1950 and 1963. Of those, nine died within three months; as of 1963, only six were living. These figures give a definite impression that the serow was a difficult mammal to maintain in captivity. Significant among the contributing factors to these embarrassing records were the conditions of their capture and the transfer methods of those animals. Many specimens that were sent to Tokyo arrived under unusual circumstances. Some of them were rescued as an emergency measure due to presumably orphaned status or dog attacks, and were already weakened or injured upon arriving at the zoo. Some were captured in a crude and primitive method, tied with rope, and carried on the collectors' back from the heart of the mountains to the inhabited localities. Without any acclimation period, they were shipped to their destination through the inadequate public transportation system.[18]

Those individuals who cared for these goat antelopes were amazed to learn that they were at ease within close proximity of humans, even after a few days of captivity. It had been believed that Japanese serows, which are alpine animals, could not tolerate summer heat. However, people were surprised to see that the lowland climate, typified by hot and humid summers, did not seem to affect them all that much. The Japanese serow, by nature, is not a difficult mammal. The turning point for them came in 1965. A female named Chiko was born on August 12 of that year at the Japan Serow Center in the west-central part of the country. This was the first full reproduction of the species in captivity. On August 26 of the same year, another serow was born at the Kobe Arboretum. Chiko became a mother at the Serow Center on May 24, 1968, marking the first multigeneration breeding of the species in captivity. Today, the birth of a serow is no longer an unusual event.

Popular zoo animals such as basic stock mammals overshadow amphibians, which rarely draw the attention of either the public or the zoo staff. In Hiroshima, Asa Zoo personnel have been working since its opening in 1971 on the giant salamander, another unique Japanese species. After observing their breeding behavior in the wild, they simulated essential environmental components for reproduction in captivity. In an off-exhibit area, the staff constructed a four-chambered tank that resembled an underwater cave system, a natural breeding site for giant salamanders. It was discovered in the wild that a group of adults congregates during breeding season, so several salamanders were put together in the tank, and in 1979 the first captive reproduction took place.[19]

The Tama Zoo opened in 1958, and in 1962 a giraffe barn was completed in the African area. Two reticulated giraffes were moved to the new facility, founding a breeding program. The first calf, a male, was born on May 14, 1964. Dozens more calves followed, and on March 11, 1986, the 100th was born, 21 years and 10 months after the first calf. Across the Pacific, Cheyenne Mountain Zoo in Colorado Springs (Colorado, United States) had 100 giraffes born between April 1960 and August 1976, according to Marvin Jones.[20] Thus, the Tama Zoo became the second zoo in the world to achieve an institutional record of 100 giraffe births, a remarkable success considering the long road that Japanese zoos struggled along. On January 1, 1998, the 150th calf, a female, was born at the Tama Zoo.

On a global level, a paramount task for zoos has become the propagation of endangered wildlife species. Japan can no longer afford to continue its isolation as an island nation on the

IMPROVED PROPAGATION SUCCESS AT JAPANESE ZOOS AND AQUARIUMS

Successful animal breeding has increased in recent decades at Japanese zoos. What follows is a sampling from a given year, taken from the brief review of the Births and Hatchings column in the 1989 *JAZGA Annual Report*.* In 1989, significant breeding of Japan's endemic taxa included births of the Ryukyu Islands dwarf wild boar, *Sus scrofa riukiuanus*, the smallest and most threatened subspecies of *S. scrofa*. Two zoos bred them: Hirakawa Zoo in the southern city of Kagoshima (three born, none survived) and Inokashira Zoo in Tokyo (five born, all raised). The Ryukyu Islands are also the home of the endemic Amami rabbit, *Pentalagus furnessi*; Hirakawa had one born and raised. Among mainland species, five Japanese serows were born and four raised, in three zoos.

After the koala craze of the mid-1980s, the koalas began to reproduce in Japanese zoos. In four zoos nine were born, and all survived. Notable births of unusual species have included Formosan serows, *Capricornis crispus swinhoei*, five born and two raised at the Japan Serow Center. Also, the Serow Center and Kyoto Zoo had a total of three gorals, *Nemorhaedus goral*, born and raised. Thirteen red pandas were born in six mostly small zoos, and ten were raised. Some zoos have bred a large number of one species. For example, a bear park in Noboribetsu on Hokkaido, the northern main island, had 36 Hokkaido brown bears born, and raised 9. And the Asa Zoo in Hiroshima had 28 Cape hyrax born, of which 16 survived.

Marine mammals have always been popular. In fissipeds, eight sea otters were born and raised in four aquariums. In pinnipeds, excluding the commonly reproducing California sea lions and harbor seals, five species had births: Steller's sea lion, with four born and three raised in three institutions; southern sea lion, five born and four raised in four institutions;

eastern edge of Asia. There have already been efforts to participate in the conservation endeavor, such as the inception of international studbooks for the Japanese serow and the red-crowned crane. In the late 1980s, the time was ripe for JAZGA to formulate a national animal-breeding program. The objective was to upgrade animal management from the individual to the species level, on a national scope, and beyond the boundaries of zoos. It was aimed at participation in international wildlife conservation efforts, as well as securing animals for exhibition.

On February 29, 1988, the JAZGA board created the Species Survival Committee of Japan (SSCJ). The JAZGA chairman was directed to head this committee, which also included some other board members, zoo and aquarium representatives of various geographical regions, and academic advisors. SSCJ was also expected to be a contact with such international organizations as the International Union for the Conservation of Nature and Natural Resources (IUCN, now the World Conservation Union), the Species Survival Commission (SSC), the Conservation Breeding Specialist Group (CBSG), the International Union of Directors of Zoological Gardens (IUDZG), and the International Species Information System (ISIS). The SSCJ chose 32 taxa for inclusion in this project. Thus JAZGA, with approximately 150 member institutions, started a program similar to the European Endangered Species Programme (EEP), the Joint Management of Species Group (JMSG) in the British Isles, the Australian Species Management Programme (ASMP) in Australia, and the Species Survival Plan (SSP) in North America. SSCJ

northern fur seal, six born and three raised in two institutions; African fur seal, one born and raised; and ringed seal, one born but did not survive. The cetacean birth data are as follows: Gill's bottle-nosed dolphin, ten born in six institutions and three raised; white-sided dolphin, two born but did not survive, in one institution; and a dolphin hybrid, Gill's bottle-nosed dolphin × false killer whale, one born but did not survive.

Breeding cranes has been one of the traditions of Japanese zoos. Kyoto Zoo's first successful breeding of the red-crowned crane was recorded in 1905; the white-naped crane also bred soon after the zoo's opening in 1903. During 1989, 17 zoos hatched a total of 60 cranes belonging to 10 species, and 45 of them were raised. Propagation of *Spheniscus* penguins is also common in Japan: 28 African penguins were hatched and 12 were raised in four institutions; a total of 149 Humboldt penguins were hatched and 73 were raised in 26 institutions. Of these penguins, 93 were hatched and 50 were raised in 11 aquariums; one of them, the Otaru Aquarium in Otaru, raised 16.

Other noteworthy hatchings include the following: Nogeyama Zoo in Yokohama hatched three kagu, two of which were raised; Omachi Alpine Museum in Nagano Prefecture hatched six rock ptarmigan and raised three; 42 ibises of six species were hatched, 28 of them in Tama Zoo, which has established a reputation as a breeding center for these birds, as well as for cranes, storks, and flamingos. Raptor breeding has been gaining ground, albeit slowly. Maruyama Zoo in Sapporo hatched and raised two great horned owls, the first breeding of this species in Japan. Other raptors that bred during 1989 include eagle owl, snowy owl, Ural owl, collared scops owl, goshawk, white-tailed sea eagle, golden eagle, common buzzard, and kestrel.

The presence of a large number of aquariums is the basis of successful breeding of marine mammals and penguins. However, reptiles and amphibians have not made an inroad in Japanese institutions.

* JAZGA, *Japanese Association of Zoological Gardens and Aquariums 1989 Annual Report*, JAZGA, Tokyo, 1990 [in Japanese]; see also Kawata, Ken, "Japanese Association of Zoological Gardens and Aquariums: some notes on the 1989 annual report," *International Zoo News*, 37, 41, 1990.

receives financial support from the Foundation for Earth Environment and the World Wildlife Fund Japan Committee.[21,22]

At its first conference in October 1988, JAZGA added one more species to the list of taxa and laid out the operational details. The SSCJ divided the taxa into seven taxonomic groups as follows: marsupials, primates, carnivores, ungulates, marine mammals, birds, and fish. Each group had a coordinator and within each group priority species were named. Each taxon had a species coordinator, who also served concurrently as a studbook keeper, and several propagation group members. The species coordinator would implement a breeding plan based on genetics and demography. Arrangement of animal transfers, coordination of loan agreements, and offering advice were a few of the tasks of the committee.

As the scope of the breeding programs expanded, the SSCJ added more taxa (71 taxa were listed as of 1996). Examples included red-crowned crane, Japanese serow, koala, black lemur, ruffed lemur, lion-tailed macaque, drill, black gibbon, orangutan, chimpanzee, gorilla, polar bear, red panda, Siberian tiger, cheetah, Grevy's zebra, Malayan tapir, Indian rhinoceros, southern white rhinoceros, black rhinoceros, Asian elephant, sea otter, California sea lion, southern sea lion, harbor seal, Kuril harbor seal, Gill's bottle-nosed dolphin, Andean condor, white-tailed sea eagle, Steller's sea eagle, golden eagle, Blakiston's fish owl, eagle owl, white-naped crane, hooded crane, wattled crane, Asian white stork, waldrapp, salmon-crested cockatoo, Bali mynah, Japanese giant salamander, and Asian arowana. In addition, 14 endemic taxa of freshwater fish were chosen for the national breeding program.

Like any other national breeding program of this scope, the activities of SSCJ rely entirely on the voluntary efforts of zoos and aquariums. JAZGA, its parent organization, has no legal or enforcement authority over its members. Therefore, the cooperation of member institutions is a lifeline for such a program. These institutions have varying degrees of awareness concerning the needs of global wildlife conservation. Views and values of superiors in the governing authorities may sharply differ from those of the zoo staff. Potential problems experienced by the SSCJ often seem to originate in this conflict between the governing authorities and those who are directly responsible for zoo operations. The issue of animal ownership is a prime example of this tension.

Since mainstream zoos in Japan are municipal, the municipal superiors may be reluctant to delegate authority to the zoo staff for moving animals off the city's premises. Animal ownership that is firmly secured within a bureaucratic system is typically incompatible with a philosophy of worldwide wildlife conservation. In an effort to alleviate the problem, during one SSCJ meeting, a coordinator of a primate species program urged strongly that the monetary value of the animals be rated modestly. He theorized that if an animal is listed in a higher price range in the municipal inventory, it will become difficult to move the animal around for a species survival program.

The willingness to transfer animals for breeding purposes differs from species to species. For example, the red panda species survival coordinator sent a questionnaire to holding institutions. One of the questions asked was, "Can you follow instructions of a species coordinator for the prevention of inbreeding?" Of the 23 zoos that responded, 19 said "Yes." Furthermore, to the question, "Can you exchange animals or participate in loan programs with overseas zoos?" 18 out of 23 responded with "Yes." This appears encouraging. As a matter of fact, moving animals between zoos for a national breeding program has already begun in species such as the black rhinoceros.

The situation involving a charismatic megamammal such as the gorilla presents a different picture. As of September 1997, there were 40 western lowland gorillas in 17 Japanese institutions. Of these, five zoos had a single animal and eight had one male and one female each. Six of the gorillas were captive born, indicating that zoos still relied mostly on animals caught in the wild. No gorilla has been born in Japan since 1986. As of December 1997, there were 662 western lowland gorillas in the ISIS inventory.[23] With the exception of 40 specimens, or 6%, the specimens ISIS listed were in North America and Europe. The captive-born ratio of the ISIS population was 69% and the number per institution was 6.6, compared with 2.3 in Japan. The high rate of captive-born animals in North America and Europe is the result of interzoo cooperation, as well as the practice of maintaining gorillas in multimember groups.

For Japanese zoos to establish a self-sustaining population, a carefully thought-out breeding plan is needed, under which gorillas would be transferred from one zoo to another, preferably into multimember groups. Japan's gorilla population is aging rapidly as more specimens are moving beyond the reproductive age class. Inaction will eventually turn the Japanese gorilla population into a modern-day dinosaur.

In every respect, the growth of the gorilla population represents a challenge. Recruitment through importation is extremely difficult because of the international animal trade treaty, and fecundity is embarrassingly poor in Japan. In addition, the ability of gorillas to attract visitors creates an ownership issue, causing political sensitivity that interferes with moving gorillas from one institution to another. In fact, most Japanese zoo directors do not have the authority to move gorillas. A request for approval for such an action may involve the city council. Chances are the local government would cling onto such a "calling card" animal as the gorilla at all cost. After all, if a gorilla is seen as a commodity, a tool to attract more zoo patrons, then the transfer of a commodity, right or wrong, becomes a matter of expediency, not of bioethics.

Indeed, every kind of difficulty that could plague a species survival program is found in the situation involving the gorilla. Even if breeding loans between zoos should become a reality, there would still be problems to overcome. In almost all Japanese zoos, the design of gorilla facilities is focused on exhibiting a pair, the "postage-stamp collection" concept, and are too small for housing breeding animals. For a zoo to accept a gorilla for breeding, there must be enough room for flexibility, for example, to introduce animals to each other. Specifically, a zoo would need several stalls and shift facilities to accommodate breeding animals. Such a basic component is nonexistent in most zoos in Japan. There have been a few cases in the past where gorillas were loaned between zoos, but none ever resulted in successful breeding. Had it ever worked, zoos may not be so hesitant to take a chance. For a more progressive program, zoo staffs will need to take a bold and comprehensive approach. However, the difficulties facing them appear overwhelming.

Under the SSCJ plan, the western lowland gorilla is one of 13 "priority species" to receive concentrated efforts toward long-term propagation. However, as of 2000, little action had been taken on gorillas since the inception of SSCJ in 1988. The gorilla issue is often conspicuously absent during SSCJ conferences. The gorilla may be regarded as an indicator, or barometer, of the success or failure of a national species survival program. The lack of action in the gorilla breeding program seems to illustrate the fact that Japanese zoos lack airtight insulation against external political interference and pressure.

Against this background, Tokyo's Ueno Zoo made an unprecedented attempt in the early 1990s when it grouped gorillas for a national breeding program. As was common in Japan, Ueno had maintained an aging pair of gorillas in a small space. After a complete overhaul of the animal management scheme, they brought in six gorillas on breeding loan between December 1992 and November 1993. Of the six new arrivals, two were from the Tama Zoo, which is also in Tokyo and under the same governing authority. The rest came from four zoos across the country. The surprising aspect is that three of these zoos were privately owned, profit-oriented operations. It is remarkable that zoos under varying ownerships and philosophies have established a solid cooperative relationship in this bold endeavor.

In May 1994, all of the gorillas — three males and five females — were housed in a new facility. Their ages ranged from 16 to 40 years for the males and 16 to 27 years for the females. Although the outcome of the gorilla grouping remains to be seen, those zoos that participated in the program are to be commended. One important factor behind this project was the investment in a sizable breeding facility. The switch of this facility's function from mere exhibitry to breeding was revolutionary by Japanese standards. However, no reproductive behavior was noted in the group. In an attempt to stimulate breeding activities, a ten-year-old male was brought in from Howletts Zoo in England on December 4, 1997.

Concerning housing facilities, a review of another species, the polar bear, is instructive. This species entered the Japanese zoo scene for the first time in 1902 at the Ueno Zoo. Between 1917 and 1992, 104 cubs were born in Japanese zoos. Of these, 11 survived at four zoos, the first one raised at the Asahiyama Zoo in Asahikawa, Hokkaido, in 1974. By comparison, in one estimate, 187 cubs were born in zoos in the United States between 1976 and 1985; 99 of them survived.

The requirements for successful captive breeding of the polar bear are far more stringent than those for other bear species. A maternity facility that ensures complete privacy and a sense of security are absolute necessities. In addition, careful planning and handling are required. The polar bear can tolerate varying captive conditions, and may live for 20 years. Japanese zoos have been content with merely exhibiting the polar bear, and have never made a serious concerted effort to breed them. The fact that only 6 of 30 zoos maintaining the species have maternity facilities is evidence of this shortcoming. Moreover, ownership, the focal issue in the gorilla dilemma, has in some instances also become a barrier in the polar bear–breeding program. A one-sentence command from a superior, "Our zoo shouldn't lose a popular beast like the polar bear," could easily impede the reproductive potential of the captive population. What is needed, not only for the polar bear, but also for other taxa, is a shift in philosophy from viewing zoo animals as a commodity to viewing them as a biological resource that requires professional management. Without such a reform, a national breeding plan may be doomed to failure.

As an organization in its infancy, the SSCJ is experiencing its share of growing pains. The SSCJ will no doubt face more challenges involving varying policies of individual institutions, external political interference such as the animal ownership conflicts, and financial difficulties. Each of these will test the strength of zoo professionalism. To be fair, however, it must be noted that, within a short period of time, the SSCJ has achieved at least partial success through efforts to increase awareness of the importance of its conservation efforts within the Japanese zoo circle and, to an extent, the news media. There is no doubt that the SSCJ has cast a ray of hope in the overall Japanese zoo scene. For Japan to be accepted as a legitimate member in the international zoo arena it is vitally important that the SSCJ be strengthened and expanded.

9.5.5 Research Activities

Despite limiting conditions, there have been achievements in the area of biology and zoo technology. In postwar Tokyo, Ueno Zoo director Tadamichi Koga used an antibiotic called Aureosrysin as an agent against aspergillosis, a disease caused by the fungus *Aspergillus,* which commonly kills penguins in captivity. By this method he succeeded in long-term captive maintenance of polar penguins.[24] He was also engaged in the study of cranes, particularly reproductive biology. "He pioneered the captive breeding of cranes," commented George Archibald.[25] During the 1956 breeding season, he was able to collect 17 eggs from a female white-naped crane, 10 of which hatched.[26] His off-scene breeding unit was no more than a wooden shack, and the incubator was an antique, yet many crane chicks were raised in this facility.

Jiro Kobara, a former director of the Asa Zoo in Hiroshima, made a breakthrough in the ecological study of the Japanese giant salamander. He led a team of young zoo staff members into fieldwork, covering such areas as habitat destruction, growth rate, migration, and reproductive behavior. This research eventually resulted in the first captive breeding of the species in the world.[27]

Collectively, however, Japanese zoos have not kept up with the rest of the industrialized world in terms of scientific research. This failure is particularly striking, considering the large number of academic institutions in the country with a high level of technological advancements. Unfortunately, the historically weak liaison between zoos and academic circles may be responsible for this poor performance. Unlike the United States, where many biologists and geneticists have working relationships with zoos, collaboration is rare in Japan.

Earlier this century, there were several opportunities for cooperation between zoos and the academic world. In 1900, Chiyomatsu Ishikawa of the University of Tokyo, one of the founders of modern zoology in Japan, was appointed zoo administrator at the Ueno Zoo. He had a clear vision of establishing a grand zoo for Japan. Some believe that, had the professor stayed in the position, Japanese zoos may have taken an entirely different course. Alas, in 1907, he was forced to resign because of an alleged unauthorized purchase of two giraffes (the first in Japan) from Hagenbeck. In 1924, the Ueno Zoo was transferred from the national government to the City of Tokyo. Meanwhile, in Kyoto, the first zoo director was a trained zoologist and a protégé of Professor Ishikawa. He was drafted into the army after a tenure of only 18 months. Then in 1934, the mayor, against stiff opposition from the city council, appointed as zoo director Tamiji Kawamura, a famed professor of zoology at the University of Kyoto. Bitterly disappointed in the municipal government and its politics, the professor left the zoo after only two years.

9.6 Internationalization and Cultural Uniqueness

9.6.1 The *Sakoku* Factor

Herman Kahn made the predictions that "Japan will pass the United States in per capita income by the year 2000," and that "Japan is very likely to become a superpower as well as a superstate."[28] Whatever the outcome of these predictions, it is undeniable that Japan has become a towering giant in international economics and finance. Yet, it does not appear it will take a major role as a superstate in the international wildlife conservation and zoo arenas anytime soon.

According to the *International Zoo Yearbook*, as of 1994, there was a total of 870 zoos in the world including 69 in Japan.[29] JAZGA lists 162 institutions (zoos and aquariums) as of 1997. Regardless of which statistics one relies upon, it is apparent that Japan is one of the most zoo-popular nations in the world, having so many zoos and aquariums in a land approximately the size of the state of Montana in the United States. Still, very little is known about Japan's zoos outside of its boundaries. A huge information and communication vacuum exists on the eastern edge of Asia. Indeed, Japan occupies a peculiar position in the global zoo scene; it belongs to neither the "third world" nor to the highly advanced zoo community of Europe and North America. For the most part, this situation has resulted from Japan's isolation from the rest of the world.

A long history of isolation, first geographical and later self-imposed (*sakoku*), has enabled Japan to develop and maintain a distinct culture. However, as Edwin O. Reischauer, a leader of Japanese studies in the United States, notes, "No people have committed themselves more enthusiastically to internationalism than the Japanese or have so specifically repudiated nationalism." According to Reischauer,

Life in Japan also is in some ways as international as anywhere in the world. Newspapers and television give good international coverage. In fact, on average the Japanese probably receive more world news than any other people. Japanese scientists are on the frontiers of science, and their scholars are well aware of the intellectual trends of the West. World fads, styles, and fashions sweep Japan as fast and fully as anywhere.[30,31]

This phenomenon of internationalization often tends to be superficial, and therefore misleads Western observers. According to sociologist Merry White, despite dramatic economic changes after the last war, the patterns of social relations and cultural values of Japan have not changed. To the Japanese, the family is *uchi* (inside); it represents the private life of the Japanese, while institutions such as the school and the workplace are *soto* (outside) and represent the public part of their lives. The line between *uchi* and *soto* is clear; what matters most is *uchi*. White maintains, "It may even be said that under pressures to internationalize, Japan itself has become a macro-*uchi*" and "generally speaking, what is being defended is the Japanese *uchi*, the preservation of a homogeneous island identity." The concept of *uchi* may be extended to the point that whatever takes place outside of the Japanese archipelago rarely generates serious concern.[32]

9.6.2 Environmental Awareness

To examine the zoos of a nation, the attitudes of its people toward nature must be reviewed, for their views on nature may be inextricably intertwined with the status of their zoos. The Japanese are often said to possess a sense of harmony with nature. For example, H. G. Summers wrote about Japanese novelist Yasunari Kawabata, winner of the 1968 Nobel prize for literature:

> I was struck, while reading Kawabata's work, with the thought that Western ecologists are dilettantes indeed. Kawabata evokes the close relationship between the natural world and the human world that has been for centuries almost second nature in the Orient. With attention to the most minute detail — the flowering of a cherry tree, the first buds of a ginko tree, the symmetry of pines in a distant forest, the rumbling of mountain — Kawabata raises ecology to its highest degree.[33]

Too often, such a romantic and unrealistic view tends to be superficial, failing to shed a deeper insight, and may face an opposing view by many Japanese themselves. In 1938, against the tide of mounting nationalism and militarism, the Japanese sociologist Nyozekan Hasegawa criticized the Japanese tendency to view nature in its parts rather than as a whole. He scolded the Japanese garden concept, stating that the Japanese tendency "is to transmute nature into forms, and to view its separate parts without any comprehensive attempt to appreciate nature as a whole. It is the same love of detail as is seen in the miniature tree and the miniature landscape on a tray." He went on to say, "The Japanese approach to nature, thus, is a literal case of not seeing the wood from the trees," and concluding, "Thus the failure of the present-day [1938] Japanese to appreciate nature means that modern Japanese culture cannot be truly creative."[34]

Concerning the issue of rapid economic growth in postwar Japan, Kahn and Pepper made the observation that until "the late 1960s, the middle class as a whole certainly took the view that, since life was constantly getting better in economic terms, a few deficiencies in the system could be tolerated." It was during this time that the middle class began to voice an antigrowth position.

> Various surveys of Japanese opinion reveal just how deeply the antigrowth position took hold in the late 1960s and early 1970s. In a well-known survey of national character taken every five years since 1953, a cross section of the population was asked to select the view about the relation of man and nature closest to its own. A sudden reversal of the trend occurred in 1973, when the traditional view that man should subordinate himself to nature reasserted itself, as against the modern view that man should conquer nature.[35]

The consequences of the reassertion of this traditional view could be misleading, in the sense that it will not necessarily evolve into an elevated environmental awareness.

Kahn and Pepper also brought up a concept called the "postindustrial marriage of machine and garden," which they define as the idea "to seek ways to use affluence and technology in an aesthetically satisfying manner to create a harmonious whole." They suggest that the American vision of the garden is one of the virgin wilderness. By contrast, they assert, the Japanese concept is one "not necessarily in untouched form, perhaps more like [one] which is artfully contrived but nevertheless still natural."[36]

Kellert's more structured and scientific approach seems to be in general agreement with these views. Kellert notes: "This apparent paradox of a presumed Japanese appreciation and respect for nature concurrent with a pattern of repeated environmental and wildlife destruction has puzzled many scholars." As his study reveals, "The Japanese appreciation and respect for nature has tended to be very narrow and idealized, primarily focusing on single species and individual aspects of the environment, and typically lacking an ecological or ethical perspective." Furthermore, he comments that "the Japanese nature-oriented traditions of bonsai, flower-arranging, and rock-gardening were typically divorced from issues of ecological functions" and that "substantially less ecological and ethical concern for nature and wildlife was found among Japanese of all age and educational groups, in considerable contrast to that encountered in the American and German surveys." For instance, "Substantially more Americans (30%) than Japanese (18%) regarded habitat destruction due to human overpopulation as a major cause of species endangerment."[37]

Such perceptions on nature, coupled with the strong inclination to treasure *uchi* more than anything else, paint a troubling picture regarding Japanese interactions with the environment. It becomes increasingly clear that whatever happens outside (*soto*) of its archipelago often makes no difference to Japan. In that sense, Japanese zoos continue to be consumers, rather than producers, of wildlife.

9.6.3 Penchant for Group Acceptance

As Merry White has observed, "The tightly-knit social relations in Japan have resulted in a strong code of behavior for its members, demanding their loyalty to the group."[32] It has been noted that most Japanese long for group approval, and acquire group membership by being part of a whole, rather than "different," independent or individualistic.

In contrast to the United States, job security is superbly and superiorly instituted in Japan. Employers practically never fire, or even lay off, employees. However, this job security comes with a price; a high achiever in a Japanese workplace is likely to be constrained. In spite of the social system, there are competent and enthusiastic young middle-level managers in Japanese zoos, but their potential seems to have little room to develop.

9.6.4 Uncharted Waters

In more recent decades, zoos in Europe, America, and, most recently, Australia have made significant strides in wildlife conservation programs and collaborative efforts. Japan, however, has been absent from this arena and seems to wish to continue to maintain the status quo instead of taking up new challenges in wildlife conservation. Early in the 1990s, George Rabb, then Chairman of the SSC, and the present author, a CBSG member, approached JAZGA in a futile attempt to involve Japan in the global wildlife conservation field. Around the same time, CBSG began work to establish a more active dialogue with JAZGA. Some expected that my background as a native of Japan would help bridge the two worlds. CBSG Chairman Ulie Seal and I hoped to at least open Japan's door.

Specifically, an effort was made to introduce the concept of the Population and Habitat Viability Analysis (PHVA) to the Japanese, and to assist them in hosting a PHVA meeting. PHVAs have been conducted worldwide, many of them in less affluent nations. By the late 1990s, it became obvious that there would be no tangible results. Evidently, some of the younger and lower-level zoo staff members were enthusiastic about PHVA, but they were stymied due to the inaction of upper-level management.

Some are probably left with a nagging question: Where does the uniqueness of the Japanese zoos come from? Why are they so different? According to one school of thought, the phenotypic features of a people play a role in the determination of thought processes and that many Japanese think differently.[38] Similar observations have been made by some Japanese themselves. One researcher pointed out that only the Japanese brain has the receptive function to discriminate the intricate and delicate insect sounds. That theory provides an explanation for the Japanese ability to appreciate the distinct insect songs, while to others these songs are merely noise produced by "bugs."[39]

However, these accounts are hardly convincing. It is unlikely that morphology, or the variation in the anatomical features between human races, causes such a decisive difference in the function of the brain. Rather, what causes the difference is more cultural and historical. In this case, the force that separates the Japanese mainly lies in the aforementioned *sakoku* factor. To some, the very thought of international isolation in this day and age may seem anachronistic. Yet, the profound sense of isolation still exists in the Japanese mind.

Language is also a problem for Japan. Since English has come to be the zoos' *lingua franca*, one needs to be familiar with a variety of printed media in English to keep up with the rest of the world. Such a practice of keeping up simply does not apply to many Japanese zoo administrators. As noted before, typical zoo directors have no background in biology, nor do they have training and experience in the zoo field. Therefore, even without the language barrier, the directors' grasp of zoo-related topics is limited.

Fundamental changes are required to reform Japan's zoos; however, the possibility of a reform largely depends on what the public wants from its zoos. Perhaps the reform could become a reality, if the people of Japan genuinely wish to have a higher level of professionalism in their zoos. Thus far, no public outcry has been noted. More recently, an animal advocacy group movement has developed in Japan, resulting in news media coverage, but it has merely proved to be a passing storm. In the absence of a groundswell of grassroots movement for zoo reform, one has to conclude that most citizens are happy with what they have now: a zoo as a showplace of exotic animals.

If that is the type of zoo Japan wants, some may argue, then so be it. Such an opinion lacks the global perspective. The large number of animals in Japanese zoos represents a valuable biological resource. Also, as Hediger put it, wild animals are a part of the heritage that belongs to all peoples of the world.[11] In this context, wild animals need to be shared and managed collectively.

Rabb noted, "Zoological parks are evolving institutions in respect to the conservation of biological diversity. From past functions in recreation as menageries and in education as living museums, they are coming to discharge these functions, plus other meaningful ones in research and conservation, as internationally oriented conservation centers." In his chart on the evolution of public zoos, the menagerie sits at the bottom left; in a straight-line diagonal succession toward upper right, the menagerie elevates itself into the zoological park, and then into the conservation center.[40] At each step, their respective themes, subjects, concerns, and exhibitry are also upgraded in terms of quality and perspective. Evaluated according to this model, Japanese zoos are still near the bottom in terms of quality, no matter how modern their facade may appear. This is odd and so out of balance, considering Japan's global status as an economic and financial giant with a highly educated citizenry. Unfortunately, Japanese zoos have yet to make a noticeable impact in the arena of science and conservation, within and outside of their country.

Some might argue that although the process has been slow and only partial, Japanese zoos have come a long way, and there is no reason to deny that they cannot make improvements in the administrative and managerial aspects. However, a nagging question still lingers: If the Japanese system were reformed by discontinuing the practice of appointing nonqualified persons as directors and replacing the paper-thin, often ineffectual middle management with a solid, professional team of curators, would Japan join the mainstream of world zoos?

The use of foreign models from the West became part of Japanese culture and modernization long ago. In 1862, the *Shogunate* (Japan's feudal government) sent a 38-member mission to Europe. According to the itinerary, it visited Jardin d'Acclimatation in Paris and zoos in London, Rotterdam, Amsterdam, and Berlin. Then in January 1867, the *Shogunate* sent 56 boxes of insect specimens to Paris for exhibition at the World Exposition. In the group accompanying this shipment was Yoshio Tanaka (1838–1916), who later became known as the father of Japanese museums. His documented recollections reveal that Tanaka also visited Jardin des Plantes, which included a menagerie. On that spring day in Paris, this famed institution must have left a profound impression on him, for he is credited with playing an important role in establishing the first zoo in his country. The new Japanese government then participated in the 1873 World Exposition held at Vienna. After collecting a large number of materials from various parts of Japan for Vienna, the government decided to exhibit some of them to the public in Tokyo, thereby building a museum. On April 15, 1873, the exhibits, including live animals, opened in an area near the current location of the Imperial Hotel. After the museum closed in July 1881, the animal collection was moved to the new location in Ueno Park, and Japan's first zoo was born on March 20 of the following year.

One can easily imagine how those pioneers were lost in wonder at the sight of animal exhibits in Europe. It is not possible to know if they had detailed plans for future zoos. However, judging from the recorded history, their concept of a zoo, including Tanaka's, was closely tied to the European model. Decades later, when Japan's zoos chose their directions, they differed from what the early-day pioneers initially had envisioned when they sailed boldly into uncharted waters. Would that today's Japanese had followed their pioneer fathers, who took a bold move to sail into uncharted waters.

Facing the rapidly changing world many Japanese, who seem quite content in "their familiar national cocoon," find their own sense of uniqueness difficult to shake.[41] It would be naive for an observer looking in from the outside to assume that the Japanese are embryonic Westerners, waiting to mature and develop. Improvements will be made gradually in the zoo field. Japan will acquire what it can afford, be it rare animals or the top-of-the-line exhibit system. However, in the most fundamental sense, it is doubtful that in the foreseeable future Japan will upgrade its zoos to the level of their Western counterparts.

References

1. Kawata, Ken, "Tea house, war and panda craze: a review of Japanese zoo history," in *AAZPA Regional Conference Proceedings,* American Association of Zoological Parks and Aquariums, Wheeling, WV, 1991, 808.
2. Osaka Tennoji Zoo, *Osaka Tennoji Zoo: The 70 Year History,* Tennoji Zoo, Osaka, 1985 [in Japanese].
3. Sasaki, Tokio, *Zoo History,* Nishida Shoten, Tokyo, 1975 [in Japanese].
4. Takizawa, Akio, *The Record of Okazaki Zoo,* Akio Takizawa, Kyoto, 1986 [in Japanese].
5. Ueno Zoo, *Ueno Zoo: The 100 Year History,* Tokyo Metropolitan Government, Tokyo, 1982 [in Japanese].

6. Yamamoto, S., "Zoos in Japan," in *The World of Zoos*, Kirchshofer, Rosl, Ed., Viking Press, New York, 1968, 269 [translated by Hilda Morris].
7. JAZGA, *Japanese Association of Zoological Gardens and Aquariums 1990 Annual Report*, JAZGA, Tokyo, 1991 [in Japanese].
8. Kawata, Ken, "Japanese Association of Zoological Gardens and Aquariums: some notes on the 1990 annual report," *International Zoo News*, 38, 42, 1991.
9. Truett, Bob, Ed., *Zoos & Aquariums in the Americas 1970–1971*, AAZPA, Washington, D.C., 1970.
10. Hahn, Emily, *Animal Gardens*, Doubleday, New York, 1967.
11. Hediger, Heini, *Man and Animal in the Zoo*, Routledge & Kegan Paul, London, 1970, 8 [translated by Gwynne Vevers and Winwood Reade].
12. Takashima, Haruo, *Animal Tales*, Yasaka Shobo, Tokyo, 1986 [in Japanese].
13. Hagenbeck, Lorenz, *Animals Are My Life*, Bodley Head, London, 1956, 162–164, 173 [translated by Alec Brown].
14. Anonymous, "Kenya animals for Japan: miniature zoo exported," *East Africa Standard*, 1933? [Undated newspaper clipping].
15. JAZGA, *Japanese Association of Zoological Gardens and Aquariums 1996 Annual Report*, JAZGA, Tokyo, 1997 [in Japanese].
16. Yajima, Minoru, "The insect ecological land at Tama Zoo," *International Zoo Yearbook*, 30, 7, 1991.
17. Yajima, Minoru, "The development of insectariums and their future," *International Zoo News*, 43, 484, 1996.
18. Komori, Atsushi, *Keeping Animals in Captivity*, Kinokuniya Publishers, Tokyo, 1964 [in Japanese].
19. Kuwabara, K., Suzuki, N., Wakabayashi, F., Ashikaga, H., Inoue, T., and Kobara, J., "Breeding the Japanese giant salamander *Andrias japonicus* at Asa Zoological Park," *International Zoo Yearbook*, 28, 22, 1989.
20. Jones, Marvin, personal communication, 1986.
21. Kawata, Ken, "Japan's species survival programme gets off the ground," *International Zoo News*, 38, 6, 1991.
22. Kawata, Ken, "Species Survival Committee of Japan," *International Zoo News*, 39, 39, 1992.
23. International Species Information System, *ISIS Mammal Abstract*, ISIS, Apple Valley, 1997.
24. Koga, Tadamichi, Fukuda, N., Asakura, S., and Nakagawa, S., "On the Aspergillosis of penguins in Ueno Zoological Gardens," *The Scientific Report of Ueno Zoological Gardens*, 1, 23, 1955 [in Japanese].
25. Archibald, George, "ICF is 25," *The ICF Bugle*, 24, 1, 1998.
26. Koga, Tadamichi, "Studies on the reproduction of cranes," *Journal of JAZGA*, 3, 51, 1961 [in Japanese].
27. Kobara, Jiro, *The Giant Salamander*, Dobutsusha, Tokyo, 1985 [in Japanese].
28. Kahn, Herman, *The Emerging Japanese Superstate: Challenge and Response*, Prentice-Hall, Englewood Cliffs, NJ, 1970.
29. Olney, P. J. S. and Fisken, Fiona A., Eds., *International Zoo Yearbook*, 34, 1995.
30. Reischauer, Edwin O., *The Japanese*, Harvard University Press, Cambridge, MA, 1981, 402.
31. Reischauer, Edwin O., *The Japanese Today: Change and Continuity*, Harvard University Press, Cambridge, MA, 1988.
32. White, Merry, *The Japanese Overseas: Can They Go Home Again?* Collier Macmillan, London, 1988.
33. Summers, H. G., "The basic difference between East and West," *The Kansas City Star*, June 28, 1970.
34. Hasegawa, Nyozekan, *The Japanese Character: A Cultural Profile*, Kodansha International, Tokyo, 1966, 127 [translated by John Bester].
35. Kahn, Herman and Pepper, Thomas, *The Japanese Challenge: The Success and Failure of Economic Success*, Thomas Crowell Publishers, New York, 1978, 25.
36. Kahn and Pepper, 1978, 14.
37. Kellert, Stephen R., "Japanese perception of wildlife," *Conservation Biology*, 5, 297, 1991, 299.
38. Keeling, Clinton, "Keeling's cognitions," *ZOO!* 1, 20, 1995.

39. Ogura, Akira, *The Japanese Ear*, Iwanami Shoten, Tokyo, 1977 [in Japanese].
40. Rabb, George B., "The changing roles of zoological parks in conserving biological diversity," *American Zoologist*, 34, 159, 1994.
41. Reischauer, 1988, 411.

Additional Sources

1. JAZGA, *Japanese Association of Zoological Gardens and Aquariums 1989 Annual Report*, JAZGA, Tokyo, 1990 [in Japanese].
2. Kawata, Ken, "Japanese Association of Zoological Gardens and Aquariums: some notes on the 1989 annual report," *International Zoo News*, 37, 41, 1990.

Camel exhibit during the mid-1920s at Ueno Zoo, Tokyo, Japan. Photograph from *Ueno Zoo: the 100 Year History*. © Ueno Zoo.

10
Zoological Gardens of Africa

Wilhelmus Labuschagne and Sally Walker*

10.1 Introduction

Africa is a large, culturally diverse region bounding, until recently, with wildlife. There were few early African civilizations, such as the Ghana kingdom during the 700s to the 1000s in what is now Mali and the Songhai kingdom of Timbuktu during the 1400s to the 1500s. These few early civilizations were short-lived, and it is not known if they had animal collections. Egypt is the only African country to have maintained animal collections throughout its history. Many African collections were not established until colonial botanical stations began holding animals in the 1600s. Colonial officials and naturalists also started private collections, as they did elsewhere. Few of these collections survived, and most extant African zoos and aquariums were established during the 1900s, many since colonial independence began in the 1960s.

10.2 Arab Republic of Egypt

Egyptian animal collections have existed since ancient times (see Chapter 1 by Kisling on Ancient Collections). Pharaohs maintained royal collections during their rule over Egypt (2700–332 B.C.). Greek Ptolemaic rulers (332–30 B.C.) and then Roman rulers (30 B.C.–ca. A.D. 619) continued the Alexandria collection. Persians, famous for their paradise gardens, then ruled Egypt, as they had on several previous occasions, until Muslim Arabs conquered the region in 642. Arab rulers, and then Ottoman Turkish rulers, continued the animal collection at Alexandria until the British took over with their occupation of Egypt in 1882. Collections were also established in Cairo, elsewhere in Egypt (which became independent in 1922), and in the Sudan territory (which Egypt controlled until it became the Democratic Republic of the Sudan in 1956).

10.2.1 Giza Zoological Gardens

Giza Zoological Gardens (Cairo) is one of the oldest of the modern zoos in Africa, and has the richest heritage. Giza Zoo had its origins in the "magnificent gardens" known as the "Haremlik" of Ismail Pasha (1830–1895), Viceroy of Egypt and then Khedive of Egypt. Stanley S. Flower, Director of the Zoological Service for Egypt from 1898 to 1924, wrote about the Egyptian and

---

* Wilhelmus Labuschagne contributed the sub-Saharan section and Sally Walker contributed the Egyptian section.

S. S. Flower

Stanley Smyth Flower (1871–1946) was director of the Zoological Service, the Giza Zoo, and the Giza Aquarium (Cairo, Egypt) from 1898 to 1924. He was also an inspector and advisor for all zoos in Egypt and the Sudan territory. Flower had been exposed to Britain's foremost naturalists during his childhood when he accompanied his father, Sir William Henry Flower (Director of the British Natural History Museum and president of the Zoological Society of London), to meetings at the London Zoo. S. S. Flower was deputed to Siam (now Thailand) in 1896 as scientific advisor to the Siamese government and traveled extensively in that country and in Malaya. He contributed detailed lists of animals, particularly reptiles and amphibians, which were published in the *Proceedings of the Zoological Society of London*. Flower remained in Siam until 1898, when he became director of the Giza Zoo."

During Flower's tenure as director of the Giza Zoo he steered the zoo into a "Golden Age," during which many "firsts" for captive management and conservation occurred. Similarly, under his management, the Zoological Service flourished and made efforts to improve the country's zoos. It provided a system of inspection for smaller Egyptian municipal menageries, which probably represented the first official mechanism for inspecting zoos on a national basis.

Flower might have been the first zoo director to be made an Officer of the Order of the British Empire (c. 1920), as was his zoo inspector, El Saghkolaghasii Malmoud Effendi Hilmi el Samma. Moreover, four zookeepers from the Zoological Service were awarded the medal of the Order of the British Empire for their service in the Camel Corps during World War I." Flower's designs for attractive and practical animal enclosures were pioneering efforts, heretofore unrecognized. His quarter century at the Giza Zoo contributed not only to zoo management in Egypt, but also in many respects to the whole world.

Sudanese zoos in his little known, but pioneering *A List of Zoological Gardens of the World*. Flower states that Ismail Pasha had created the Giza Zoo's precursor sometime between 1867 and 1872. This fledgling collection had several aviaries for birds and a few mammals, but it was a private collection and did not admit the public.[1-3]

Giza Zoo's centenary brochure relates that Ismail Pasha conceived the idea of having a public zoological garden in Giza to mark the opening of the Suez Canal in 1869. While there was not enough time to develop the zoo, he did collect some animals, keeping them at his Gizerah Palace (now the Hotel Marriott Zamalek). The private gardens in Gizerah also contained an aquarium, which later opened to the public, along with a couple of aviaries. This centenary brochure also claims that, in 1889, Khedive Tewfik Pasha (Ismail Pasha's son who took over the government in 1879) set aside some 50 feddans (approximately an acre) of his mother's palace gardens as a holding area for animals, and that the collection at this palace, El Walda Pasha near El Orman, was the beginning of the zoological gardens.

Other reports on the founding of Giza Zoo differ. Flower states that in 1891, when the authorities decided to start a public zoological garden, the British Government "allowed" the Pasha's garden to be used. Later, in 1898, the Khedive's "Selamlik" Garden was added, doubling

Flower was scrupulous in his record keeping and attention to detail, both scientifically and administratively. As a result, there exists a systematic record of his achievements in the form of numerous and regular yearly reports, tour reports, lists of animals, scientific articles, and notes written during his tenure at Giza, a total of 36 publications for the Zoological Service. Unfortunately, copies of these publications have become difficult to locate, and for this reason, Flower's achievements are no longer well known. Of particular importance are Flower's comprehensive list of zoos of the world and his report on the zoos of India and Asia.***

Flower took up correspondence with other zoos, museums, and zoological institutions around the world, as evidenced by the publications received at the Giza Zoo, all carefully recorded in the zoo's annual reports. This correspondence, combined with Flower's travels to zoos in three continents, paid dividends in the form of his *A List of Zoological Gardens of the World*, which the Zoological Service published and later updated (it was reprinted in *The Zoologist* in 1909).† In January 1912, this list was brought up to date with the help of "correspondents in many countries." According to Flower, this list was reported to be "very useful" to zoo personnel of other institutions, as well as to animal traders all over the world. It was translated into French, German, and Russian when published independently in Paris, Frankfurt, and Moscow, respectively. The list, which was very comprehensive, was probably the first effort of its kind in the zoo community.††

The Zoological Service sent Flower throughout South Asia to tour the zoos there for the purpose of studying how their management might be used to improve Egyptian zoos. His objects were to study the management and housing of large animals, the use of forage materials, accounting methods, staffing patterns, collection registration and record keeping, and management of large crowds. These are interesting objectives for a foreign study tour, particularly the last. Flower visited 10 zoos during his visit to South Asia and made detailed notes on all aspects related to the objectives. He "looked at every animal in each zoo, noted its condition, food, and accommodation compared to the menagerie at Giza." He also visited stables for horses and elephants to learn new and different techniques of managing these ungulates. However, he received no insight (at least none that was recorded) on how to

continued

the area. Some years later, when Gustave Loisel, the French zoo historian, visited the "Government Zoological Gardens Giza," he credited the British for its establishment. The seeming contradiction in these reports is not difficult to understand, as inconsistent reports of this type frequently show up from previously colonized countries. It is probably safe to presume Ismail Pasha had the initial idea for the zoological garden, while the British government made it happen and took full credit. The truth is probably that the Khedive and the colonial government both contributed to the project. In any event, on March 1, 1891 the zoological garden opened to the public. There were about 24,000 visitors the first year (compared with 5,000,000 during the centennial year).[4,5]

Giza Zoo is to be distinguished from the palace collection at Gizerah (the similarity between Giza and Gezira, as it was sometimes spelled, has led to some confusion), which also had an extensive collection. Sir William Flower (S. S. Flower's father) recorded in his diary of April 2, 1874 that he saw in this private menagerie at Gizerah, "two African Elephants, seven Giraffes, sixteen Lions (of all ages), three Leopards, two Servals, one Spotted Hyena, three Nylgales [sic], four Hartebeeste, two Leucoryx, smaller Antelopes, Deer, Kangaroos, Secretary Birds, Flamingos, good collection of Pheasants and fowls, Emu, etc." The Gizerah collection was not open

> ## S. S. FLOWER
>
> manage large crowds of people. His report is the best (and almost the only) source of information on Indian and Asian zoos of that period.[†††]
>
> Flower helped draft legislation for the Game Preservation Department of the Government of Sudan and was active in the protection of Egypt's wildlife. He is credited with having ensured the survival of egrets in Egypt, which plumage hunters had reduced to just two small colonies by 1912. As a result of protective legislation and Flower's efforts, egrets multiplied. As part of this effort to save egrets, Flower established one of the first conservation breeding programs in the world. This consisted of several aviaries constructed throughout Egypt for the captive breeding of the highly endangered egret. Hundreds of birds were bred and released to the wild between 1912 and World War I.
>
> Little else is known about S. S. Flower. Unlike his father, Sir William Henry Flower, who is often mentioned in biographical references for his natural history work and official positions in Britain, S. S. Flower is not found in these same references for his equally significant natural history work and his official positions in Africa and Asia.
>
> [*] See Flower, S. S., "Reptiles of the Malay Peninsula and Siam," *Proceedings of the Zoological Society of London*, 609, 1900; Flower, S. S., "Mammals of Siam and the Malay Peninsula," *Proceedings of the Zoological Society of London*, 306–309, 332–333, 368–369, 378–379, 1900; and Hindle, E., "Major S. S. Flower, O.B.E. Obituary," *Nature*, 157, 327, 1946.
>
> [**] Flower, S. S., *Report on the Zoological Service for the year 1920 in which is included the 22nd Annual Report of the Giza Zoological Gardens*, Pub. No. 34, Zoological Service and Ministry of Public Works/Government Press, Cairo, 1921.
>
> [***] Flower, S. S., *A List of Zoological Gardens of the World*, reprinted in *The Zoologist*, May, 1909; see also Flower, S. S., *Report on a Zoological Mission to India in 1913*, Pub. No. 26, Zoological Service and Ministry of Public Works/Government Press, Cairo, 1914.
>
> [†] Flower, 1909.
>
> [††] Flower, S. S., *Report on the Zoological Service for the Year 1913 in which is included the 15th Annual Report on the Giza Zoological Gardens*, Pub. No. 27, Ministry of Public Works/Government Press, Cairo, 1914; Flower, S. S., *Report on the Zoological Service for the Year 1912 in which is included the 14th Annual Report on the Giza Zoological Gardens*, Pub. No. 25, Ministry of Public Works/Government Press, Cairo, 1913.
>
> [†††] Flower, 1914.

to the public and, after some years, only the aquarium remained. In 1902, the Public Works Department repaired and remodeled this structure for public visitation.[6]

Gustave Loisel waxed eloquent about the streams in Giza Zoo and the "wonderful vegetation of fig trees, papyrus, lotus and other plants of tropical flora." He also commented that "the colored mosaic and marble footpaths were the 'favorites' of the Khedive." And that the zoo is "still embellished by the grottoes constructed by Ismail Pasha." This supports the contention in the zoo's centenary brochure that the Khedive made preparations for the zoo, even though he could not complete it. And an insight into where the inspiration for open enclosures originated may be found in Loisel's comment that "in this country, where rain is so rare, and where the thermometer never goes down to zero, it was unnecessary to build large buildings; accordingly, the animals in this garden live mostly out of doors, sometimes in a state of semi-liberty." Zoos in tropical countries had open, moated enclosures long before the days of Hagenbeck.[5]

It is typical of Flower's efficiency that, having become director of the Zoological Service in 1898, he published the first report on the Giza Zoo in 1900, which covered his first full year of activities. During his tenure Flower compiled and authored many official and scientific publications on behalf of the Zoological Service. From these reports we know that Flower increased the zoo's animal collection considerably. In 1898, there were 270 specimens of 98 species of mammals, birds, and reptiles in the zoo; by 1899, the number had increased to 473 specimens of 132 species. There was a high death rate, as the number of registered additions to the zoo was 671. Of these, 515 were purchased. Flower also made regular collecting trips to the Sudan to obtain African species. Donations were another source, and even Flower's wife was a donor, having contributed lizards, small birds, and other small animals. Flower also added the first amphibians (then called Batrachians) to the collection.[7,8]

Flower initiated and supervised the reconstruction of many structures during his first year, several of which dated to the Pasha's early attempts to establish the animal collection. Judging from Flower's reports, he spent as much time during his first few years repairing, repainting, landscaping, and cleaning as he did developing new features. The gazelle paddocks were fenced to prevent the animals from falling into the exhibit's canal, which previously had led to several drowning deaths. Flower also rebuilt a new entrance gatehouse with materials Victor A. Flower, presumably a relative, had donated. Victor Flower was also responsible for a detailed plan for an Elephant House. During 1899 there were 44,296 visitors, of which students and soldiers were admitted free (except Fridays and Sundays).[7]

On March 20, 1901, Giza Zoo was transferred from the Ministry of Finance to the Department of Public Works. The menagerie and its surrounding gardens, which had been separate, were amalgamated into one administrative unit. Flower reported that Giza Zoo covered 50 acres and the animal collection had, again, increased considerably to 700 "live animals, birds and reptiles" and a large number of plant varieties. There were three miles of walks paved with colored mosaic stones and more than 20 bridges. Rules for visitors in the zoo were adopted in 1901. At this time the zoo employed 9 staff, 81 workers on monthly wages, and up to 30 laborers on daily wages. There was no paid veterinarian, but rather an "Honorary Veterinary Surgeon," whose services may or may not have been available in any given year. With the exception of the director and veterinary surgeon, all staff were native Egyptians.[9]

Loisel, visiting in 1911, observed that the zoo and the aquarium collection contained 1,761 animals, representing 115 mammal species, 207 bird species, 42 reptile species, 10 amphibian species, and 27 fish species. Loisel made special mention of the fact that the collection was particularly well stocked with African species and thus "of particular interest to the zoologist, as well as instructive to visitors." Loisel also remarked on the availability of a book that listed every animal in the Giza Zoo.[5]

A detailed *Plan and Guide to the Zoological Gardens Giza near Cairo, Egypt* was published in 1902 in English, French, and German. This guide stated that the objectives of the zoo were to "stimulate and foster in the people a love for animals and plants and to promote the science of biology" both for the education of laypeople, and for the needs of naturalists and artists. The collection was described as representative of African species, but did not exclude other exotic species which "the natives of Egypt are anxious to see ... tiger, bear, kangaroo, two humped camel, etc." Plants were collected from all over the world with "special attention" to certain Egyptian plants such as lotus and papyrus, which Western visitors wanted to see. The guide also contained a map of the zoo with information on the animals, including scientific name, common name (in three languages), distribution, the animal donor's name, and an occasional interesting bit of trivia. For example, Ranee, the Tigress was caught in central India and was received from the Calcutta Zoo in 1901; a genet was captured on the battlefield whereupon Captain W. W. Cordeaux presented it to the gardens in 1898; and Hathi, a male elephant, was a donation from the government of India, while Chang, a female elephant, was purchased.[10]

Notable exhibits and buildings in the early years of the zoo included a Tropical House, constructed in 1902 with heating apparatus using coal. Special admission was charged for visiting this tropical house. In 1904, the Giza Zoo expanded along its entire length on the south side of the gardens.[11,12] According to Flower, 1905 was the most important year in the short history of the zoo because the government subsidy was increased and many changes were put into effect, including the lowering of the admission fee. As a result, the number of visitors increased by more than 100,000. Flower also felt the monkey and lemur collection, at that time numbering 33 species, was "one of the finest collections ever brought together alive."[13]

Another important exhibit, initiated in 1905, was a series of paddocks for zebras and large antelopes, which Flower thought was one of the finest exhibits of its kind in the world. Since Flower was an intrepid traveler, his opinion probably meant quite a lot, despite the fact that he was commenting on his own zoo. The design of this exhibit considered the comfort of the animals and reflected a vision of future zoo architecture. Each paddock was separated from the others in the series with scrupulous landscaping. This arrangement prevented animals in neighboring cages from fighting, controlled dust raised by galloping ungulates, and provided an aesthetic, naturalistic appearance. Special attention was also paid to the iron fences between the animals and people. These, although tall, were constructed of materials that were of "light appearance" and painted dark green for the least possible obstruction of viewing. Care was taken to "break" the straight lines of the fence with "curves, trees and sunk fences."[13]

Flower's cassowary paddock, constructed in 1906, was another example of advanced design. It was a relatively large space for a bird in those days and used special wire-mesh fencing imported from the United States. Flower took care to provide a pool in which the bird loved to bathe in hot weather and to plant a green lawn with clumps of other shrubs scattered throughout. Wire fence supports were concealed with a purple-flowered climber so that visitors could see the cassowary "to great advantage wandering amongst greenery and flowers."[14]

Flower had a flair for experimenting and improving exhibit designs. Because he was constantly traveling and making contacts with other zoos, the Giza Zoo carried on an active gift and exchange program with zoos around the world. Sometimes the "removals" from the zoo were in dramatic numbers, such as when, in 1906, "ten lions [were] sent to India for H. H. Maharajah of Gwalior; thirty-five animals sent to the Stellingen Zoological Gardens and one hundred and seven sent to the New York Zoological Society's Menagerie at Bronx Park."[14]

In 1908, hippopotamus and Nubian bustard are mentioned as having been exhibited in Giza Zoo for the first time. This year was also cited as the most successful in breeding animals, including zebra, kudu, addax, and waterbuck. During 1908, the zoo administration paid special attention to venomous snakes because of the great interest of visitors, with five of the six recorded Egyptian species exhibited.[6] Native Egyptians became more involved in the zoo, donating animals and writing impressions of their visits to the zoo. H.H. Prince Youssef Kamel Pasha captured a rhinoceros on the Kit River and presented it to the Giza Zoo in 1910, the first of its species to be exhibited at the zoo. The number of visitors had reached 200,000 and the number of animals nearly 1,500 specimens of almost 400 species.[15]

In 1910, William C. Beebe of the New York Zoological Society visited Egypt and described the Giza Zoological Gardens as a "delight to the lover of beauty as well as to the zoologist" and said Captain Flower "deserves the greatest credit for having brought order out of chaos, and with the able help of his assistant, Mr. Nicoll, is building up a most valuable and interesting collection of African animals." Beebe went on to comment that "it would be well if a similar segregation of the indigenous fauna could be made in all zoological gardens, as it possesses a peculiar interest both for tourists and natives." Beebe also noted that "the cages of the birds and monkeys are open and portable. This makes it possible to remove an infected cage and replace it at once by a fresh one."[16]

Construction was finished on the giraffe paddocks, which had been started in 1907. This giraffe installation seems to have been a matter of some pride, having "two large paddocks with a shelter and small yard in each." According to Flower, the giraffe paddock at Giza Zoo was the "largest and most effective of its kind in any menagerie in the world." Its circumference was 942 feet with paths on two sides and in the center from which visitors could view these animals. It is interesting to note that Giza Zoo kept pigs, since they are neither a wild animal nor a popular animal in a Muslim country. Yet, "the principal work during 1913 was the repairing and repainting of the Pigs' Cages which was the original Lion House." Outdoor cages for local reptiles were also constructed, these having plate-glass viewing areas.[17]

Among the civil works completed in 1914 were 24 new seats for use in the zoo, making a grand total of 450. This number seems excessive, even for an 80-acre zoo, but it seems there were many public complaints about the lack of seats and about seats with broken legs. Clearly, the zoo, at that time, was a place where the public spent its leisure time sitting and socializing.[17]

An old building known as the Giza Zoological Museum was used as a kind of early "education building," with its collection supplementing the zoo's collection. After the museum began about 1912, the Cairo School of Medicine and the Giza Zoo donated local specimens. In these early years of the modern zoo, the number of species in a collection was of primary interest. Late nineteenth- and early twentieth-century zoos were consumed with taxonomic curiosity, and the number of species (and subspecies or forms) that could be collected and displayed determined the quality of the collection.[17]

The mammal collection at the Giza Zoo included the obligatory large species, such as apes, elephants, rhinoceroses, hippos, giraffes, zebras, and kangaroos, but also a representative sample of smaller species one would not necessarily expect, such as bats, rodents, hyrax, and insectivores. A reason for the inclusion of so many noncharismatic, minivertebrates was the contemporary interest in natural history. Early twentieth-century motivations for developing a living museum, which had inspired Raffles to start the London Zoo and inspired other colonial officials to begin collections, would have affected Flower as well. Flower was a naturalist who was interested in all lifeforms, and whose father was one of the notables in the Zoological Society of London.

Flower's administration was successful in increasing the Giza Zoo collection through captive animal breeding and other means. Of the 486 animals added to the collection in 1913, nearly 200 were born in the zoo (some not so easily bred in captivity, such as fruit bats and spiny tailed gerbils), the public donated 130, the zoo staff collected 118, 9 were from other zoos, and 6 were deposited for safekeeping. Others were procured from the Sudan Game Preservation Department, Cairo bird shops, European animal dealers, and individuals not in the trade.[17]

Also notable were the departures from the menagerie during 1913. Of the 1,600-odd individuals disposed of, 78 animals were "removed, sent away alive," some "set at liberty," and others sent to various zoos around the world: Washington, Berlin, Stellingen (Hagenbeck's zoo in Germany), and various private collections. Three animals were removed to feed carnivores and 17 were destroyed as "undesirable." And 31 animals disappeared, wild animals killed 26, and there were 32 accidental deaths, many of which were attacks by conspecifics in the same enclosure. Death from natural causes claimed 246 others.[17]

Flower inspected the zoos of Egypt and the Sudan once or twice a year and gave suggestions for their improvement. Municipalities with a zoo would invite the Giza Zoo Director to inspect its collection and suggest improvements. It was customary to apply to the Ministry of Public Works for this privilege, with the municipality paying the inspection costs. These small municipal zoos did not have directors; instead, governors of the provinces provided supervision. Flower, as director of the Giza Zoo and the Zoological Service, or one of his assistants (later called "Inspectors"), performed the inspections, along with occasional visits to sort out more complex problems.[17]

In those years of plentiful wildlife, Giza Zoo authorities had no reason to feel embarrassed about their collecting expeditions. Nearly all of Flower's reports for the early 1900s devote a section to "Expeditions," in which are listed the date, place, and purpose of the expedition. In 1910, for example, Flower made his eleventh annual Sudan tour, visiting live animal collections and collecting along the Blue Nile. He returned with 170 live animals, 140 bird skins, and a few others preserved in spirits. And in 1911, Flower and seven keepers visited Suakin, Erkowit, Khartoum, El Obeid, Sennar, and Singa, returning with 183 live animals and some preserved specimens. The success of this expedition was, according to the report, due to the kindness and help of "every officer of the Egyptian Army and officials of the Sudan Government," just to name a few.[8,15,18]

In 1912, Flower proudly reported in the preface to his report that the menagerie at Giza held "1,608 individual animals ... the largest number ... so far maintained." He also mentioned that in July the zoo received a "valuable collection of 183 live animals and over 150 museum specimens" from the Sudan, and that legislation against hunting was passed. The irony of this mixture of achievements was lost on Flower and, undoubtedly, others of that era. These regulations were just in time to save the cattle egret from total extermination. The birds were in demand for their lovely feathers, which European ladies liked for decorating their hats and to wear in their hair.[19]

Even officials of the Giza Zoo were called upon to monitor the situation. The Director himself, as well as the assistant director took "Bird Protection Duty in the provinces for a month in periods of 2–6 days during the year." In 1921, when the laws and enforcement measures had been in place for nearly a decade, it was reported that the cattle egret population had recovered to a very large extent. During the summer of 1920 it was estimated there were at least 100,000 of "these useful birds in the country," of which a great number (64,000) were in wild breeding colonies which the Egyptian Zoological Service managed. There were also an estimated 10,000 birds in a free-ranging colony at the Giza Zoo itself.[20]

This egret conservation scheme even included a captive breeding program. The Egyptian Zoological Service made an attempt to initiate artificial breeding colonies, maintaining birds in large aviaries and ultimately releasing them into the wild. Flower comments that the aviaries were also useful for "studying the habits of the birds and their needs." During 1912, the government constructed two of these aviaries, one at Giza and one at Gizerah. A third aviary was completed during 1913 in Gharbia Province, and another at Luxor in Upper Egypt. The government of Sharqia Province also started an aviary for egrets at its Zagazig Zoo and the government of Beheira proposed to create one in their public gardens at Damanhur. Results of the breeding colonies, both natural and artificial, were encouraging. Flower remarks, in fact, that "the success ... has exceeded the most sanguine expectations."[17,19]

During World War I no annual reports were issued and it is presumed that many activities were put on hold for the duration. In 1921, Flower reports that the zoo grounds were restored to their "pre-war condition of cleanliness" and the collection of animals increased. No mention is made of the aviaries and egret colonies that had been started with such enthusiasm for "artificial breeding." In 1920, the Zoological Service continued its involvement with wildlife protection and also began to provide wildlife control, "controlling beasts of prey" that threatened human life.[20]

Flower retired from the Giza Zoo and Zoological Service in 1924. After his departure, information about Giza and other Egyptian zoos is not easily found. During these years the history of Egypt is fraught with violence and economic difficulties. The British Army remained in Egypt until 1956. Since the 1960s Egypt has suffered through wars, religious fundamentalism, political coups and assassination, and economic depression. Since Flower retired from the directorship, officials posted as director of the Zoological Service have been transferred frequently and have not had an opportunity, as did Flower, to settle into the job and build projects and contacts. Nor did they have such a zoologically illustrious background as Flower's zoological "blue blood." Moreover, as is appropriate, records and reports since Flower's time are more likely to have been written in the Egyptian script.

One is, therefore, dependent on those official brochures and handouts from the zoo that are in English, as well as occasional articles by visitors and the press, for further information on the years after Flower retired. In 1949, the Survey of Egypt published a map of the zoo with brief descriptions of the chief exhibits in Arabic, French, and English. The chief exhibits at that time consisted of many more native species, but also an impressive number of animals from South America, Asia, and Europe.[21]

Geoffrey Schomberg visited the Giza Zoo in 1961 and found an attractive setting with a series of large enclosures for herbivores.[22] In 1982, J. C. Suares wrote about the severe air pollution in Cairo, with particular reference to its effect on the zoo. Another problem in Cairo that badly affected the zoo was the large population of stray animals. Stray dogs and cats in the zoo learned keepers' feeding schedules and followed them around to snatch a portion of the food for themselves.[23] In 1986, Giza Zoo reported an attendance of 5,000,000 and a collection of 7,850 specimens of 395 species.[24]

With the help of the International Center for Bird Preservation and the Egyptian National Wildlife Service, Giza Zoo started a Conservation Education Center in 1988. This center was the first environmental education facility of its kind in the country. It had a trained education staff, produced educational literature and public awareness materials, and held the first regional workshop on conservation education. The main aim of the project was to enhance the existing educational capacity of the zoo to encourage support for the conservation of Egyptian wildlife and habitats. Unfortunately, the center — a model facility — was closed as a public facility and is used now only on special occasions, such as conferences.[25,26]

In 1991, the zoo celebrated its centennial with a full-day celebration. There was talk of improving enclosures, including an air-conditioned den and pool for the two polar bears. Four nongovernmental societies associated with the Giza zoo — the Egyptian Zoological Society, Society for the Protection of Natural Resources, Egyptian Wildlife Society, and Friends of Animals Club — participated in the historic occasion.[27] At the time of the centenary, the Giza zoo held 12,000 animals of 370 species from all over the world.[4]

More recently, Peter Klaver, veterinarian at the Amsterdam Zoo, visited Giza Zoo in 1995 to assist the zoo's veterinarians. Klaver found limited instruments and drugs available at the zoo, but a very willing group of veterinarians. He reported that the veterinarians had good theoretical knowledge about surgery, but no routine for doing it, and that supplies necessary for operations are unavailable or difficult to get. Such is the current state of affairs.[28]

10.2.2 Other Egyptian and Sudanese Territory Zoos

Regarding the other zoos in Egypt and the Sudan territory, information is very scarce. Many zoos started at the turn of the twentieth century no longer exist. Flower included several Egyptian and Sudanese zoos in his annual reports for the Zoological Service, but many of these "zoos" seem to have served largely as collection points and holding areas for animals intended for the Giza Zoo or for export.

In addition to eight or so municipal menageries, there was the Gizerah Aquarium, which was part of Khedive Ismail Pasha's royal collection at the Gizerah palace. The Khedive did not maintain the aquarium after 1877, and in 1902 the Egyptian Public Works Department renovated the aquarium and opened it to the public.[6] In 1904, the aquarium boasted 30 fish species, of which 29 were local Nile species the remaining 1 the goldfish. Two keepers looked after the aquarium and Flower, director of the Zoological Service and the Giza Zoo, was also head of the aquarium. G. A. Boulenger, a naturalist well known for his natural history studies in Asia, examined a small collection of fish from the aquarium, which Flower had sent to him, and pronounced the identification as perfectly correct.[18]

William Beebe, of the New York Zoological Society, commented in 1910 on the aquarium, which contained 17 tanks placed in an "artistic artificial grotto" made of cement. Beebe was impressed with the "huge electric catfish [which was] capable of giving a fatal shock to a man."[16] The facility at Gizerah became known as the "Gezira Aquarium and Grotto Garden" and employed, in addition to the two keepers, a head gardener, gatekeeper, night watchman, 13 gardeners, and a garden boy.[19] In 1912, an aviary was built on the palace grounds near the aquarium. This aviary was filled with cattle egrets in 1913 (as part of the effort to breed this endangered species) and a second aviary was built to hold the other bird species.[17]

Khartoum Zoological Gardens was founded in 1901 in the center of Khartoum (Sudan), but in 1903 was moved to a site on a tongue of land between the White and Blue Niles. The gardens were under the control of the municipality, but the animal collection was under the control of the Sudan Game Preservation Department.[6] When Flower made his eleventh annual Sudan tour in 1911, he visited not only the Khartoum Zoo, but the zoos at Merowe, Port Sudan, and Singa.[18] During visits in 1920 and 1921, Flower commented that this zoo was kept in a "far better and cleaner condition than I have ever seen it formerly. The garden is very neat and attractive and the collection of the animals varied, valuable, and in good case, and well illustrating the fauna of the Sudan."[20]

Singa Zoological Garden (Sudan) was begun as a private collection of wild animals in 1906. Major C. E. Wilson, East Lancashire Regiment and Governor of Sennar, started the collection, which was located close to the governor's house on the left bank of the Blue Nile. It was surrounded with a hedge, which Flower described as "strong enough not only to prevent the animals from escaping but also to keep the Spotted Hyaenas, which prowl round at night, from entering and attempting to devastate the collection."[20]

Governor Wilson's successor as governor, Captain G. S. Nickerson, took "great interest" in the Singa Zoo. When Nickerson passed away in 1911, Captain A. A. C. Taylor, Royal Dublin Fusiliers, became governor and continued the interest of his predecessor. These three governors rendered much assistance in collecting animals in the Sennar Province and in looking after the collection. Singa Zoo served as a "way station" or "branch establishment" for the Giza Zoo's collecting ventures and was, as such, financed by the Zoological Service. Flower describes it as a "most valuable depot." Two keepers, a Sudanese man and his wife, cared for Singa Zoo's animals.[20]

Flower wrote that it was "unique among Zoological Gardens of the world in that the animals, with exception of the carnivora, are nearly all allowed to run loose over the whole area of the institution." He further commented that it was a "most picturesque arrangement" but had its difficulties, such as the time an elephant destroyed the zoo's gate and the entire collection wandered into the neighborhood. Flower continues, saying visitors are admitted free to the zoo but "at their own risk."[20] In the middle of 1912, the entire collection at Singa Zoo, except for some monkeys and geese, were taken to the zoological gardens at Khartoum and Giza. A few months later, in February 1913, there were only a serval, a spotted hyena, porcupine, waterbuck, roan antelope, four giraffes, a hippopotamus, and two ostriches.[19] The record of the eventual demise of this zoo is not available.

Alexandria Zoological Garden, which began in 1907 as a "small menagerie," is the only zoo besides Giza Zoo to survive well into the twentieth century. It was located at Nuzha in a beautiful park a bit outside and east of Alexandria. "Cages were built and a considerable number of mammals and birds obtained by presentation and purchase."[6,18] By 1912, Alexandria Zoo had a collection of 57 mammals of 23 species, 157 birds of 31 species, and 5 tortoises of 1 species, or about 208 animals, not including domestic animals. By 1913, the collection had lost nearly

25% of its animals, although the report does not mention any problems at the zoo. Some of the animals were the sacred baboon (*Papio hamadryas*) which, according to Flower, bred well at the zoo, a female Grant's zebra (*Equus burchelli granti*), a "fine old kangaroo (*Macropus giganteus*) which had been more than 12 years in captivity," and a beautiful specimen of American heron (*Ardea cocoi*).[19] In 1921, Flower visited the Alexandria zoo twice and maintained that the "cages were clean, the animals in good condition, and the friendliness to their keepers exhibited by the large carnivora very pleasing to see."[29] The Alexandria Zoo continued, and in 1989 contained 2,620 animals of 255 species and had 1,840,000 visitors.[30]

Zagazig Zoological Garden (Egypt) was a municipal institution founded by H. E. Hassan Hassib Pasha, Mundir of Sharqia, in 1911. Flower described it as a "very pretty garden," located along with the city's water pumping and electric power station, on the south bank of the Bahr Moes, west of the city of Zagazig in Sharqia Province. An entrance fee was levied, band performances were held twice a week in the garden, and there was a café for refreshments. Flower reported that it was "deservedly popular with the inhabitants of Zagazig and of great interest and instruction to visitors from all parts of the Province of Sharqia."[19]

Zagazig, like the other menageries in Egypt, contracted with the Egyptian government for the services of the Giza Zoo director to inspect its facility and make suggestions. Flower comments that, for a zoologist, the main feature of this zoo was its collection of local animals from the Sharqia Province, which consisted of wild cats, foxes, and other small mammals, although the casual observer would find the monkeys, lion, leopard, saber-horned antelope, ibex adjutant stork, pelicans, and flamingos more spectacular. Flower also states that the Zagazig Zoo distinguished itself with cages, ponds, and walks that were particularly clean and neat, as well as a wealth of flowering plants in well-arranged beds. He said "out of more than forty public gardens containing menageries which I have visited in Africa, Asia and Europe, few have made such a pleasant impression on my memory as that of Zagazig." Lack of signage was a problem, however, and he felt the value of the animal collection would be improved if "names of the animals, the places of origin and the dates of arrival were written up in both Arabic and European characters." By 1912, the collection at Zagazig consisted of 66 mammals of 34 species, 208 birds of 78 species, and 2 tortoises of 1 species, a total of 366 animals plus some domestic rabbits and poultry.[19] Zagazig Zoo is not mentioned in the 1922 annual report of the Zoological Service, and it is possible the war years took their toll on the collection.[31]

In autumn 1916, a collection of wild animals was kept in the public garden at Tanta (Egypt). Ibrahim Kadry inspected Tanta Zoological Gardens in June 1922 and reported having seen 12 mammals of 8 species, 38 birds of 15 species (the most important being an emu and four ostriches), and 2 tortoises. Nothing is known of its fate.[31]

Merowe Zoological Gardens was located in the Mudiriya Gardens at Merowe, which was the capital of Dongola Province (Sudan). Flower refers to it often as it formed a regular part of his itinerary when he visited Sudan. Mudiriya Gardens, by 1920, consisted of approximately 50 acres. Flower, much impressed by this garden, stated that of the "many beautiful gardens watered from the Nile from Alexandria and Ismailia in the north, to Roseires and Mongalla in the south ... those of Merowe are the most attractive as gardens by themselves, and are the best laid out and the neatest and tidiest." As an efficient government servant, Flower also appreciated the fact that these gardens were self-supporting from the sale of fruit and other plant products grown there.[20]

Merowe Zoo exhibited all the wild bovine species of Dongola Province in a series of paddocks, and had particularly good success with these ungulates. In June 1920, there were "fifteen Gazelles of the local form of the Gazella dorcas group, three Addra or Ryl Gazelles, Gazella ruficollis, four White Oryx, four Addax Antelopes, and eight Arui Wild Sheep." Other mammals exhibited were grivet and red hussar monkeys. Birds included spur-winged geese, blue-wattled guineafowls, Indian peafowls, large Sudan bustards, and crowned cranes. There were also large land tortoises, *Testudo calcaraia*.[20] Flower credits General Sir Herbert Jackson Pasha, governor of Dongola with the "wonderful success" of these gardens. Jackson gave them the "greatest personal attention."[20] Unfortunately, Merowe Zoological Gardens was completely washed away during a flood on August 7, 1921.[31]

Port Sudan Zoological Gardens was a small menagerie kept in the public garden at Port Sudan (Sudan). Flower was unable to visit this zoo during the war years and the animal collection at Port Sudan was dispersed before Flower's 1920 Sudan visit.[20] Other municipal zoos that received the director of the Zoological Service's official inspections were those at Benha (Qalyubiya Province, Egypt) and at Mind el Qamh (Sharkiya Province, Egypt). There was also an aviary for egrets constructed in 1913 at Sakka (Gharbia Province, Egypt).[17,20]

Today, there are a number of zoos throughout Egypt, in addition to those the Zoological Service (Ministry of Agriculture) manages, mostly government-managed facilities. There are also restaurants and other establishments that have animals, as is prevalent in the Middle East. And there are other private zoos. Two of the most famous restaurants with zoos in the 1990s were the Seagull Restaurant in Alexandria and Omar's Farm Restaurant on the Alexandria–Cairo Desert Road. According to visitors and expatriates, as well as the press, these various zoos seem to occupy the bottom of the Ministry of Agriculture's priorities list, and there seems little interest in making the situation better. Today, Egyptian zoos are not educational or conservation facilities, but merely gardens or recreation grounds with animals. Entrance fees to the zoos are very low and funds for management and maintenance are insufficient. What had been a golden age of animals in a magnificent culture has now deteriorated.

10.3 Sub-Saharan Africa

10.3.1 Kenya

Louis Leakey's interest in primate behavior as a source of clues to early human behavior led him to establish the Primate Research Centre (Tigoni) in 1960, now known as the Institute of Primate Research (Nairobi). The Institute has 476 primates and is now managed by Louis' son Richard and the National Museums of Kenya. Current research focuses on infectious diseases, reproductive health, primate medicine, and primate conservation. Another facility of the National Museums of Kenya is the Nairobi Snake Park, established in 1962. The Snake Park now has 103 reptiles of 34 species and 300 fish of 125 species. Nairobi Safari Walk began in 1964 as the Nairobi Animal Orphanage, a facility of the Kenya National Parks. The Safari Walk is home to orphaned, injured, and abandoned Kenya wildlife; provides wildlife education about indigenous species and Kenya ecology; and provides an introduction to the Kenya park system. Mount Kenya Game Ranch (Nanyuki) is a private collection of African mammals with a few birds and reptiles.

10.3.2 Republic of South Africa

Port Elizabeth was in its infancy when the idea for a museum was considered. In 1854, the commissioners of the municipality acquired land and began construction on a building to house the town hall, municipal offices, library, and an athaeneum. One small room of the athaeneum was used to house a collection of natural history specimens. This small collection was expanded in 1856 into a museum, which received a constitution and officially became the Port Elizabeth Museum in 1897. In 1906, F. W. FitzSimons, the well-known herpetologist, became director of the museum. FitzSimons revitalized the museum exhibits and added live snake exhibits. A seal pool was opened in December 1933, but did not exist for long. Although the seals were popular with the public, their barking and bickering at night intruded upon the lifestyle of Bird Street's "merchant prince" residents, and, following a court case, the seals were banished.

During World War II, the museum received an appeal from the South African Institute for Medical Research for larger supplies of venom. The serum was urgently required for the Allied forces serving in snake-infested parts of Africa. John Pringle, director since 1936, undertook snake-collecting trips throughout the Eastern Cape to meet this demand. By 1958, the museum had grown considerably and a much-needed move was undertaken. A new museum building was constructed, as was a new Snake Park facility. During the following year an oceanarium was completed. In 1968, a dolphin lake was added, as was a seal pool (after a 35-year banishment). A tropical house, a combined horticultural and bird exhibit, opened in 1972. Since 1975, museum research has expanded considerably and includes the study of marine mammals, gannet biology and ecology, marine predator–prey relationships, and herpetology. An education center has also been established, as education has become an even more important function of the museum.

East London Zoological Gardens began as a menagerie at the Botanical Gardens some time after the East London municipality established the gardens in 1890. This zoo now has 175 mammals of 49 species, 122 birds of 41 species, and 29 reptiles of 8 species. East London Aquarium opened to the public on December 2, 1931. Established as a public amenity and tourist attraction, the aquarium has developed into a public education facility and scientific reference center for other institutions at the national and international level. In addition to displaying a wide variety of local and exotic marine specimens, the aquarium has a performing seal display and a breeding colony of the endangered Cape jackass penguin. Marine animal and bird rescue, recovery, and rehabilitation are also undertaken by the aquarium. The aquarium currently has 526 fish of 119 species in 34 tank displays, along with a few mammals, birds, reptiles, and invertebrates.

On October 21, 1899, J. W. B. Gunning, director of the State Museum, moved a collection of animals from the museum's backyard to the farm "Rus in Urbe" on the banks of the Apies River. This was the start of the National Zoological Gardens of South Africa.[32] Animals donated to the State Museum (now the Transvaal Museum) for the purpose of being stuffed for display in the museum were kept in the backyard as Gunning did not have the heart to kill them. Instead, he instigated people living in the vicinity of the museum to complain about the smell and noise of the animals; this led to permission to move the animals from Gunning to the farm with the aim of establishing a zoo. This animal collection consisted of a serval, striped polecat, a leopard, a jackal, a genet, baboons, vervet monkeys, various antelope species, dormice, a bat, a spotted eagle owl, 50 other birds, a monitor lizard, a python, and a tortoise. Fish were added to the collection in 1910 when the City Council of Pretoria donated the Sammy Marks fountain and fish pool.

Independence from the museum came in 1913, at which time the zoo became the Transvaal Zoological Gardens and Gunning was appointed director. During his tenure with the zoo, he oversaw the building of the lion house, buffalo and zebra exhibit, elephant and rhinoceros house, bear house, and raptor aviaries. Gunning died in 1913 and A. K. Haagner served as director from 1914 to 1926. Although the zoo could not afford many animals, it served as an intermediate home for animals on their way to Europe and the United States. The zoo was given national status in 1916 (the only African zoo with national status) and its name was changed to the National Zoological Gardens of South Africa. Under Haagner's management, the primate house, giraffe house, ostrich house, and reptile vivarium were built.

When Rudolph Bigalke took over as director in 1927, the zoo was in dire straits. In 1928, he had a tea room and bandstand built to attract more visitors. And in 1930, he convinced the Department of Public Works that the national government was responsible for maintaining the zoo, as it was for maintaining the national museums. Bigalke was the first zoo director in Africa to become a member of the International Union of Directors of Zoological Gardens, receiving his membership shortly after World War II. Under his supervision, the zoo constructed the lion and tiger enclosures on the other side of the Apies River (the mountain area north of the river which the zoo had acquired in 1909), the Eileen Orpen bird aviary, bear enclosures, and the veterinary hospital (named the Rudolph Bigalke Institute).

In 1960, Frank Brand took over as the fourth director of the zoo. Brand believed the zoo should represent animals in their natural environment and he undertook the renovation of the exhibits to accommodate this philosophy. Brand also believed it was important to develop captive breeding programs for endangered species. However, providing larger natural exhibits and providing facilities for captive reproduction was difficult in the zoo's limited space. In 1974, the Lichtenburg municipality made available some 14,800 acres of western Transvaal high veld for the National Zoo to use as an endangered species breeding center, named Lichtenburg Game Breeding Centre. Père David's deer and pygmy hippopotamuses are housed in the center's wetland area, while large herds of impala, springbok, zebra, blesbok, and red hartebeest roam the vast grassland area. The collection also contains populations of the endangered addax, mhorr gazelle, scimitar-horned oryx, and Arabian oryx. Part of the wetlands area has also been developed into a system of dams and pans, which serves as a natural haven for thousands of waterbirds such as spoonbills, a variety of ducks, kingfishers, ibis, and herons.

Potgietersrus Game Breeding Centre was established in 1981. It is located on 3,700 acres of bush veld and its purpose is to supplement the zoo's collection and to help with its endangered species breeding programs. Indigenous bird species can also be found throughout the center and in its walk-through aviary. The zoo has also been able to procure the use of the De Wildt Breeding Centre, property of Ann van Dyk, to assist the zoo with its breeding programs. This center was responsible for the first captive breeding of the king cheetah (once considered a separate species, but now considered a rare variety of the normal cheetah). The zoo also completed its Zoo Farm in 1982, and an aquarium and a reptile park were opened in 1974. The zoo's education building was constructed in 1982 and new elephant quarters were built in 1984.

After Brand's retirement in 1985, Wilhelmus Labuschagne, with a background as curator of the Johannesburg Zoo for 13 years, was appointed director of the zoo. In 1989, the Department of Public Works commissioned a master plan for the zoo, a long-term project that is to be implemented in phases over the following 30 to 40 years. The focus will be bioexhibits, each of which will display multiple species in natural habitat enclosures. Thus, the progress that has been made since 1985, including the first insectarium in Africa (which opened in 1990), a parrot enclosure, modernization of the veterinary clinic (the Bigalke Institute), a beaver enclosure, upgrading the chimpanzee enclosure, opening of an animal nursery, upgrading the hippo and pygmy hippo enclosure, improving the gibbon enclosure, improving the bear enclosure to house two Kodiak bears, renovation of the aquarium, and creation of the zoomobile (another first for an African zoo), will be continued. In time for the zoo's centenary celebrations, a koala exhibit will be opened. It will be the first time a koala has been exhibited in Africa. In 1999, the zoo housed 110 mammal species, 175 bird species, 270 fish species, and 90 reptile species, for a total of 5,770 specimens.

Johannesburg Zoological Gardens began in 1904 as a small, privately owned collection of African wildlife belonging to Sir Percy FitzPatrick. This collection and its site were donated to the Johannesburg City Council to establish the zoo. Over the years, to 1912, and under

FitzPatrick's guidance, numerous animals were added to the collection. This necessitated an expansion, and in 1912 the zoo expanded onto adjacent land held in trust as a war memorial to the soldiers who fell during the South African War (1899–1902). In 1947, the South African National Museum of Military History was added to the ground adjacent to the War Memorial. As with most zoos, the Johannesburg Zoo has evolved from iron and concrete cages into modern, multispecies, open enclosures. Landscaped animal enclosures, hidden moats, and shady public lawns typify the Johannesburg Zoo. The zoo emphasizes education and views itself as a living classroom.

In 1947, the local Durban Wildlife Society undertook a scientific expedition to a pristine coastal region of the Province of KwaZulu Natal, just south of the Mozambique border with South Africa. Impressed with the diversity of marine life on that stretch of the coast and struck with its complexity of fragile and unresearched ecosystems, those present recognized the need for more knowledge about this region. One night, around a campfire near the beach, the expedition members resolved to establish a marine biological station in KwaZulu Natal. On January 30, 1951, following a great deal of work by conservationists, academics, and anglers, the South African Association for Marine Biological Research was founded as an independent, nonprofit, nongovernmental organization. The association recognized the need for an aquarium to assist their research and public education efforts, and in June 1959 an aquarium was officially opened. Because Durban was in the midst of celebrating its centenary year, the aquarium became known as the Durban Centenary Aquarium. It consisted of a single, circular marine tank, an entrance hall, and two offices. Simultaneously with the opening of the aquarium, the Oceanographic Research Institute was established and became affiliated with the University of Natal.

During the late 1950s and early 1960s, a series of shark attacks plagued Durban and affected the city's tourist trade. The province commissioned the Oceanographic Research Institute to investigate this situation and to recommend the best way to protect bathers from shark attacks. In 1961, an experimental shark tank was opened at the aquarium for public viewing whenever researchers were not using it. It remains today as the aquarium's shark exhibit. During the late 1960s further additions were made, including the construction of new exhibits, a restaurant, gift shop, lecture hall, and additional laboratories for the researchers. In 1976, the dolphinarium was built, adding marine mammals to the aquarium collection. True to the originators' concept, the aquarium emphasizes the complexity of the marine life and ecosystems of the region's coast. Since the very beginning, the aquarium has been open to all sectors of the community, never bowing to the apartheid laws of the time. Today, the association manages three facilities: the aquarium renamed Sea World, the Institute, and the Sea World Education Centre.

Walter Mangold founded World of Birds Wildlife Sanctuary in 1973. It emphasizes large, landscaped walk-through aviaries that contain over 300 species of birds and small animals. Alan Abrey, a veterinarian with a passion for psittacines, founded Umgeni River Bird Park in 1983. Taking over a disused quarry, he built a series of walk-through aviaries and bird exhibits in a well-landscaped environment. Today, the Bird Park continues to emphasize psittacines and cranes.

Cango Wildlife Ranch began as the Cango Croc Ranch. Andrew and Glenn Eriksen established the ranch in 1976 and opened it to the public in 1977, presenting South Africa's first crocodile show and farm. Today, while the ranch continues to house a large crocodilian collection, it has turned its attention to the plight of endangered species. An expansion of the ranch allowed it to include a number of big cats in its collection, with an emphasis on the cheetah. In 1988, the ranch developed a facility, Cheetahland, in an effort to promote its goal of public awareness of endangered species. The ranch also developed its center for breeding cheetah, as well as other endangered species, including the Cape wild dog, pygmy hippo, serval cat, and Bengal tiger. The collection at the ranch includes other African animals, cats, birds, and a snake park.

Cape Town is the original link between the land and sea for all of southern Africa. It is also here that the cold water of the Atlantic Ocean meets the warmer waters of the Indian Ocean. Few places in the world feature aquatic life so different in their habitat requirements. It is remarkable that it took so long before a major aquarium was conceived and built here. Finally, in 1994, the Two Oceans Aquarium Trust was chartered for the purpose of establishing a privately owned commercial aquarium, which opened to the public in 1996. This aquarium was designed to exhibit, compare, and contrast the habitats and lifeforms of the Atlantic and Indian Oceans, as well as to educate the public about the ocean environment in general.

10.3.3 Other African Nations

Colonial botanical stations began in the major West African ports (such as Madeira, São Tomé, and Fernando Po) in the 1400s, although most botanical stations were not established and did not start holding animals until the 1600s. Cape Town, in what was to become the Republic of South Africa, had one of the most important of these stations, which the Dutch East India Company established there in 1652. Throughout the rest of Africa, colonial collections were established, although very little is known about these either because they were not documented or because the documentation is difficult to locate. For example, there were the private collections of Cecil Rhodes in South Africa and the French acclimatization stations in Algeria. However, for most of Africa, these collections do not seem to have had a connection with the modern African zoos and aquariums of the twentieth century, Egypt being a notable exception.

Malawi

A group of enthusiasts, formed as a zoological society, started the Blantyre Zoo in 1967 to house animals in need of shelter. Eventually, the society disbanded and handed the zoo over to the Blantyre City Council, which now manages the zoo through its Department of Parks and Recreation. The collection contains a few large animals along with a number of birds and small mammals. In a country with national parks teeming with wildlife, it might seem unnecessary to have a zoo, but if it were not for the zoo, many urban dwellers would never see the native animals of their own country.[33]

Morocco

Although a small zoo existed in Casablanca, the government of Morocco decided to create a larger, more modern zoo, which it did in 1969. This new zoo, the Parc Zoologique National de Rabat, was located on the main highway between Casablanca and the capital city of Rabat. To start the collection, the royal family presented the zoo with 20 lions from its palace collection. The zoo was officially opened to the public in 1973. Part of the Ministry of Interior during its construction, it was turned over to the Ministry of Agriculture upon its opening. In 1999, the zoo had 693 mammals of 95 species, 974 birds of 143 species, and 96 reptiles of 19 species.

Mozambique

Both zoos of Mozambique are located in Maputo: the Jardim Zoologique and the Maputo Zoo. In the case of the latter zoo, veterinarians formed the Associacao do Jardim Zoológico de Moçambique and erected a small zoo and animal reserve in 1929, originally known as the Lourenco Marques Zoo. Many have considered the Maputo Zoo one of the most beautiful in Africa. With the independence of Mozambique in 1975, the management and most members of the association left the country, while the remaining members turned the zoo over to the city council. The carnivores in the collection were the first to perish since there was little to eat, other than the cabbage and fish that both the people and the zoo animals had to subsist on, a result of the turmoil during the civil war that led to independence. Unbelievably, six lions managed to survive on their new diet of cabbage, becoming the first vegetarian lions. As times worsened under independence, the zoo lost 90% of its animal collection.

In 1989, veterinarians again came to the service of the zoo. After investigating the situation at the zoo, these veterinarians formed the Maputo Zoo Fund and employed Joanna Dalton to manage the zoo with the aid of Cobus Raath of Kruger Park. This fund eased the plight of the zoo between 1989 and 1993, but in 1993 the fund was closed, culminating in the reactivation of the association in 1994. Although a private zoo by law, Maputo Zoo is still co-managed with the city council and the city employs the zoo staff. The association feeds the animals and pays for repairs. Maputo Zoo also has a sponsorship from the Royal Dutch Embassy and has a close relationship with the Johannesburg Zoo. With all of this support, the zoo has been able to sustain a slow recovery.

Ghana, Senegal, and Nigeria

Ghana's Department of Game and Wildlife manages Kumasi Zoological Gardens (Kumasi, Ghana). It is a small zoo with 52 mammals of 15 species, 76 birds of 11 species, and 55 reptiles of 13 species. Hann Zoological Garden (Dakar, Senegal) operates under the direction of the Ministére de la Protection de la Nature, Service des Eaux, Forêts et Chasses.

Ibadan Zoo (Ibadan, Federal Republic of Nigeria) is an operation of the University of Ibadan. It is a small zoo that the University Zoology Department operates. In 1999, it had 71 mammals of 26 species, 68 birds of 24 species, and 82 reptiles of 25 species. The Zoological Society of Jos manages, with financial aid from the national government, the Jos Museum Zoo (next to the National Museum in Jos), which specializes in West African fauna. In 1999, it had 62 mammals of 25 species, 60 birds of 18 species, and 39 reptiles of 9 species.[34]

10.4 African Region

10.4.1 Malagasy Republic (Madagascar)

Tsimbazaza Zoo (Parc Botanique et Zoologique de Tsimbazaza), founded in 1925, is now a botanical and zoological garden with a center for captive breeding of rare and endangered Malagasy fauna. The zoo's goal is the conservation of these species and the education of the public about protecting wildlife and natural habitat. An education program exists for school children and a zoomobile reaches rural people who do not get a chance to come into town (Antananarivo) to visit the zoo.[35–38]

10.4.2 Mauritius

Mauritius has lost much of its forests, which at one time covered 90% of the island. Only small areas of indigenous forest remain in the inaccessible mountain regions. In cooperation with the government of Mauritius, the Mauritian Wildlife Foundation was established to conserve and manage the indigenous fauna and flora of Mauritius; to preserve genetic, species, and ecosystem biodiversity; to coordinate and administer conservation projects; and to educate and involve Mauritian people in these conservation efforts.

To carry out these conservation projects, it was necessary to manage captive breeding programs. Thus, the Gerald Durrell Endemic Wildlife Sanctuary was established in 1984 on the western coast of the island to undertake these captive breeding programs. The sanctuary is not open to the public, but is the facility most closely resembling a zoo on the island. Current breeding programs involve the Mauritius kestrel, pink pigeons, and echo parakeets, among others. The Mauritius kestrel is the only endemic bird of prey on the island. Due to habitat loss and pesticide contamination, there were only four known individuals in 1974, but by 1994 there were 500 birds. The pink pigeon captive breeding and release program has raised population numbers from about 20 in 1985 to around 300 in 1998. The echo parakeet is the only endemic parakeet on the island and its total population is estimated at about 100 individuals.

References

1. Walker, Sally, personal correspondence file, 1999.
2. Flower, S. S., *A List of Zoological Gardens of the World*, reprinted in *The Zoologist*, May, 1909.
3. Hadden, Nellie, "The Giza Zoological Gardens, Cairo," *Windsor Magazine*, 29, 630, 1909.
4. Anonymous, *1 March 1991 100th Anniversary, Giza Zoo Centenary*, Giza Zoo, Giza, 1991.
5. Loisel, Gustave, *Histoire des menageries de l'antiquite a nos jours*, Octave Doin et Fils and Henri Laurens, Paris, 1912.
6. Anonymous, *Zoological Gardens, Giza near Cairo, Report for the Year 1908: Tenth Annual Report*, Government of Egypt Public Works Department, Cairo, 1909.
7. Flower, S. S., *Ghizeh Zoological Gardens, Report for the Year 1899*, Ministry of Finance, Cairo, 1900.
8. Hadden, Nellie, "The Sudan as a collecting ground: stocking the Giza Zoological Gardens," *Windsor Magazine*, 33, 307, 1911.
9. Flower, S. S., *Zoological Gardens Ghizeh, near Cairo, Report for the Year 1901: Third Annual Report*, Government of Egypt Public Works Department, Cairo, 1902.
10. Flower, S. S., *Plan and Guide in English, French and German. Zoological Gardens Giza, near Cairo, Egypt*, Al-Mokattam Printing Office, Cairo, 1902.
11. Flower, S. S., *Zoological Gardens, Giza near Cairo, Report for the Year 1902: Fourth Annual Report*, Government of Egypt Public Works Department, Cairo, 1903.
12. Flower, S. S., *Zoological Gardens, Giza near Cairo, Report for the Year 1903: Fifth Annual Report*, Government of Egypt Public Works Department, Cairo, 1904.
13. Flower, S. S., *Zoological Gardens, Giza near Cairo, Report for the Year 1905: Seventh Annual Report*, Government of Egypt Public Works Department, Cairo, 1906.
14. Flower, S. S., *Zoological Gardens, Giza near Cairo, Report for the Year 1906: Eighth Annual Report*, Government of Egypt Public Works Department, Cairo, 1907.
15. Anonymous, *Zoological Gardens, Giza near Cairo, Report for the Year 1910: Twelfth Annual Report*, Government of Egypt Public Works Department, Cairo, 1911.
16. Beebe, William, "The Ghizeh Zoological Gardens," *Zoological Society Bulletin*, 39, 660, 1910.
17. Flower, S. S., *Report on the Zoological Service for the Year 1913 in which is included the 15th Annual Report on the Giza Zoological Gardens*, Pub. No. 27, Ministry of Public Works/Government Press, Cairo, 1914.
18. Anonymous, *Zoological Gardens, Giza near Cairo, Report for the Year 1911: Thirteenth Annual Report*, Government of Egypt Public Works Department, Cairo, 1912.
19. Flower, S. S., *Report on the Zoological Service for the Year 1912 in which is included the 14th Annual Report on the Giza Zoological Gardens*, Pub. No. 25, Ministry of Public Works/Government Press, Cairo, 1913.
20. Flower, S. S., *Report on the Zoological Service for the year 1920 in which is included the 22nd Annual Report of the Giza Zoological Gardens*, Pub. No. 34, Zoological Service and Ministry of Public Works/Government Press, Cairo, 1921.
21. Anonymous, *Giza Zoological Gardens Guide Plan*, Survey of Egypt, Giza, 1949.
22. Schomberg, G., "A Visit to the Gezira Zoo," *International Zoo News*, 8, 36, 1961.
23. Suares, J. C., "Supply-side Economics goes to the Zoo," *Harper's Magazine*, September, 48, 1982.
24. Olney, P. J. S., Ed., "Zoos and aquariums of the world," *International Zoo Yearbook*, 24/25, 651, 1986.
25. Holt, Peter, personal communication, 1990.

26. el Din, Mindy Baha, personal communication, 1999.
27. Schmidt, William C., "The trouble says a zoo is a herd of humans," *New York Times*, March 18, 1991, Section B 1, 5.
28. Klaver, P., "Giza Zoo, Cairo, Egypt," *International Zoo News*, 42, 449, 1995.
29. Flower, S. S., *Report on the Zoological Service for the Year 1921 in Which is Included the 23rd Annual Report of the Giza Zoological Gardens*, Pub. No. 35, Zoological Service and Ministry of Public Works/Government Press, Cairo, 1922.
30. Olney, R. J. S. and Ellis, P., Eds., "Zoos and aquaria of the world," *International Zoo Yearbook*, 28, 557, 1989.
31. Flower, S. S., *Report on the Zoological Service for the Year 1922 in which is included the 24th Annual Report of the Giza Zoological Gardens*, Pub. No. 36, Zoological Service and Ministry of Public Works/Government Press, Cairo, 1923.
32. Oberholster, J. M. M., *Die Geskiedenis van die Nasionale Dieretuin van Suid-Afrika, Pretoria, 1899–1984*, Universiteit van Pretoria, Pretoria, 1992.
33. Garland, V., *Blantyre and the Southern Region of Malawi*, Central Africana Ltd., Blantyre, 1991.
34. Nason, I., *Enjoy Nigeria*, Spectrum Books, Ibadan, 1991.
35. Jean, *Problèmes et perspectives du Parc Zoologique de Tsimbazaza de 1980–l'an 2000*, IMATEP, Antananarivo, 1993.
36. Millot, J. and Paulian, R., *Le Parc Botanique et Zoologique de Tananarive*, Société des Amis du Parc Botanique et Zoologique, Tananarive, 1949.
37. ORSTOM, *Parc ORSTOM de Tsimbazaza: Guide de 1970*, ORSTOM, Tananarive, 1970.
38. Rakotondraparany, Félix, *Recherche au sein du Département Faune*, P.B.Z.T., Tsimbazaza, 1997.

Additional Sources

1. Flower, S. S., "Reptiles of the Malay Peninsula and Siam," *Proceedings of the Zoological Society of London*, 609, 1900.
2. Flower, S. S., "Mammals of Siam and the Malay Peninsula," *Proceedings of the Zoological Society of London*, 306–309, 332–333, 368–369, 378–379, 1900.
3. Flower, S. S., *Report on a Zoological Mission to India in 1913*, Pub. No. 26, Zoological Service and Ministry of Public Works/Government Press, Cairo, 1914.
4. Hindle, E., "Major S. S. Flower, O.B.E. Obituary," *Nature*, 157, 327, 1946.

Tiger painted from a specimen in the Raja Serfagee menagerie (early to mid-1880s), the precursor to the Shivaganga Gardens Zoo, Tamil Nadu, India. From the Raja Serfagee Collection. © The British Library.

11 Zoological Gardens of South America

James F. Ellis, Jr. and Georgeann A. Ellis

11.1 Introduction

Destruction of the ancient Aztec and Inca animal collections in the early 1500s (see Chapter 1 by Kisling on Ancient Collections) left a 350-year vacuum in Central and South America. Nor until 1888 was the Buenos Aires Zoo (Jardin Zoológico Municipal de Buenos Aires), according to recent censuses the oldest zoological garden in the region, founded, although its roots can be traced to 1874. Since 1888 only a few zoos have been listed in published censuses. A review of the literature on the history of zoos gives one the mistaken impression that zoological gardens are nonexistent, or at least very few in number, in South America. Outside of South America, published information on these facilities is not readily available; however, the *International Zoo Yearbook* census of zoos of the world lists more than 30 facilities in South American countries between the 1970s and 1990s. These facilities vary greatly in size, total attendance, and total specimens held.[1] But even this list of zoos does not reveal the full number of zoos that exist in these countries: the Sociedade de Zoológicos do Brasil lists 121 zoos in Brazil alone in 1998.[2]

Economic differences among the nations of the region, together with differences in urban population concentrations, has undoubtedly affected zoo development in South America. When comparing the number of zoos across the South American countries, Brazil again ranks first, based on 1980s and 1990s data, which gives credence to the significant correlation among economic conditions, population distribution, and the number of zoological gardens. Also, Brazilian curators and biologists are foremost among their South American colleagues in the number of published articles in the *International Zoo Yearbook*.[3] Because of Brazil's leadership position, its zoos are treated here in some detail.

11.2 Brazilian Zoos

Brazil is divided into "macro regions."[4-7] The North (Norte) incorporates the Amazon River Basin, the Northeast (Nordeste) is characterized primarily by its arid *sertao* region, the Southeast (Sudeste) is a major agricultural region, the Center-West (Centro-Oeste) is dominated by high plains, and the South (Sul) is a temperate region of high plateaus. These macro regions are reviewed to examine the external forces that have influenced the historical development and distribution of zoological gardens within Brazil.

Number of South American Zoos by Country

Country	1959	1962	1968	1972	1974	1978	1980	1993–95
Argentina	3	3	7	8	7	7	7	3
Bolivia	2	2	2	2	2	1	4	1
Brazil	2	3	5	3	6	7	25	10
Chile	3	5	4	4	4	2	2	1
Colombia	1	NA	2	3	3	4	4	5
Ecuador	NA	NA	NA	1	1	NA	1	NA
French Guiana	NA	NA	NA	NA	NA	NA	NA	NA
Guyana	1	1	1	1	1	1	1	1
Paraguay	NA	1	1	1	1	1	NA	NA
Peru	1	1	1	1	1	1	2	1
Suriname	NA	NA	NA	NA	NA	NA	NA	NA
Uruguay	1	1	1	1	2	2	3	3
Venezuela	1	4	5	6	6	0	6	6
South America	15	21	29	31	34	34	55	31
United States	93	104	245	258	230	199	176	165

Note: Number of zoos listed in the *International Zoo Yearbook* for South America, by country since 1959. Also included for comparison purposes are *International Zoo Yearbook* figures for U.S. zoos. NA = not available.
Source: *International Zoo Yearbook*, Zoological Society of London, London, 1960+.

Historically, the economy of Brazil has exhibited product cycles, dependent on world markets and generally based on dominant individual products: brasilwood (1500–1550), sugar cane (1550–1700), gold (1700–1775), and then eventually rubber (1850–1930). The era of industrialization began in the late 1800s, and by 1940 three quarters of the country's employment and industrial output was concentrated in the states of São Paulo, Rio Grande do Sul, Minas Gerais, Rio de Janeiro, and Pernambuco. This concentration continues to the present with these southern states serving as agricultural, industrial, and financial centers for the country.[8–10]

The distribution of zoo membership and the number of zoos reporting animal census data vary greatly among Brazil's macro regions.[11] Much of this data is reported through the Sociedade de Zoológicos do Brasil. At the time the Sociedade formed in 1979, it was the only South American professional zoological garden organization.

Unfortunately, historical records for individual Brazilian zoos are generally unwritten, unpublished, or difficult to find and examine. These histories of the facilities would be the best vehicles for understanding the problems faced by Brazilian and other South American zoos,

Regional Distribution of Brazilian Zoos, Museums, and Parks

Region	Museums	State Parks	National Parks	Zoos (1980)	Zoos (1998)
Northern	10	0	10	6	6
Northeast	88	0	7	3	5
Central-West	19	1	4	4	6
Southeast	317	27	11	34	85
Southern	108	21	4	5	16
Totals	542	49	36	52	118

Note: Number of museums and parks in Brazil based on 1980–1987 data. Data on the zoos are for 1980 and 1998.
Source: Instituto Brasileiro de Geografia e Estatistica,[1,64] Puglia,[11] and Giacomini.[39]

problems generally related to the geographic and socioeconomic conditions that have set the stage for the historical development at the zoological gardens.

11.2.1 The North (Norte)

A number of zoological gardens have existed in the north,[12] including several in Manaus. However, it is fitting to start in the State of Para, whose zoo opened in 1986 as the most modern facility in the country. This was the Parque Natural de Carajas, which the Companhia Vale do Rio Doce (CVRD) owned and operated. Its main goal has been to exhibit the native flora and fauna of the Amazonian region. Carajas is a CVRD-owned and -managed town built to support the local mining district, located 380 miles southwest of Belem.

At the time of its opening, this zoo showed potential for excellence unparalleled in the Brazilian zoo profession. This potential was due to its location in an extensive natural area and due to its association with one of the largest mining operations in the country. In the short period of seven months the construction of the zoo cost U.S.$1,000,000,[13] and its alliance with the CVRD mining project had the benefit of access to unlimited construction, engineering, and technological expertise. Unfortunately, its remote location, favorable for the goal of working with native animals, is removed from the mainstream of the region's population, which is located 84 miles away at Maraba, in the center of the Serra Norte mining district.

One of Brazil's older zoos, now closed, was Manaus, the capital city of the State of Amazonas. The municipal zoo, Aviaquario Municipal, was built in 1938 at Plaza Osvaldo Cruz in front of the city's cathedral. This facility was easily accessible on foot and by boat on the river, as the park is but a minute's walk up from the docks. The major benefit of the zoo was its accessibility to school groups. The site was approximately 2.5 acres, surrounded with a protective wall that had large grated openings allowing people to view the animals from outside the wall. Following its opening, the zoo, including the trees within the park, was covered entirely with a chicken wire mesh in an attempt to create a free-flight area. The zoo housed only native fauna such as manatees, giant otter, Amazon river turtles, caiman, numerous native fish including arapaima, and many avian species. Breeding was reported for the giant amazon turtle and a number of aquatic avian species. The zoo sold and exported many of its species, especially fish, and these sales were the only funds available to the zoo to increase its cages and exhibits, and to provide for its maintenance. Plans for a larger park were considered early on; in his annual report of 1940, Antonio Maia suggests that the municipal zoo could move to a park then known as "10 of November" where a "proper" Amazonian zoo could be built.[14]

Silvio Barros and Robin Best report that the Manaus zoo, from its inception, never received proper attention and reached its nadir in 1977.[15,16] According to the March 30, 1981 issue of the Manaus newspaper, *A Critica*, the zoo was converted between 1977 and 1979 into a children's zoo, but then returned to its prior status in 1979 under pressure from the public.[17] The zoo also considered increasing the number of animals exhibited even though it was having financial difficulties in maintaining those it had. Education was a major objective and there was a classroom that the municipal secretary of education operated. On March 17, 1984 the newspaper, *A Noticia*, reported the near total disappearance of the animals, the closure of the administrative offices, and the presence of trash in almost all the still-existing animal enclosures. *A Noticia* further mentions that the highly active classroom and educational facilities were transferred to another locale because of a lack of security in the park.[18] In a follow-up article on March 21, the newspaper reported the closure of the facility, which had only three birds remaining in the collection.[19] The *Jornal de Commercio* on the same date mentioned the rapid growth of prostitution associated with the park's restaurant as the main reason for the closure.[20]

During the 1980s there were also indications that a major zoo would be built for Manaus that would contain rare regional wildlife, as well as African species. This concept had been in the minds of municipal leaders and was probably based on several reports, such as the report Wallauer, Bruck de Andrade, and Grieger presented in 1980. This report studied the possibility of such a zoological garden and suggested a park of regional animals exhibited in large natural settings. To avoid financial difficulties, they suggested the park have special admission fees to the costly exhibits, that the park sell plants and captive-born animals, and that the park have souvenir sales, restaurants, and other sources of income.[19] Silvio Barros of the State Tourism Board followed this with his report in 1981. Barros had traveled extensively in the United States visiting a number of facilities, many of which were tourism oriented. These facilities included San Diego Wild Animal Park, Marine World Africa U.S.A., Sea World, Lion Country Safari, Monkey Jungle, Metro Toronto Zoo, and Vancouver Aquarium. His study concluded that the major stumbling block for development of a "proper" major zoo was the cost of construction and maintenance. Again, suggestions were made that involved tourism-oriented sales, rental of animals for filming commercials, and the creation of a support society, among many others.[21] In the end, however, the development of this new zoo has still not progressed beyond the talking and idea stage.[15]

Another zoo in the city, and the one to which animals from the original Manaus Zoo were transferred, is the Center for Instruction in Jungle Warfare (Centro de Instrucao de Guerra na Selva). The center officially opened in 1964 for the purpose of training military forces in jungle warfare.[22] In 1966, this zoo began to house animals that the military encountered in their operations and used for training staff at the center.[23] The zoo, which has a military veterinarian as an administrator and soldiers in training as staff members, has always been open to the public.[17]

In 1976, as part of the ever-growing tourism activities in the city, the Tropical Hotel (a subsidiary of Brazil's national airlines, Varig) built a small zoo on the outskirts of town for its patrons' use.[24] The zoo, whose goal at the time was to exhibit native fauna, had grown to the point where, in 1985, funding for a newer facility adjacent to the hotel was approved. Access to the hotel's kitchen and financial resources has made this facility one of the best kept and maintained in the region.

What is perhaps the westernmost zoo in Brazil is located in the city of Rio Branco, the capital city of the State of Acre. Rio Branco is located approximately 860 miles southwest of Manaus. Historical information about the Parque Zoobotânico of the University of Acre is based on accounts described in its 1981 and 1982 annual reports.[25,26] In 1979, the university's Department of Natural Sciences took over the supervision of a biological reserve on campus and began the construction of a zoological garden the following year. The goals of this facility were the study, management, preservation, and exhibition of regional flora and fauna. The reserve and zoo served the university by providing a practical classroom for its courses. Additional goals included local education, tourism, and development of nurseries for regional flora, among others. In 1981, a division of animal care was created to develop and improve the small zoo. The older cages lacked keeper safety areas and were built with inadequate materials (allowing animals to escape); exhibits did not have shelters for the animals, and there was inadequate security for the public. Another problem was that so many animals were acquired through donations from the Instituto Brasileiro de Desenvolvimento Florestal and from the local residents that the accumulation of these animals led to improper housing conditions.[26] Diet was also a problem, for the animals depended on local daily donations of food items, with variations in type and quantity preventing development of stable and well-defined diets for the animals in the collection.

In 1982, the zoo reviewed its conditions with the assistance of Ladislau Deutsch, then curator of mammals at the São Paulo Zoo. The review reemphasized the inadequacies of the facility that were already known and proposed that an area for the quarantine of animals be established, as well as the implementation of diet controls, veterinary and animal records, and other management controls. Although the zoo is still in operation, it is likely that technical or financial support for its improvement is still unavailable within the region.[27]

In addition to the major facilities of Amazonia discussed above, there are also other zoos that exist, or have existed, in the region; however, as is the nature of underdeveloped frontier regions, these smaller zoos are ephemeral, do not cooperate with the formal zoo networks, and information on them is virtually nonexistent outside of each zoo's immediate area. Development projects, such as dams, often build zoos to ameliorate their intensive impacts on the landscape and wildlife. However, the long-term existence of these zoos is questionable; their initial goals are usually to hold animals only long enough for translocation to other habitats or shipment to other zoos. As the region continues to develop and cities grow, these zoos could become permanent or new zoos could replace them.

11.2.2 The Northeast (Nordeste)

There were only three known zoological gardens in the Northeast in 1985: Horto Zoobotânico de Dois Irmaos (Recife), Parque Zoobotânico Getulio Vargas (Salvador), and Fundaqão Zoobotânico de Piaui (Teresina). The zoo at Recife, which began when its botanical garden (begun in 1916) was converted to a combined zoo and garden in 1938,[28,29] is the focus here. This zoo was the first state effort to protect the fauna of the Northeast region. For more than 10 years, efforts were made to harmonize the flora and fauna with the exhibits in its 250 acres of secondary-growth forest. The botanical garden was originally operated under the administration of the municipality of Recife, passing in 1935 to the state foundation known as the Agronomy Research Institute (Fundaqão Instituto de Pesquisas Agronomicas). The Jardim Zoobotânico de Dois Irmaos opened in 1939 under the joint administration of the city of Recife and the Agronomy Research Institute with both agencies ceding the facility over to the State of Pernambuco's Department of Tourism (Empresa Pernambucana de Turismo) in 1969. It was at this point that the zoo no longer functioned as it was originally planned, losing its native and regional fauna character. This last move in administration was an effort to revive the zoo's declining facilities, quality, and attendance because tourism was thought to be the only possible source of revenue in this already economically depressed region. As of the mid-1980s, the infusion of tourist revenue was insufficient and the zoo still lacked both direction and resources.

11.2.3 The Center-West (Centro-Oeste)

Zoological gardens in this region include Parque Zoológico de Goiania (Goiania), Jardim Zoológico do Distrito Federal (Brasilia), Zoológico da Universidade Federal de Mato Grosso (Cuiaba), and Jardim Zoológico Municipal (Alta Floresta). As is the case with other zoos in Brazil, relatively little information is available on these facilities. The least amount of information is available for the Parque Zoológico de Goiania, built in 1935. Goiania was a planned city built to be the Goias state capital, whose more cosmopolitan population perhaps explains the zoo's more exotic collection, composed of roughly 50% non-Brazilian species.

Jardim Zoológico do Distrito Federal in Brasilia, the nation's capital, was founded at the same time as the city in 1960.[30] Novacap, the agency in charge of building Brasilia, originally operated the zoo. The roughly 200-acre zoo was planned to be on the outskirts of the new town, but found itself located next to a *favela* (shantytown) that grew up soon after the start of construction of the city, a situation that has plagued the facility.

Zoológico da Universidade Federal de Mato Grosso was founded in 1977 by the staff of the university's biology department. Raul Neto, zoo director in the mid-1980s, furthered the development of the zoo through his many contacts and his many visits to the zoo at the Museu Goeldi in Belem, as well as other facilities in the country.[31] His vision for the zoo was to build a facility with open enclosures exhibiting regional fauna. Financially, the zoo was more stable than most other zoos, as it was linked with a federal university having a relatively stable budget structure.

Jardim Zoológico Municipal in Alta Floresta was founded in 1978 when a private individual relocated from the city of São Paulo as a participant in a colonization program. The zoo began as a private facility, but shortly after its opening it was turned over to the city and its director became a city employee in charge of the parks and streets. The animals in its collection are specimens received from the public living in the region.

11.2.4 The Southeast (Sudeste)

By the 1980s there were 34 zoos located in the Southeast with 28 of these in the State of São Paulo alone. These numbers have increased dramatically since then, to over 85 in the region and over 50 in the State of São Paulo, by the mid-1990s. A great deal more information exists on the major city facilities of Rio de Janeiro, São Paulo and Sorocaba, with little information on the rest of the zoos of the region. One of the youngest zoos in Brazil is located in the city of Americana, São Paulo, which opened in October 1984.[32] The zoo at Ilha Solteira has perhaps the most indigenous emphasis in its collection and goals. Founded in 1979 in association with a nearby hydroelectric dam, the zoo is operated by the Electrical Company of São Paulo (Companhia Electrica de São Paulo), which is completely state owned. The zoo's association with the hydroelectric project was probably related to the displacement and relocation of animals generally associated with these types of projects.

The Jardim Zoológico da Cidade de Rio de Janeiro opened in 1945 and is located adjacent to the historic Imperial Palace, which houses the Museu Nacional. This zoo had its origins in another zoo that Barao de Drummond at Vila Izabel established in 1888. The new zoo has always had financial and attendance difficulties, and gambling has become a major attraction drawing visitors. From the beginning, the zoo was conceived piecemeal without a master plan and it continues to be troubled by this lack of organization. Plans were made to move the zoo to a better location where an "ultra-modern" zoo would be built. As of the 1980s, however, the zoo remained at its original location and continued to suffer from a lack of funding and attention.[30,33,34]

The first of Sorocaba's zoos was a collection maintained in 1916 in Praca Frei Barauna, one of the city's extensive parks. This facility displayed only native species, many of which local residents brought to the zoo. During the 1920s, and until its disappearance in 1930, the zoo was subject to political harassment and neglect, which eventually led to its demise. It was not until 1965 that a resurgence of the city's parks and their gardens, along with an increased donation of small birds, capuchin monkeys, and other animals by residents, created a need for a new city zoo. This second zoo, opened with much celebration in 1966, became known as the Jardim da Margem. This zoo grew rapidly with the addition of many larger, native animals and continued to exist until 1971 when, due to a community improvement project, its site was paved over as part the construction of Avenida Marginal, a new avenue serving the city. Most of the animals were moved to the third and current zoo.

In 1968, with the acquisition of the Parque Municipal Quinzinho de Barros, the third and current zoo at Sorocaba came into existence. The Prestes de Barros family had formerly owned the newest zoo site, which had previously served the community as a general meeting place, sports center, and city park for many years before to the development of the zoo. Planning for the zoo led to discussions with the zoo in São Paulo about loans of native animals for exhibition in the new facility. By 1968, the new zoo was ready for inauguration and officially opened to the public as the Parque Zoológico Municipal "Quinzinho de Barros." Close collaboration continued with the director of the São Paulo Zoo, Mario Autuori, who made it possible for the park to receive its first large animal, a Kodiak bear, in 1969. From all indications, this marked the beginning of exotic animal exhibition at the new zoo. In 1979, Lazaro R. R. Puglia took over as director of the zoo and continued to build on the cultural and educational aspects of the zoo. Puglia, along with the supportive mayor and politicians of the city, has fostered an excellent reputation for the zoo, both regionally and nationally. The zoo has even served as headquarters for the Sociedade de Zoológicos do Brasil. This newest Sorocaba Zoo has very high standards of management and has been in the forefront of Brazilian zoo educational programs.[35]

The zoo in the city of São Paulo, the Fundaqão Parque Zoológico de São Paulo, was the idea of the governor of the State of São Paulo, Janio da Silva Quadros. His desire was to have a center for the study of native and foreign fauna. His first act as governor was to order a study commission composed of leading consultants and scientists from the region. By March 1958, the new zoo was officially inaugurated with Mario Paulo Autuori as its first director. He served in this position until his death in 1982.[36,37] In December 1958, The Fundaqão Parque Zoológico de São Paulo was officially created under state law, with its goals as follows: to maintain live animals of all species for education, recreation, and research; to establish a biological station for research on regional fauna; and to furnish facilities for national and international researchers in zoology through the use of agreements, contracts, or grants.

The foundation and its goals were overseen by a high council (Conselho Superior), an advisory council (Conselho Orientador), a fiscal council, and a directorship (Diretoria) for the zoo. All council members are specialists in zoology. The high council is composed of representatives nominated by the following institutions: Instituto Biologico, Instituto Butantan, Faculdade de Medicina Veterinaria e Zootecnia da Universidade de São Paulo, Instituto de Pesca, Museu de Zoologia da USP, and Instituto de Biosciencias da Universidade de São Paulo. The advisory council is composed of three individuals skilled in the science of general biology who meet monthly with the director. The fiscal council is composed of three board members from the high council. In 1959, this foundation was linked to the office of the secretary of state for São Paulo, but later in that same year was transferred to the state secretary of agriculture, and then in 1975 to the state secretary of sports and tourism. The zoo operates with financing from the state government, and is supervised by the above councils, which are generally considered independent of political interference. In 1964, the zoo began a major expansion and renovation program. In 1979, the zoo received a 1,400-acre farm on the outskirts of town from the state, which is being used for the production of produce, meats, and other items for use in the zoo. Attendance at the zoo reached over two million in 1995.

The State of São Paulo was the birthplace of Brazil's professional organization of zoos, known as the Sociedade de Zoológicos do Brasil. A group of zoo professionals from the region started the organization in 1979, and it has been active over the years in providing technical assistance both locally and to the federal government's forestry and wildlife agency, the Instituto Brasileiro de Desenvolvimento Florestal. Annual and regional conferences are held and annual censuses of zoos and their collections are provided.[12] São Paulo is also the home of the only private zoo consulting firm in Brazil, known as Implantacao de Parques de Preservacao, which was created by Ladislau Deutsch, formerly mammal curator of the São Paulo Zoo, in collaboration with several other local professionals. This firm provides architectural planning, husbandry programs, and supervisory services for the construction and long-term administration of zoos and parks.

11.2.5 The South (Sul)

Sixteen zoos now appear in the inventory of South Brazilian zoos according to the Sociedade de Zoológicos do Brasil, as of 1998. The largest facility in this region in 1985 was the Parque Zoológico da Fundaqão Zoobotânico do Rio Grande do Sul in Porto Alegre. This zoo, actually located in the suburb of Sapucaia do Sul approximately 14 miles from Porto Alegre, was built in 1962.[38] Since its inception it has suffered seriously from a lack of financial support, which led to the creation of a foundation in 1972.[38,39] In December 1972, the State of Rio Grande do Sul government authorized under state law the establishment of the Fundaqão Zoobotânico do Rio Grande do Sul. This foundation was established as a private entity under the supervision of the secretary of the interior, regional development, and public works. The foundation had administrative and financial autonomy and its objectives were to administer and maintain areas designated for the preservation and conservation of natural resources of the State of Rio Grande do Sul, as well as to contribute technical support to political efforts to preserve the region's environment.[40]

A master plan for development of the zoo was prepared in 1976 based on the infrastructure and support facilities identified as lacking or in need of rebuilding.[41] In 1979, the foundation was transferred to the supervision of the office of the state secretary of culture, sports, and tourism.[40] The foundation has also been responsible for the supervision and management of the botanical garden and the Natural Science Museum, founded in 1955. As of 1985 the zoo was the most functional of the three foundation-supervised facilities, and the museum was closed for repairs. According to Claudio Giacomini, then director of the zoo and president of the Sociedade de Zoológicos do Brasil, the region was experiencing major economic problems which severely impaired the operation of the foundation. The foundation even discontinued publication of the outstanding educational and informational magazine *Natureza em Revista* in 1980. It had been hoped that establishing the foundation, much the same as the one in São Paulo, would financially improve the zoo operation. The foundation was able to stabilize the zoo's situation, but linking the foundation with the other facilities in the region has caused a detrimental division of already meager resources.

11.2.6 Summary

Although direct links between the economic situation in each region and zoo development may not be immediately obvious, the inference can be made based on zoo distribution alone. The State of São Paulo, at the core of the Brazilian economy, is also at the center of the zoo profession. A number of larger zoos in the South depend on the operation of large farms for the production of staples required for feeding their animals. Others, especially smaller facilities, rely on the donation of food materials from local farmers' markets. Some facilities are able to acquire animals via outright purchase, but many depend exclusively on the donation of regional fauna from the general public and the hydroelectric dam construction projects.

Budget information available on Brazilian zoos is scant and confusing. These figures, taken in the context of the time and in comparison with U.S. currency of the period, clearly raise the question of how these facilities survived on such meager funding resources. Given this meager funding, it is impressive that the Brazilian zoo profession has been able to take root and to grow. Some zoos were operating on funds provided without an actual budget, as was the case with the zoo in Belem up until the early 1980s. Other factors that must be taken into consideration are the variations in cost of living from region to region, which can drastically hamper operations. The contrast between funding for U.S. and European zoos and funding for Brazilian zoos is significant, and should provide a proper perspective when analyzing or criticizing situations in developing countries such as Brazil.[42,43]

11.3 Parque Zoobotânico Museu Paraense Emilio Goeldi

A detailed description of the Parque Zoobotânico Museu Paraense Emilio Goeldi in Belem is available only because of the extensive records the authors located in nooks and crannies of the museum library, curator's bookcases, under the kitchen sink, and in long-forgotten closets during a stay in Brazil between 1985 and 1989. These details not only provide a rich understanding of the Museu Goeldi zoo in particular, but conversations and visits with many other colleagues in Brazil and reviews of information on other South American zoos indicated that other zoos in the Amazonian region, as well as those throughout South America, have experienced a similar history.

In 1894, the governor of the State of Para, Laudro Sodre, appointed Emilio Augusto Goeldi to reorganize and direct the Museu Paraense.[44] Goeldi, in his first annual report (1894) to the governor, noted the need for appropriate facilities to house the live animal collection at the museum. It is in this report that he first acknowledged a zoo facility when he incorporated it as part of the museum's administrative regulations. This was also the beginning of the process whereby the zoo and the botanical garden became more influential in the future of the museum.

Goeldi, in his report on the activities of 1894, recognized that the acquisition of a new building on the outskirts of Belem (which today is in the heart of the city) would allow for the inclusion of the zoo and botanical garden, as well as for their future growth. Goeldi wanted to create an attractive educational facility where the public could learn about the natural wonders of Amazonia. Goeldi believed that it should not be the intent of the institution to imitate the great zoos and gardens overseas that were collecting notable specimens of the animal and plant kingdom from all over the world; rather, his wish was to exhibit that which was Amazonian.

Based on the style of the zoo's enclosures and layout, however, it is obvious that European gardens had their influence on Goeldi, who was born in 1859 in the city of Ennetbòhl, Switzerland.[45] Educated at the Universities of Jena and Leipzig, Germany, he graduated with a doctorate degree in 1883 and was exposed to the numerous European parks and zoos. Goeldi's references in his 1894 annual report to iron-barred enclosures, wire cages, and concrete tanks for aquatic animals, as well as his comments on the maintenance and feeding of the collection, all indicate this early exposure to the styles and methods used at the time in European zoological gardens. Goeldi even mentions in his annual report for 1895 that the wire for the large flight aviary built in that year was fabricated in Paris and was similar to that used at Parc St. Germain. Then, in his 1898 annual report, Goeldi further refers to his European connections with the acquisition of signs fabricated in Germany.

Management difficulties at the zoological garden appeared early. In 1895, Goeldi notes that the construction costs in the region are extremely expensive — "E incontestavelmente carissima a mao de obra aqui na Amazonia" ["Labor is incontestably expensive here in Amazonia"] — which to this day continues to present problems.[46] In 1895, Goeldi continued to relate the frustration of constructing two lakes and large feline cages with co-workers who had no understanding of the work and who had never seen anything like the enclosures being built.

In 1897, Goeldi reports a worsening financial crisis as the budgets authorized became insufficient for the operation of the museum and the depreciation of the currency required the budget to be augmented by one third over the previous year. Goeldi attempted to maintain the status quo in the zoo and wrote that it is repulsive to even think of diminishing the feed rations of the animals. It is also during this period that he offers his strongest defense for the existence of the zoo and botanical garden. In this defense of the park he suggests it would be poor practice, indeed, to offer the visitor a postage-stamp collection of minuscule exhibits. Apparently this defense was offered at a time when there was a growing fear by some that the zoo was following a concept of unlimited growth, in no certain direction. In Goeldi's words, those so thinking were "in error" and the work of the park culminates in two words, "fauna amazonica." Beginning in 1899, one can see a decline in the philosophical discussion of the problems of the zoo and botanical garden, with the annual reports only highlighting events rather than their political and conceptual concerns. It is possible that by this time Goeldi's views had been accepted and no longer needed to be noted in his reports.

The zoo received negative publicity in a 1907 newspaper article, which a disgruntled employee apparently instigated, suggesting that the animals were starving to death. Other husbandry problems included the death of several jaguars and the sudden blindness of a four-month-old jaguar kitten purchased on the island of Marajo. Goeldi resigned as director of the museum due to poor health and returned to Europe on March 21, 1907, becoming a professor of zoogeography and biology at the university in Bern, Switzerland.[45–49] Jacques Huber, whom Goeldi had brought in to head the botany department in 1895, took over as director.[45]

Huber noted in 1908 that the number of aquatic birds was so high in the aviary they were inhibiting other species, and he ordered the sacrifice of a number of night heron to remedy the situation. Huber, in his 1909 annual report, notes that a chimpanzee house had been built and an aquarium was under construction. Additionally, it is at this time that one notes an interest in higher numbers of species and individuals in the collection. The inventory at that time was 183 species, with about 600 to 700 specimens. The aquarium was completed in 1910.

Between 1921 and 1930 the museum was in state of decline, with its zoological garden the only active division during that period. This deterioration coincided with the period when the State of Para was also experiencing a declining economy as a result of the collapse of the rubber industry upon which it so heavily relied.[50,51] The year 1930, however, brought a radical change in the zoological garden, and in fact it was between 1930 and 1945 that the zoo was considered at its zenith, becoming known as the best zoo in Brazil.[50] The facility grew rapidly and reached a peak of 2,000 animals in the collection.[51] It was during this period that the zoo's efforts were directed toward the breeding of *Podocnemys* and many other fish species, including pirarucu and tucunare.[51,52]

Unfortunately, the years 1946 through 1955 brought another decline, and by 1955 the museum was essentially abandoned. Then, in 1955 the Conselho Nacional de Pesquisas (National Council for Research), which had been founded in 1951, took over the operations of the museum under an agreement with the State of Para in an effort to revitalize scientific research in the region.[44,50,51,53] From this time on, the museum research areas began to grow anew and flourish, but the zoological and botanical gardens continued to be maintained with great sacrifices and in a deficient manner.[50] From 1962 to 1968 the director of the museum, Dalcy Albuquerque, charged with meeting the very real needs of the other museum facilities, tried to downsize and, if possible, to close the zoo and botanical garden. The closures never took place. Three generations of residents of the city had grown up with these annexes to the museum, and there was public pressure to keep them open. In May 1967, the municipal government of Belem backed the continued existence of a zoological garden within the city by authorizing preservation of an area for the zoo's development.

There was a major effort in 1984 to redesign the existing facilities of the zoological and botanical gardens of the museum. With private financial support, the museum received funding to upgrade its park facilities under a project known as "Projeto Consciencia Ecologica da Amazonia." The effort to rebuild began with the development of a master plan for the botanical portion of the park. The project initially allowed time for development of a master plan for the exhibits in the zoo portion of the project as well. However, there were technical problems, and the project developed piecemeal rather than following an overall master plan. Various consultants to the project offered a wide range of solutions, going so far as to suggest that because of its small size (six acres) the zoo facility would be unable to support large breeding groups of animals, and so it should emphasize fewer individuals of as many different species as possible.

The master plan called for the zoological and botanical gardens to continue maintaining an Amazonian focus in their exhibits and goals. Plans to remove the larger animals (especially the jaguars) from the collection due to the lack of space at the site stimulated discussions regarding the future needs for such animal exhibits. The master plan, which was completed in October 1986, brought the facility and exhibits into the modern age of zoological garden management within the Brazilian community. Discussions also led to a cooperative agreement among city, state, and federal agencies for the preservation of an area known as Utinga, where a new, more modern zoological garden could be built specifically for larger animals that were not easily exhibited in the museum facility.[27] This new facility would complement, enhance, and increase existing recreational and educational areas available to the citizens of Belem.

A series of animal inventories extensively document the animal species and numbers of individuals that were in the zoo. These inventories indicate the zoological garden has always been a truly Amazonian facility. The occasional non-native species, such as chimpanzees in the early 1900s, were brought in from overseas, were circus donations, or were brought up from the southern parts of the country. The records for the animal collection exist, for the most part, as handwritten logbooks. These logbooks contain data dating from the 1930s and, when incorporated with the animal collection notes Goeldi left in his annual reports, provide a fairly good overview of the collection's activities.

Inventory totals of the animal collection for the years 1895 through 1909 were published in the annual reports of Goeldi and Huber and demonstrate that the zoo's collection leveled off at close to 1,000 individuals and over 150 species. These numbers have been the norm to the present day. Additionally, these inventories lend credence to the figure mentioned earlier of 2,000 individuals for the 1930s, which includes the numerous fish held in the zoo's ponds. An analysis of the logs shows a sharp rise in the number of individuals brought into the facility during 1967, followed by another rise between 1975 and 1978. The first entry acknowledging the arrival of confiscated wildlife coincides with the sharp rise in 1967.

Data from 1969 through 1974 indicate efforts took place to acquire new stock through purchase; however, this commercialization of animals did not continue. Beginning in 1975 purchases were no longer used as a method for stock acquisition, which holds true to this date. This change in acquisition policy is in accordance with Brazilian federal laws passed in 1967 prohibiting commercial traffic in native species. Of interest as well has been a trend toward larger numbers of animals being donated to the zoo during the early months of the year, which coincides with the period of high rainfall during the South American summer months.[54,55] Many animals are displaced to higher ground and areas of human habitation as the water levels rise.

A number of animal species donated to the museum zoo were impacted by the development activities in the region during the 1960s and 1970s. The peak in donations in the mid-1970s may again be due to the confiscation activities of the Instituto Brasileiro de Desenvolvimento Florestal.[56] Sloths in particular were a problem and most of them were brought in by local residents rather than by law enforcement agents. These animals may be an indicator species for the level of deforestation in the area, as their arboreal adaptation and the preference of some species to a diet of *Cecropia* sp. make it highly unlikely they would be found on the ground. However, most of the people bringing in these animals reported finding them on the ground on the outskirts of Belem where deforestation was the greatest.

The high rate of animal donations continued through the late 1980s and had an impact on the facility. Donations filled every enclosure, causing overcrowding. Disease in the collection was difficult to isolate, identify, and treat, thus leading to high levels of mortality. Expenses for feed continued to escalate as did malnutrition, due in part to a lack of quarantine areas for the adequate separation of these animals, and due in part to poor diet control.

During this same time, exhibits were transformed to improve keeper safety, to provide enclosures with isolation and separation cages for the more dangerous animals, to reduce the problem of animal escapes, to improve veterinary care, and to increase the visual appeal of the exhibits for the public, while at the same time reducing the stress on the animals. Restricting the public through the use of special viewing areas and not allowing 360° access to the exhibits, in combination with improved husbandry standards, gave the visitor a much more positive experience. Animals became more active, behaving and interacting in a more natural manner.

By 1989, approximately one third of the rebuilding had been completed. However, economic difficulties in the country continued to hamper the construction of planned work, surplus native wildlife became an increasing problem throughout the region due to continued development, and the problem of potential disease continued. Although positive changes in training, exhibition, and increased professional management were occurring, access to needed support infrastructure was still unavailable within the region.

A positive step was a renewed emphasis on the education programs of the museum. Major efforts to educate the local public both on site and off site were developed. A program to take the museum to the beach (Museu vai a Praia) during summer holiday months was an acclaimed success. With international funding, the museum developed liaison programs with local public schools and a rather extensive specimen loan collection for school children to use in class projects. Meetings with teachers and local administrators have shown strong interest. The museum and its zoo continue to survive to this date in the heart of a city that has reached over one million inhabitants in a rapidly changing region.

11.4 South American Zoos in Other Countries

11.4.1 Argentina

Historical information on the zoos of Argentina is scant; however, that of the Jardin Zoológico Municipal de Buenos Aires is well documented before 1979. The zoo's 45 acres was situated in the Palermo neighborhood of Buenos Aires near the botanical garden. Its signature landmark was a golden iron fence forged in Europe. The history of this zoo has been closely linked to Europe since 1840 when a small zoo was started in Buenos Aires in association with the residency of the British diplomat Henry Southern, and then later with the mayor of Buenos Aires who further expanded the collection to entertain guests, especially those from abroad. In 1874, a presidential decree created the Parque Tres de Febrero, and within this park a zoological section was created that is considered to be the first record of the Jardin Zoológico Municipal. The park was inaugurated in 1875 with over 30,000 visitors attending (out of a total of approximately 180,000 living in Buenos Aires at the time). The zoo developed in this area through 1888 and consisted principally of native animals, which various government leaders and notable citizens donated. Of interest were meetings in 1883 between Don Carlos Pellegrini and Carl Hagenbeck in Hamburg, Germany. The former was very closely associated with the mayor of Buenos Aires, had traveled throughout Europe, and was familiar with European zoological gardens. He felt the zoo must have enough attractions to induce workers of the city to pay for the cable car ride to the zoo, or to make the walk to the zoo. The latter, Carl Hagenbeck, provided rare and valuable animals to the collection on behalf of the Emperor of Brazil in Rio de Janeiro.

Another connection of this zoo with Europe is the Belgian architect, Julio Dormal, who designed the zoo, as well as other government houses and the well-known Teatro Colon. Furthermore, the zoological section received help from an eminent Russian zoologist and director of the Museo Nacional de Buenos Aires, Carlos Berg. The zoo, after suffering the deaths of many of its animals due to dogs running wild and other problems, was moved in 1888 to a new location in the park and remained there through 1979, at which time it was scheduled to move to a more ample location in Parque Almirante Brown, still within the urban area of Buenos Aires.

From 1888 to 1903, under the direction of Eduardo Holmberg, the zoo began to take on a scientific mission. Holmberg, in 1893, stated that a zoo is not a luxury, nor a reflection of vanity, nor superfluous, but rather it is a reflection of the national laws relative to the instruction of the public. Based on this philosophy, the regulations of the Jardin Zoológico were established. The "new" zoo would not be free to the public, but would cost 10 centavos. On Thursdays it cost 20 centavos for entire school groups with their teachers, and the first Sunday of each month and holidays were free. The zoo provided a clean and hygienic place for visitors to come and learn about nature.

From 1904 to 1924 Clemente Onelli directed the zoo. Onelli was born and educated in Rome and traveled to Buenos Aires to see the world and work at the Museo de La Plata. Onelli continued the philosophy of education and popularizing the study of natural sciences. By the end of 1904 attendance at the zoo had reached 150,000, which was 10 times that of the year before.

The period 1924 through 1944, under the directorship of Adolfo Holmberg, began an era of change. He opened enclosures using moats and other means, and removed or reduced the number of cages and other typical enclosures of the time. The documented history of the zoo then jumps to the early 1970s when, in 1973, the zoo was forced to close temporarily because of a massive die-off of animals (32 in all), which was attributed to food contaminated with phosphorus; however, no definitive proof that this was the cause was ever reported. Attendance had reached over two million and by 1977 the City of Buenos Aires asked for an international study to develop a Parque Zoofitogeographico. This new "biopark" was to occupy 370 acres in the Parque Almirante Brown. However, the zoo has yet to relocate.[57]

11.4.2 Bolivia

Opened in 1979 following four years of construction, the Zoológico de La Ciudad de Santa Cruz, once one of several Bolivian zoos, may now be the country's only zoo. Located in Santa Cruz de la Sierra, the zoo is situated in what was a remote part of the country in the Amazonian rain forests of Bolivia. Noel Kemp Mercado, the zoo director in 1986, was heavily involved in studies of the flora and fauna of the region and a leader in understanding the natural areas of the country. He was active in the elevation of a regional reserve, German Busch, to national park status and compared the development of this national park as a tourism destination to that of the Galapagos Islands. He also felt Bolivian environmental laws were poorly legislated, nonfunctional, protected nonexistent species, and transferred control of nonrenewable natural resources to corporations. This background is important in the discussion of this zoo and of Mercado in particular. Mercado, responsible for the construction of the zoo, was also responsible for the construction and reconstruction of the Jardin Botânico (which a storm destroyed in the early 1980s) and all of the green spaces of the city of Santa Cruz. He predicted at the time of its opening that this new zoo would become one of the major tourist attractions of the country. According to Mercado over 120,000 visitors came to the park in the first 20 days after its opening. The park was built on 16 acres and the animal collection at the zoo was exclusively regional. Mercado was unaware of any other zoos or related institutions breeding South American species in the area. The zoo in 1985 was little more than a place to visit and spend the day since no special education programs existed at the time for the public or school groups.

However, the zoo was a significant regional facility and Mercado was heavily involved in the conservation movement of the country.[58,59]

Unfortunately, events took a sad turn in 1986 according to reports on September 7, 8, and 9 in the Santa Cruz de la Sierra newspaper *El Mundo*. On Saturday, September 6, Mercado, two other biologists (one from Spain), and a pilot were performing a natural history survey of the Serrania de Huanchaca on the Bolivian frontier with Brazil, and in the northern sector of the Province of Velasco part of the Santa Cruz district. On this day, he and two others of the group were killed after they landed on an undocumented airstrip they had located in the forest. According to the Spanish biologist, who survived by running into the jungle, they were shot shortly after walking up a trail off the runway. Upon hearing of the murders, the army was mobilized, but the air force could not operate over the weekend, and the Ministry of the Interior could not explain why it took over 30 hours to get forces into the region. Final reports indicate that by the time the military arrived all that was left were the burned bodies of those who died and the plane. The group had stumbled upon one of the largest narcotics operations in Bolivia, and possibly in the entire region; however, no one was apprehended, as the killers had ample time to leave the area. On September 9, 1986, the 1,337,00-acre area known as the Serrania de Huanchaca was renamed the Parque Nacional Kempff Mercado.[59–61]

11.4.3 Colombia

Zoológico de Cali is one of the more significant of the five zoos in Colombia. Cali Zoo exhibits comprise a total of 20 acres in a 50-acre park. Park attendance increased from under 40,000 in 1982 to 240,000 in 1995.[1] In 1989, the equivalent of 9% of the population of Cali visited the zoo and the entry fees from these visitors accounted for approximately 38% of the zoo income. In 1991, a private zoo foundation managed the zoo, although the zoo was a municipal facility. Members of the foundation represent regional organizations involved in education, recreation, and conservation. Funding was still over 46% municipal, with other sources including parking lot revenues, restaurant concessions, donations, and rental of the park for special events. As in other Brazilian facilities, much of the animal collection comes from donations, confiscations of wildlife, and from pet owners who no longer want their pets. The condition of these animals at the time of receipt is poor and probably accounts for the low survival rates of these donated animals.[62]

Of significance as well is the fact that the zoos of Colombia formed a professional association in 1976, the Asociacion Colombiano de Parques Zoológicos, which in 1991 was composed of the country's five zoos: the Zoológico Santacruz in Bogota, the Zoológico de Barranquilla in Barranquilla, the Zoológico de Cali in Cali, the Zoológico Santa Fe in Medellin, and the Zoológico Matacafia in Pereira. The association has as its priority the professional exchange of information and collaboration between institutions. Lack of funding, animal mortality, mistrust between institutions, and misplaced priorities (such as striving to have the largest and best collection in Colombia) are factors that were identified as major impediments to professional growth. By 1991, awareness of the conservation and educational roles of the zoos in Colombia was taking hold, and as contact with external institutions has increased so has the professionalism within these facilities.

11.4.4 Venezuela

Jardin Zoológico Las Delicias in Maracay was established in 1973, having its origins in a collection that General Gomez established with animals given by friends, workers, and other individuals from around the country. Although mostly native species, this collection continued to grow and in the early 1970s began to acquire a number of African and Asian species. Another of the country's major zoos is that found in the Parque Caricuao, which was built in a corner of Caracas where citizens could satisfy the need for contact with nature. This park was part of a much larger natural area of over 740 acres. This large natural area was established as a national park in 1973, and the first stage of development of the park was the creation of a zoo in which the animals found in the park could be exhibited. Venezuela has been home to the only large-scale commercial venture in capybara production and has had numerous experimental efforts at commercial breeding of crocodilians, in particular caimans.[63]

11.4.5 Summary

While scant, what information there is about the history of each of these South American facilities reflects the history of South America, one that is intricately intertwined in the evolution of the culture of each country, as well as of the world. All zoos of the region, having come from a rich heritage, have demonstrated an incredible ability to survive and grow, even in the light of serious economic hazards and professional impediments. It is important to note that in virtually all the zoos there has been a strong commitment both to education and to conservation of the environment. The typical tourist may not recognize this commitment, given the menagerie-like conditions that exist for economic reasons at some of the facilities. However, these zoos are the only natural areas where the local citizenry can go to spend the day, a holiday, or a weekend, so they occupy a valued position in the society.

Western European and U.S. definitions have guided the development of zoos in South America, which has often meant having the most, the largest, and the rarest. The zoos of South America have found a most exciting and novel approach to this development, which should be encouraged — that of the regional zoo intertwined with the local natural context, having the goal of educating visitors about the natural heritage and the history of the region and the country. In contrast, U.S. zoos have established as a professional priority the conservation of endangered species from many parts of the world, and the development of breeding programs to implement this priority. South American zoos, and Brazilian zoos in particular, have been unable to achieve this goal to any significant extent (outside of three or four major zoos), but have made an effort to conserve their local species.

Finally, it is important to encourage a certain creativity in zoological garden philosophy and to look at the local or regional history to understand why some zoos have succeeded, why others have failed, and why some have simply remained in a limbo of uncertainty. A key to the future of zoos of South America lies in their own communities and the ability of their facilities and their staff to establish themselves as a vital part of their particular communities. By becoming a community resource for local and national education, as science or natural history laboratories, sociocultural study areas, and centers for environmental education, the zoos add to their vitality and their ability to survive. Through these linkages, zoos can establish themselves professionally and gain the support for conservation programs that will inevitably become a part of their activities.

References

1. *International Zoo Yearbook*, Zoological Society of London, London, 1960+ [an annual publication of the ZSL; some volumes contain a directory of zoos and aquaria of the world and of national and regional zoo associations].
2. SZB (Sociedade de Zoológicos do Brasil), *Base de Dados Tropical*, Fundaqão Tropical de Pesquisas e Technologia "Andre Tosello," 1998 [Directory of Brazilian Zoos].
3. Ellis, J., *Brazilian Zoological Gardens: Their Status and Conservation Potential*, Master's thesis, Center for Latin American Studies, University of Florida, Gainesville, 1987.

4. Adas, Melhem, *Panorama Geografico do Brasil*, Editora Moderna, São Paulo, Brazil, 1985 [*Geographic Panorama of Brazil*].
5. Coelho, Marcos de Amorim and Soncin, N. Bueno, *Geografia do Brasil*, Editora Moderna, São Paulo, Brazil, 1985 [*Geography of Brazil*].
6. Vesentini, William J., *Geografia do Brasil*, Editora Atica, São Paulo, Brazil, 1986 [*Geography of Brazil*].
7. Instituto Brasileiro de Geografia e Estatistica, *IX Recenseamento Geral do Brasil — 1980*, Fundaqão Instituto Brasileiro de Geografia e Estatistica, Rio de Janeiro, Brazil, 1982 [*IX General Census of Brazil — 1980*].
8. Burns, E. Bradford, *A History of Brazil*, Columbia University Press, New York, 1980.
9. Guimaraes, Maria A. de Alencastro, *An Outline of Brazilian History*, Cultural Division of the Ministry of Foreign Relations, Rio de Janeiro, Brazil, 1952.
10. Anonymous, *25 Anos de Economia Brasileira*, Grafica Record Editora, Rio de Janeiro, Brazil, 1965 [*25 Years of Brazilian Economy*].
11. Puglia, Lazaro Ronaldo Ribeiro, Director Parque Zoológico Municipal "Quinzinho de Barros," Sorocaba, Brasil, personal communication, 1985.
12. Fundaqão Tropical de Pesquisas e Technologia "Andre Tosello," *Base de Dados Tropical, Directory of Brazilian Zoos*, 1998 [*Tropical Database and Directory of Zoos*].
13. Freitas, Davies, Director, the Parque Natural de Carajas (CVRD), Carajas, Brazil, personal communication, 1985.
14. Maia, Antonio B., *Relatorio — Prefeitura Municipal de Manaus*, Tip. Fenix de Sergio Cardo, Manaus, Amazonas, Brazil, 1940 [*Report — Municipal district of Manaus*].
15. Barros, Silvio, Amazonas Tourism Board, Manaus, Brazil, personal communication, 1985.
16. Best, Robin, Biologist, Instituto Nacional de Pesquisas, Manaus, Brazil, personal communication, 1985.
17. Barboza, Claudio, "Manaus: esta faltando sombra no coracao da floresta," *A Critica*, March 30, 1981, 5 ["Manaus: lacking shade in the heart of the forest"].
18. Anonymous, "Aviaquario perde animais e 'conquista' um boate," *A Noticia*, March 17, 1984 ["Aviary/Aquarium loses and wins a round"].
19. Wallauer, Jordan P., Andrade, Glenio Bruck de, and Grieger, Paulo Alceu, Parque Zoobotânico de Manaus, Unpublished Report, Manaus, Amazonas, Brazil, 1981 [Zoobotanical Park of Manaus].
20. Anonymous, "Aviaquario lacrado hoje pela Semusp," *Jornal de Commercio*, March 21, 1984 ["Aviary/Aquarium complimented today by Semusp"].
21. Barros, Silvio M., Estudo Preliminar para Implantacao de um Parque Ecológico, Unpublished Report, Manaus, Brazil, 1981 [Preliminary Study for the Development of an Ecological Park].
22. Anonymous, "CIGS completa hoje 20 anos de criacao," *Jornal do Commercio*, March 2, 1984 ["CIGS completes 20 years of existence today"].
23. Benitez, Luiz Mutti, Director/Veterinarian, CIGS Zoo, Manaus, Brazil, personal communication, 1985.
24. Silva, Dr., Director, Hotel Tropical Zoo, Manaus, Brazil, personal communication, 1985.
25. Edegard de Deus, Carlos, *Relatorio de Atividades — 1982*, Universidade Federal do Acre, Rio Branco, Acre, Brazil, 1982 [*Activities Report — 1982*].
26. Pita, Antonio de Oliveira, *Relatorio de Atividade — 1981*, Universidade Federal do Acre, Rio Branco, Acre, Brazil, 1982 [*Activity Report — 1981*].
27. Soares, Antonio Carlos Lobo, Director, Parque Zoobotânico Museum Paraense Emilio Goeldi, Belem, Brazil, personal communication, 1986.
28. Sobrinho, J. V., de Morais, J. B., and Tenorio, D. de Oliveira, "Levantamento Regional das Organizacoes Estaduais, Municipais e Empresariais," in *F.B.C.N. Encontros Regionais Sobre Conservacao da Fauna e Recursos Faunisticos — Recife*, Ministerio de Agricultura–I.B.D.F., Brasilia, D.F., Brazil, 1977, 227 ["Regional Study of the State Municipal and Industrial Organizations"].

29. Cavalcanti de Sa, Mauro, Director, Jardim Zoobotânico de Dois Irmaos, Recife, Brazil, personal communication, 1985.
30. Levy, Dennis, Zoological Gardens in Argentina, Brazil and Uruguay, Unpublished report to the Zoological Society of London, London, England, 1961.
31. Neto, Raul, Director, Zoológico da Universidade Federal de Mato Grosso, Cuiaba, Brazil, personal communication, 1985.
32. Simon, Faical, Assistant Director, Fundaqão Parque Zoológico de São Paulo, São Paulo, Brazil, personal communication, 1985.
33. Hilton, R., *The Scientific Institutions of Latin America*, California Institute of International Studies, Stanford, 1970.
34. Coimbra-Filho, A. F., Director, Centro de Primatas, Rio de Janeiro, Brazil, personal communication, 1985.
35. Frioli, Adolfo, Parque Municipal "Quinzinho de Barros" e o Terceiro Zoologico de Sorocaba e do Brasil, Unpublished Report, Parque Zoológico "Quinzinho de Barros," Sorocaba, São Paulo, Brazil, n. d. [Municipal Park "Quinzinho de Barros" is the third zoo of Sorocaba and of Brazil].
36. Saliba, Adayr Mfuz, Director, Fundaqão Parque Zoológico de São Paulo, São Paulo, Brazil, personal communication, 1985.
37. Anonymous, O Parque Zoológico de São Paulo — Parque Professor Mario Paulo Autuori, Unpublished Report, Parque Zoológico de São Paulo, São Paulo, Brazil, n. d. [Zoological garden of São Paulo — Professor Mario Paulo Autuori].
38. Martins da Silva, Elisabete M., "Editorial," *Natureza em Revista*, 7, 2, 1980.
39. Giacomini, Claudio, Director, Parque Zoológico do rio Grande do Sul, Porto Alegre, Brazil, personal communication, 1986.
40. Martins da Silva, Elisabete M., "Editorial," *Natureza em Revista*, 6, 2, 1979.
41. Madureira, M. Saint-Pastous and Schmitt, Maria T. V., "Plano piloto do Parque Zoológico da Fundaqão Zoobotânico do Rio Grande do Sul," *Natureza em Revista*, 1, 36, 1976 ["Master plan of the Zoological Garden of the Rio Grande do Sul Zoobotanical Foundation"].
42. Boyd, Linda, Ed., *Zoological Parks and Aquariums in the Americas 1984–1985*, American Association of Zoological Parks and Aquariums, Wheeling, WV, 1984.
43. Boyd, Linda, Ed., *Zoological Parks and Aquariums in the Americas 1994–1995*, American Zoo and Aquarium Association, Wheeling, WV, 1994.
44. Cavalcanti, Paulo B., *Guia Botânico do Museu Goeldi*, Museu Paraense Emilio Goeldi, Belem, Brazil, 1983 [*Botanical Guide to the Museum Goeldi*].
45. Rodrigues da Cunha, Osvaldo, "Emilio Augusto Goeldi (1859–1917)," *Ciencia e Cultura*, 35, 1965, 1983.
46. Goeldi, Emilio A., *Relatorio 1894–1895*, Boletim do Museu Paraense Tomo I (fasc 1–4), Belem, Brazil, 1894–1896, 221 [*1894–1895 Report*].
47. Goeldi, Emilio A., *Relatorio 1896*, Boletim do Museu Paraense Tomo II (fasc. 1–4), Belem, Brazil, 1897–1898 [*1896 Report*].
48. Goeldi, Emilio A., *Relatorio 1897–1900*, Boletim do Museu Paraense Tomo III (fasc. 1–4), 1900–1902 [*1897–1900 Report*].
49. Goeldi, Emilio A., *Relatorio 1903–1904*, Boletim do Museu Paraense Tomo V (fasc. 1–2), Belem, Brazil, 1909 [*1903–1904 Report*].
50. Rodrigues da Cunha, Osvaldo, Sugestoes para um Projeto de Installacao de um Novo e Moderno Parque Zoológico na Cidade de Belem, Museu Goeldi Unpublished Report, Belem, Para, Brazil, 1971 [Suggestions for a Project Developing a New Modern Zoological Garden in the City of Belem].
51. Rodrigues da Cunha, Osvaldo, "108 Anniversario do Museu Paraense Emilio Goeldi," *Revista de Cultura do Para*, 16–17, 151, 1974.
52. Neto, Paulo Nogueira, *A Criacao de Animais Indigenas Vertebrados*, Edicoes Tecnapis, São Paulo, Brazil, 1973 [*The Breeding of Indigenous Vertebrate Animals*].

53. Coordenacao Editoral, *CNPq — Origens e Perspectivas*, Grafica CNPq, Rio de Janeiro, Brazil, 1980 [*CNPq — Origins and Perspectives*].
54. Moreira, Eidorfe, *Belem e Sua Expressao Geografica*, Imprensa Universitaria, Belem, Para, Brazil, 1966 [*Belem and Its Geography*].
55. Conway, H. McKinley and Liston, L. L., *The Weather Handbook*, Conway Research, Inc., Atlanta, 1974.
56. Instituto Brasileiro de Desenvolvimento Florestal, *Codigo Florestal, Protecao a Fauna, Criacao do IBDF, Regulamento dos Parques Nacionais Brasileiros*, Ministerio da Agricultura, Brasilia, Distrito Federal, Brasilia, n. d [*Forestry Laws, Protection of Wildlife, Creation of IBDF, Regulations of the Brazilian National Parks*].
57. Del Pino, Diego A., *Historia del Jardin Zoológico Municipal*, Municipalidad de la Ciudad de Buenos Aires, Buenos Aires, Argentina, 1979 [*History of the Municipal Zoological Garden*].
58. Mercado, Noel Kempff, "El zoológico de la ciudad de Santa Cruz de la Sierra," *Bolivia Turistica*, 1, 24, 1979 ["The zoo of the city of Santa Cruz de la Sierra"].
59. Anonymous, "Los recursos naturales no renovables, el zoológico, el nuevo botánico preocupan a Noel Kempff Mercado," *El Mundo*, September 7, 1986, 12 ["Nonrenewable natural resources, the zoo, the new botanical garden preoccupy Noel Kempff Mercado"].
60. Anonymous, "Sobreviviente relata la tragedia de Huanchaca," *El Mundo*, September 8, 1986, 1 ["Survivor tells about the tragedy of Huanchaca"].
61. Anonymous, "El mas grande laboartorio de cocaina," *El Mundo*, September 9, 1986, 1 ["The greatest cocaine laboratory"].
62. White, Teresa Gutierrez de, "ACOPAZOO and the status of zoos in Colombia," *International Zoo News*, 38, 19, 1991.
63. Luxmoore, Richard, Wildlife Trade Monitoring Unit, International Union for the Conservation of Nature and Natural Resources, Cambridge, U.K., personal communication, 1985.
64. Instituto Brasileiro de Geografia e Estatistica, *Anuario Estatistico do Brasil — 1982*, Fundaqão Brasileiro de Geografia e Estatistica, Rio de Janeiro, Brazil, 1982 [*1982 Annual Statistics of Brazil*].

Appendix

Zoos and Aquariums of the World

Zoos and aquariums of the world are listed chronologically according to the regions presented in the chapters. This listing of zoos and aquariums is not meant to be comprehensive since it only lists those institutions with a known date of establishment. For information on extant zoos and aquariums, refer to these international directories: *International Zoo Yearbook* (Zoological Society of London, London, published annually since 1960) or *Global Zoo Directory* (Captive Breeding Specialist Group in cooperation with the Zoological Society of London and International Species Information System, Apple Valley, MN, 1993, 1996).

Zoological Gardens of Great Britain and Ireland

1245–1832	Tower Menagerie (London, England)
1800–1828	Exeter 'Change (London, England)
1828	Zoological Society of London Zoo (London, England)
1831	Royal Zoological Society of Ireland Zoo (Dublin, Ireland)
1831–1856	Surrey Zoo (London, England)
1832–1863	Liverpool Zoo (Liverpool, England)
1835	Bristol Zoo (Bristol, England)
1836–1977	Belle Vue Zoo (Manchester, England)
1837–1900	Rosherville Zoo (Rosherville, England)
1838–1842	Manchester Zoo (Manchester, England)
1838–1844	Cheltenham Zoo (Cheltenham, England)
1838–1844	York Zoo (York, England)
1839–1855	Edinburgh Zoological Gardens (Edinburgh, Scotland)
1840–1860	Hull Zoo (Hull, England)
1853	London Zoo Aquarium (London, England)
1865–1870	Preston Zoo (Preston, England)
1871–1936	Crystal Palace Aquarium (London, England)
1872	Brighton Aquarium (Brighton, England)
1874–1878	Manchester Aquarium (Manchester, England)
1874–1900	Southport Aquarium (Southport, England)
1875–1963	Blackpool Tower Zoo (Blackpool, England)
1876–1888	Yarmouth Aquarium (Yarmouth, England)
1876–1890	Westminster Aquarium (Westminster, England)
1878–1880	Edinburgh Aquarium (Edinburgh, Scotland)
1888	Plymouth Aquarium-Marine Biological Association of the United Kingdom (Plymouth, England)

1890–1910	Eastham Zoo (Eastham, England)
1895	Woburn Abbey Deer Park (Woburn, England)
1896–1899	Oldham Zoo (Oldham, England)
1900–1910	New Brighton Zoo (Brighton, England)
1900–1910	Sutton Coldfield Zoo (Sutton Coldfield, England)
1900–1941	Cardiff Zoo (Cardiff, Wales)
1909–1913	Edinburgh — Portobello Zoo (Edinburgh, Scotland)
1909–1918	Halifax Zoo (Halifax, England)
1913	Royal Zoological Society of Scotland Zoo (Edinburgh, Scotland)
1923	Paignton Zoological and Botanical Gardens (Paignton, England)
1927–1940	Skegness Zoo (Skegness, England)
1928–1939	Grimsby Zoo (Grimsby, England)
1931	Chessington Zoo (Chessington, England)
	Chester Zoo (Chester, England)
	Whipsnade (London, England)
1932–1937	Oxford Zoo (Oxford, England)
1932–1938	Liverpool Zoo (Liverpool, England)
1933	Belfast Zoo (Belfast, Ireland)
1933–1939	Sheringham Zoo (Sheringham, England)
1933–1940	Southend Zoo (Southend, England)
1933–1962	Maidstone Zoo (Maidstone, England)
1937	Dudley and West Midlands Zoological Society Zoo (Dudley, England)
1942–1973	Wellingborough Zoo (Wellingborough, England)
1948–1969	Ilfracombe Zoo (Ilfracombe, England)
1949–1984	Bognor Regis (England)
1951–1955	Ferndown Zoo (Ferndown, England)
1952	Southport Zoo (Southport, England)
1955	Sandown Zoo (Isle of Wight)
1955–1971	Ashover Zoo (Ashover, England)
1956–1960	Doncaster Zoo (Doncaster, England)
1959	Jersey Wildlife Preservation Trust Zoo (Jersey, Channel Islands)
1961	Flamingo Gardens and Zoological Park (Olney, England)
1962	Norfolk Wildlife Park (Great Witchingham, England)
1962–1973	Aberdeen Zoo (Aberdeen, Scotland)
1962–1975	Cromer Zoo (Norfolk, England)
1962–1978	Exmouth Zoo (Exmouth, England)
1962–1980	Cardiff — Barry Zoo (Cardiff, Wales)
1963	Twycross Zoo (Twycross, England)
1963–1985	Southampton Zoo (Southampton, England)
1963–1994	Poole Zoo (Poole, England)
1964	Mole Hall Wildlife Park (Widdington, England)
	Shaldon Wildlife Trust Zoo (Shaldon, England)
1964–1974	Birmingham Zoo (Birmingham, England)
1965	Curraghs Wildlife Park (Ballaugh, Isle of Man)
	Skegness — Natureland (Skegness, England)
1965–1983	Plymouth Zoo (Plymouth, England)
1966	Howletts Zoo Park (Bekesbourne, England)
1966–1972	Bideford Zoo (Bideford, England)
1966–1980	Coventry Zoo (Coventry, England)

1966–1980 Knaresborough Zoo (Knaresborough, England)
1966–1982 Southam Zoo (Southam, England)
1967   Cricket Wildlife Park (Cricket St. Thomas, England)
           Drayton Manor Park Zoo (Tamworth, England)
           Glasgow Zoo (Glasgow, Scotland)
1967–1974 Weyhill Zoo (Weyhill, England)
1967–1990 Zoological Trust of Guernsey Zoo (Guernsey, Channel Islands)
1968   Banham Zoo (Banham, England)
           Penscynor Zoo (Penscynor, Wales)
1968–1973 Sherwood Zoo (Sherwood, England)
1969   Linton Zoological Gardens (Linton, England)
1969–1982 Thorney Zoo (Thorney, England)
1970   Basildon Zoo (Basildon, England)
           Cotswold Wildlife Park (Burford, England)
           Highland Wildlife Park (Kingussie, Scotland)
1971   Blackpool Municipal Zoological Gardens (Blackpool, England)
1972   Marwell Zoological Park (Marwell, England)
1978   Port Lympne Zoo Park (Lympne, England)

Zoological Gardens of Western Europe

Austria

1752	Menagerie Schönbrunn/Tiergarten Schönbrunn (Vienna)
1961	Salzburger Tiergarten Hellbrunn (Salzburg)
1962	Alpenzoo (Innsbruck)

Belgium

1843	Royal Zoological Society of Antwerp (Antwerp)
1851–1879	Zoo Bruxelles (Brussels)
1956	Dierenpark Plackendael (Mechelen)

Denmark

1859	Zoologisk Have (Copenhagen)
1950	Odense Zoo (Odense)
1953	Aalborg Zoologiske Have (Aalborg)
1969	Safaripark (Knuthenborg)

Finland

1889	Helsinki Zoo (Helsinki)
1973	Zoo Ähtäri (Ähtäri)

France

1793	Menagerie du Jardin des Plantes (Paris)
1855	Jardin Zoologique Marseille (Marseille)
1858	Jardin Zoologique des la Ville de Lyon (Lyon)
1860	Jardin d'Acclimatation (Paris)

1868	Parc Zoologique et Botanique de Mulhouse (Mulhouse)
1920	Parc Zoologique, Fondation Jean Delacour (Cleres)
1934	Parc Zoologique de Paris (Paris)

Germany

1844	Zoologischer Garten (Berlin)
1858	Zoologischer Garten der Stadt Frankfurt (Frankfurt-on-Main)
1860	Zoologischer Garten Köln (Cologne)
1861	Zoologischer Garten Dresden (Dresden)
1863–1930	Zoologischer Garten Hamburg (Hamburg)
1865	Zoologischer Garten Breslau (Breslau, now Wroclaw, Poland)
	Zoologischer Garten Hannover (Hannover)
1866	Zoologischer Garten Karlsruhe (Karlsruhe)
1869–1910	Berliner Aquarium unter den Linden (Berlin)
1871–1906	Nills Thiergarten (Stuttgart)
1874–1945	Zoologischer Garten Düsseldorf (Düsseldorf)
1875	Westfälischer Zoologischer Garten Münster (Münster)
1878	Zoologischer Garten Leipzig (Leipzig)
1881	Zoologischer Garten (Elbenfeld, now Wuppertal)
1882–1905	Zoologischer Garten (Aachen)
1896	Zoologischer Garten Königsberg (Königsberg, now Kaliningrad, Russia)
1901	Zoologischer Garten Halle (Halle)
1907	Carl Hagenbeck's Tierpark at Stellingen (Hamburg)
1911–1923	Zoologischer Garten München (Munich)
1912	Tiergarten der Stadt Nürnberg (Nuremberg)
	Zoologischer Garten Landau (Landau)
1928	Münchener Tierpark Hellabrunn (Munich)
	Tiergrotten Zoo am Meer (Bremerhaven)
1932	Zoologischer Garten Saarbrücken (Saarbrücken)
1933	Tierpark Bochum (Bochum)
1934	Krefelder Zoo (Krefeld)
	Tiergarten Heidelberg (Heidelberg)
	Zoo Duisburg (Duisburg)
1935–1944	Tierpark (Aachen)
1936	Zoo Osnabrück (Osnabrück)
1937	Tiergarten der Stadt Straubing (Straubing)
	Thierpark Rheine NaturZoo Rheine (Rheine)
	Zoologischer Garten Augsburg (Augsburg)
1948	Aquazoo (Düsseldorf)
1949	RuhrZoo (Gelsenkirchen)
	Wilhelma, Zoologischer und Botanischer Garten (Stuttgart)
1950	Heimattierpark Neumünster (Neumünster)
	Zoologischer Garten (Magdeburg)
1953	Zoologischer Garten Dortmund (Dortmund)
1955	Tierpark Berlin-Friedrichsfelde (Berlin)
1956	Opel-Zoo (Kronberg/Taunus)
	Zoologischer Garten Rostock (Rostock)
	Zoologischer Garten Schwerin (Schwerin)
1958	Thüringer Zoopark Erfurt (Erfurt)
	Vogelpark Walsrode (Walsrode)

1960	Aachener Tierpark (Aachen)
1961	Vivarium Darmstadt (Darmstadt)
1966–1973	Bremer Tierpark (Bremen)
1974	Serengeti Safaripark Hodenhagen (Hodenhagen)

Italy

1911	Giardino Zoologico del Commune de Roma (Rome)
1930	Giardino Zoologica (Milan)
1950	Giardino Zoologica di Napoli (Naples)
1955	Giardino Zoologico (Turin)

The Netherlands

1838	Natura Artis Magistra (Amsterdam)
1858	Rotterdam Zoo (Rotterdam)
1863–1943	Haagse Dierenpark (The Hague)
1913	Burgers' Zoo and Safari (Arnheim)
1932	Ouwehands Dierenpark (Rhenen)
1935	Noorder Dierenpark (Emmen)
1936	Wassenaar Zoo/Wassenaar Wildlife Breeding Center (Wassenaar)
1947	Dierenpark Amersfoort (Amersfoort)
1956	Alphen Birdpark Avifauna (Alphen)
1968	Safaripark Beekse Bergen (Beekse Bergen)
1971	Apenheul (Apeldoorn)

Portugal

1884	Jardim Zoológico (Lisbon)

Spain

1869	Jardin Zoológico (Madrid)
1894	Parque Zoológico de Barcelona (Barcelona)
1972	Loro Parque (Tenerife)

Sweden

1891	Skansen (Stockholm)
1901–1916	Furuvikparken (Gävle)
1936	Furuvikparken (Gävle)
1950	Parken Zoo (Eskilstuna)
1962	Borasparken (Boras)
1965	Kolmardens Djurpark (Kolmarden)

Switzerland

1874	Zoologischer Garten Basel (Basel)
1892	Wildpark Peter und Paul (Saint Gallen)
1923	Tierpark Goldau (Goldau)
1929	Zoologischer Garten Zürich (Zurich)
1939	Städtischer Tierpark Dählhölzli (Bern)

Zoological Gardens of Eastern Europe

Bulgaria

1888	Zoologicheska Gradina Sofia (Sofia)
1932	Akvarium Varna (Varna)
1965	Zoologicheska Gradina Varna (Varna)

Czech and Slovak Republics

1919	Severoceska Zoologicka Zahrada Liberec (Liberec, Czech)
1931	Zoologicka Zahrada Praha (Prague, Czech)
1946	Vychodoceska Zoo-Safari, Dvur Kralove nad Labem (Dvur Kralove, Czech)
1948	Zoologicka Zahrada Usti nad Labem (Usti nad Labem, Czech)
1949	Zoologicka Zahrada Decin (Decin, Czech)
1950	Zoologicka Zahrada Mesta Brna (Brno, Czech)
1951	Zoologicka Zahrada Bojnice (Bojnice, Slovakia)
1953	Zoologicka Zahrada a Zamek Lesna-Zlin (Zlin-Lesna [Gottvaldov], Czech)
	Zoologicka Zahrada Ostrava (Ostrava, Czech)
1956	Zoologicka Zahrada Olomouc (Olomouc, Czech)
1957	Zoologicka Zahrada Jihlava (Jihlava, Czech)
1960	Zoologicka Zahrada Bratislava (Bratislava, Slovakia)
1963	Zoologicka a Botanicka Zahrada Mesta Plzne (Pilsen, Czech)
1972	Zoologicka Zahrada Ohrada (Hluboka nad Vltavou, Czech)
1979	Zoologicka Zahrada Kosice (Kosice, Slovakia)

Hungary

1866	Budapest Fovaros Allat-Es Novenykertje (Budapest)
1958	Kittenberger Allat-Es Vadaspark Veszprem (Veszprem)
195?	Nagyerdei Kulturpark Allat-Esnovenykertje (Debrecen)
1961	Mecseki Kulturpark Zoo (Pecs)
196?	Xantus Zoo (Gyor)
1989	Szegedi Vadaspark Zoo Szeged (Szeged)

Poland

1865/1948	Miejski Ogrod Zoologiczny (Breslau, Germany/now Wroclaw)
1871	Wielkopolski Park Zoologiczny (Poznan/Old)
1917	Miejski Ogrod Zoologiczny (Zamosc/Old)
1928	Miejski Ogrod Zoologiczny w Warszawie (Warsaw)
1929	Miejski Park i Ogrod Zoologiczny w Krakowie (Krakow)
1938	Miejski Ogrod Zoologiczny w Lodzi (Lodz)
1951	Miejski Ogrod Zoologiczny (Plock)
1953	Miejski Ogrod Zoologiczny Opole (Opole)
1954	Miejski Ogrod Zoologiczny Wybrzeza (Gdansk)
1958	Slaski Ogrod Zoologiczny (Katowice)
1970	Bialystok Zoo (Bialystok)
1971	Akwarium Morskiego Instytutu Rybactwa (Gdynia)
1974	Ogrod Zoologiczny w Poznaniu (Poznan/New)
1978	Ogrod Fauny Polskiej (Bydgoszcz)
1982	Miejski Ogrod Zoologiczny (Zamosc/New)
1996	Swierkocin Safari Zoo (Swierkocin)

Russia and Former Soviet Union

1864	Moscow Zoo (Moscow, CIS-Russia)
1865	Sankt Peterburg Zoo (St. Petersburg [Leningrad], CIS-Russia)*
1892	Zoofarm of the State Steppe Reserve, Askaniya-Nova (Askaniya-Nova, Ukraine)
1895	Charkov Zoo (Charkov, Ukraine)
1896/1947	Kaliningrad Zoo (Königsberg, Germany/now Kaliningrad, CIS-Russia)
1897	Sevastopolskii Morskoi Akvarium (Sevastopol, CIS-Ukraine)
1901	Nikolaev Zoo (Nikolaev, Ukraine)
1908	Kiev Zoo (Kiev, Ukraine)
1912	Rigas Zoologiskais Darzs (Riga, Latvia)
1924	Kazan Zoo (Kazan, CIS-Tatar)
	Tashkent Zoo (Tashkent, CIS-Uzbekistan)
	Termez Zoo (Termez, CIS-Uzbekistan)
1927	Rostov-on-Don Zoo (Rostov na Donu, CIS-Russia)
	Tbilisi Zoopark (Tbilisi, Georgia)
1929	Ashkabad Zoo (Ashkhabad, CIS-Turkmenistan)
	Grodno Zoo (Grodno, Belorussia)
1930	Sverdlovsk Zoo (Sverdlovsk, CIS-Russia)
1933	Perm Zoopark (Perm, CIS-Russia)
1936	Odessa Zoo (Odessa, Ukraine)
1937	Alma-Atinskii Zoopark (Alma-Ata, Kazakhstan)
	Tallin Zoo (Tallin, Estonia)
1938	Karaganda Zoo (Karaganda, Kazakhstan)
	Lietuvos Zoologijos Sodas (Kaunas, Lithuania)
1941	Erevan Zoo (Erevan, Armenia)
1945	Baku Zoo (Baku, Azerbaijan)
1947	Novosibirsk Zoo (Novosibirsk, CIS-Russia)
195?	Kishinev Zoo (Kishinev, Moldavia)
1961	Dushanbe Zoo (Dushanbe, Tadzhikistan)
196?	Batumskii Okeanarium (Batumi, Georgia)
197?	Center of Marine Culture, Juru Lietuva (Klaipeda, Lithuania)
1981	Penza Zoopark (Penza, CIS-Russia)

Zoological Gardens of the United States

1859/1874	Philadelphia Zoological Garden (Philadelphia, Pennsylvania)**
1861	Central Park Zoo (New York, New York)
1868	Lincoln Park Zoological Gardens (Chicago, Illinois)
1872	Roger Williams Park Zoo (Providence, Rhode Island)
1873	National Aquarium (Washington, D.C.)
1875	Buffalo Zoological Gardens (Buffalo, New York)
	Cincinnati Zoo (Cincinnati, Ohio)
	Ross Park Zoo (Binghamton, New York)
1876	Baltimore Zoo (Baltimore, Maryland)
1882	Cleveland Metroparks Zoo (Cleveland, Ohio)
1887	Metro Washington Park Zoo (Portland, Oregon)
1888	Dallas Zoo (Dallas, Texas)

---

* CIS = Commonwealth of Independent States
** Chartered/opened

1889	National Zoological Park (Washington, D.C.)
	San Francisco Zoological Gardens (San Francisco, California)
	Zoo Atlanta (Atlanta, Georgia)
1890	Dickerson Park Zoo (Springfield, Missouri)
	St. Louis Zoological Park (St. Louis, Missouri)
1891	John Ball Zoological Garden (Grand Rapids, Michigan)
	Miller Park Zoo (Bloomington, Illinois)
1892	Milwaukee County Zoological Gardens (Milwaukee, Wisconsin)
1893	Prospect Park Zoo (New York, New York)
	St. Augustine Alligator Farm (St. Augustine, Florida)
1894	Seneca Park Zoo (Rochester, New York)
	Zoo at Buttonwood (New Bedford, Massachusetts)
1896	Aquarium for Wildlife Conservation (New York, New York)
	Denver Zoological Park (Denver, Colorado)
1897	St. Paul's Como Zoo (St. Paul, Minnesota)
1898	Alameda Park Zoo (Alamogordo, New Mexico)
	Omaha's Henry Doorly Zoo (Omaha, Nebraska)
	Pittsburgh Zoo (Pittsburgh, Pennsylvania)
1899	Wildlife Conservation Park/Bronx Zoo (New York, New York)
1900	Toledo Zoological Gardens (Toledo, Ohio)
	Virginia Zoological Park (Norfolk, Virginia)
1901	Glen Oak Zoo (Peoria, Illinois)
1904	Belle Isle Aquarium (Detroit, Michigan)
	Oklahoma City Zoo (Oklahoma City, Oklahoma)
	Waikiki Aquarium (Honolulu, Hawaii)
	Woodland Park Zoological Gardens (Seattle, Washington)
1905	Brandywine Zoo (Wilmington, Delaware)
	Lafayette Zoological Park (Lafayette, Indiana)
	Orange County Zoo (Orange, California)
	Point Defiance Zoo & Aquarium (Tacoma, Washington)
	Walter D. Stone Memorial Zoo (Stoneham, Massachusetts)
1906	Memphis Zoological Garden (Memphis, Tennessee)
1907	Sequoia Park Zoo (Eureka, California)
1908	Chaffee Zoological Gardens of Fresno (Fresno, California)
1909	Fort Worth Zoological Park (Fort Worth, Texas)
	Kansas City Zoological Gardens (Kansas City, Missouri)
1910	Franklin Park Zoo (Boston, Massachusetts)
1911	Henry Vilas Zoo (Madison, Wisconsin)
1912	Bramble Park Zoo (Watertown, South Dakota)
	Los Angeles Zoo (Los Angeles, California)
	Utah's Hogle Zoo (Salt Lake City, Utah)
1914	Audubon Park & Zoological Garden (New Orleans, Louisiana)
	Burnet Park Zoo (Syracuse, New York)
	Jacksonville Zoological Park (Jacksonville, Florida)
	San Antonio Zoological Gardens (San Antonio, Texas)
	Utica Zoo (Utica, New York)
1915	Sacramento Zoo (Sacramento, California)
1916	Boise City Zoo (Boise, Idaho)
	San Diego Zoological Park (San Diego, California)
1917	Potawatomi Zoo (South Bend, Indiana)
	Potter Park Zoological Gardens (Lansing, Michigan)

1919	Abilene Zoological Gardens (Abilene, Texas)
	Jackson Zoological Park (Jackson, Mississippi)
1920	Beardsley Zoological Gardens (Bridgeport, Connecticut)
	Pueblo Zoo (Pueblo, Colorado)
	Roosevelt Park Zoo (Minot, North Dakota)
1921	Houston Zoological Gardens (Houston, Texas)
1923	Lake Superior Zoological Gardens (Duluth, Minesota)
	Louisiana Purchase Gardens & Zoo (Monroe, Louisiana)
	Racine Zoological Gardens (Racine, Wisconsin)
	Steinhart Aquarium (San Francisco, California)
1924	Erie Zoo (Erie, Pennsylvania)
1925	Albuquerque Biological Park (Albuquerque, New Mexico)
	Washington Park Zoological Garden (Michigan City, Indiana)
1926	Alexandria Zoological Park (Alexandria, Louisiana)
	Cheyenne Mountain Zoological Park (Colorado Springs, Colorado)
	Little Rock Zoological Gardens (Little Rock, Arkansas)
1927	Columbus Zoological Gardens (Powell, Ohio)
	Lee Richardson Zoo (Garden City, Kansas)
	Tulsa Zoo & Living Museum (Tulsa, Oklahoma)
1928	Detroit Zoological Park (Detroit, Michigan)
1929	Mesker Park Zoo (Evansville, Indiana)
1930	John G. Shedd Aquarium (Chicago, Illinois)
	Marshfield Wildwood Park & Zoo (Marshfield, Wisconsin)
1930s	Hillcrest Park Zoo (Clovis, New Mexico)
	Lowry Park Zoological Garden (Tampa, Florida)
1931	Brookgreen Gardens (Murrells Inlet, South Carolina)
1933	Catoctin Mountain Zoological Park (Thurmont, Maryland)
	Catskill Game Farm (Catskill, New York)
	Chahinkapa Zoo (Wahpeton, North Dakota)
	Monkey Jungle (Miami, Florida)
	Sunset Zoological Park (Manhattan, Kansas)
	Topeka Zoological Park (Topeka, Kansas)
1934	Chicago Zoological Park (Brookfield, Illinois)
	Cohanzick Zoological Garden (Bridgeton, New Jersey)
	Emporia Zoo (Emporia, Kansas)
	Oakland Zoo (Oakland, California)
1935	Lincoln Park Zoo (Manitowoc, Wisconsin)
	Tautphaus Park Zoo (Idaho Falls, Idaho)
1936	Dallas Aquarium (Dallas, Texas)
	Parrot Jungle & Gardens (Miami, Florida)
	Staten Island Zoo (Staten Island, New York)
	Trevor Zoo (Millbrook, New York)
1937	Black Hills Reptile Gardens (Rapid City, South Dakota)
	Warner Park Zoo (Chattanooga, Tennessee)
1938	Marineland (Marineland, Florida)
	Tracy Aviary (Salt Lake City, Utah)
1941	El Paso Zoo (El Paso, Texas)
1947	Brit Spaugh Zoo (Great Bend, Kansas)
	Honolulu Zoo (Honolulu, Hawaii)
	Knoxville Zoological Gardens (Knoxville, Tennessee)
1948	Miami Metrozoo (Miami, Florida)

1950	Hattiesburg Zoo (Hattiesburg, Mississippi)
	Riverside Zoo (Scottsbluff, Nebraska)
1952	Arizona-Sonora Desert Museum (Tucson, Arizona)
	Fejervary Zoo (Davenport, Iowa)
	Mill Mountain Zoo (Roanoke, Virginia)
	National Aviary (Pittsburgh, Pennsylvania)
	Santa Ana Zoo (Santa Ana, California)
1953	Akron Zoological Park (Akron, Ohio)
	Caldwell Zoo (Tyler, Texas)
	Marineland (Rancho Palos Verdes, California)
	North Eastern Wisconsin Zoo (Green Bay, Wisconsin)
1954	Salisbury Zoological Park (Salisbury, Maryland)
1955	Birmingham Zoo (Birmingham, Alabama)
	Cameron Park Zoo (Waco, Texas)
	Charles Paddock Zoo (Atascadero, California)
	Seaquarium (Miami, Florida)
1957	Micke Grove Zoo (Lodi, California)
1959	Busch Gardens (Tampa, Florida)
	Folsom Children's Zoo (Lincoln, Nebraska)
1960	Bergen County Zoological Park (Paramus, New Jersey)
	Greenville Zoo (Greenville, South Carolina)
	Morro Bay Aquarium (Morro Bay, California)
1961	Dakota Zoo (Bismarck, North Dakota)
	Gettysburg Game Park (Fairfield, Pennsylvania)
1962	Phoenix Zoo (Phoenix, Arizona)
1963	Great Plains Zoo (Sioux Falls, South Dakota)
	Santa Barbara Zoo (Santa Barbara, California)
	Turtle Back Zoo (South Mt. Reservation, New Jersey)
1964	Clyde Peelings Reptiland (Allenwood, Pennsylvania)
	Indianapolis Zoo (Indianapolis, Indiana)
	Sea Life Park Hawaii (Waimanalo, Hawaii)
	Sea World of California (San Diego, California)
1965	Blank Park Zoo (Des Moines, Iowa)
	Ft. Wayne Children's Zoo (Ft. Wayne, Indiana)
	Reid Park Zoo (Tucson, Arizona)
	Wildlife World Zoo (Litchfield Park, Arizona)
1966	Happy Hollow Zoo (San Jose, California)
	Spring River Park Zoo (Roswell, New Mexico)
1967	Ellen Trout Zoo (Lufkin, Texas)
	Parkman Zoological Gardens (South Euclid, Ohio)
	Saginaw Children's Zoo (Saginaw, Michigan)
	Scovill Children's Zoo (Decatur, Illinois)
1968	Louisville Zoological Garden (Louisville, Kentucky)
	Marine World Africa USA (Vallejo, California)
	Queens Wildlife Conservation Center (New York, New York)
1969	Alaska Zoo (Anchorage, Alaska)
	New England Aquarium (Boston, Massachusetts)
1970	African Safari Wildlife Park (Port Clinton, Ohio)
	Greater Baton Rouge Zoo (Baton Rouge, Louisiana)
	Henson Robinson Zoo (Springfield, Illinois)
	Living Desert (Palm Desert, California)
	Sea World of Ohio (Aurora, Ohio)

1971	Gladys Porter Zoo (Brownsville, Texas)
	Sedgwick County Zoo (Wichita, Kansas)
1972	Bear Country USA (Rapid City, South Dakota)
	Montgomery Zoo (Montgomery, Alabama)
	San Diego Wild Animal Park (Escondido, California)
	Santa Fe Teaching Zoo (Gainesville, Florida)
	Wildlife Safari (Winston, Oregon)
1973	International Crane Foundation (Baraboo, Wisconsin)
	Mystic Marinelife Aquarium (Mystic, Connecticut)
	Sea World of Florida (Orlando, Florida)
1974	North Carolina Zoological Park (Asheboro, North Carolina)
	Riverbanks Zoological Park (Columbia, South Carolina)
	St. Catherines Wildlife Conservation Center (Midway, Georgia)
	Wild Animal Habitat (Kings Island, Ohio)
1975	Central Florida Zoological Park (Lake Monroe, Florida)
	Northwest Trek Wildlife Park (Eatonville, Washington)
	NZP Conservation & Research Center (Front Royal, Virginia)
1976	Discovery Island Zoological Park/Disney World (Orlando, Florida)
	North Carolina Aquarium at Fort Fisher (Kure Beach, North Carolina)
	North Carolina Aquarium at Pine Knoll Shores (Atlantic Beach, North Carolina)
	North Carolina Aquarium at Roanoke Island (Manteo, North Carolina)
	Texas Zoo (Victoria, Texas)
1977	Binder Park Zoo (Battle Creek, Michigan)
	Cape May County Park Zoo (Cape May Court House, New Jersey)
	Chehaw Wild Animal Park (Albany, Georgia)
	Oglebay's Good Children's Zoo (Wheeling, West Virginia)
	Popcorn Park Zoo (Forked River, New Jersey)
	Seattle Aquarium (Seattle, Washington)
1978	Minnesota Zoological Garden (Minneapolis, Minnesota)
	ZooAmerica (Hershey, Pennsylvania)
1980	Belle Isle Zoo (Detroit, Michigan)
1981	National Aquarium (Baltimore, Maryland)
	Thousand Islands Zoo (Redwood, New York)
1983	Heritage Zoo (Grand Island, Nebraska)
1984	Applegate Park Zoo (Merced, California)
	Fossil Rim Wildlife Center (Glen Rose, Texas)
	Gulf Breeze Zoo (Gulf Breeze, Florida)
	Monterey Bay Aquarium (Monterey, California)
1986	Hutchinson Zoo (Hutchinson, Kansas)
	Living Seas/Disney World (Orlando, Florida)
1988	Sea World of Texas (San Antonio, Texas)
1990	Aquarium of the Americas (New Orleans, Louisiana)
	Grassmere Wildlife Park (Nashville, Tennessee)
	Texas State Aquarium (Corpus Christi, Texas)
1992	Birch Aquarium — Scripps Institution of Oceanography (La Jolla, California)
	Dallas World Aquarium (Dallas, Texas)
	New Jersey State Aquarium (Camden, New Jersey)
	Oregon Coast Aquarium (Newport, Oregon)
	Tennessee Aquarium (Chattanooga, Tennessee)
	Zoo of Acadiana (Broussard, Louisiana)
1994	Brevard Zoo (Melbourne, Florida)

1995	Florida Aquarium (Tampa, Florida)
1997	Albuquerque Biological Park Aquarium (Albuquerque, New Mexico)
1998	Alaska SeaLife Center (Seward, Alaska)
	Aquarium of the Pacific (Long Beach, California)
	Disney's Animal Kingdom Park/Disney World (Orlando, Florida)
1999	Newport Aquarium (Newport, Kentucky)
	Ocean Journey [Aquarium] (Denver, Colorado)
2000	South Carolina Aquarium (Charleston, South Carolina)

Zoological Gardens of Australia

1861	Royal Melbourne Zoological Gardens (Melbourne)
1883	Adelaide Zoological Gardens (Adelaide)
	Moore Park Zoo (Sydney)
1898	Perth Zoological Gardens (Perth)
1916	Taronga Zoo (Sydney)
1922–1953	Hobart Municipal Zoo (Hobart)
1934	Healesville Sanctuary (Healesville)
1977	Western Plains Zoo (Dubbo)
1983	Victoria's Open Range Zoo at Werribee (Werribee)
1989	Territory Wildlife Park (Darwin)
1993	Monarto Zoological Park (Monarto)

Zoological Gardens of Asia

Southwest Asia (Middle East)

1938–1980	Tel Aviv Zoo (Tel Aviv, Israel)
1939	Istanbul Belediyesi Gulhane Zoo (Izmir, Turkey)
1940	Ankara Zoo (Ankara, Turkey)
	Tisch Family Zoological Gardens (New Biblical Zoo) (Jerusalem, Israel)
1950	Limassol Municipal Zoo (Limassol, Cyprus)
	Stavros tis Psokas Forest Station (Cyprus)
	Zoological Garden of the Biblical Institute (Mount Carmel, Israel)
1954	Salwa Zoo (Kuwait)
1957	Eilat Aquarium (Eilat, Israel)
1958	Tehran Zoo (Shemiran, Iran)
1960s	Mashhad Zoo (Iran)
1961	Meir Segals Garden for Zoological Research (Tel Aviv, Israel)
1962	Eilat Zoo (Eilat, Israel)
1964–1981	Ramat-Gan Zoo (Ramat-Gan, Israel)
1965	Riyadh Zoological Gardens (Riyadh, Saudi Arabia)
1966	Petah-Tiqya Zoo (Israel)
	Robert College Zoo (Istanbul, Turkey)
1967	Dubai Zoological Gardens and Aquarium (Dubai, United Arab Emirates)
	Kabul Zoo (Kabul, Afganistan)
	Tuzcu Zoological Gardens (Istanbul, Turkey)
1968	Kuwait Zoological Garden (Kuwait)
1970	Al Ain Zoo and Aquarium — National Zoological Gardens (Abu Dhabi, United Arab Emirates)

1971	Hai Bar Camel Reserve (Mount Carmel, Israel)
	Negev Desert Chai Bar Reserve (Israel)
1976	Oman Mammal Breeding Center (Bait al Barakh, Oman)
1978	Shaumari Wildlife Reserve (Jordan)
1981	Zoological Centre Tel Aviv — Ramat Gan (Ramat Gan, Israel)
1986	National Wildlife Research Centre (Taif, Saudi Arabia)
1987	King Khalid Wildlife Research Centre (Riyadh, Saudi Arabia)
1991	Al-Wabra (Doha, Qatar)
	Doha Zoological Garden (Doha, Qatar)

South Asia

1872	Lahore Zoological Gardens (Lahore, Pakistan)
1881	Karachi Zoological Gardens (Karachi, Pakistan)
1909	Peshawar Zoological Garden (Pakistan)
1909–1939	Colombo Museum Zoo (Colombo, Sri Lanka)
190?-1936	Zoological Gardens Company (Colombo, Sri Lanka)
192?-1944	Osman Hill Zoo (Colombo, Sri Lanka)
1932	Central Zoo (Katmandu, Nepal)
1936	Deliwala Zoo — National Zoological Gardens (Colombo, Sri Lanka)
1962	Karachi Municipal Aquarium (Karachi, Pakistan)
1964	Dacca Zoological Gardens (Dacca, Bangladesh)
1970s	Pinnawala Elephant Orphanage (Pinnawala, Sri Lanka)
1980s	Gampaha Snake Farm (Gampaha, Sri Lanka)
1990s	Udawalawe Elephant Orphanage (Udawalawe, Sri Lanka)
1992	Jungle Kingdom Theme Park (Rawalpindi, Pakistan)

Southeast Asia

1864	Thao Cam Vien Saigon — Saigon Zoological–Botanical Garden (Ho Chi Minh City, Vietnam)
1870s-1903	Singapore Botanic Garden Zoo (Singapore)
1906	Rangoon Zoo (Rangoon, Myanmar)
1923	Pasteur Institute Snake Farm (Bangkok, Thailand)
1928	Johor Baru Zoo (Johor Baru, Malaysia)
1930	Bangkok Reptile Grove (Bangkok, Thailand)
1938	Dusit Zoo (Bangkok, Thailand)
1954	Chiang Mai Zoo (Chiang Mai, Thailand)
1955–1993	Van Kleef Aquarium (Singapore)
1957	Zoo Negara Malaysia (Kuala Lumpur, Malaysia)
1959	Manila Zoological and Botanical Garden (Manila, Philippines)
1960	Crocodile Farm (Samut Prakarn Province, Thailand)
1961	Sarawak Museum Aquarium (Kuching, Sarawak)
1962	Taiping Zoo (Taiping, Malaysia)
1964	Melaka Zoological Garden (Melaka, Malaysia)
1971	Jurong Bird Park (Singapore)
	Singapore Zoological Gardens (Singapore)
1972	Hassanal Bolkiah Aquarium (Brunei)
1974	Khao Kheow Open Zoo (Chonburi, Thailand)
1976	Hanoi Zoo (Hanoi, Vietnam)
1980s	Louis Mini Zoo (Brunei)
	Safari World (Thailand)

1986	Batang Duri Mini Zoo (Batu Apoi National Park, Brunei)
	Eakao Breeding Station (Dac Lac Province, Vietnam)
1989	Nakorn Ratchasima Zoo (Korat, Thailand)
	Songkla Zoo (Songkla, Thailand)
1990	Bavi Breeding Station (Hanoi, Vietnam)
1991	Underwater World (Singapore)
1992	Gibbon Rehabilitation Project (Bang Pae Forest Reserve, Thailand)
	Vientiane Zoo (Vientiane, Laos)
1993	Endangered Primate Rescue Centre (Cuc Phuong National Park, Vietnam)
1995	Phnom Tamao Zoo (Phnom Penh, Cambodia)
1997	Hanoi Rescue Centre (Hanoi, Vietnam)

Indonesia

1864	Kebun Binatang Ragunan Zoo (Jakarta, Java)
1878	Taman Wisata Satwa Taru Jurug (Surakarta, Java)
1916	Kebun Binatung Surabaya (Surabaya, Java)
1929	Taman Marga Satwa dan Budaya Kinantan (Bukittinggi, Sumatra)
1933	Kebun Binatang Tamansari Bandung (Bandung, Java)
1936	Taman Hewan Pematangsiantar (Pematangsiantar, Sumatra)
1951	Taman Margaraya Tinjomoyo Semarang (Semarang, Java)
1953	Gembira Loka Zoological Garden (Yogyakarta, Java)
1973	Kebun Binatang Medan (Medan, Sumatra)
	Kebun Binatang Taman Ria Aneka Rimba Jambi (Jambi, Sumatra)
1974	Oceanarium Gelanggang Samudra Jaya Ancol (Ancol, Java)
1976	Taman Mini Indonesia Indah (Jakarta, Java)
1986	Taman Safari Indonesia (Jakarta, Java)
1995	Bali Butterfly Park (Bali)
	Bali Reptile Park (Bali)
	Citra Bali Bird Park (Bali)
1997	Indonesia Jaya Reptile and Crocodile Park (Bali)

East Asia

1871/1945	Hong Kong Zoological and Botanical Gardens (Hong Kong)
1906	Beijing [Peking] Zoological Gardens (Beijing, China)
1909	Chang-gyeong Weon Zoological and Botanical Gardens (Seoul, South Korea)
1915	Taipei Zoo (Taipei, Taiwan)
1951–1993	Lai Chi Kok Zoo (Hong Kong)
1953	Shanghai Zoo (Shanghai, China)
1954	Chengdu Zoological Gardens (Chengdu, China)
1956	Guangzhou [Canton] Zoo (Guangzhou, China)
1960	Pyongyang Zoological Gardens (Pyongyang, North Korea)
1977	Ocan Park (Hong Kong)
1979	Edward Youde Aviary (Hong Kong)
1994	Kadoorie Farm and Botanic Garden (Hong Kong)

Zoological Gardens of India

1801–1879	Barrackpore Park (Barrackpore)
1854–1998	Marble Palace Zoo (Calcutta)

1855–1983	People's Park (Chennai, formerly Madras)
1857	Trivandrum Zoo (Thiruvananthapuram)
1863	Sakkarbaug Zoo (Junagadh)
1873	Veermata Jijabhai Bhosle Udyan Zoo (Mumbai, formerly Bombay)
1875	Alipore Zoo (Calcutta)
1877	Ram Nivas Gardens Zoo (Jaipur)
1878	Udaipur Zoo (Udaipur)
1879	Sayyaji Baug Zoo (Baroda)
1882–1994	Shivaganga Gardens Mini Zoo (Tanjore)
1885	Thrissur State Museum and Zoo (Thrissur)
1892	Sri Chamarajendra Zoological Gardens (Mysore)
1893	Maharaja Shivaji Chattrapathy Zoo (Kholapur)
1894	Maharaj Baug Zoo (Nagpur)
1909	Madras Aquarium (Chennai, formerly Madras)
1921	Gandhi Zoological Park (Gwalior)
	Prince of Wales Zoological Park (Lucknow)
1922	Bikaner Zoo (Bikaner)
1936	Jodphur Zoo (Jodphur)
1936–1994	Shikardadi Deer Park (Udaipur)
1951	Kamala Nehru Zoological Garden (Ahmedabad)
1953	Peshwa Park Zoological Garden (Pune)
1956	Sarnath Deer Park (Varanasi)
1958	Assam State Zoo and Botanical Garden (Guwahati)
	National Zoological Park (New Delhi)
	Padmaja Naidu Himalayan Zoological Park (Darjeeling)
1959	Nehru Zoological Park (Hyderabad)
1960	Nandankanan Biological Park (Nandankanan)
1965	V.O.C. Park Mini Zoo (Coimbatore)
1969	Bondla Zoo (Usgaon)
	Sanjay Gandhi National Park (Borivilli Mumbai)
1970	Indira Gandhi Zoological Park (Visakhapatnam)
1971	Chennai [Madras] Snake Park Trust (Guindy)
1972	Maitri Baug Zoo (Bhilai)
1973	Lady Hydari Park and Animal Land (Shillong)
	Sanjay Gandhi Biological Park (Patna)
	Sepahijala Zoological Park (Sepahijala)
1974	Calcutta Snake Park and Zoological Garden (Calcutta)
	Kamala Nehru Prani Sangrahalaya (Indore)
	Kanpur Zoological Park (Kanpur)
1975	Mahavir Harina Vanasthali Deer Park (Vanasthalipuram)
1976	Madras Croc Bank/Centre for Herpetology (Mamallapuram)
	Manipur Zoological Garden (Imphal)
1977	Mahendra Chaudhury Zoological Park (Kholapur)
1978	Indroda Nature Park (Gandhinagar)
1980	Itanagar Zoological Park (Itanagar)
1981	Bellary Children's Park Zoo (Bellary)
	Van Vihar National Park (Bhopal)
1984	Surat Nature Park (Surat)
1985	Arignar Anna Zoological Park (Vandalur)
	Aurangabad Municipal Zoo (Aurangabad)
	Bannerghatta Zoological Garden (Bannerghatta)

1986	Rohtak Zoo (Rohtak)
1987	Bhagwan Birsa Biological Park (Ranchi)
	Rajkot Municipal Zoo (Rajkot)
	Sri Venkateswara Zoological Park (Tirupati)
1989	Jawaharlal Nehru Biological Park (Bokaro)
1990	Himalayan Zoological Park (Gangtok)
1992	Tata Steel Zoological Park (Jamshedpur)
2000	Nilgiri Biosphere Conservation Park (Coimbatore)

Zoological Gardens of Japan[*]

1882	Gifu Park
	Ueno Zoological Gardens (Tokyo)
1903	Kyoto Municipal Zoo (Kyoto)
1912	Hirakata Park
1913	Uozu Aquarium
1915	Osaka Municipal Tennoji Zoo (Osaka)
1919	Kofu Yuki Park Zoo
1924	Takarazuka Zoological & Botanical Gardens
1926	Ayameike Zoo (Nara)
	Komoro Zoo
1927	Marinepia Matsushima Aquarium
1929	Kumamoto Zoological Park
	Ritsurin Park Zoo
1929/1950	Hanshin Park[**]
1930	Shirahama Aquarium
1931	Katsurahama Aquarium
1932	Itozu Zoological Park
1934	Hiyoriyama Park
1935	Oshima Park Zoo
1937	Higashiyama Zoological & Botanical Gardens (Nagoya)
1941	Izu Mito Sea Paradise
1942	Inokashira Park Zoo (Tokyo)
	Omuta Zoo
1948	Beppu Cable Rakutenchi
1950	Hamamatsu Municipal Zoo
	Kochi Zoo
	Odawara Zoo
1951	Himeji Municipal Zoo
	Kobe Oji Zoo (Kobe)
	Nogeyama Zoological Gardens (Yokohama)
	Omachi Alpine Museum
	Sapporo Maruyama Zoo (Sapporo)
1952	Mishima City Park Rakujuen

[*] Zoos and aquariums listed by Japanese Association of Zoological Gardens and Aquariums. Municipal zoos and aquariums have the city listed. Regional authorities or private companies manage the others.

[**] Originally opened 1929, reopened 1950.

1953	Fukuoka Municipal Zoological & Botanical Gardens (Fukuoka)
	Iida Zoo
	Ikeda Zoo
	Kiryugaoka Park Zoo
	Muroran Aquarium
	Saitama Omiya Park Zoo
	Tamano Marine Museum
1954	Enoshima Aquarium
	Kurume City Bird Center
	Toyohashi Municipal Zoo
1955	Toba Aquarium
1956	Okhotsk Aquarium
	Primates Zoo/Japan Monkey Centre
	Shimonoseki Municipal Aquarium
	Takeshima Aquarium
1957	Hitachi City Kamine Zoo
	Misaki Park Zoo & Aquarium
	Tokushima Zoo
1958	Atagawa Banana–Crocodile Gardens
	Kanazawa Zoo
	Noboribetsu Bear Park
	Tama Zoological Park (Tokyo)
1959	Etizen Matusima Aquarium
	Izu Cactus Park
	Nagasaki Aquarium (Nagasaki)
1960	Takaoka Kojo Park Zoo
	Tokuyama Zoo
1961	Biwako Bunkakan
	Sasebo Subtropical Zoological & Botanical Garden
	Ube Tokiwa Park
1962	Suzaka Zoo
1963	Kanazawa Aquarium
	Obihiro Zoo
1964	Awashima Marine Park
	Namegawa Island
	Oita Ecological Aquarium
	Shonai Beach Kamo Aquarium
	Yomiuri Land Marine Aquarium
1965	Futami Seaparadise
	Unesco-Mura Pet Zoo
	Yagiyama Zoological Park (Sendai)
1966	Amakusa Natural Aquarium
	Himeji City Aquarium
	Nagasakibana Parking Garden
	Toyota Kuragaike Park
1967	Asahikawa Asahiyama Zoo
	Oga Aquarium
	Shimoda Floating Aquarium
	Utsunomiya Zoo
1968	Aburatsubo Marine Park Aquarium
	Noshappu Kanryu Aquarium

1969	Kushimoto Marine Park Center
	Shizuoka Municipal Nihondaira Zoo
	Taiji Whales Museum
	Yashima Sea Palace
1970	Kamogawa Sea World
	Marine Science Museum
	Oarai Aquarium
	Shima Marineland
	Wakayama Park Zoo
1971	Asa Zoological Park (Hiroshima)
	Phoenix Natural Zoo
1972	Hirakawa Zoological Park
	Korankei Snake Center
	Okinawa Kodomonokuni Zoo & Aquarium
	Yumemigasaki Zoological Park
1973	Akita Omoriyama Zoo
	Japan Serow Center
	Otaru Aquarium
1974	Inubosaki Marine Park
1975	Ashizuri Kaiyokan
	Kushiro Zoo
	Sabae Nishiyama Park Zoo
1976	Kyushu African Safari
	Okinawa Expo Aquarium
1977	Akiyoshidai Safari Park
	Walking Safari Izu Bio Park
1978	Fukuyama Zoo
	Hamura Zoological Park (Tokyo)
	Nanki Shirahama Adventure World
	Sunshine International Aquarium (Tokyo)
1979	Gunma Safari Park
1980	Fuji Safari Park
	Hiroo Aquarium/Seaside Park Hiroo
	Joetsu City Aquarium
	Minamichita Beach Land
	Nagasaki Bio Park
	Saitama Children's Zoo
1981	Marine Plaza Miyajima
	Tobu Zoological Park
	Uminonakamichi Beach Park Zoo
1982	Hekinan Beach Aquarium
	Kanazawa Zoological Garden
	Notojima Beach Park Aquarium
	Sun Piaza Aquarium
	Wakayama Municipal Natural Science Museum
1983	Aomori Prefectural Asamushi Aquarium
	Nagano Chauasuyama Zoo
	Saitama Municipal Aquarium
	Teradomari Aquarium
1984	Himeji Safari Park
	Toyama Family Park Zoo

1985	Awaji Farm Park Zoo
	Chiba Zoological Park
1986	Izu Andyland
	Sayama Chikosan Park Children's Zoo
1987	Ichikawa Zoological & Botanical Gardens
	Suma Aqualife Park (Kobe)
	Yambaru Wildlife Park
1988	Ehime Prefectural Tobe Zoo
	Sakaigahama Marine Park Aquarium
1989	Morioka Zoological Park
	Tokyo Sealife Park (Tokyo)
1990	Marineworld Uminonakamichi
	Miyazu E. L. Aquarium
1991	Niigata Aquarium
1992	Port of Nagoya Public Aquarium (Nagoya)

Zoological Gardens of Africa

1856	Port Elizabeth Museum (Humewood, South Africa)
1890	East London Zoological Gardens (East London, South Africa)
1891	Giza Zoological Gardens (Cairo, Arab Republic of Egypt)
1899	National Zoological Gardens of South Africa (Pretoria, South Africa)
1901	Khartoum Zoological Gardens (Khartoum, Democratic Republic of the Sudan)
1902–19??	Gizerah Aquarium (Cairo, Egypt)
1904	Johannesburg Zoological Gardens (Parkview, South Africa)
1906–19??	Singa Zoological Garden (Singa, Sudan)
1907	Alexandria Zoological Garden (Alexandria, Arab Republic of Egypt)
1911–19??	Zagazig Zoological Garden (Zagazig, Egypt)
1916–19??	Tanta Zoological Gardens (Tanta, Egypt)
1925	Parc Botanique et Zoologique de Tsimbazaza (Antananarivo, Malagasy Republic)
1929	Maputo Zoo (Maputo, Moçambique)
1931	East London Aquarium (East London, South Africa)
1959	Sea World Durban (Durban, South Africa)
1960	Institute of Primate Research (Nairobi, Kenya)
1962	Nairobi Snake Park (Nairobi, Kenya)
1964	Nairobi Safari Walk (Nairobi, Kenya)
1967	Blantyre Zoo (Blantyre, Malawi)
1969	Parc Zoologique National de Rabat (Temara, Morocco)
1973	World of Birds Wildlife Sanctuary (Hout Bay, South Africa)
1976	Cango Wildlife Ranch (Somerset West, South Africa)
1983	Umgeni River Bird Park (Northway, South Africa)
1984	Gerald Durrell Endemic Wildlife Sanctuary (Mauritius)
	Two Oceans Aquarium (Cape Town, South Africa)

Zoological Gardens of South and Central America

Argentina

1874	Jardin Zoológico Municipal de Buenos Aires (Buenos Aires)
1945	Acuario Municipal (Rosario)
1966	Jardin Zoológico de Bahia Blanca (Bahia Blanca)

Bolivia

1979	Zoo Santa Cruz (Santa Cruz)

Brazil

1895	Jardim Zoológico Museu Paraense Emilio Goeldi (Belem)
1916	Jardim Zoo-Botânico do Dois Irmaos (Recife)
1931	Fundaqão Hermann Weege
1935	Parque Zoológico de Goiania (Goiania)
1938	Horto Zoobotânico de Dois Irmaos (Recife)
	Parque Zoológico Municipal (Manaus)
1945	Acuario Municipal de Santos (Santos)
	Jardim Zoológico (Rio de Janeiro)
	Prefeitura Municipal de Santos (Santos)
1949	Zoológico Municipal Antonio J. M. Andrade (Andradina)
1951	Parque Zoológico de Goiania (Goiania)
1955	Parque Zoobotânico Getulio Vargas (Salvador)
1956	Parque Zoológico Municipal (Ondina)
1958	Fundaqão Parque Zoológico de S.P. (São Paulo)
1959	Cebus Zoológico da Usipa (Ipatinga)
1960	Jardim Zoológico (Brasilia)
1962	Fundaqão Zoobotânica do Rio Grande do Sul (Sapucaia do Sul)
1963	Zoológico de Uberaba — Parque Jacaranda (Uberaba)
1964	Clube de Caza d Pesca Alvorada (Ipatinga)
	Zoológico de Centro de Instrucao de Guerra na Selva (Manaus)
1965	Parque Zoológico Dr. Mario Frota (Varginia)
1966	Jardim da Margem (Sorocaba)
1967	Zoológico Municipal de Amparo (Amparo)
1968	Parque Zoológico Municipal de Limeira (Limeira)
	Parque Zoológico Municipal Quinzinho de Barros (Sorocaba)
1970	Zoo Municipal do Leme — Parque Municipal Mourao (Leme)
1971	Parque do Inga (Maringa)
1972	Simba Safari LTDA S/C (São Paulo)
	Zoológico Municipal de Piracicaba (Piracicaba)
1973	Parque Zoológico (Teresina)
	Parque Zoológico de Londrina (Londrina)
	Zoológico Municipal de São Jose do Rio Preto (S. Jose do Rio Preto)
1974	Parque Zoobotânico do Piaui (Piaui)
1976	Parque Ecológico de São Carlos (São Carlos)
	Parque Zoológico Municipal de Paulinia (Paulinia)
	Zoológico Hotel Tropical (Manaus)
1977	Zoológico da Universidade Federal de Mato Grosso (Mato Grosso)
1978	Jardim Zoológico Municipal (Alta Floresta)
	Parque Municipal Danilo Galafassi (Cascavel)
	Zoológico Ilha de São Pedro (S. Jose do Rio Pardo)
1979	Zoológico de Matelandia (Matelandia)
1980	Parque das Hortensias Zoológico Municipal (Taboa da Serra)
	Parque Zoológico Municipal de Bauru (Bauru)
1981	Santur Parque Balneário Camboriu (Balneário Camboriu)
	Zoológico Municipal de Volta Redonda (Volta Redonda)

1982	Bosque/Zoológico Municipal de Pedreira (Pedreira)
	Parque Zoológico Municipal Guarulhos (Guarulhos)
	Zoológico de Cubatao (Cubatao)
	Zoológico de Curitiba (Curitiba)
	Zoológico Parque do Sabia (Uberlandia)
1983	Zoológico Municipal de Andradas (Andradas)
	Zoológico Municipal de Lins (Lins)
1984	Parque Balneário Turístico Oasis (Santa Maria)
	Parque Ecológico Municipal de Americana (Americana)
1985	Fundaqão Rio Zoo (Rio de Janeiro)
	Parque Governador Jose Rolemberg Leite (Aracaju-SE)
	Parque Natural da CBMM (Araxa)
	Parque Zoobotânico de Carajas (Carajas)
	Parque Zoológico Municipal Chico Mendes (S. Bernardo do Campo)
1986	Parque Natural de Carajas (Para)
1988	Zoológico de Dourado (Dourado)
1989	Zoológico Municipal de Mogi Mirim (Mogi Mirim)
1990	Bosque Municipal Nestor Bologna (Vargem Grande do Sul)
	Mini Zoo de Sete Lagoas (Sete Lagoas)
	Parque Ecológico Municipal Boituva (Boituva)
	Parque Municipal do Mocambo (Patos Minas)
1991	Zoológico de Araraquara (Araraquara)
	Zoológico de Belo Horizonte (Belo Horizonte)
1992	Centro de Educaqão Ambiental Francisco Mendes
	Parque Ecológico Zoobotânico Brusque
	Zoológico do Litoral — Contaregis (Osorio)
	Zoológico Municipal de Santa Barbara D'Oeste (Sta. Barbara D'Oeste)
1993	Zoológico Paraiso das Aves (Jarinu)

Chile

1924/1925	Jardim Zoológico Nacional (Santiago de Chile)
1959	Jardin Zoológico de Chillan (Chillan Viejo)

Colombia

1965	Jardin Zoológico de Sociedad de Mejoras Publicas (Barranquilla)

Guyana

1947	Guyana Zoological Park (Georgetown)

Peru

1968	Parque de Las Leyendas (Lima)

Venezuela

1966?	Parque del Este (Caracas)
1973	Jardin Zoológico Maracaibo (Maracaibo)
1976	Parque Sur Maracaibo (Maracaibo)
1977	Parque Zoológico Caricuao (Caracas)

Central America

1883	Parque Zoológico Nacional (San Salvador, El Salvador)
1892	Zoológico Nacional La Aurora (Guatemala City, Guatemala)
1916	Zoo Nacional Simon Bolivar (San Jose, Costa Rica)
1926	Parque Zoológico de Chapultepec (Mexico City, Mexico)
1958	Parque Zoológico Cdte Edgar Lang Sacasa (Managua, Nicaragua)
1982	Belize Zoo (Belize City, Belize)
1985	Zoo Ivan Montenegro Baez (Tegucigalpa, Honduras)
	Centro de Producion de Animales en Peligro de Extinction (Panama City, Panama)

Index

A

AAZPA (American Association of Zoological Parks and Aquariums), 168, 171, 173–174
Abderrahman III, 21
Abrey, Alan, 345
Acclimatisation Society of Victoria, Australia, 182, 184
Acclimatization programs
    Australia, 182, 184
    colonial times, 36–37
    England, 36
    France, 36, 90
    Hungary, 142
    Prague, 141
    United States, 154
Adams, Martha, 147
Adamson, Joy, 202
Adelaide Zoo, Australia
    breeding programs, 208
    educational role, 205, 206
    enclosure design, 187
    establishment, 184, 186
    modernization period, 203
    recreational role, 210
    reptile collection, 192
    rhinoceros in, 191
    slow period, 200
    visitors to, 197, 198
Affenberg Salem monkey house, 110
Afghanistan leopard, 218
Afghanistan zoological gardens, 217–218
African zoological gardens
    Egypt, *see* Egyptian zoological gardens
    Kenya, 342
    other collections, 346–348
    Republic of South Africa
        aquariums, 346
        bird sanctuary, 345
        breeding programs, 344
        cheetah collection, 345–346
        Durban aquarium, 345
        early collections, 342–343
        East London Zoo, 343
        Johannesburg Zoo, 344–345
        National Zoological Garden, 343–344
        Port Elizabeth Museum, 343
        Transvaal zoo establishment, 343

Agri-Horticultural Society, Singapore, 232
Ahmad, M.F., 225
Ahmedabad Zoo, India, 278–279
Ahungalla Zoo, Sri Lanka, 228
Ain-e-Akbari, 254, 255
AIPE (American Institute of Park Executives), 165
Akbar (naturalist), 254, 255
Akbar Nama, 254
Akkad, 8
Al Ain Zoo and Aquarium, United Arab Emirates, 223
Albuquerque, Dalcy, 360
Aldrovandi, Ulisse, 76
Alexander the Great, 18
Alexander VI, Pope, 29
Alexandria Zoological Garden, Egypt, 340–341
Alfred Brehm House, Germany, 109
Ali Mahomed, 226
Ali Murad, 226
Ali Sadikin, 237
Allwetterzoo Münster, 108
Alma-Ata zoo, Soviet Union, 139
Alpenzoo Innsbruck, Austria, 109
American Association of Zoological Parks and Aquariums (AAZPA), 168, 171, 173–174
American Association of Zoo Veterinarians, 171
American bison, 144, 166
American Bison Society, 166
American Institute of Park Executives (AIPE), 165
American Zoo and Aquarium Association (AZA), 174
Ameshoff, Arnoldus, 82
Amphitheatrales magistri, 39
Amsterdam Zoological Society, 91
Anaconda, 55
Ancient animal collections, *see also* Menageries
    appeal of exotic collections, 1–2
    Arabian countries, 21
    Asian
        attitudes toward animals, 16
        types of collections, 16–17
    Aztec
        arts and sciences, 25
        gardens, 25–26
        types of collections, 26
    collection of wild animals, 7–8
    domestication of wild animals, 5–7

Egyptian
- animal husbandry, 13
- animal-oriented sports, 14
- garden development, 13
- Punt expeditions, 15
- religion and sciences, 13
- society history, 12–13
- types of animals, 14–15

environmental knowledge of animals
- folk systematics, 3–5
- historical periods, 2, 3

Greco-Roman
- attitudes toward nature, 17–18
- capture techniques, 20
- exotic animals, 18–19
- keeping of animals, 18
- types of collections, 20
- uses of animals, 19

Incan, 26–27

Mayan, 27

medieval period
- Chinese royal collections, 24–25
- European royal collections, 22–24

Mesopotamian
- garden development, 9
- habitat re-creation, 10, 12
- luxury item trade, 9
- religion and sciences, 8–9
- royal collections, 10, 11
- servant/animal keepers, 12
- society history, 8

Persian, 21

Ancol Oceanarium, Java, 238
Anderson, John, 263, 264
Andhra Pradesh Forest Department, India, 281
Animal Grottos, Germany, 105
Animal husbandry
- advances in capture techniques, 172
- in Australia, 209
- beginnings of, 6
- in China, 242
- in Egypt, 13
- feeding of animals in England, 53

Animal trading
- in Australia, 188–189
- for colonial menageries, 35–36
- in India, 271, 280
- in Japan, 312–313
- post-medieval Europe, 82–83
- in Taiwan, 240
- through Hong Kong, 239
- in Western Europe, 81, 82

Animal Welfare Act of 1966, United States, 173
Ankara Zoo, Turkey, 222
Anomadassi, Aluthnuwara, 228

Anschütz, O., 124
Antelope in collections
- Giza Zoo, Egypt, 336
- Jardin des Plantes, France, 89
- Knowsley Hall, England, 60
- Tower of London menagerie, 54

Antwerp Zoo, 91, 92–94, 100
Apenheul monkey house, 109
Appendix, 369–390
Aquariums
- Amsterdam, 92
- Australia, 198
- background, 40–41
- Brunei, 229
- chronological listing, 369–390
- definition, 40
- Egypt, 339–340
- etymology, 38–40
- first public, 70, 71
- first walk-through enclosure, 98
- India, 267–268
- Israel, 219
- Japan, 298, 303–304
- Java, 238
- Pakistan, 226
- Russia, 135
- Singapore, 233
- South Africa, 345, 346
- United Arab Emirates, 223
- United States, 19th century, 155, 157, 164
- United States, 20th century, 168, 175–176

Arabian countries, 21, 32; *see also* Southwest Asian zoological gardens
Arabian oryx, 216–217, 221, 222; *see also* Père David's deer in collections
ARAZPA (Australasian Regional Association of Zoological Parks and Aquaria), 204
Arbel (French traveler), 273
Architecture, *see also* Enclosure design
- in Australian zoos, 187
- Barrackpore Menagerie, India, 259
- Egyptian style, 93
- exotic designs, 100–101
- Incan gardens, 27
- Liverpool Zoo, England, 63–64
- Panorama design, 103–104
- *paradeisos* in Persia, 21
- Sayyaji Baug Zoo, India, 273
- Schönbrunn Zoo, Austria, 84–85
- systematic approach to Berlin Zoo, 95
- Versailles menagerie, 82
- Western European zoos, 97–99, 100–101

Argentina zoological gardens, 362–363
Arignar Anna Zoological Park, India, 268–269
Aristotle, 4, 18

Arizona-Sonora Desert Museum, 172
Artis, 91–92
On the Art of Hunting with Birds, 77
Ashari, D., 228
Ashurbanipal, 11, 12
Ashurnasirpal II, 11
Asian ancient animal collections
    attitudes toward animals, 16
    types of collections, 16–17
Asian zoological gardens, see also East Asian zoological gardens; South Asian zoological gardens; Southeast Asian zoological gardens
Askaniya Nova collection, Russia
    establishment, 134–135
    nationalization of, 137–138
ASMP (Australasian Species Management Plan), 204
Asoka, 253
Assam State Zoo, India, 279
Association of Indian Zoo Directors, 287
Association of Zoo Directors of Australia and New Zealand (AZDANZ), 204
ASZK (Australian Society of Zoo Keepers), 208
Atkins (animal keeper), 62
Auckland, Lord, 260
Audubon, John James, 156
Aurochs, 118
Australasian Regional Association of Zoological Parks and Aquaria (ARAZPA), 204
Australasian Species Management Plan (ASMP), 204
Australian Society of Zoo Keepers (ASZK), 208
Australian zoological gardens
    background, 181–182
    foundation and development
        acclimatization movement, 182, 183–184
        animal acquisition and transportation difficulties, 190
        animal acquisition purposes, 182–183
        animal exchanges, 188–189
        aviaries, 192
        conservation efforts, 195–196
        educational role, 193
        enclosure design, 186–188
        establishment of public zoos, 184–185
        limited research efforts, 196–197
        native animal use, 191–192
        popularity of exhibits, 190–191
        recreational role, 193–195
        reptile collections, 192
        sources of animals, 189–190
        visitors, 197–198
    mid-twentieth century
        additions to Taronga Zoo, 200–201
        animal collections, 198–200
        aquariums, 198
        conservation efforts, 201–202
        Depression effects, 199
        educational role, 201
        enclosure design, 198
        recreational role, 201
        WWII effects, 200
    modernization period
        breeding programs, 207–208
        conservation efforts, 204, 206–207, 208
        educational role, 205–206
        individual zoo programs, 202–204
        recreational role, 210
        research center role, 209
        training programs, 208–209
        visitors, 211
    objectives of zoos, 181
Australian Zoologist, 196
Austrian zoological gardens
    damage from wars, 86, 87
    design, 84–85
    establishment, 84, 141
    popularity of, 86
    pre-WWII improvements, 87
    sources of animals, 85
    transformation to a zoo, 86
    types of specimens, 85
    wartime effects on, 106
Autuori, Mario, 357
Aviaquario Municipal, Brazil, 353
Aviaries, see also Birds in collections
    in Australia, 192
    books about, 39
    in England, 64
    in Hong Kong, 239
    Loro Parque, Germany, 109
    in Malaysia, 231
    in Taiwan, 240
Aviarium/aviaria, 39
Aviary, 39
AZA (American Zoo and Aquarium Association), 174
AZDANZ (Association of Zoo Directors of Australia and New Zealand), 204
Aztec civilization animal collections
    arts and sciences, 25
    gardens, 25–26
    types of collections, 26

B

Babolsar Zoo, Iran, 219
Baboons in collections, 341
Babylon-Nineveh hanging gardens, 11
Bacon, Francis, 30
Bactrian wapiti, 218

Bahadur Rana, Judha Sumser Janga, 224
Bailey, George, 150
Baker, Frank, 160
Balfour, Edward, 265
Bandung Zoo, Java, 238
Bangkok Reptile Grove, Thailand, 234
Bangladesh zoological gardens, 223–224
Barbar, 255
Bärengraben (Bern bear pit), Germany, 79
Barnum, Phineas T., 155
Barrackpore Menagerie, India
    decline in status, 260–261
    establishment of, 257–258
    financial situation, 259
    scientific focus, 258–259
    types of animals, 259–260
Barros, Silvio, 353, 354
Bartlett, Abraham D., 68, 71
Basham, A.L., 252
Batang Duri Mini Zoo, Brunei, 229
Bayely, John, 51
Bears in collections
    in Australia, 187
    in Poland, 119
    in Russia, 133
    Tower menagerie, 53–54
    in Western Europe, 79–80
Beebe, William C., 336, 340
Beijing Zoo, China, 243–244
Belle Vue Zoo, England, 65–66
Benchley, Belle J., 168, 170
Benedict XII, Pope, 23
Bennett, E.T., 51
Bentinck, Lord, 260
Berlin Tiergarten, 76, 77, 78
Berlin Zoo, Germany
    breeding programs, 95
    design, 87–8, 100
    effect of wars on, 104
    establishment, 87–88, 94
    financial support for, 94–95
    improvements and successes, 95
    post-WWI specimen numbers, 106
    specimen additions, 88
    systematic approach use, 95, 101–102
    WWII destruction, 107
Bern bear pit (Bärengraben), Germany, 79
Bernier, François, 256
Bern Zoo, Belgium, 108
Best, Robin, 353
Bestiarii, 39
Bharals (blue sheep), 90
Bhutan zoological gardens, 224
Bible and explanation of biodiversity, 31–32
Bigalke, Rudolph, 344

Bin Abdul Aziz, Saud, 216
Bin Abdullah, Mohd Louis, 229
Bin Abdullah al-Sabah, Shaikh Jaber, 216
Bin Said, Qaboos, 221
Biological classification systems, 4–5
Biopark definition, 40
Birds in collections, *see also* Aviaries
    Australian importation of, 183, 192
    Rotterdam Zoo, 102
    in South Africa, 345
    specialty parks, 109
    Tower of London menagerie, 54–55
Birds International, Philippines, 232
The Birds of Middle Europe, 98
Bison in collections
    in Australia, 200
    in Bulgaria, 144
    conservation efforts in America, 166
    European, 104, 120, 132, 143
    live animal shipments, 35
Blackburn, William, 160
Blair, W. Reid, 169
Blaszkiewitz, Bernhard, 80, 109
Blauwe Jan, 82
Blijdorp Zoo, 104–105
Blue sheep (bharals), 90
Bodinus, Heinrich, 94–95
Bogelpark Walsrode, Germany, 109
Bolivia zoological gardens, 363–364
Bombay Natural History Society, India, 271
Bombay Wildlife Act, 1951, 285
Bombay Zoo, India, 270–271
Boone and Crockett Club, United States, 162
Bornean orangutan, 81
Boston Aquarial and Zoological Gardens, United States, 157
Botanic and Zoological Gardens of the Society for the Protection of Nature in Israel, 220
Boulenger, G.A., 339–340
Bourez, Albert, 107
Boy's Town Hendela, Sri Lanka, 228
Boyle, Robert, 32
Brachetka, Julius, 87, 107
Brand, Frank, 344
Brazilian zoological gardens
    in Center-west, 355–356
    educational efforts, 353, 362
    historical aspects, 351–353
    in the North
        Carajas zoo, 353
        Manaus zoo, 353
        Rio Branco zoo, 354–355
    in Northeast, 355
    private collections, 357
    the South, 358

the Southeast
    prevalence of, 356
    in Rio de Janeiro, 356
    in São Paulo, 357
    in Sorocaba, 356–357
  status of, 358
Breeding programs
  Arabian oryx, 221
  in Australian zoos, 207–208
  Berlin Zoo, 95
  Budapest Zoo, Hungary, 143
  in China, 244
  in Indian zoos, 278–279, 280, 282, 283–284
  in Japan
    amphibians, 317–318
    animal transfers, 320–322
    conservation efforts, 319–320
    examples of successes, 318–319
    giraffes, 318
    serows, 317
    SSCJ challenges, 322
  in Java, 238
  in Jordan, 221
  Kabul Zoo, Afghanistan, 218
  Leningrad Zoo, 138
  London Zoo, England, 68–72
  Moscow Zoopark, 138–139
  in Philippines, 232
  pigmy hog, 279
  polar bears, 142, 322
  Poznan Zoo, Poland, 132
  in Saudi Arabia, 222
  Sofia Zoo, Bulgaria, 145
  in South Africa, 344
  in United Arab Emirates, 223
  in Vietnam, 235–236
  Warsaw Zoo, Poland, 121
  in Western Europe, 110
  writings about, 265–266
Brehm, Alfred, 154
Brehm, Wolf, 109
Brehm Foundation, 109
Breslauer Zoologischer Garten, Poland, 123–124, 126
Brisbin, James, 155
Bristol Zoological Garden, England, 66–67
British colonies, 36
Bronx Zoo, New York, United States, 163
Brooke, Charles, 230, 231
Brookfield Zoo, Chicago, United States, 169
Brown, Allen, 269
Brunei zoological gardens, 229
Buchanan, Francis, 259
Budapest Zoo, Hungary, 143
Buddha, 253
Buddhist Theyboo Monastery, Myanmar, 231

Buffalo Zoological Gardens, New York, United States, 157
Buffon (Jardin director), 89
Bulart, O.J., 223
Bulgaria, 143–145
Burchell, William, 53
Burckhardt, Jacob, 30
Burdzinski, Wenanty, 121, 135
Bureau of Fisheries Aquarium, United States, 157
Burger Park, Germany, 110
Burma, *see* Myanmar
Burnes (traveler), 254
Bustard, 195, 222, 336
Butcher, Alfred Dunbavin, 202, 207, 211
Butterfly House, Australia, 207
Bydgoszcz Zoo, Poland, 129

C

Cabinet collections, 30, 31
Calcutta Zoological Gardens, India
  establishment, 262, 263–264
  popularity, 264–265
  proposal for, 263
  sources of animals, 35–36
Camac, William, 151
Cambodia zoological gardens, 229
Camels in collections
  in Poland, 118, 119
  for rides in Australian zoos, 194
Cameron (zoo director), 274
Canadian Association of Zoological Parks and Aquariums (CAZPA), 173
Cango Wildlife Ranch, South Africa, 345
Canton Zoo, China, 244
Captive Breeding Centre, Philippines, 232
Captive breeding programs, *see also* Breeding programs
  Berlin Zoo, 95
  Philippines, 232
Captive wildlife management writings, 174–176; *see also* Literature and writings on zoos
Captivity studies, 175
Caracal in collections, 53
Care of animals, *see also* Enclosure design; Veterinarians; Wartime effects on zoos
  Artis, Amsterdam, 92
  attitudes toward animals in Yemen, 223
  in Australia, 191, 192
  Berlin Zoo, Germany, 95, 100
  in Brazil, 354
  capture techniques in Rome, 20
  Central Park Zoo, United States, 154
  domestication, 5–7
  habitat re-creation in Mesopotamia, 10, 12

live animal shipments, 32, 35, 60, 85, 190
London Zoo, England, 68, 70–71
Schönbrunn Zoo, Austria, 85–86, 87
in Sri Lanka, 228
Tower of London menagerie, 50, 51–54
traveling menageries, 83
at Windsor Great Park, England, 57–58
Carolina parakeet, 155
Carr-Harley (animal dealer), 220
CAZPA (Canadian Association of Zoological Parks and Aquariums), 173
Ceby City Zoo, Philippines, 232
Center for Tropical Conservation Studies, Philippines, 232
Central-Eastern Europe zoological gardens
   Bulgaria, 143–145
   Czech and Slovak Republics
      private royal collections, 140–141
      public zoo establishments, 141–142
   general comments, 145
   Hungary, 142–143
   Poland, *see* Polish zoological gardens
   Russia and Soviet Union, *see* Russian zoological gardens
Central Park Zoo, New York, United States, 154
Central Station Aquarium, United States, 157
Central Zoo, Nepal, 224
Centre for Ecological Zoology, Israel, 219
Ceylon, *see* Sri Lanka
Chamarajendra Zoological Gardens, Mysore, India, 274–275
Chandragupta, 253
Chandran, P.R., 269
Chang-gyeong Weon Zoological and Botanical Gardens, Korea, 240
Charkov Zoo, Russia, 135
Charlemagne, 22
Charles V, 22
Chatterjee, Dinabandhu, 261
Cheetahs in collections
   in India, 255, 259
   in South Africa, 345–346
   Tower of London menagerie, 53
Chengdu Zoological Gardens, China, 243, 244
Chester Zoological Garden, England, 68
Chhatbir Zoo, India, 280
Chiang Mai Zoo, Thailand, 234
Chimpanzees in collections, *see also* Gorillas in collections; Monkeys in collections
   Australian zoos, 190–191, 198
   Japanese zoos, 307, 308, 313–314
   menageries in Western Europe, 81, 84
Chinese Association of Zoological Gardens, 242
Chinese Zoo Association, 241
Chinese zoological gardens
   ancient animal collections, 16–17
   background, 241–244
   breeding programs, 244
   etymology of animal collections, 38
   royal animal collections, 24–25
Christie, Sarah, 289
The Chronicles of Matthew Paris, 50
Chronological listing of zoos, 369–390
Cicero, Johann, 77
Cikini Zoo, Jakarta, 237
Cincinnati Zoo, Ohio, United States, 154–155, 157
CITES (Convention on International Trade in Endangered Species of Wild Fauna and Flora), 132, 140, 173, 311
Civil Works Administration, United States, 169
Clapperton, Captain, 61
Clark Zoo, Philippines, 232
Classification systems, 4–5, 9
Cleveland Zoo, Ohio, United States, 158
Cohen, Pinchas, 219
The Cologne Zoological Garden, 100
Colombia zoological gardens, 364
Colombo Zoo, Sri Lanka, 227
Colonial menageries
   acclimatization programs, 36–37
   animal transactions, 35–36
   collecting stations, 34–35
   European interest in new species, 33–34
   live animal shipments, 34–35
Conklin, William, 154
Conservation and research efforts of zoos
   after WWI, 104
   American bison, 166
   Arabian oryx, 216–217
   in Australia, 195–197, 201–202, 204, 206–207, 208, 209
   in China, 244
   conservation park concept, 40
   egrets, 338
   European bison, 104, 132
   Giant Panda, 242–243
   Giza Zoo, Egypt, 338
   in India, 253, 258–259, 270, 278–279, 280–281, 284–286
   in Japan, 322–323
   in Pakistan, 226
   Père David's deer, 241–242
   in Philippines, 232
   in Poland, 131–132
   in Saudi Arabia, 222
   in Sri Lanka, 227
   Sumatra rehabilitation centers, 239
   in Taiwan, 240
   in Thailand, 235

Index 397

in United States, 164, 166, 173–174, 176
in Vietnam, 235, 236
Convention on International Trade in Endangered Species of Wild Fauna and Flora (CITES), 132, 140, 173, 311
Cops, Alfred, 51–52, 56
Cops, James, 56
Corbett, Jim, 284
Cortes, Hernando, 26
Crawford Market, Bombay, India, 271
Crocodiles in collections
    Indian Crocodile Project, 283
    in Pakistan, 226
    in Thailand, 234
Cross, Edward, 58, 59, 62
Crowcroft, Peter, 210
Crowninshield, Jacob, 148, 149
Curzon, Lord, 258
Cutting, James, 157
Cuvier, Georges, 90
Cyprus mouflon, 218
Cyprus zoological gardens, 218
Czech and Slovak Republics
    private royal collections, 140–141
    public zoo establishments, 141–142

D

D'Oyly, Charles, 261
Dacca Zoological Gardens, Bangladesh, 224
Dankworth, Margaret A., 169, 171
Dathe, Heinrich, 109
David, Armand, 241
David, Reuben, 278
de Alwis, Lyn, 227, 233
Deer moats in Western Europe, 79
de Fabeck (zoo builder), 271
Dehiwala Zoo, Sri Lanka, 227
Delacour, Jean, 78
de la Vega, Garcilaso, 27
Delgeur, Lodewijk H., 93
Denver Zoo, Colorado, 160
Department of Conservation and Land Management, Australia, 204
Department of Tropical Research, United States, 164
Der Zoologische Garten zu Coeln, 100
Desai, J.H., 287
de Tournefort, Joseph P., 4
Deutsch, Ladislau, 354, 357
Diaz del Castillo, Bernal, 26
Die Praxis der Naturgeschichte, 100
Die Vögel Mitteleuropas, 98
Dioscorides, 4
Disease in Captive Wild Mammals and Birds, 164
Dogs in collections, 53

Doi Suthep National Park, Thailand, 234
Dolphinariums, 110
Domestication of wild animals, 1, 5–7
Dom Manuel I, 23, 29
Dorcas gazelle, 222
Dormal, Julio, 362
Draak, Rijndert, 91
Dresden Zoological Garden, Germany, 109
Dubai Zoological Gardens and Aquarium, United Arab Emirates, 223
Dudley Zoological Garden, England, 68
Dugong in collections, 315
Duisburg dolphinarium, 110
Durban Centenary Aquarium, South Africa, 345
Durrell, Gerald, 218
Dusit Zoo, Thailand, 234
Düsseldorf Zoo, 105
Dvur Kralove zoo, Czech Republic, 142

E

East Asian zoological gardens
    China, 241–242
    Hong Kong, 239
    Korea, 240
    Macau, 240
    Taiwan, 240–241
East London Zoo, South Africa, 343
EAZA (European Association of Zoos and Aquaria), 110
Echidna in collections, 120
Ecosystems Research and Development Bureau Centre, Philippines, 232
Edinburgh Zoological Garden, 66
Educational efforts of zoos
    Australian zoos, 193, 201, 205–206
    in Brazil, 353, 362
    Calcutta Zoo, 264
    Giza Zoo, Egypt, 337, 338
    in Japan, 311–312, 316–317
    Jardin des Plantes, France, 89
    in Java, 237
    in Poland, 130–131
    primary aims of zoos, 100
    Sofia Zoo, Bulgaria, 145
    in Soviet Union, 137
    in Sri Lanka, 227
    in United States, 158, 169, 172–173
    Western European zoos, 104–105
Edwards, Arthur M., 41
Edwards, John C., 252
Edward VII, 58
Edward Youde Aviary, Hong Kong, 239
EEP (European Endangered Species Program), 110
Eggenschwyler, Urs, 103

Egrets conservation, 338
Egyptian zoological gardens
    Alexandria Zoo, 340–341
    ancient animal collections
        animal husbandry, 13
        animal-oriented sports, 14
        garden development, 13
        Punt expeditions, 15
        religion and sciences, 13
        society history, 12–13
        types of collections, 14–15
    etymology of animal collections, 38
    Giza Zoo
        conservation efforts, 338
        current status, 338
        educational role, 337, 338
        establishment, 331–333
        exchange programs, 336
        exhibits and buildings, 335–336
        objectives of, 335
        types of animals, 335, 336–337
    Gizerah Aquarium, 339–340
    Khartoum Zoo, 340
    Merowe Zoo, 341–342
    Port Sudan Zoo, 342
    Singa Zoo, 340
    Tanta Zoo, 341
    Zagazig Zoo, 341
Eilat Aquarium, Israel, 219
Elephants in collections
    in ancient Rome, 19
    Antwerp Zoo, 93
    Berlin Zoo, 106
    Budapest Zoo, 143
    in Japan, 300, 305
    Jardin des Plantes, France, 89
    Liverpool Zoo, England, 62
    Manchester Zoo, England, 65
    medieval period, 23, 29
    in Poland, 118, 120, 121
    for rides in Australian zoos, 190, 194
    in Russia, 133
    Schönbrunn Zoo, Austria, 85
    in Siam/Thailand, 234
    in Sri Lanka, 228
    St. Petersburg Zoo, Russia, 134
    Tower of London menagerie, 50, 54
    in United States, 148–149
Eliat Zoo, Israel, 219
Elitch, Mary, 170
Elitch Amusement Gardens, United States, 170
Elliott, John, 30
Emanuel I, 81
Enclosure design, *see also* Architecture
    in ancient China, 16–17
    in ancient Rome, 19
    Antwerp Zoo, 89–90
    Australian zoos, 186–188, 198
    in Aztec collections, 26
    barless concept, 103–104, 164
    Berlin Zoo, Germany, 87–88, 100
    bird cages in ancient Egypt, 14
    bird cages in Medieval period, 22
    Central Park Zoo, United States, 154
    developments in, 188
    Dudley Zoo, England, 68
    in Exeter 'Change, England, 58
    flooded pits in England, 63
    in Incan collections, 27
    lion cages in Mesopotamia, 10
    modern Western Europe, 110–111
    Panorama design, 103–104
    pits and moats in Western Europe, 79–80
    Schönbrunn Zoo, Austria, 86, 87
    in United States, 164
    Versailles, France, 82
    Western European zoos, 104–105
    Whipsnade Zoo, England, 73
    writings about, 265
Endangered Species Act of 1973, United States, 173
English zoological gardens
    Belle Vue Zoo, 65–66
    Bristol Zoo, 66–67
    Chester Zoo, 68
    Dudley Zoo, 68
    Edinburgh Zoo, 66
    estate collections, 56–57
    Exeter 'Change, 58–59
    first permanent collection, 49
    Hull Zoo, 67
    impetus behind zoos, 90–91
    Jersey Zoo, 73
    Knowsley Hall, 60
    Liverpool Zoo
        closing of, 65
        types of animals, 62–64
    London Zoo
        bear pits in, 80
        breeding and housing successes, 68–72
        financial status, 72
        first public aquarium, 70, 71
        Windsor collection transferred to, 58
    Manchester Zoo, 65
    Paignton Zoo, 67–68
    popularity of exotic specimens, 49
    Preston Zoo, 67
    Rotherham Museum, 61
    Stubton Hall, 60
    Surrey Zoo, 59
        sale of collection, 62

Index

types of animals, 61–62
Tower menagerie
  care of animals, 51–52
  establishment, 50
  specimen descriptions, 52–55
  transfer of collection, 55–56
  types of animals, 50–51
Wentworth collection, 61
Whipsnade Zoo, 72–73
Windsor Great Park, 57–58
Eriksen, Andrew and Glenn, 345
Erkenbrecher, Andrew, 154
Esarhaddon, 11, 12
Ethology, 96–99
Etymology, 38–40
Europe, see also Western European zoological gardens
  etymology of animal collections, 39
  interest in new species, 33–34
  menageries
    biodiversity explanations, 30–32
    cabinet collections, 30, 31
    origins of animals, 28, 30
    shipments of animals, 32–33
  royal animal collections, 22–24
European Association of Zoos and Aquaria (EAZA), 110
European bison
  Budapest Zoo, 143
  conservation of, 104, 120, 132
European Endangered Species Program (EEP), 110
Evolution of zoos
  conservation park concept, 39
  etymology, 38–40
  proposal for zoological study, 37–38, 91
  shift from private to public funding, 37
Exeter 'Change, England, 58–59

F

Falconry, 22, 77
Falz-Fein, Friedrich, 135
Far East Animal Farm, Taiwan, 240
Farrell, H.P., 225
Fayrer, Joseph, 263
Faz'l, Abul, 255
Federal Emergency Relief Administration, 169
Feng-Huang-Ku Bird Park, Taiwan, 240
Ferdinand I, 140
Ferdinand II, 81
Fiedler, A., 120
Fiedler, Walter, 87
Field Naturalist's Club, Australia, 195
Fintelmann, Gustav Adolph, 87, 88
Fish House, London Zoo, England, 70

Fitzinger, Leopold Joseph, 76
FitzPatrick, Percy, 344–345
FitzSimons, F.W., 342–343
Fitzwilliam, Earl, 61
Fleay, David, 199, 200
Flower, Stanley S.
  biography, 332–334
  trips to India, 254, 267, 269, 271, 272, 273, 274, 275, 276–277
  trips to Pakistan, 225, 226
  work at Giza zoo, 334–335, 337
  writings about Egyptian zoos, 333
Flower, Vincent, 335
Flower, William, 36, 334
Folk systematics, 3–5
Forbes, Rosita, 256
Fox, Herbert, 164
France
  acclimatization programs, 36
  Jardin des Plantes, 89–90
  menageries in, 82
Frankfurt-on-Main Zoo, Germany, 110
Franz I, 84, 85
Franz II, 85
Franz Josef, 86
Frederick I, 77, 78
Frederick II, 22, 77, 80
Frederick William I, 77, 78
Frederick William III, 87, 88
Frederick William IV, 88
French zoological gardens
  background, 89
  bear pits in, 80
  Parc Zoologique de Paris, 90
  scientific activities, 90
  sources of animals, 89–90
  Versailles, 82, 89
  wartime effects on, 106
From Amoeba to Gorilla, 125
Fugger, Hans, 83
The Fundamental of Zoological Garden Work, 143
Fundaqão Parque Zoológico de São Paulo, Brazil, 357
Fundaqão Zoobotânico do Rio Grande do Sul, Brazil, 358

G

Gaekwad, Sei Fatesingh Rao, 256
Gaekward, Fatesingh Rao P., 273
Gaekwar of Baroda, 272
Galstaun, Benjamin, 237
Gampaha Snake Farm, Sri Lanka, 228
Gangetic dolphins, 224
Gazelles in collections, 222
Gdansk Zoo, Poland, 129

Gebbing, Johannes, 104
Gebhardt, Sophia and Jules, 134
Gelder, Duke of, 81
Gembira Loka Zoological Garden, Java, 238
George, Wilma, 32
George IV, 57
Gerenday, Jozsef, 142
Germain, Louis A., 235
German zoological gardens, *see also* Western European zoological gardens
    Berlin Zoo
        breeding programs, 95
        design, 100
        effect of wars on, 104
        establishment and design, 87–88
        establishment of, 94
        financial support for, 94–95
        improvements and successes, 95
        post-WWI specimen numbers, 106
        specimen additions, 88
        systematic approach, 95
        systematic approach use, 101–102
        WWII destruction, 107
    menageries in, 82
    post-WWII zoos, 109
    Tierpark Berlin-Friedrichsfelde, 80, 109
    Tierpark Dählhölzi, 79
    Tierpark Hellabrunn, 105
    Tierpark Rheine, 110
Gesner, Conrad, 76
Gewalt, Wolfgang, 110
Ghana, Africa, 347
Giacomini, Claudio, 358
Gibbon Rehabilitation Projects, Thailand, 235
Gibbons, William, 259
Gilgamesh epic, 9
Giraffes in collections
    ancient Rome, 19
    Antwerp Zoo, 93
    Giza Zoo, Egypt, 336
    in Japan, 313
    Jardin des Plantes, France, 89
    medieval period, 25
    menageries in Western Europe, 80
    Schönbrunn Zoo, Austria, 85
    St. Petersburg Zoo, Russia, 134
    in United States, 150
    at Windsor Great Park, 57
Gir lions, 270
Giza Zoo, Egypt
    conservation efforts, 338
    current status, 338
    educational role, 337, 338
    establishment, 331–333
    exchange programs, 336
    exhibits and buildings, 335–336
    objectives of, 335
    types of animals, 335, 336–337
Gizerah Aquarium, Egypt, 339–340
Glass tanks, 41
Goeldi, Emilio, 359–360
Goitered gazelle, 218
Golden pheasants, 78
Goldfish collections, 24
Goral, 224
Gorillas in collections, *see also* Chimpanzees in collections; Monkeys in collections
    Berlin Zoo, 95
    breeding programs in Japan, 320–322
Gosse, Philip H., 40
Gowda, C.D. Krishne, 275
Great Britain, *see* English zoological gardens
Great Depression effects on zoos, 169, 199
Greater bilby conservation, 207
Great New York Aquarium, 157
Grecian ancient animal collections
    attitudes toward nature, 17–18
    capture techniques, 20
    etymology of animal collections, 39
    exotic animals, 18–19
    keeping of animals, 18
    types of collections, 20
    uses of animals, 19
Griffon vulture, 221
Ground parrot, 65
Groznoj, Ivan, 132–133
Grzimek, Bernhard, 110
Grzimek's Animal Life Encyclopedia, 110
Grzimeks Tierleben, 110
Guangzhou Zoo, China, 244
Gucwinski, Antoni and Hanna, 130–131
Guelre, Dukes of, 23
A Guide to the Marine Aquarium, 267
Guinea pig, 35
Gunning, J.W.B., 343
Gwalior Zoo, India, 282

H

Haagner, A.K., 343
Habsburgs menageries, 83–84
Hagenbeck, Carl, 62
    animal donations to Japan, 313
    in Argentina, 362
    consulting roles, 141
    contribution to zoo design, 102–104
    design of Indian zoo, 282–283
    popularity in Australia, 187
Hagenbeck, John, 227
Hagenbeck, Lorenz, 313

Hahn, Emily, 307
Hai Bar Carmel, Israel, 219
Hai-Bar Nature Reserve, Israel, 219
Haifa Educational Zoo, Israel, 220
Hallstrom, Edward, 200–201
Hancocks, David, 202
A Handbook of the Management of Wild Animals in Captivity in Lower Bengal, 264, 265
Han dynasty, 17
Hanging gardens of Babylon-Nineveh, 11
Hanoi Zoo, Vietnam, 235
Harappa, India, 16
Harkness, Anna M., 169
Harkness, Ruth, 172–173, 243
Harkness, William, 172, 173
Haroun-el-Raschid, 22
Harrison, Bernard, 233, 281
Harrisson, Barbara, 239, 257
Harrisson, Tom, 231
Hasegawa, Nyozekan, 324
Hassanal Bolkiah Aquarium, Brunei, 229
Hatchepsut, 15
Hawking, 77
Hayvanat Bahcesi-Izmir Zoo, Turkey, 222
Healesville Sanctuary, Australia, 203
Heber, Reginald, 259, 277
Heck, Heinz, 105
Heck, Ludwig, 95, 100, 104
Heck, Lutz, 95
Hediger, Heini, 105, 108
Heinroth, Katharina Berger, 95, 98
Heinroth, Oskar, 96–99
Hellenistic Greek period, 18
Hendrik Engel's Alphabetical List of Dutch Zoological Cabinets and Menageries, 76
Henke, Judith, 206
Henry I, 22, 49
Henry III, 22
Henry VI, 80
Heron, Robert, 60
Herons in collections, 341
Hertwig (veternarian), 88
Het Kleine Loo menagerie, 81, 89
Hetzgarten (hunting garden), 77
Highland Park Zoo, Pittsburgh, United States, 162
Hill Garden Zoo, India, 278–279
Hippopotamus in collections
    Australian zoos, 191
    Giza Zoo, Egypt, 336
    at Hull Zoo, 67
    in South Africa, 344
    in United States, 150
Histoire des ménageries de l'antiquité à nos jours, 76
Histoire naturelle de l'elephant, 89
Histoire naturelle des mammiféres, 90

History of Menageries from Antiquity to the Present, 76
History of the British Empire in India, 258
Hittite collections, 12
Hoare, M.E., 196
Hobart Municipal Zoo, Australia, 181
Hogle Zoo, Utah, United States, 300
Holmber, Eduardo, 363
Holmberg, Adolfo, 363
Hong Kong Zoological and Botanical Gardens, 239
Hornaday, William T., 158, 163, 166, 195
Horto Zoobotânico de Dois Irmaos, Brazil, 355
Housing of animals, *see* Enclosure design
Hsinchu City Zoo, Taiwan, 240
Huber, Jacques, 360
Hulein, Ernst, 144
Hull Zoological Garden, 67
Hungary zoological gardens, 142–143
Hutson, V.M., 228, 230
Hybridization attempts, 314
Hyder Ali, 274
Hyenas in collections, 53

I

Ibis, 222
IBWL (Indian Board for Wildlife), 282, 285–286
Iguanas, 34
I-Lan Wild Animal Rescue Centre, Taiwan, 240
Implantacao de Parques, Brazil, 357
Incan ancient animal collections, 26–27
Indian Board for Wildlife (IBWL), 282, 285–286
Indian Crocodile Project, 283
Indian Forest Service, India, 281
Indian Wildlife Protection Acts, 286
Indian zoological gardens
    ancient animal collections, 16–17
    animal trading in, 271, 280
    attitude toward nature, 253, 256
    Baroda zoos, 273
    Bombay Zoo, 270–271
    breeding programs, 278–279, 280, 282, 283–284
    Calcutta zoos, 257–258
        Barrackpore Menagerie, 257–261
        focus of menageries, 257
        Marble Palace, 261–262
        new zoo establishment, 263–264
        new zoo popularity, 264–265
        new zoo proposal, 263
        Wajid Ali Shah menagerie, 262–263
    conservation efforts, 253, 270, 278–279, 284–286
    early menageries and collections, 251–252
    future of, 289–290
    historical periods
        European, 256–257
        Gupta, 253–254

Mogol, 254–256
post-independence stage, 257
Vedic, 252–253
Jaipur Zoo, 271–272
Kerala Trivandrum, 269
management of zoos
coordination goal, 286–287
funding help, 289
government legislation, 287–288
standards setting, 288–289
modern zoos
Ahmedabad Zoo, 278–279
Assam State Zoo, 279
biological park invention, 281
Chhatbir Zoo, 280
condition of, 283
Crocodile Project, 283
deer park, 281–282
Hill Garden, 278
Indira Gandhi Zoo, 281
Kanpur Zoo, 280–281
Lucknow, 277–278
Madhya Pradesh zoos, 282
Nandankanan Biological Park, 279
National Zoological Park, 282–283
Nehru Zoo, 281
Sanjay Gandhi Biological Park, 279–280
specialist zoos, 284
Sri Venkateshwara Zoo, 281
traveling menageries, 280
old Madras State zoo
aquarium in, 267–268
Arignar Anna Zoo, 268–269
People's Park, 265–267
types of animals, 267
old Mysore's zoos, 274–275
other collections, 276–277
private menageries, 256, 272, 273
Sakkarbaug Zoo, 270
Sayyaji Baug Zoo, 272–273
Shivaganga Gardens Zoo, 272–273
Trichur, 269
types of animals
in Gupta period, 254
in Lalbaug, 274
in Mogol period, 255, 256
in Vedic period, 252–253
Udaipur Zoo, 272
wildlife and zoo conservation
All India Convention, 284–285
government support, 285–286
Indira Gandhi Zoological Park, India, 281
Indo-Aryan societies, 16
Indonesian Parks Association, 228

Indonesian zoological gardens
diversity of, 237
early royal collections, 236
Java, 237–238
rhinoceros in collections, 236
Sumatra, 238–239
Indonesian Zoological Parks Association, 236–237
Indore Zoo, India, 282
Indus civilization, 16
Insect collections
in England, 70
in India, 254
in Japan, 316
in Java, 238
Institute of Economical and Biological Resources, Vietnam, 235
International Society for Saving the European Bison, 104, 120, 132
International Union of Directors of Zoological Gardens (IUDZG), 122
Iranian zoological gardens, 218–219
Iraqi zoological gardens, 219
Ishikawa, Chiyomatsu, 323
Islamic caliphate, 21
Israeli zoological gardens
establishment of zoos, 219–220
reintroduction programs, 221
wartime effects on zoos, 220
Istanbul Belediyesi Gulhane Zoo, Turkey, 222
Italian menageries, 81
IUDZG (International Union of Directors of Zoological Gardens), 122

J

Jacobi, E.F., 92
Jacquin, Nicolaus, 85
Jafry, Hasan Abid, 284
Jagiello, Wladyslaw, 117
Jaguars in collections, 53
Jahangir (naturalist), 254, 255–256
Jahangir the Naturalist, 255
Jaipur Zoo, 271–272
James I, 57
Janda, Jiriho, 141, 142
Japanese Association of Zoological Gardens and Aquariums (JAZGA), 298, 302, 303, 305–306, 319
Japanese zoological gardens
animal exchanges, 314–315
animals as commodities, 312–313
breeding programs
amphibians, 317–318
animal transfers, 320–322
conservation efforts, 319–320

examples of successes, 318–319
  giraffes, 318
  serows, 317
  SSCJ challenges, 322
environmental awareness, 324–325
exhibit reconstruction, 315–316
first zoo, 327
governing authorities
  complacency in regulations, 310–311
  privatization movement, 306–307
  railway company influences, 306, 307
graphics and signage, 316–317
historical overview
  modern zoos emergence, 296–298
  post-war rebuilding, 300–302
  pre-restoration era, 289
  wartime effects on zoos, 298–300
hybridization attempts, 314
insectariums, 316
institutional overview
  accreditation, 302
  aquariums, 303–304
  attendance at zoos, 302
  categories of zoos, 302–303
  safari parks, 304–305
  traveling menageries, 304
isolation from worldwide zoos, 323–324, 326–327
marketing and events, 311–312
philanthropy, 312
popularity of animals, 313–315
publications, 311
reform consideration inputs, 325–326
research activities, 322–323
staffing
  managers and directors, 309–310
  positions and scope of duties, 308–309
  statistics, 307
  veterinarians, 309, 310
  zoological society, 305–306
Japan Monkey Centre (JMC), 307
Jaquemont, 260
Jardim da Margem, Brazil, 356
Jardim Zoológico da Cidade de Rio de Janeiro, Brazil, 356
Jardin des Plantes, France
  background, 89
  bear pits in, 80
  scientific activities, 90
  sources of animals, 89–90
  wartime effects on, 106
Jardin Zoologique d'Acclimatation, France, 36, 90
Java zoological gardens, 237–238
JAZGA (Japanese Association of Zoological Gardens and Aquariums), 298, 302, 303, 305–306, 319

Jennison, John, 66
Jenny (elephant), 106
Jersey Wildlife Preservation Trust, England, 73
Jerusalem Biblical Zoological Garden, Israel, 220
Jijabhai Bhosle Udyan Zoo, India, 270–271
JMC (Japan Monkey Centre), 307
Joachim II, 77
Johannesburg Zoological Garden, South Africa, 344–345
Johor Baru Zoo, Malaysia, 230
Jordanian Royal Society for the Conservation of Nature, 221
Jordan zoological gardens, 221
Jungle Kingdom Theme Park, Pakistan, 226
Jurong Bird Park, Singapore, 233
Justinian I, 24

K

Kabul Faculty of Science, Afghanistan, 217, 218
Kabul Zoo, Afghanistan, 217–218
Kadoorie Farm and Botanic Garden, Hong Kong, 239
Kagawa, Isamu, 313
Kahn, Herman, 323
Kangaroo in collections
  in Egypt, 341
  Tower of London menagerie, 54
Kanpur Zoo, India, 280–281
Kaohsiung Zoo, Taiwan, 240
Karachi Municipal Aquarium, 226
Karachi Zoological Gardens, Pakistan, 225–226
Katowice Zoo, Poland, 129
Kawabata, Yasunari, 324
Kawamura, Tamiji, 323
Kebun Binatang Cikini, Java, 237
Kebun Binatang Medan, Sumatra, 238
Kebun Binatang Taman Ria Aneka Rimba Jambi, Sumatra, 238
Kebun Binatang Tamansari Bandung, Java, 238
Kebun Binatung Surabya, Java, 237
Kenya Zoological Gardens, 342
Kerala Trivandrum Zoo, India, 269
Kets, Jacques, 93
Khao Kheow Open Zoo, Thailand, 234
Khao Kheow Wildlife Sanctuary, Thailand, 234
Khartoum Zoological Garden, Egypt, 340
King, George, 264
King Khalid Wildlife Research Centre, Saudi Arabia, 222
King Mahendra Trust for Nature Conservation, Nepal, 224–225
Kipling, Lockwood, 273
Kircher, Athanasius, 31, 32
Kish Island Zoo, Iran, 219
Klaver, Peter, 338

Kleiner, Salomon, 84
Klös, Heinz-Georg, 95, 107
Knowsley Hall, 60
Knox, Robert, 66, 226
Knudson, Gus, 171
Koalas in collections, 199–200, 315
Kobara, Jiro, 323
Koga, Tadamichi, 322
Kommer, H.F., 237
Komodo monitors in collections, 314
Korea zoological gardens, 240
Krakow Zoo, Poland, 121–122, 123
Krause, Alois, 86
Kubinyi, Agoston, 142
Kublai Khan, 24
Kukrail Park, India, 278
Kumar, Sri Pushp, 281
Kuwait zoological gardens, 221
Kyoto Zoo, Japan, 298, 300

L

Laackmann, H., 120
Labuschagne, Wilhelmus, 344
The Lady and the Panda, 172
Lahore Zoological Gardens, Pakistan, 225
Lai Chi Kok Zoo, Hong Kong, 239
Lalbaug Botanical Gardens, India, 274
Laos zoological gardens, 229
Law, Ong Swee, 233
Lawenson, W.R., 285
Leakey, Louis, 342
Leefoo Zoo, Taiwan, 240
Leningrad Zoo, Soviet Union, 138; *see also* St. Petersburg Zoo
Lenné, Peter Joseph, 87, 88, 94
Lensink, B.M., 92
Leopards in collections, 240
Leo X, Pope, 23, 29, 81
Leporarium, 39
Le Souef, Albert, 184, 186, 190
Le Souef, Caroline, 192
Le Souef, Dudley, 190, 195
Le Souef, Sherbourne, 186, 188
Le Souef family, 183–184, 190
Leverkuhn, Paul, 144
Liberec zoo, Czech Republic, 141
Library of Entertaining Knowledge, 51
Lichtenburg Game Breeding Centre, South Africa, 344
Lichtenstein, Martin Hinrich, 88
Life Among the Wild Beasts in the Zoo, 71
Life Beneath the Waters, 41, 157
Limassol Municipal Zoo, Cyprus, 218
Lincoln Park Zoo, Chicago, 154
"Lingyou," 17
Linnaeus, Carolus, 3, 4
Lions in collections
    in Czech Republic, 140
    in Israel, 220
    medieval period, 23
    in Mesopotamia, 10
    Sakkarbaug Zoo, India, 270
    Tower of London menagerie, 50, 52–53
    in United States, 147
    in Western Europe, 80, 81
A List of Zoological Gardens of the World, 333
Literature and writings on zoos
    aquariums, 155
    captive wildlife management, 174–176
    enclosure design, 265
    ethology, 99
    exotic styles, 100
    on history of zoos, 126–127
    from India, 266–267
    from Japan, 305–306, 311
    from Poland, 123, 125
    professionalization of zoo management, 174–176
    Tower of London menagerie, 52–55
    treatment of animals, 254–255
    wartime effects on zoos, 107
    in Western Europe, 76
Liverpool Zoological Garden, England
    closing of, 65
    types of animals, 62–64
Llamas in collections, 54
Lodrick, D.O., 254
Lodz Zoo, Poland, 122, 123
Loisel, Gustave, 76, 258, 272, 334, 335
London Zoological Garden, England
    bear pits in, 80
    breeding and housing successes, 68–72
    financial status, 72
    first public aquarium, 70, 71
    Windsor collection transferred to, 58
Lord McAlpine's Pearl Coast Zoological Gardens, Australia, 181
Lorenz, Konrad, 98
Loro Parque, Germany, 109
Louis IX, 22
Lubetkin, Berthold, 68
Lucknow: The Last Phase of an Oriental Culture, 262
Lucknow Zoo, India, 277
Ludlow, F., 225
Lukaszewicz, Karol, 122, 124, 125–127
Lut'f'ullah, 273
Lytton, Lord, 260

M

Macau zoological gardens, 240
MacKenzie, Colin, 201, 203
Macomber, Zebedee, 150
Madagascar, Africa, 347
Madras Government Central Museum, India, 265–267
Madras Snake Park, India, 284
Magistri, 39
Maharaja Fatesingh Zoo, India, 273
Mahavir, 253
Mahazat Assayd Nature Reserve, Saudi Arabia, 222
Mahebatkhanji Babi II, 270
Mahendra Chaudhary Zoological Park, India, 280
Mahraj Baug Zoo, India, 276
Maia, Antonio, 353
Malabon Zoo, Philippines, 232
Malagasy Republic, Africa, 347
Malawi, Africa, 346
Malaysian Association of Zoological Parks and Aquaria, 229
Malaysian-Borneo Exhibition, Singapore, 233
Malaysian zoological gardens, 229–231
Malaysian Zoological Society, 230
Mallorca, Ornis, 109
Management of zoos, *see also* Scientific societies; Zoological societies
  in Afghanistan, 217–218
  in Argentina, 363
  in Australia
    emphasis on recreational role, 193–195
    financial status, 199
    lack of research efforts, 201–202
    modernization period, 202–208
    Taronga Zoo, 200–201
    visitors' associations, 211
  Berlin Zoo financial status, 94
  in Bolivia, 363
  in Brazil, 353, 354, 356–358, 359–362
  Budapest Zoo, Hungary, 143
  in Bulgaria, 145
  in China, 242–244
  in Colombia, 364
  in Egypt
    general comments, 342
    Giza Zoo, 334–335, 336, 337, 338
    Zagazig Zoo, 341
  in India
    coordination goal, 286–287
    funding help, 289
    government legislation, 287–288
    standards setting, 288–289
  in Indonesia, 236–237
  in Japan
    accreditation, 302
    complacency in regulations, 310–311
    privatization movement, 306–307
    railway company influences, 306, 307
    staffing, 309–310
  in Java, 237–238
  London Zoo, England, 68, 71, 72
  in Malaysia, 229–230
  Museu Paraense, Brazil, 359–362
  nationalization of Russian zoos, 136–137
  in Philippines, 232
  post-war Germany, 109–111, 112
  post-war Poland, 124, 126, 128–129
  pre-war Polish zoos, 122
  professionalization of, 173–176, 264
  in Republic of South Africa, 345
  in Russia, 134
  Schönbrunn Zoo, Austria, 87
  in Soviet Union, 140
  in Sri Lanka, 227–228
  in United States
    Bronx Zoo, New York, 163
    coordination of executives, 165, 168
    modernization period, 169
    national zoo establishment, 158–160
    Philadelphia Zoo plans, 151
    visitors' associations, 172
    during wartime, 164
  in Western Europe, 78
  zoo biology efforts at Bern Zoo, 105, 108
  zoo establishment in Thailand, 234
Man and Nature, 18
Manasang (circus owner), 238
Manaus zoo, Brazil, 353
Manchester Zoo, 65
Mandelbaum Gate, Israel, 220
Mangold, Walter, 345
Manila Zoological and Botanical Garden, Philippines, 231–232
Marble Palace Zoo, Calcutta, India, 261–262
Margaraya Tinjomoyo Semarang, Java, 238
Maria Stuart, 81
Maria Theresa, 84
Marigowda, M.H., 275
Mari-it Conservation Park, Philippines, 232
Marineland, Florida, United States, 41, 169
Marine Mammal Protection Act of 1972, United States, 173
Marine Science Institute, Russia, 135
Marseille, 91
Marsh, George, 18
Martin, Philipp Leopold, 100
Mashhad Zoo, Iran, 219
Mauritius, Africa, 347–348
Mauritius dodo, 141
Maximilian, 76

Maximilian II, 84, 140
Mayan ancient animal collections, 27
McCoy, Frederick, 183
Mearns (grounds keeper), 65
Medieval period ancient animal collections
    Chinese royal collections, 24–25
    European royal collections, 22–24
Meir Segals Garden, Israel, 219
Meissner, M., 120
Melaka Zoological Garden, Malaysia, 230
Melbourne Zoo, Australia
    conservation efforts, 207
    educational role, 205
    enclosure design, 198
    establishment, 184
    modern period, 202
    recreational role, 210
    reptile collection, 192
    rhinoceros in, 191
    visitors to, 197
Menagerie, 39
Menageries, *see also* Ancient animal collections
    colonial
        acclimatization programs, 36–37
        animal transactions, 35–36
        collecting stations, 34–35
        European interest in new species, 33–34
        live animal shipments, 34–35
    definition, 257
    European
        biodiversity explanations, 30–32
        cabinet collections, 30, 31
        origins of animals, 28, 30
        shipments of animals, 32–33
    traveling, in Japan, 304
    in Western Europe
        educational benefits, 83
        popularity among royals, 80–81
        private collections, 82
        purpose of, 80
        trading of animals, 81, 82
From the Menagerie to the Animal Paradise: 125 Years of the Berlin Zoo, 107
Mercado, Noel Kemp, 363–364
Merodach-Baladan II, 11
Merowe Zoological Garden, Egypt, 341–342
Mesopotamian ancient animal collections
    etymology of animal collections, 38
    garden development, 9
    habitat re-creation, 10, 12
    luxury item trade, 9
    religion and sciences, 8–9
    royal collections, 10, 11
    servant/animal keepers, 12
    society history, 8

Michailovitch, Aleksei, 133
Middle East, *see* Southwest Asian zoological gardens
Miller, Stefan, 122
Milne-Edwards, Henri, 241
Milus in collections, 95
Minchin family, 186, 190, 191
Mir Shikar Tola, India, 280
Mitchell, D.W., 71
Mogol period in India, 254–256
Mohenjo-daro, India, 16
Monarto Zoological Park, Australia, 204
Mongoose in collections, 53, 54
Monkeys in collections, *see also* Chimpanzees in collections; Gorillas in collections
    in England, 54, 65
    in Germany, 109, 110
    in Japan, 307
    in Western Europe, 100, 105
Montezuma, 26
Moore Park Zoo, *see* Taronga Zoo, Australia
Mörner, Count, 196
Morocco, Africa, 346
Morrin, T.D., 235
Morris, Graham, 205
Moscow Acclimatization Society, 134
Moscow Zoological Gardens, Soviet Union
    breeding programs, 138–139
    nationalization of, 137
    opening of, 134
    wartime effects on, 138
Motithang Zoo, Bhutan, 224
Mottershead, George S., 68
Mozambique, Africa, 346–347
Mugger Pir, Pakistan, 226
Müller, Karl, 87
Müller, Richard, 122
Mullick Bahadur, Rajah Rajendro, 261–262
Münster All-Weather-Zoo, 108
Muscovy duck in live animal shipments, 35
Muséum d'Histoire Naturelle, 82
Museum of Ethnology, 92
Museu Paraense, Belem, Brazil
    background, 359–360
    directorship changes, 360
    European influences, 359
    ownership changes, 360
    renovations, 360–361, 362
    types of animals, 361
Mutamid Khan, 255
Myanmar zoological gardens, 231

N

Nair, P. Kesavan, 269
Nairobi Safari Walk, 342

Nairobi Snake Park, 342
Nakorn Ratchasima Zoo, Thailand, 234
Nandankanan Biological Park, India, 279
National Aquarium, United States, 155, 157
National Fisheries Center and Aquarium, United States, 157
National Museum of Marine Biology/Aquarium, Taiwan, 241
National Wildlife Research Centre, Saudi Arabia, 222
National Zoological Park, India, 282–283
National Zoological Park, United States, 160
Natura Artis Magistra, 91
Natural History of the Elephant, 89
Natural History of the Mammals, 90
Natural History Project, Barrackpore Menagerie, India, 259
Nature id the Master of Arts, 91
Nebuchadnezzar II, 11
Negev Desert Chai Bar Reserve, Israel, 219
Negotiator ursorum, 39
Nehring, Arnold, 77
Nehru Zoological Park, India, 281
Nepal zoological gardens, 224–225
Netherlands menageries, 81
Neugebäu, Austria, 78
New Biblical Zoo, Israel, 220
New South Wales Zoological Society, England, 186
New York Aquarium, United States, 164
New York Zoological Society, United States, 162, 164
NFEFI Conservation Center, Philippines, 232
Nickerson, G.S., 340
Nigeria, Africa, 347
Nightjar breeding, 98
Nikolaev Zoo, Russia, 135
Nogge, Gunther, 218
Novikov, S.N., 134
Nubian ibex, 222
Numbat, 208

O

O'Connor, Patricia, 168, 171
Oceanarium
　definition, 40
　in Hong Kong, 239
　in Java, 238
　in Taiwan, 240–241
Oceanarium Gelanggang Samudra Jaya Ancol, Java, 238
Ocean Park, Hong Kong, 239
Ocean World, Taiwan, 240–241
Oelschlaeger, Max, 2
Oeming, Al, 241
Ogrody Zoologiczne-Wczoraj-Dzis-Jutro, 127
Oliwa Zoo, Poland, 129
Olmsted, Frederick Law, 159
Oman Mammal Breeding Center, 221–222
Oman zoological gardens, 221–222
Onager, 221, 222
Onelli, Clemente, 363
Open Range Zoo at Werribee, Australia, 202
Opole Zoo, Poland, 129
Opossums in live animal shipments, 34
Orangutans in collections
　Australian zoos, 190–191, 198
　in India, 269
　in Lalbaug zoo, India, 274
　in Malaysia, 231
　Schönbrunn Zoo, Austria, 85
　St. Petersburg Zoo, Russia, 134
　Sumatra rehabilitation centers, 239
　in United States, 148
　in Western Europe, 81
Orligh, Leopold von, 276
Osman Hill, W.C., 227
Ossendowski, F., 120
Ostriches in collections, 148
Ostrorog, Jan, 118
Our Vanishing Wild Life, 195
Outline of the History of Menageries, 76

P

Padmaja Naidu Himalayan Zoological Park, India, 284
Paignton Zoological Garden, England, 67–68
Pakistan zoological gardens, 225–226
Pal, A.C., 261
Palermo menagerie, Sicily, 80
Pandas in collections
　Giant Panda conservation, 242–243
　interzoo exchanges, 139
　in Japan, 314–315
　in United States, 169, 172–173
Panorama design, 103–104
Paradeisos, 21, 38
Parc Zoologique de Paris, France, 90
Pardisan Nature Park, Iran, 219
Paris, Matthew, 50
The Park at Barrackpore, 261
Parks, Fanny, 259, 261, 276, 277
Parque Municipal Quinzinho de Barros, Brazil, 356–357
Parque Natural de Carajas, Brazil, 353
Parque Zoobotânico, Brazil, 354–355
Parque Zoobotânico Museu Paraense Emilio Goeldi, *see* Museu Paraense
Parrot Park, Germany, 109
Parrots in live animal shipments, 34
Pasha, H.E. Hassan Hassib, 341

Pasha, Herbert Jackson, 342
Pasha, Ismail (Khedive), 331, 333, 334
Pasha, Tewfik (Khedive), 332
Passenger pigeon, 60, 155, 156–157
Pasteur Institute Snake Farm, Thailand, 234
Patnaik, S.K., 288
Patna Zoo, India, 279–280
Peacock Island (Pfaueninsel), Germany, 87
Pechlaner, Helmut, 87
Peking Zoo, China, 241, 243–244
Pelican, 55
Penrose Research Laboratory, United States, 164, 171–172
Père David's deer in collections, *see also* Arabian oryx
  in Australia, 200
  conservation efforts, 241–242
  reintroduction program in Asia, 216–217
  in South Africa, 344
Persia, *see also* Iran
  ancient animal collections, 21
  etymology of animal collections, 38
Persian fallow deer, 221
Perth Zoo, Australia
  educational role, 205–206
  establishment, 186
  modernization period, 204
  visitors to, 197
Peshawar Zoological Garden, Pakistan, 226
Petah-Tiqya Zoo, Israel, 220
Petit, D.M., 271
Pfaueninsel (Peacock Island), Germany, 87
Pheasantries, 77–78, 82, 218
Philadelphia Zoo, United States, 151–152
Philippine Eagle Foundation, Philippines, 232
Philippine zoological gardens, 231–232
Philip VI, 22
Phnom Tamao Zoo, Cambodia, 229
PHVA (Population and Habitat Viability Analysis), 326
Pidcock, Gilbert, 58
Pierre, J.L., 235
Pigmy hog, 279
Pilawin on Wolyn collection, Poland, 119
Pinjarapoles in India, 253–254
Pinnawala Elephant Orphanage, Sri Lanka, 228
Pintung Rescue Centre, Taiwan, 240
Piscinae, 39
Piscinarii, 39
Pisit na Patalung, Thailand, 234
Pizarro, Francisco, 26, 27
Plant collections, 1
Platypus in collections, 199–200
Plock Zoo, Poland, 128
Poinsett, Joel R., 150

Polar bears
  breeding programs in Japan, 322
  in Prague zoo, 142
  in Russian collections, 133
  Tower menagerie, 50
  in United States, 148
Polish zoological gardens
  during and after WWII
    animal transfers, 122–123
    Breslauer opening, 123–124, 126
    municipal management, 126
    new zoo establishments, 126, 128–129
    rebuilding efforts, 123–124
  characteristics and comments
    attendance records, 130
    education and publicity attempts, 130–131
    science and conservation, 131–132
  flood effects, 128
  modern zoos origins
    Krakow, 121–122
    Lodz, 122
    other small collections, 122
    Poznan, 119–120
    Warsaw, 120–121
  wild animal keeping
    popularity among royals, 117–118
    private menageries, 118–119
    sources of animals, 118
    types of animals, 118
Polito, S., 58
Polo, Marco, 24, 25
Poniatowski, Stanislaw August, 119
Population and Habitat Viability Analysis (PHVA), 326
Port Elizabeth Museum, South Africa, 342–343
Portielje, Anton F.J., 92
Port Sudan Zoological Garden, Egypt, 342
Portugal, 29, 81
Potgietersrus Game Breeding Centre, South Africa, 344
Potocki, J., 119
Poznan Zoo, Poland
  animal losses from WWII, 122, 123
  breeding programs, 132
  origins of, 119–120, 129
Praca Frei Baraun, Brazil, 356
The Practice of Natural History, 100
Praga Park Zoo, Poland, 122
Prague zoo, 142
Preston Zoological Garden, England, 67
Prince of Wales Zoo, India, 277–278
Pringle, John, 343
Procurator ad elephantos, 39
Professionalization of zoo management, 173–176, 264

Przekroj przez Zoo, 123
Psenner, Hans, 109
Psittacines, 345
Ptolemy II, 18
Puglia, Lazaro, R.R., 357
Punt expeditions, 15
Purdy, Eisenhart, 150
Pygmy lori, 132
Pyongyang Zoological Gardens, Korea, 240

Q

Qin dynasty, 17
Qureshi, Z.A., 226

R

Rabb, George, 325, 326
Rabbits in Australia, 184
Radziwill, Karol, 119
Radziwill, Mikolaj, 118
Ragunan Zoo, Jakarta, 237
Railway company zoos in Japan, 306, 307
Raja of Dantheal, 256
Rajasinghe of Kandy, 226–227
Ramat-Gan Zoo, Israel, 220
Ramesses II, 14
Rangoon Zoo, Myanmar, 231
Recreational role of zoos
    in ancient Rome, 19
    aquariums popularity, 40–41
    Australian zoos, 190, 193–195, 201, 210
    in England, 49
    game parks in Western Europe, 76, 77
    in Germany, 86
    in India, 264–265, 279
    in Japan, 308
    Manchester Zoo, England, 66
    Polish zoos, 130
    post-war Germany, 110
    primary aims of zoos, 100
    in Republic of South Africa, 343, 344, 345
    in United States, 151, 152, 169
    West European zoos efforts, 110
Red Panda Project, India, 284
Reintroduction programs, *see also* Conservation and research efforts
    in Israel, 221
    Southwest Asia, 216–217
Reischauer, Edwin O., 323–324
René, Count of Anjou, 23
Reptile collections
    in England, 69, 71
    in India, 269
    in Nairobi, 342
    Republic of South Africa, 343

    in Thailand, 234
Republic of South Africa zoological gardens
    aquariums, 346
    bird sanctuary, 345
    breeding programs, 344
    cheetah collection, 345–346
    Durban aquarium, 345
    early collections, 342–343
    East London Zoo, 343
    Johannesburg Zoo, 344–345
    National Zoological Garden, 343–344
    Port Elizabeth Museum, 343
    Transvaal zoo establishment, 343
Rhinoceros in collections
    Assam State Zoo, India, 279
    Australian zoos, 191
    Budapest Zoo, Hungary, 143
    Giza Zoo, Egypt, 336
    in Indonesia, 236
    Kanpur Zoo, India, 280–281
    London Zoo, England, 70
    Manchester Zoo, England, 65
    medieval Europe, 29
    in Nepal, 224
    People's Park, Madras, India, 267
    in United States, 150
Rhodes, Cecil, 34
Richardson, Harry, 65
Rix, Cecil, 194
Riyadh Zoological Gardens, Saudi Arabia, 222
Robert College Zoo and Natural History Museum, Turkey, 222
Roboraria, 39
Rodent control, 53
Roger Williams Park Zoo, Rhode Island, United States, 154
Rome
    ancient animal collections
        attitudes toward nature, 17–18
        capture techniques, 20
        exotic animals, 18–19
        keeping of animals, 18
        types of collections, 20
        uses of animals, 19
    etymology of animal collections, 39
Rookmaaker, Kees, 62
Rost, E.A., 134
Rostov-na-Donu Zoo, Soviet Union, 139
Rotenbeck, Michael, 76
Rotherham Museum, 61
Rotterdam, 91
Rousselet, Louis, 272
Royal Zoological Society of New South Wales, Australia, 197
Rudolf II, 140

Ruhe, Hermann, 108
Ruhr Zoo, Germany, 108
Rusa deer in collections, 54
Russian zoological gardens
    origin of modern zoos to 1917
        Askaniya Nova collection, 134–135
        financial problems, 134–135
        private menageries, 133–134
        public collections, 134
    Soviet Union
        characteristics and comments, 140
        political and economic considerations, 135–137
        Russian zoo nationalizations, 137–139
        small zoos, 139
    wild animal keeping, 132–133

S

al-Sabah, Jaber Abdullah, 221
Saddam Hussain Park, Iraq, 219
Safari Park, Java, 238
Safari parks
    Germany, 110
    India, 281
    Indonesia, 238
    Japan, 304–305
    Java, 238
    Kenya, 342
    Singapore, 233
    Thailand, 234–235
Safari World, Thailand, 234–235
Saigon Zoo, Vietnam, 235
Saint-Hilaire, Albert Geoffroy, 90
Saint-Hilaire, Etienne Geoffroy, 90
Saint-Hilaire, Isidore Geoffroy, 90
Saint-Pierre, Bernhardin de, 106
Sakkarbaug Zoo, India, 270
Saleh, Raden, 237
Salwa Zoo, Kuwait, 221
Samut Prakarn Province, Thailand, 234
Sand gazelle, 222
San Diego Zoo, United States, 165
San Francisco Zoo, United States, 158
Sanjay Gandhi Biological Park, India, 279–280
Sanyal, Ram Brahma, 264–267
Sarawak Museum, Malaysia, 231
Sargon II, 11
Satwataru Zoological Park, Java, 237
Saudi Arabia zoological gardens, 222
Savage, Arthur, 147
Saxe-Coburg-Gotha, Ferdinand, 143
Sayyaji Baug Zoo, India, 272–273
Schilling, Diebold, 79
Schmidt, Maximilian, 95
Schomberg, Geoffrey, 339
Schön, Alexander, 86
Schönbrunn and Its Zoo, 107
Schönbrunn Zoo, Austria
    damage from wars, 86, 87
    design, 84–85
    establishment, 84, 141
    popularity of, 86
    pre-WWII improvements, 87
    sources of animals, 85
    transformation to a zoo, 86
    types of specimens, 85
    wartime effects on, 106
Schornstein, Mordechai, 219
Schouman, Aart, 81
Schumann, Adolf, 144
Schwendler, Carl Louis, 263
Science and conservation, *see* Conservation and research efforts
Scientific societies, 94; *see also* Management of zoos; Zoological societies
Seal, Ulie, 325
Sea otters in collections, 315
Semarang Zoo, Java, 238
Sen, R. M., 231
Senanayake, Ranil, 228
Senegal, Africa, 347
Sennacherib, 10, 11, 12
Serfagee, Raja, 272–273
Servais, Charles, 93
Severoceska Zoologicka Zahrada Liberec, Czech Republic, 141
Shang dynasty, 16
Shanghai Zoo, China, 244
Sharar, Abdul Halim, 262, 263, 277–278
Shariff, Z. A., 225
Sharma, S.C., 287
Shaumari Wildlife Reserve, Jordan, 221
Shipments of animals
    colonial menageries, 34–35
    to/from Europe, 32–33
Shiraz Zoo, Iran, 219
Shivaganga Gardens Zoo, India, 272–273
Shulov, Professor, 220
Siam, *see* Thailand
Sieber, August, 88
Siesian Zoological Park, Poland, 129
Silver pheasants, 78
Singapore Naturalists Society, 232, 233
Singapore Night Safari, 233
Singapore zoological gardens, 232–233
Singa Zoological Garden, Egypt, 340
Singh, Jaipur Maharaja Savai Ram, 271
Singh, Maharaja Sajjan, 272
Sir Colin MacKenzie Sanctuary, 203

Slovak Republics
    private royal collections, 140–141
    public zoo establishments, 141–142
Smith, Tangier, 172, 173, 243
Smithsonian Institute, United States, 150, 158
Snake collections, *see* Reptile collections
Snow leopards, 218
Sociedade de Zoológicos do Brasil, 357
Societé de Zoologie d'Anvers, Antwerp, 92, 94
Societé Royal de Zoologie d'Horticulture et d'Agrément de la Ville de Bruxelles, Germany, 97
Société Zoologique d'Acclimatation, France, 36, 90
Society for the Acclimatization of Birds, Cincinnati, United States, 154
Society for the Acclimatization of Plants and Animals, Hungary, 142
Society for Zoo and Acclimatization, Prague, 141
Sofia Zoo, Bulgaria, 144–145
Song dynasty, 17
Songkla Zoo, Thailand, 234
Sosnowskij, I.P., 137
South Africa, *see* Republic of South Africa zoological gardens
South American zoological gardens
    Argentina, 362–363
    Bolivia, 363–364
    Brazil
        Belem zoo, *see* Museu Paraense
        Carajas zoo, 353
        in Center-west, 355–356
        historical aspects, 351–353
        Manaus zoo, 353
        private collections, 357
        Rio Branco zoo, 354–355
        the South, 358
        the Southeast, 356–357
        status of, 358
    Colombia, 364
    focus of regional zoos, 365
    Venezuela, 364–365
South Asian zoological gardens
    Bangladesh, 223–224
    Bhutan, 224
    Nepal, 224–225
    Pakistan, 225–226
    Sri Lanka, 226–227
Southeast Asian zoological gardens
    Brunei, 229
    Cambodia, 229
    Laos, 229
    Malaysia, 229–231
    Myanmar, 231
    Philippines, 231–232
    Singapore, 232–233
    Thailand, 233–235
    Vietnam, 235–236
    zoological societies formed, 228
South East Asian Zoological Parks Association, 228
Southwest Asian zoological gardens
    Afghanistan, 217–218
    background of zoos, 215–217
    Cyprus, 218
    Iran, 218–219
    Iraq, 219
    Israel
        establishment of zoos, 219–220
        reintroduction programs, 221
        wartime effects on zoos, 220
    Jordan, 221
    Kuwait, 221
    Oman, 221–222
    reintroduction program, 216–217
    Saudi Arabia, 222
    Turkey, 222–223
    United Arab Emirates, 223
    Yemen, 223
Soviet Union zoological gardens, *see also* Russian zoological gardens
    characteristics and comments, 140
    political and economic considerations, 135–137
    Russian zoo nationalizations, 137–139
    small zoos, 139
Spain, 29, 81
Species Survival Committee of Japan (SSCJ), 319, 320, 321
Sports with kept animals
    in ancient China, 17
    in ancient Rome, 20
    in Egypt, 14
    Indian attitudes against, 253
    in Indian Mogol period, 255
    Lucknow Zoo, India, 277–278
    in Russia, 133
    Sayyaji Baug Zoo, India, 272–273
    Udaipur Zoo, India, 272
Sri Lanka zoological gardens, 226–227
Sri Mahavir Harina Vanasthali Deer Park, India, 281–282
Sri Venkateshwara Zoo, India, 281
SSCJ (Species Survival Committee of Japan), 319, 320, 321
St. Petersburg Zoo, 134
    Leningrad Zoo breeding program, 138
    nationalization of, 137
Stanley, Earl of Derby, 60
Stanton, H.R., 313
Staten Island Zoo, New York, United States, 171
Stavros tis Psokas Forest Station, Cyprus, 218
Stevens, Edward, 65

Strachan, W., 226
Strack, Heinrich, 94
Strahan, Ronald, 201, 203, 205, 207
Strype, John, 50
Stubton Hall, 60
Suares, J.C., 338
Sumatran rhino, 143
Sumatra zoological gardens, 238–239
Sumer, 8
Summers, H.G., 324
Sunier, A.L.J., 92
Surabaya Zoological Gardens, Java, 237
Surrey Zoological Garden, England, 59, 61–62
Sweyn II, 23
Sykes, William Henry, 224
Systema Naturae, 5
Systematic approach to design, 101–102
Systematics, 3–5
Szabo, Jozsef, 142
Szczerkowski, K.W., 120, 122

T

Taipei Zoo, Taiwan, 240
Taiping Zoo, Malaysia, 230
Taiwan zoological gardens, 240–241
Takarazuka Zoo, Japan, 316
Takin, 224
Tallin Zoo, Soviet Union, 139
Taman Hewan Pematangsiantar, Sumatra, 238
Taman Marga Satwa dan Budaya Kinantan, Sumatra, 238
Taman Marga Satwa Jakarta, Java, 237
Taman Mini Indonesia Indah, Java, 238
Taman Safari Indonesia, 238
Taman Wisata Satwa Taru Jurug, Java, 237
Tama Zoo, Japan, 307, 316, 318
Tamburong Zoo, Brunei, 229
Tanaka, Yoshio, 296, 327
Tang dynasty, 17
Tanta Zoological Garden, Egypt, 341
Taronga Zoo, Australia
  additions to, 200–201
  educational role, 205
  enclosure design, 187, 198
  establishment, 186
  modernization period, 203
  new research emphasis, 197
  recreational role, 210
  visitors, 198
Tasmanian tiger, *see* Thylacine in Australia
Taworski, Tadeusz, 128
Taxonomy, 101–102
Taylor, A.A.C., 340
Tehran Zoo, Iran, 218–219

Tel Aviv University Research Zoo, Israel, 219, 220
Temminck, Jacob, 82, 91
Tennoji Zoo, Japan, 297
Terrariums, 41
Territory Conservation Commission, Australia, 202
Territory Wildlife Park, Australia, 202, 206, 207
Thaiarry, Philphat, 234
Thailand zoological gardens, 233–235
Thao Cam Vien Saigon, 235
Theatrum Naturae, 76
Theophrastus, 4
Therotrophium, 39
Thompson, Frank J., 152
Thu Le Park Zoo, Vietnam, 235
Thurneisser, Leonhard, 83
Thylacine in Australia, 189, 195
Tiergarten, *see* Berlin Tiergarten; Berlin Zoo, Germany
Tiergarten Schönbrunn, *see* Schönbrunn Zoo, Austria
Tiergrotten, 105
Tierpark Berlin-Friedrichsfelde, Germany, 80, 109
Tierpark Dählhölzi, Germany, 79
Tierpark Hellabrunn, Germany, 105
Tierpark Rheine, Germany, 110
Tigers in collections
  in India, 269
  Tower of London menagerie, 53
  in United States, 148
Tiglath-Pileser I, 11
Tigrine frog, 269
Tikell, S.R., 224
Tilson, Ron, 234
Tisch Family Zoological Gardens, Israel, 220
The Tower Menagerie, 52–55
Tower of London menagerie
  care of animals, 51–52
  establishment, 50
  specimen descriptions, 52–55
  transfer of collection, 55–56
  types of animals, 50–51
Training programs
  in Australia, 208–209
  in Japan, 309, 310
  in Poland, 130
  in United States, 174–175
  in Western Europe, 110
Trichur Zoo, India, 269
Turkeys in live animal shipments, 35
Turkey zoological gardens, 222–223
Tuthmosis IV, 14
Tuzcu Zoological Garden, Turkey, 223
Two Oceans Aquarium, South Africa, 346
Tyler, William, 62
Tyroler House, Germany, 87

U

U.S.S.R. Red Data Book, 140
Über die Kunst, mit Vögeln zu Jagen, 77
Udaipur Zoo, India, 272
Ueno Zoo, Japan
  acreage, 303
  breeding programs, 320
  opening of, 296, 327
  rebuilding efforts, 301
  specimen numbers, 302
  wartime effects on, 300
Undawalawe Elephant Orphanage, Sri Lanka, 228
Underwater Observatory Marine Park, Israel, 219
Unicorns, 25
United Arab Emirates zoological gardens, 223
United States zoological gardens
  animal shows, colonial times, 147
  aquariums in 19th century, 152, 155, 157, 164
  aquariums in 20th century, 168, 175–176
  menageries in 18th century
    exotic animals shown, 147–148
    natural science emergence, 150
    traveling shows and menageries, 148–149
  menageries in 19th century, 150
  zoos in 19th century
    Central Park Zoo, 154
    Cincinnati Zoo, 154–155, 157
    education and science role, 158
    European influences, 152
    Lincoln Park Zoo, 154
    Mid-Atlantic to Western zoos, 157–158
    national zoo establishment, 158–160
    national zoo suggestion, 150
    New York Zoo establishment, 162–163
    Philadelphia zoo establishment, 152
    Philadelphia zoo plans, 151–152
    popularity of urban parks, 152, 153
    Roger Williams Park Zoo, 154
    spread of openings, 160, 162
  zoos in 20th century
    captivity studies, 175
    conservation efforts, 166, 173–174, 176
    coordination of executives, 165, 168
    education advancements, 172–173
    modernization with federal funds, 169
    municipal zoos, 176
    open, moated exhibit use, 164
    post-WWII expansions, 170–171
    research center role, 164
    scientific and research activities, 171–172
    training programs, 174–175
    trends in, 177
    women zookeepers, 168–169
    WWI effects, 164–165
    WWII effects, 169
Urbanization influence on animal collecting, 8

V

Vaclav VI, 140
Vagner, Josef, 142
van Aken, Cornelis, 92
van der Schot, Richard, 85
Van Kleef Aquarium, Singapore, 233
van Ravensteyn, Sybold, 104
van Steckhoven, Adrian, 84
Van Vihar National Park, India, 282
Velten, Jan, 82
Venezuela zoological gardens, 364–365
Versailles, France, 82, 89
Versuch einer Geschichte der Menagerien, 76
Veterinarians and zoo keepers
  in ancient Egypt, 13
  in ancient Rome, 20
  in Australian zoos, 208
  Alfred Cops, Tower of London menagerie, 51–52
  first veterinary clinic, 164
  in German zoos, 96–99
  Giza Zoo, Egypt, 335
  Oskar Heinroth, 96–99
  Hertwig, Berlin Zoo, 88
  hospital in Pakistan, 225
  in Japanese zoos, 309, 310
  London Zoo, England, 68, 72
  Karol Lukaszewicz, Krakow Zoo, 125–127
  in Mesopotamia, 9
  Karl Müller, Schönbrunn Zoo, 87
  research in United States, 171–172
  Ram Brahma Sanyal, India, 264–267
  in United States zoos, 170–172
Veterinarius, 39
Victoria, 58
Vientiane Zoo, Laos, 229
Vietnam zoological gardens, 235–236
Visitors to zoos study, 266
Vivariology, 127
Vivarium, 20, 39, 41
Vivarium/vivaria, 39
Von der Menagerie zum Tierparadies: 125 Jahre Zoo Berlin, 107
von Holzing (forester), 82
von Linné, Carl, 3, 4
von Mataram, Soesoehoenan, 236
von Mengden, Mrs. Ulrike, 237
von Mueller, Ferdinand, 188
von Orligh, Leopold, 277
Vosmaer, Arnout, 81

W

Wajid Ali Shah menagerie, India, 262–263
A Walk through the Zoo, 123
Wallace Line, 238

Ward, Nathaniel, B., 41
Wardian Cases, 41
Warington, Robert, 40
Warrington Cases, 40
Warsaw Zoo, Poland
    animal losses from WWII, 123
    origins of, 120–121
Wartime effects on zoos
    in Afghanistan, 218
    Civil War in America, 152
    damage to Schönbrunn Zoo, Austria, 87
    in Iraq, 219
    in Israel, 220
    in Japan, 298–300
    in Kuwait, 221
    in Philippines, 231
    in Poland
        animal transfers, 122–123
        Breslauer opening, 123–124, 126
        municipal management, 126
        new zoo establishments, 126, 128–129
        Poznan Zoo, 120
        pre WWII, 122
        rebuilding efforts, 123–124
    Soviet Union zoos, 138
    in Vietnam, 235
    in Western Europe, 106–108
    WWII effects in Australia, 200
    WWII in America, 169
Wegeforth, Harry M., 165
Weinman, Aubrey, 227
Welch, Rufus, 150
Wellesley, Arthur, 56
Wellesley, Richard, 257–259
Welser, Marcus, 77
Wentworth collection, 61
Wen Wang, 16
Werribee Zoological Park, Australia, 202
Westerhof, Jan Barentze, 82
Westerman, Gerardus Frederick, 91
Western Australian Acclimatisation Committee, 186
Western European zoological gardens
    early modern zoos
        Berlin Zoo, 94–95
        exotic designs, 100–101
        exotic style popularity, 97–99
        location of zoos, 91
        Panorama design, 103–104
        prevalence in England, 90–91
        primary aims, 100
        systematic approach, 101
        zoological societies formed, 91, 92
    menageries transition to zoos
        Berlin Zoo, 87–89
        public menagerie collections, 83
        Schönbrunn, *see* Schönbrunn Zoo, Austria
    modern zoos
        changes in exhibit designs, 104–105
        education and breeding programs, 110
        effect of wars on, 104, 105, 106–108
        post-WWII rebuilding, 108–109
        specialty parks, 109–110
        zoogeographic concept, 105
    post-medieval
        bear pits, 79–80
        deer moats, 79
        falconry, 77
        game parks, 76–77
        lion cages, 80
        menageries, 80–83
        pheasantries, 77–78
        writings about zoos, 76
Western Plains Zoo, Australia, 203, 207
Whales in collections, 155, 314, 315
Wheeler, Margaret, 171
Whipsnade Zoo, England, 72–73
White, Merry, 324, 325
Whitley, Herbert, 67
Whittaker, David W., 157
Wild Animals in Captivity, 71
Wildlife Conservation Society, United States, 162
Wildlife Rescue Centre DENR, Philippines, 232
Wilhelma Botanical Garden, Germany, 108
Wilkie, Andrew, 198, 202
William III, 81
William IV, 23, 58, 81
William V, 81
William the Conqueror, 22
Wilson, C.E., 340
Windsor Great Park, England, 57–58
Winsent, *see* European bison
Wodeyar, H.H. Chamarajendra, 275
Women zookeepers, 168–169, 170–171
Woodward, John, 32
Works Progress Administration, United States, 169
Worlds of Birds Wildlife Sanctuary, South Africa, 345
Wroclaw Zoo (Breslauer Zoologischer Garten), Germany, 123–124, 126

X

Xantus, Janos, 142

Y

Yalooni Camp, Oman, 221
Yellow-footed rock wallaby, 208
Yemen zoological gardens, 223
Yi Lung Hau Gung Yuen, Macau, 240
Yogyakarta Zoo, Java, 238

Yongle (Ming ruler), 25
Yu, 17, 38
Yuan, 17, 38

Z

Zabinski, Jan, 121, 123
Zagazig Zoological Garden, Egypt, 341
Zamosc Zoo, Poland, 122, 123
Zebras in collections
    Antwerp Zoo, 93
    in Egypt, 341
    in Germany, 82
    Giza Zoo, Egypt, 336
Zheng He, 25
Zhou dynasty, 16
Ziggurats, 12
Zoo Animals: Friends in a Foreign Country, 107
Zoo biology, 96, 265
Zoogeographic concept, 105
Zoological Board of Victoria, Australia, 202
Zoological Centre Tel Aviv-Ramat Gan, Israel, 220
Zoological garden definition, 40
Zoological Garden of the Biblical Institute, Israel, 219
Zoological Gardens Government Society, Czech Republic, 141
Zoological Gardens of Ceylon, Sri Lanka, 227
Zoological Gardens - Yesterday, Today, Tomorrow, 127
Zoological Park of Plackendael, 94
Zoological Parks and Gardens Board of Victoria, Australia, 202
Zoological Parks Organisation of Thailand, 234
Zoological societies, *see also* Management of zoos; Scientific societies
    Amsterdam, 91
    Antwerp, 92–94
    Australia, 182, 186, 197
    Belgium, 97
    for Berlin Zoo, 94
    China, 241, 242
    common attributes, 93
    England
        acclimatization programs, 36
        conservation efforts, 242
        first public aquarium, 41
        interest in science of animals, 52
        proposal for zoological study, 37–38
        Tower menagerie given to, 55–56
    formation of, 91
    France, 36, 90
    Germany, 97
    Hungary, 142
    Indonesia, 236–237
    in Japan, 298, 302, 303, 305–306, 319
    Java, 237
    late twentieth century, 110
    Moscow, 134
    Prague, 141
    Singapore, 232
    Southeast Asia, 228
    United States, 150, 151, 162, 165
Zoological Society of Antwerp, 92
Zoological Society of Frankfurt-on-Main, 97
Zoological Society of London
    acclimatization programs, 36
    conservation efforts, 242
    first public aquarium, 41
    interest in science of animals, 52
    proposal for zoological study, 37–38
    Tower menagerie given to, 55–56
Zoological Society of San Diego, United States, 165
Zoological Society of the Natura Artis Magistra, 91–92
Zoological Society of Victoria, Australia, 182
Zoologischer Garten bei Berlin, *see* Berlin Zoo, Germany
Zoo Negara Malaysia, 230
Zoopark, *see* Moscow Zoological Gardens, Soviet Union
Zoo-technics discipline in Russia, 136–137
Zootiere Freunde in der Fremde, 107
Zuckerman, Lord, 72